SpringerBriefs in Applied Sciences and Technology

More information about this series at http://www.springer.com/series/8884

Sissi Closs

DITA – the Topic-Based XML Standard

A Quick Start

Springer

Sissi Closs
Hochschule Karlsruhe Technik und
    Wirtschaft
Karlsruhe
Germany

ISSN 2191-530X   ISSN 2191-5318   (electronic)
SpringerBriefs in Applied Sciences and Technology
ISBN 978-3-319-28348-7   ISBN 978-3-319-28349-4   (eBook)
DOI 10.1007/978-3-319-28349-4

Library of Congress Control Number: 2016934007

Printed on acid-free paper

This Springer imprint is published by Springer Nature
The registered company is Springer International Publishing AG Switzerland

Contents

Introduction

The XML standard DITA (Darwin Information Typing Architecture) has established itself in technical communications in recent years. The systematic use of topic-based structures is a necessary precondition for efficient single sourcing and is also best suited to the modern style of content presentation and usage, particularly on mobile devices.

Therefore, DITA has recently been finding its way also into areas such as marketing, training, and corporate communications [4].

Chapter 1
What Is DITA?

DITA is often compared to the Lego system. Just as you can use Lego bricks to build the most varied replicas of the real world such as houses, cars, and landscapes, you can use DITA *topics* to create any information products necessary.

With both Lego and DITA, this modular system gives you flexibility and enables you to create all sorts of different objects from the same repertoire of pieces.

Example

Figure 1.1 shows an excerpt from a topic repertoire on the subject of *Coffee*.

From this you can build a quick start guide for *Making Coffee* (Fig. 1.2), which describes only one of the many ways to make coffee and keep it warm. Prerequisites, notes, and all other topics are deliberately omitted from the quick start guide.

You can also compile a complete coffee book from the coffee-topic repertoire (Fig. 1.3), containing the same topics as in the quick start guide but with their complete content and also other topics.

This example shows that in the quick start guide and the coffee book, not only the selection and combination of the topics can differ but also the content shown from a single topic, as well as the way the topics are presented.

1.1 DITA Meets Zeitgeist

Classical book-oriented production has reached its limits because requirements placed on information products have changed enormously. Users no longer want comprehensive books: instead, they want information packages that are individually tailored precisely to their profile, their situation, and their current needs. In addition, it must be possible to present the content optimally on the respective display medium.

© The Author(s) 2016
S. Closs, *DITA – the Topic-Based XML Standard*,
SpringerBriefs in Applied Sciences and Technology,
DOI 10.1007/978-3-319-28349-4_1

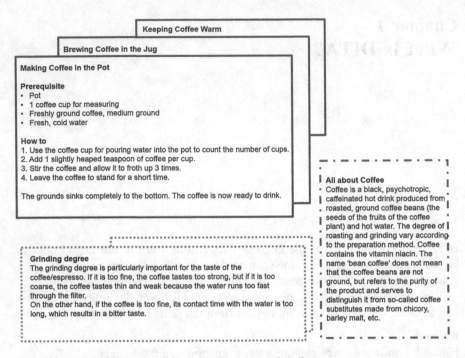

Fig. 1.1 Excerpt from a topic repertoire on the subject of *Coffee*

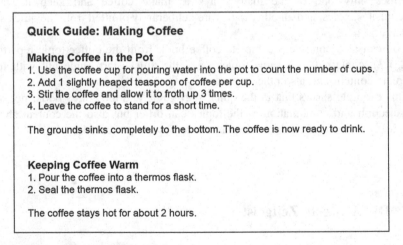

Fig. 1.2 Quick start guide, *Making Coffee*

At the same time, the immense diversity of the products is growing continuously for the information provider, and their period of validity is getting shorter and shorter. This means that with traditional documents and redundant content

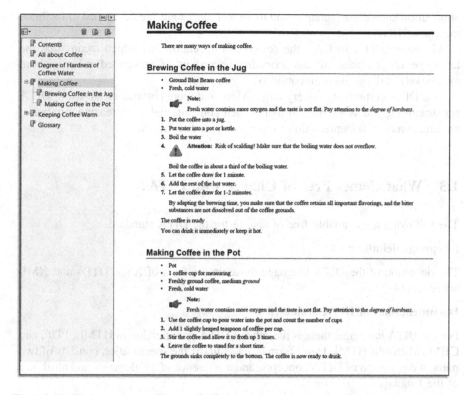

Fig. 1.3 Coffee book

maintenance, production is becoming ever more expensive and error-prone, particularly if the user expectations of today are to be met.

Topic-oriented structuring offers methods and techniques for meeting these challenges.

DITA builds on modularization, minimalism, reuse, and single sourcing, and provides an XML framework in which tried and tested methods and techniques can be implemented. These methods and techniques include information mapping® from Horn [10], class concept method® from Closs [2], function design® from Muthig and Schäflein-Armbruster [12], and reuse methods from Rockley [15]. This creates confidence and offers a chance for good, sustainable solutions.

1.2   What Makes DITA so Special?

One reason why DITA is so widely accepted is its focus on topic-oriented structuring. Another is the fact that it is an open standard and is therefore not subject to license fees. This encourages its widespread use, which in turn means that application know-how increases and more and more tools are developed to facilitate the use of DITA.

Moreover, DITA includes the concept of specialization, which means that the language can be adapted and extended according to prescribed rules without excessively endangering compatibility.

The DITA community is very large. More and more companies are using DITA for structuring and as a source format of their contents, and an increasing number of manufacturers of documentation tools are supporting DITA.

1.3   What Comes Free of Charge with DITA?

The following are available free of charge for the DITA standard.

Language definition

The definition of the DITA language exists in the form of XML DTDs and XML schemas [14].

Documentation of DITA standard

For the DITA language, there is the official DITA specification in HTML, PDF, and CHM (Microsoft HTML Help) format. Essentially, the specification consists of two parts: a description of DITA concepts, and a reference of all elements and attributes of the language.

The document sources are structured with DITA and thus serve as examples that frequently occur in a similar form in practice.

DITA Open Toolkit

With the standard, you already have a suitable, cost-free DITA Open Toolkit (DITA OT) [6]. This contains a collection of scripts and programs that use DITA sources to generate different output formats. Using the DITA OT and further plugins where necessary, you can for example generate HTML, XHTML, PDF, EPUB, HTML5 with JQuery mobile, HTML Help, JavaHelp, EclipseHelp, DocBook, Troff, and RTF. DITA OT is being continuously developed and expanded. It is the reference implementation for the standard, which means that the functionality is implemented in the way intended by the standard.

DITA Open Toolkit also offers sample data for testing the scripts. To get an initial impression of the DITA world, the best and fastest way is to start with DITA Open Toolkit.

For DITA Open Toolkit, an installation manual and other documentation are provided.

Sample applications

In harmony with the open source spirit, many DITA users provide their DITA applications free of charge.

Articles

The DITA concepts are described in numerous articles to be found on the OASIS site [7].

1.4   DITA History

In the late 1990s, IBM first defined DITA for its own documentation requirements.

To promote its spread and further development, IBM handed DITA over to OASIS (Organization for the Advancement of Structured Information Standards) in 2004 as an open-source architecture. In May 2005, DITA Version 1.0 was released as standard by OASIS; DITA Version 1.1 followed in August 2007, and Version 1.2 in November 2010. Version 1.3 is expected to appear at the end of 2015.

Where DITA details are mentioned in this book, they refer to DITA Version 1.3.

1.5   What Does *DITA* Mean?

Darwin Information Typing Architecture stands for core concepts used in DITA [5].

Darwin refers to Charles Darwin, the originator of the Theory of Evolution, which represents heredity and adaptability.

Information Typing refers to the typing of topics and maps.

Architecture means that there is a framework for topic and map types in the standard. This framework defines what they are intended for, and how they are to be used. DITA also defines rules and procedures for adapting and extending the architecture.

1.6   Where Can DITA Be Used?

DITA originally came from a technical documentation area, where it was used particularly in software documentation. This initially led to its reputation of being a software documentation standard only. But that is not true at all. The topic approach, combined with the flexibility of using suitable topic types for each kind of content, makes DITA a generic standard that can be used in any area for content structuring and as a source format. Successful DITA applications show that content reuse is possible on a broad scale and that for the first time synergies between classically separate fields such as marketing, training, and product information development can be achieved [4].

Chapter 2
Basic Principle for DITA: Topic-Oriented Structuring

DITA builds on topic-oriented structuring. The basic idea of this structuring principle is the division of content into pieces known as *topics* with the aim of assembling and reusing them flexibly. This structuring principle has a long history and was used in classical book production for lexicons and glossaries [2].

Example

A term definition is a good topic example. A term is defined only once. Its definition can then be used anywhere where the term occurs and an explanation is needed. The topic advantages can be seen immediately:

- If the term definition has to be translated, it has to be translated only once. There is no redundant content that would cause multiple translations.
- Any changes are made in a single place only, and they are then consistently available everywhere.
- A term can be defined independently of the content in which it occurs.
- Creation and maintenance can be carried out by special experts.

But topic-oriented structuring really first came to life when content could be created and displayed digitally, and when important functions, in particular linking, could be technically implemented in an effective manner.

Topic-oriented structuring experienced a first heyday with the arrival of online help for software. Tools known as help authoring tools were developed both to support the creation of topics and to integrate technical functionality without any programming effort. However, the sources created with these tools are to a high degree tool-specific. In contrast, DITA offers a largely tool- and manufacturer-independent XML basis for the sources.

© The Author(s) 2016
S. Closs, *DITA – the Topic-Based XML Standard*,
SpringerBriefs in Applied Sciences and Technology,
DOI 10.1007/978-3-319-28349-4_2

2.1 What Is a Topic?

Not every snippet of content is a good topic. As the example of the term definition shows, a topic should be a self-contained piece of content as context-independent as possible, containing a key statement and making sense on its own. The division of content into topics does not have to be just content- and usage-related: it can also have technical or organizational reasons. There are no rules governing the size of a topic. However, a topic should not be too large, for two reasons: first, there is a risk that the topic contains more than one key statement; second, it is difficult to display it on small devices. On the other hand, a topic should not be too small: it should be large enough to contain meaningful content that can be properly managed.

In the case of traditional book-oriented writing, content is created in context. Hierarchy and order of subjects are predefined. Explanation and instruction are often mixed up together, and the same content occurs redundantly but not identically at different places. In contrast, topics should be created separately from a concrete publication and context-independent so that they can be used in multiple ways. Each subject should be described only once (single point of truth). Just when a term is defined we do not necessarily know where that term is going to be used. The rule that applies when creating a topic is that its context is first revealed when topics are assembled for a special purpose and for a particular target group.

When presenting a topic on screen to the user, in the simplest case a topic is also displayed as a separate page. But this is not necessary. Content that is divided into several topics in the source can be presented contiguously if this is considered appropriate for the user.

For newcomers to topic-oriented structuring, dividing contents into topics is unfamiliar. The topic rules of the class concept method$^{®}$ help to find the right modularization [2].

DITA topic

DITA topics should conform to the topic rules and they must have a title and a particular topic type. In file-based management, each DITA topic is usually stored as a separate file.

2.2 Why Topic Types?

Topics offer more flexibility, but they also require a lot of organizational effort because the number of topics can grow very fast and numerous topics have to be planned, managed, and organized properly in order to be found again. Classification or typing is a methodical approach to keep the amount of topics under control. With suitable classification criteria, you divide topics into different types characterized by features such as heading style, content type, etc. Instead of having to plan each topic individually, you just have to design a few topic types. Usually, you only need up to ten topic types for the architecture of a specific documentation environment. For each topic type, as many topics as necessary can be created, in a controlled and consistent manner.

The design of topic types is a central task for information architects [3]. The class concept method® supports the iterative, agile development of topic types and their characteristic features [2].

Example

A typical classification criterion is the content type. According to a proven modularization rule, explanations and background information should be separated from instructions ("separate what from how"). Therefore, DITA has contained the topic types concept and task right from the beginning [1]. The optimal structure of a task has been extensively researched: first the prerequisite, then the individual action steps, then the result. DITA has defined the suitable XML elements and structures for this.

Advantages of classification

The topic types create a framework that ensures efficiency and quality and can guarantee the long-term stability of a topic pool.

DITA topic types

Finding suitable topic types is not a simple task. This is where we see another advantage of DITA. As the name says, the standard supports typing. For the commonest content types such as step-by-step instructions, descriptions, and glossary entries, suitable topic types have become established throughout the years. DITA takes these up and, starting from the generic topic type, offers predefined specializations for established topic base types. Table 2.1 shows the base types in DITA 1.3.

Additionally, DITA offers topic types for specific applications. These include a series of topic types for the learning environment such as LearningAssessment, LearningOverview, LearningPlan, and LearningSummary.

The topic types have common elements such as the title element for the title, whereas specific elements characterize the type and structure of the content for which they are intended.

Table 2.1 DITA topic types

DITA topic type	For
concept	Background information, concept, interdependencies, overview
glossentry	Glossary entry
machinery task	Instruction in engineering
reference	Facts, description of functions, commands, parameters
task	Instruction, procedure
topic	Content that does not suit any other topic type and basic type for specializations
troubleshooting (DITA 1.3)	Error message and removal

Chapter 3
Sample Class Concept

The development of an initial class concept is shown in the following using the coffee example.

3.1 Raw Material for *Making Coffee*

There often exists raw material (Fig. 3.1) that contains correct information but is neither meaningfully structured nor well formulated.

When you first look at the text in Fig. 3.1 for this well-known subject, the information first seems enough for making coffee. But if you look a bit closer, you see that important details are missing (such as how much coffee you need), and that the contents are not clearly structured (e.g., keeping the coffee hot is mentioned during the brewing phase). Moreover, there is no consistent style of writing. With unknown and difficult subjects, weak points like these result in the contents not being understood or not providing enough information to solve the task satisfactorily.

3.2 Developing a Class Concept

A topic-based solution can remove these weak points, and the class concept method helps to find systematically suitable topic types and define their features [2].

Classifying contents

The existing contents are analyzed and the different content types are marked. Figure 3.2 shows a meaningful modularization.

© The Author(s) 2016
S. Closs, *DITA – the Topic-Based XML Standard*,
SpringerBriefs in Applied Sciences and Technology,
DOI 10.1007/978-3-319-28349-4_3

Brewing in the jug
Put ground coffee in a jug. Cover it with about a third of the boiling water. Caution! Acute risk of scalding! After about a minute, add the rest of the hot water. Leave the coffee to stand for one or two minutes. Pour the coffee into a preheated thermos flask. The adjusted brewing time ensures that the coffee retails all its flavors and the bitter substances are not released from the coffee-grounds.

The water
Use only fresh water and heat it from cold. The water then contains more oxygen and the taste is not flat, as it would be with low-oxygen water.
You should use water with a hardness grade of five to six. If the water is softer, adding a pinch of salt can help. If the water is too hard, it spoils mainly the appearance of the drink. If necessary, you can use special filters to soften the water.

Brewing coffee in the pot
Simply measure how many coffee cups of water will go into your 5-liter pot. Use only fresh water! Add a slightly heaped teaspoon of coffee per cup. In the Arabian variant, add a few pinches of spice (cardamom or cinnamon) and 1 tsp of sugar.
Bring it to the boil and stir it to make it foam 3x (the pot should therefore be only 2/3 full). Then leave it to stand for a short time (1 min). The grounds fall completely to the bottom, and you can then offer the finest and best-flavored coffee hot and to all guests simultaneously!

Fig. 3.1 Raw material for *Making Coffee*

Brewing in the jug

Put ground coffee in a jug. Cover it with about a third of the boiling water. Caution! Acute risk of scalding! After about a minute, add the rest of the hot water. Leave the coffee to stand for one or two minutes. Pour the coffee into a preheated thermos flask. The adjusted brewing time ensures that the coffee retails all its flavors and the bitter substances are not released from the coffee grounds.

The water

Use only fresh water and heat it from cold. The water then contains more oxygen and the taste is not flat, as it would be with low-oxygen water.

You should use water with a hardness grade of five to six. If the water is softer, adding a pinch of salt can help. If the water is too hard, it spoils mainly the appearance of the drink. If necessary, you can use special filters to soften the water.

Brewing coffee in the pot

Simply measure how many coffee cups of water will go into your 5-liter pot. Use only fresh water! Add a slightly heaped teaspoon of coffee per cup. In the Arabian variant, add a few pinches of spice (cardamom or cinnamon) and 1 tsp of sugar.
Bring it to the boil and stir it to make it foam 3x (the pot should therefore be only 2/3 full). Then leave it to stand for a short time (1 min). The grounds fall completely to the bottom, and you can then offer the finest and best-flavored coffee hot and to all guests simultaneously!

Fig. 3.2 Modularizing

Table 3.1 Class concept

Topic type	Label	Title	Writing style	Mapping → DITA
Instruction	t	Verbal: present participle plus object	Action step: imperative	task
Background information	c	**All about** *subject*		concept
Glossary entry	g	*Term* that is defined		glossentry
Facts	r	*Subject* substantive		reference
Note	n	None		note

The sample text contains instructions (frame), although these are not in the optimal shape as step-by-step instructions. There are notes. These are also marked (dotted frames) since the same note is intended to appear in several places. And there are some facts (double frame). Additionally, it makes sense to provide for background information and term definitions.

Defining an initial class concept

With the class concept method®, we produce an initial class concept (Table 3.1).

You can successively refine the existing topic types by defining further features and add new topic types if necessary.

Designing topic models

The important, DITA-independent work consists in designing a suitable topic model for every topic type defined in the class concept. This topic model should serve authors as a template. Existing good examples can be used as a starting point and adapted and extended with methods and techniques of minimalism, class concept method®, function design®, information mapping® as well as rules for comprehensible writing.

Figure 3.3 shows the task *Keeping Coffee warm in a Thermos Flask* according to a suitable task model.

Often it is not enough just to reformulate or make small structural adaptations to the raw material: you have to completely rewrite the contents.

In our example, this applies to the contents concerning the hardness of the water. It turns out that the raw material is incomplete. This is where the editorial work starts—i.e., you have to do some research to find missing information. Figure 3.4 shows the still incomplete topic.

Even if a class concept is still incomplete, you can implement it with DITA. If you later change the class concept, you can systematically revise the existing topic pool according to the changed class concept.

Keeping Coffee Warm in a Thermos Flask

PREREQUISITE

- Thermos flask
- Warm hotplate

NOTE: *Candles heaters can't be used.*

TASK

1. Pour the coffee into a thermos flask.
2. Seal the thermos flask.

RESULT:

The coffee stays hot for about 2 hours.

Fig. 3.3 Topic: *Brewing Coffee in the Jug*

Hardness of Coffee Water

The *degree of hardness* of the coffee water is a decisive factor in the quality of the coffee. The following table shows the effects of the water hardness:

Degree of hardness	Effect	What to do
5-6	Coffee tastes good	Nothing
<5 (softer)	???	Add a pinch of salt to the water
>6 (harder)	Coffee becomes cloudy	Soften the water???with filters

Fig. 3.4 Topic: *Hardness of Coffee Water*

Chapter 4
Implementation in DITA

For marking structures without making commitments to a particular layout or tool, XML languages have been used successfully for many years. DITA too is an XML language and it is specially equipped for marking topic structures in a suitable and display-neutral manner.

The implementation in DITA is shown in the following using the coffee example. Only the central DITA elements that occur in the topics are presented. The current, complete language description can be seen on the DITA page of OASIS [14].

> The examples show that for a particular application, a small subset of the standard (which now consists of several hundred elements) is sufficient to mark the needed structures. That is reassuring but also means that defining the required DITA subset is a central initial task prior to implementation.

4.1 Elements for Block Structures and Inline Elements

For marking fundamental block and inline structures, DITA uses elements that are intentionally derived from the corresponding HTML elements.

DITA block elements

Table 4.1 shows the DITA elements for typical block elements in alphabetical order.

© The Author(s) 2016
S. Closs, *DITA – the Topic-Based XML Standard*,
SpringerBriefs in Applied Sciences and Technology,
DOI 10.1007/978-3-319-28349-4_4

Table 4.1 Selection of typical DITA block elements

Element	Marks
fig	Figure with optional caption
li	List item
note	Note
ol	Ordered list
p	Paragraph
section	Section
simpletable	Simple table
table	Table
title	Title
ul	Unordered list

Table 4.2 Selection of typical DITA inline elements

Element	Marks
image	Image (graphic)
keyword	Keyword (significant word)
term	Term
xref	Cross-reference

DITA inline elements

Table 4.2 shows the DITA elements for typical inline elements in alphabetical order.

4.2 DITA Task

Table 4.3 shows specific task elements in alphabetical order.

Table 4.3 Selection of specific task elements

Element	Marks
cmd	Command (step description)
context	Purpose of task
info	Additional information about the step description
prereq	Prerequisites to be met before carrying out the task described in the topic
result	Result of task
step	Single step
stepresult	Result of a single step
steps	Container for the single steps of the task
task	Root of a task topic
taskbody	Body of a task topic

```
- <task xmlns:ditaarch="http://dita.oasis-open.org/architecture/2005/"
  id="id146QIOW005Z">
    <title>Brewing Coffee in the Jug</title>
  - <taskbody>
      + <prereq product="book">
      - <steps>
          - <step>
              <cmd>Put the coffee into a jug.</cmd>
            </step>
          + <step>
          + <step>
          + <step>
          + <step>
          + <step>
        </steps>
      - <result>
          <p>The coffee is ready</p>
          <p>You can drink it immediately or keep it hot.</p>
        </result>
    </taskbody>
  </task>
```

Fig. 4.1 DITA task topic

Example of a task
You can tag the topic *Brewing Coffee in the Jug* with DITA as shown in Fig. 4.1.

4.3   DITA Concept

Table 4.4 shows specific concept elements in alphabetical order.

Example of a concept
You can tag the topic *All about Coffee* with DITA as shown in Fig. 4.2.

Table 4.4 Selection of specific concept elements

Element	Marks
conbody	Body of a concept topic
concept	Root of a concept topic

```
- <concept xmlns:ditaarch="http://dita.oasis-open.org/architecture/2005/" id="id158B0030401">
    <title>All about Coffee</title>
  - <conbody>
      <p>Coffee is a black, psychotropic, caffeinated hot drink produced from roasted,
         ground coffee beans (the seeds of the fruits of the coffee plant) and hot water. The
         degree of roasting and grinding vary according to the preparation method. Coffee
         contains the vitamin niacin. The name 'bean coffee' does not mean that the coffee
         beans are not ground, but refers to the purity of the product and serves to
         distinguish it from so-called coffee substitutes made from chicory, barley malt,
         etc.</p>
    </conbody>
  </concept>
```

Fig. 4.2 DITA concept topic

4.4 DITA Reference

Table 4.5 shows specific `reference` elements in alphabetical order.

Example of a reference
You can tag the topic *Hardness of Coffee Water* with DITA as shown in Fig. 4.3.

Table 4.5 Selection of specific `reference` elements

Element	Marks
propdeschd	Header of the output column for the column with the brief description
properties	List of properties or parameters
property	A property or a parameter
refbody	Body of a reference topic
reference	Root of a reference topic

```
- <reference id="id158DG0T0HNA">
    <title>Hardness of Coffee Water</title>
  - <refbody>
    - <section>
      - <p>
          The
          <term keyref="HG">degree of hardness</term>
          of the coffee water is a decisive factor in the quality of the coffee. The
          following table shows the effects of the water hardness:
        </p>
      - <simpletable frame="all" relcolwidth="33* 33* 33*">
        - <sthead>
          - <stentry>
              <p>Degree of hardness</p>
            </stentry>
          - <stentry>
              <p>Effect</p>
            </stentry>
          - <stentry>
              <p>What to do</p>
            </stentry>
          </sthead>
        - <strow>
            <stentry>5-6</stentry>
          - <stentry>
              <p>Coffee tastes good</p>
            </stentry>
            <stentry>Nothing</stentry>
          </strow>
        + <strow>
        + <strow>
        </simpletable>
      </section>
    </refbody>
  </reference>
```

Fig. 4.3 DITA reference topic

4.5   DITA Glossentry

Table 4.6 shows specific `glossentry` elements in alphabetical order.

Example of a glossentry
You can tag the term definition for the degree of water hardness in DITA as shown in Fig. 4.4.

Table 4.6 Selection of specific DITA `glossentry` elements

Element	Marks
glossbody	Details for the glossentry
glossdef	Definition of term (glossentry)
glossentry	Root of a glossentry topic
glossterm	Term that is defined

```
- <glossentry id="id146R9800RZW">
      <glossterm>Degree of hardness</glossterm>
   - <glossdef>
         Specifies the hardness of the water.
      - <note>
            <p>Ask your local council for the hardness of your water.</p>
         </note>
      </glossdef>
   </glossentry>
```

Fig. 4.4 DITA `glossentry` topic

Chapter 5
Assembling Topics

DITA provides several ways of assembling individual topics. You can nest topics or combine them in maps.

5.1 Nesting Topics

You can nest topics to form larger, continuous content blocks. The nesting defines the sequence and hierarchy of the participating topics. However, nesting can only take place outside the topic body.

The DITA configuration can define how and which topic types can be nested (if at all). The standard DITA configuration allows nesting with all topic types only for the generic root element `dita`.

> Generally it is recommended that you manage each topic in a separate file. In some cases however, nesting can make sense in reducing management effort. With a file-based organization, the authors' work may be more efficient if for example all topics of one type are physically in one file, because in the event of a general revision, only one file has to be opened and functions like search and replace can be conducted more quickly. Also, the migration of contents can be easier if the topics are stored in a single file during the migration process.

© The Author(s) 2016
S. Closs, *DITA – the Topic-Based XML Standard*,
SpringerBriefs in Applied Sciences and Technology,
DOI 10.1007/978-3-319-28349-4_5

5.2 DITA Map

The most flexible way of combining contents is provided by the DITA map. This resembles the well-known table of contents in online help programs and in the same way represents the framework for a complete information product. In a DITA map, you can combine topics according to different organization patterns and specify links for the topics listed in the map.

However, the DITA map has far more functions than the classical table of contents. The map defines the topics to be included in an information product, specifies their sequence, grouping, and hierarchy, and defines the relationships between the topics. Additionally, you use the map to define variables as well as metadata for characterizing and managing the information product. You can use a map to plan projects and information products and produce and adapt information products.

Figure 5.1 shows an extract from the DITA map for a coffee book.

Submaps

Maps can contain submaps. You use submaps for giving a meaningful structure to maps for comprehensive and more complex information products, which makes them easier to handle.

Example

It can be very useful to organize all glossary entries for a particular subject in a separate map, and to reference this as a submap in the maps for the information products.

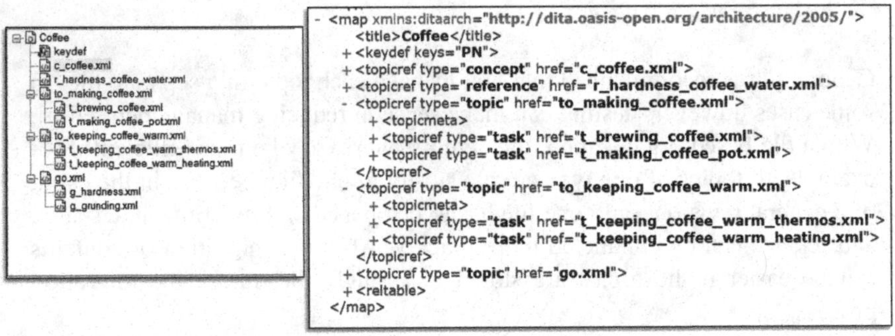

```
- <map xmlns:ditaarch="http://dita.oasis-open.org/architecture/2005/">
    <title>Coffee</title>
  + <keydef keys="PN">
  + <topicref type="concept" href="c_coffee.xml">
  + <topicref type="reference" href="r_hardness_coffee_water.xml">
  - <topicref type="topic" href="to_making_coffee.xml">
      + <topicmeta>
      + <topicref type="task" href="t_brewing_coffee.xml">
      + <topicref type="task" href="t_making_coffee_pot.xml">
    </topicref>
  - <topicref type="topic" href="to_keeping_coffee_warm.xml">
      + <topicmeta>
      + <topicref type="task" href="t_keeping_coffee_warm_thermos.xml">
      + <topicref type="task" href="t_keeping_coffee_warm_heating.xml">
    </topicref>
  + <topicref type="topic" href="go.xml">
  + <reltable>
</map>
```

Fig. 5.1 DITA map

Chapter 6
Defining Relations

You place topics in relation to each other so as to show interrelated and therefore interesting content for the user. The best-known relations in documentation are cross references, which relate content items to each other either within a document or between different documents. A cross reference leads the reader from the current place in the text to other places that the author considers to be related to the current place, to explain the current place better, or to provide further information on the current place. In the electronic world, references are known as links. They are visible as such to the user and occur in different forms.

In general, DITA supports the principle that topic content should be kept as link-free as possible. This principle has advantages for both the reader and the author:

- Readers are not distracted and can absorb link-free topics better; they can navigate more effectively via systematically arranged links.
- For the author, the separation of contents and links simplifies all processes from creation to management and from maintenance to translation.

One of DITA's great strengths is the wide variety of ways of defining relations and links. Apart from the links that can automatically be generated from the map, you can set links explicitly in the topic or in the map.

6.1 Links in the Map

The most flexible method is to define the linking in the map and therefore to first define relations between topics when the topics are assembled in their respective combination.

© The Author(s) 2016
S. Closs, *DITA – the Topic-Based XML Standard*,
SpringerBriefs in Applied Sciences and Technology,
DOI 10.1007/978-3-319-28349-4_6

These options are very interesting when it comes to the reuse factor. Since a topic can have different relations according to its context, a link in the topic would restrict its reusability. But if the link is defined in the map, you can combine the topic flexibly with other topics in different maps.

Generated links

For relations that result from the arrangement of topics in the map, the standard provides that links can be automatically generated using suitable metadata (attributes)—for example, from a topic to its subordinate (child) topics.

Example

If there are several topics on a single subject, it can help orientation according to the leveling principle to provide an overview.

For example, there are many ways of making coffee. For a fast overview, you provide an overview topic. If you get DITA to generate the links, you only need an introductory sentence in the overview topic. Figure 6.1 shows the DITA source of the overview topic.

The links to the various task topics result from the map and are generated if you set the `linking` attribute accordingly. Figure 6.2 shows a generated MS HTML help output in CHM format.

The advantages are obvious: the overview topic always automatically contains the correct and complete link list.

Linking via a relationship table (`reltable`)

A map also offers the possibility to specify relations explicitly by means of a relationship table (`reltable`).

```
- <topic id="id158AN900UT6">
      <title>Making Coffee</title>
  - <body>
        <p>There are many ways of making coffee.</p>
    </body>
  </topic>
```

Fig. 6.1 DITA source for overview topic

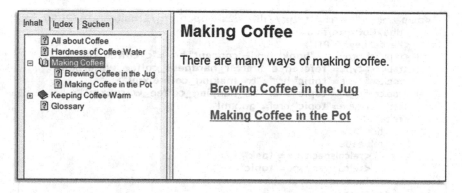

Fig. 6.2 Published overview topic with generated links

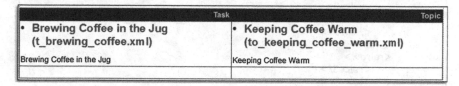

Fig. 6.3 Relationship table

Example

You can specify the link from the task *Brewing Coffee in the Jug* to the overview *Keeping Coffee Warm* in the coffee map by means of a relationship table.

In the WYSIWYG view (Fig. 6.3), the table looks like this.

Figure 6.4 shows the DITA source.

6.2 Links in the Topic

DITA also allows to set links in a topic:

- You can use `xref` to specify the classic cross-references that are well-known in paper documents and that can be used anywhere in topic content. They are often used to refer to a figure or table in the topic.
- You can also specify links in a separate `related-links` section following the topic body and separate from the actual topic contents. However, this may reduce the reusability of the topic since the referenced destinations may not suit every context.

```
- <map xmlns:ditaarch="http://dita.oasis-open.org/architecture/2005/">
    <title>Coffee</title>
  + <keydef keys="PN">
  + <topicref product="book" type="concept" href="c_coffee.xml">
  + <topicref type="reference" href="r_hardness_coffee_water.xml">
  + <topicref type="topic" href="to_making_coffee.xml">
  + <topicref type="topic" href="to_keeping_coffee_warm.xml">
  + <topicref type="topic" href="go.xml">
  - <reltable>
        <title/>
      - <relheader>
            <relcolspec type="task"/>
            <relcolspec type="topic"/>
        </relheader>
      - <relrow>
          - <relcell>
            + <topicref type="task" href="t_brewing_coffee.xml">
            </relcell>
          - <relcell>
            + <topicref type="topic" href="to_keeping_coffee_warm.xml"
            </relcell>
        </relrow>
      - <relrow>
            <relcell/>
            <relcell/>
        </relrow>
    </reltable>
  </map>
```

Fig. 6.4 DITA relationship table (reltable)

Chapter 7
Reusing Contents by Embedding

A central aspect in a DITA-based documentation environment is the orientation to single sourcing and content reuse. You can reuse material from complete topics down to individual sentences.

Example
You want to use a note at different places, not as a separate topic but embedded in the content of another topic.

DITA offers various ways of embedding.

The `conref` mechanism, well known from SGML, enables you to reuse the content of an element or an element group both within a topic and between different topics. Figure 7.1 illustrates the principle.

The `conref` mechanism in DITA exists as both a pull and a push variant. In the pull variant, you reference the target content to be embedded at the place where the content is to be inserted. In the push variant, you specify in the content to be embedded where and how the content is to be inserted.

With the `conref` mechanism, you can reuse any element or whole element groups with a unique identification (`id` attribute). However, you can only insert structurally equivalent content. If the element types are not compatible, no embedding takes place.

> Reuse at the topic level is much easier to handle and should therefore always be the first choice. However, there are good reasons for embedding, such as with safety warnings. But clear rules are needed. You have to define the type of content to be embedded in this manner and where embedding can take place. The rules must always be obeyed: otherwise, there is a danger that the organization becomes too complex and unclear.

S. Closs, *DITA – the Topic-Based XML Standard*,
SpringerBriefs in Applied Sciences and Technology,
DOI 10.1007/978-3-319-28349-4_7

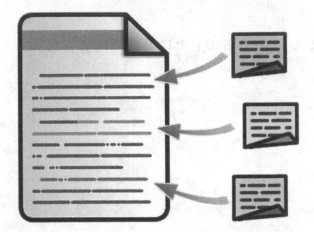

Fig. 7.1 Embedding contents

Example

In the coffee tasks, the note about fresh water is relevant in several coffee topics. Figure 7.2 shows two examples.

To avoid redundancy in the sources, we use the `conref` mechanism in the pull variant. To make it easier to manage, you create a collective topic for all notes, also called a warehouse topic. For the purpose of identification, every note receives an ID. The conventions for building the ID must be defined. Figure 7.3 shows the IDs assigned in the `id` attribute. The ID for notes starts with $N\_$, and the ID for safety warnings with $SN\_$.

The warehouse topic is the single point of truth for notes where a note is created and edited. This avoids redundancy, inconsistency, and errors, and keeps the size of

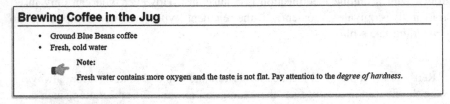

Fig. 7.2 A repeatedly needed note

```
- <topic id="id146QK0608YK">
    <title>Note Warehouse Topic</title>
  - <body>
    - <note id="N_water" type="note">
      - <p>
            Fresh water contains more oxygen and the taste is not flat. Pay attention
            to the
            <term keyref="HG">degree of hardness</term>
            .
        </p>
      </note>
      <note id="SN_scald" type="attention">Risk of scalding! Make sure that the boiling
            water does not overflow.</note>
  </body>
</topic>
```

Fig. 7.3 Warehouse topic for notes

```
- <task id="id146QI0W005Z">
    <title>Brewing Coffee in the Jug</title>
  - <taskbody>
    - <prereq product="book">
      - <ul>
          + <li>
          - <li>
              <p>Fresh, cold water</p>
              <note conref="n_notes.xml#id146QK0608YK/N_water"/>
          </li>
        </ul>
      </prereq>
    + <steps>
    + <result>
  </taskbody>
</task>
```

Fig. 7.4 Embedding a note with `conref`

the sources down. It also enables the work to be distributed easily. For example, one person can write the notes while another writes the topics containing the notes.

Wherever a note is needed, it can be taken from the warehouse topic and embedded using `conref`. Figure 7.4 shows the embedding of the note about fresh water in the `note` element in the topic *Brewing Coffee in the Jug*.

Chapter 8
Addressing

For referencing, DITA since Version 1.2 has supported not only direct but also indirect addressing.

8.1 Direct Addressing

With direct addressing, you specify the destination with its actual address. But this results in considerable dependency. If the destination is renamed, moved, or deleted, the reference has to be edited to remain intact.

8.2 More Freedom Through Indirect Addressing

DITA 1.2 introduced indirect addressing to solve the problems of direct addressing. Instead of addressing the destination directly, you just specify a freely definable key for the destination. Then, when you create the map, you assign suitable destinations to the keys and thus define the target of the reference.

Ways of using indirect addressing have been improved and extended in DITA 1.3.

Example
The simplest way to demonstrate indirect addressing is with references to glossary entries.

Our coffee subject contains the phrase *degree of hardness*. For this phrase, you create in DITA a term definition as a glossentry topic (Fig. 8.1).

© The Author(s) 2016
S. Closs, *DITA – the Topic-Based XML Standard*,
SpringerBriefs in Applied Sciences and Technology,
DOI 10.1007/978-3-319-28349-4_8

```
- <glossentry id="id146R9800RZW">
     <glossterm>Degree of hardness</glossterm>
   - <glossdef>
        Specifies the hardness of the water.
      - <note>
           <p>Ask your local council for the hardness of your water.</p>
        </note>
     </glossdef>
  </glossentry>
```

Fig. 8.1 Glossentry topic for degree of hardness

```
- <topic id="id146QK0608YK">
     <title>Note Warehouse Topic</title>
   - <body>
      - <note id="N_water" type="note">
         - <p>
              Fresh water contains more oxygen and the taste is not flat. Pay attention
              to the
              <term keyref="HG">degree of hardness</term>.
           </p>
        </note>
        <note id="SN_scald" type="attention">Risk of scalding! Make sure that the
           boiling water does not overflow.</note>
     </body>
  </topic>
```

Fig. 8.2 Indirect addressing using the key *HG* in the term element

The note about fresh water uses the term *degree of hardness*. As shown in Fig. 8.2, the key *HG* is assigned to the keyref attribute instead of the fixed address of the glossentry topic to make a reference to the definition of *degree of hardness*.

In a map for the coffee topics, the term definitions are referenced with glossref. The destination topic containing the term definition to be used in this map is assigned in the href attribute to the corresponding key in the keys attribute (Fig. 8.3).

Indirect addressing is a powerful feature. The topics in which references are used are independent of fixed addresses and do not need to be adapted if these addresses change. Moreover, from the same combination of topics, different variants can be produced if the keys in the map are assigned to different destinations.

```
- <map xmlns:ditaarch="http://dita.oasis-open.org/architecture/2005/">
     <title>Coffee</title>
   + <keydef keys="PN">
   + <topicref type="concept" href="c_coffee.xml">
   + <topicref type="reference" href="r_hardness_coffee_water.xml">
   + <topicref type="topic" href="to_making_coffee.xml">
   + <topicref type="topic" href="to_keeping_coffee_warm.xml">
   - <topicref type="topic" href="go.xml">
      - <topicmeta>
           <navtitle>Glossar</navtitle>
        </topicmeta>
      - <glossref keys="HG" type="glossentry" href="g_hardness.xml" print="yes">
         + <topicmeta>
        </glossref>
      + <glossref keys="MG" type="glossentry" href="g_grunding.xml" print="yes">
     </topicref>
   + <reltable>
  </map>
```

Fig. 8.3 Destination assignment in the map in `glossref`

DITA 1.3 has extended the assignment options for keys in a map. You can now define validities for keys, and different branches of a map can have different assignments for the same key.

Chapter 9
Variants

DITA provides very good options for setting up an efficient variant management, including if the sources are to be managed only in the file system and without a content management system. These include the options of combining different maps from the same topic pool, and using variables and filtering. The basis is the map via which these methods can be used to produce the variants.

9.1 Different Maps

If required, you can produce different maps from a pool of topics for different purposes, target groups, and output media. If maps share only a few topics and only the topic content is changed but the combination in the maps is very stable, it can make sense to make several maps.

Example
From the coffee topics, you can produce a map for the quick start guide and a map for the complete coffee book.

9.2 Variables

You can use variables in DITA, typically for product names, version numbers, etc. In this case, you use keys as with indirect addressing. For a variable, you define a specific key. In the topics, you use only keys instead of fixed names. It is the map that combines the topics that defines the values of the variables. For this reason, you do not have to define the value until the information product is produced,

© The Author(s) 2016
S. Closs, *DITA – the Topic-Based XML Standard*,
SpringerBriefs in Applied Sciences and Technology,
DOI 10.1007/978-3-319-28349-4_9

and you can generate different variants from the same topics by assigning different values to the variables in the maps.

Example

If a coffee producer wants to use the coffee topics for coffee books that are branded with his or her coffee brands, a variable can be used in the topics for the coffee brand (Fig. 9.1).

In the map, a name such as *Blue Beans* is assigned to the variable *PN* (Fig. 9.2).

```
- <task id="id146QI0W005Z">
     <title>Brewing Coffee in the Jug</title>
   - <taskbody>
      - <prereq product="book">
         - <ul>
            - <li>
               - <p>
                  Ground
                  <keyword keyref="PN"/>
                  coffee
               </p>
            </li>
            + <li>
         </ul>
      </prereq>
      + <steps>
      + <result>
   </taskbody>
   </task>
```

Fig. 9.1 Variable *PN* for coffee brand

```
- <map xmlns:ditaarch="http://dita.oasis-open.org/architecture/2005/">
     <title>Coffee</title>
   - <keydef keys="PN">
      - <topicmeta>
           <navtitle/>
         - <keywords>
              <keyword>Blue Beans</keyword>
           </keywords>
        </topicmeta>
     </keydef>
   + <topicref type="concept" href="c_coffee.xml">
   + <topicref type="reference" href="r_hardness_coffee_water.xml">
   + <topicref type="topic" href="to_making_coffee.xml">
   + <topicref type="topic" href="to_keeping_coffee_warm.xml">
   + <topicref type="topic" href="go.xml">
   + <reltable>
   </map>
```

Fig. 9.2 Defining variable *PN* in the map

9.3 Filtering

Use filtering for producing different variants from the same map. Attributes and a filter file determine which content appears in the respectively produced information product.

If the maps are very large and the variants have a large number of common topics, and a lot of changes occur (i.e., many topics are added, deleted, or moved), it is better to generate the different variants using filtering instead of using different maps. This reduces maintenance efforts.

Attributes for filtering
DITA uses the attributes listed in Table 9.1 for filtering.

You define possible attribute values—e.g. for the `audience` attribute, the values for *beginner, normal, expert*. An attribute can have several values separated by spaces.

Filter file ditaval
For production, the filter file defines the contents to be included or excluded (with `include` or `exclude`), depending on the filter attributes and their values. The filter file is an XML file with the suffix `.ditaval`.

Example
The filter attribute `product` is used, and the possible values `book` and `quickGuide` are defined. The attributes are set accordingly in the topics (Fig. 9.3) and in the map (Fig. 9.4).

Table 9.1 Filter attributes

Attribute	For specifying
`audience`	Target group
`deliveryTarget` (DITA 1.3)	Output format in the map
`otherprops`	Individually defined filter criteria
`platform`	Area
`product`	Product

Fig. 9.3 `prereq` is to be used only in the coffee book (`product=``book''`)

```
- <task id="id146OI0W005Z">
      <title>Brewing Coffee in the Jug</title>
  - <taskbody>
      + <prereq product="book">
      + <steps>
      + <result>
    </taskbody>
  </task>
```

```
-  <map xmlns:ditaarch="http://dita.oasis-open.org/architecture/2005/">
      <title>Coffee</title>
   +  <keydef keys="PN">
   +  <topicref product="book" type="concept" href="c_coffee.xml">
   +  <topicref type="reference" href="r_hardness_coffee_water.xml">
   +  <topicref type="topic" href="to_making_coffee.xml">
   +  <topicref type="topic" href="to_keeping_coffee_warm.xml">
   +  <topicref type="topic" href="go.xml">
   +  <reltable>
   </map>
```

Fig. 9.4 The topic c_coffee.xml is to be used only in the coffee book (product=``book'')

```
<val>
      <prop val="book" att="product" action="exclude"/>
</val>
```

Fig. 9.5 Filter file quickguide.ditaval for producing the quick start guide

For the production of the quick start guide, you create a filter file for excluding all contents that are to appear in the coffee book only. Figure 9.5 shows the XML source.

DITA 1.3 has introduced branch filtering, which enables you to set different filter criteria for different branches of a map.

Chapter 10
Collaboration Under Full Control

Successful collaboration is an important success factor in any environment, but it needs suitable rules so that it does not get out of control. Since Version 1.2, DITA has provided good support for successful collaboration.

You can use constraints to restrict syntax without specialization. In this way, you can define conventions in a project environment more clearly for a team. Thus you can remove elements or set them as mandatory, and specify the element sequence.

Subject schemas provide a promising way of making clear rules. For a particular environment, they allow you to define specific attribute values and names for metadata without having to edit the DTDs.

Supplements for the terminology enable you to define the terminology right up to complete taxonomies. This enables you to lay the basis for semantic web functions: systematical, standardized, and integrated in the source content.

© The Author(s) 2016   39
S. Closs, *DITA – the Topic-Based XML Standard*,
SpringerBriefs in Applied Sciences and Technology,
DOI 10.1007/978-3-319-28349-4_10

Chapter 11
How Is an Information Product Produced?

An information product is produced by combining topics in a map and defining an output format for them.

From a collection of topics, you can put together information products flexibly and according to your requirements on the basis of suitable maps: for different target groups, purposes, and output media. DITA is an XML language, so information products are generated from DITA sources according to the normal XML production process. Figure 11.1 shows the principal process and the tools required.

Creation

Basically, you can use any editor for creating DITA sources. But if you use a simple text editor, it is more difficult. There are now many XML editors that are pre-configured for DITA and provide authors with good support: input assistance for selecting suitable elements, attributes and attribute values; and preconfigured layout views for writing, generating, and displaying in different formats via a connection to the DITA OT or your own output transformations.

Management

In the simplest case, you can use the file system for managing DITA sources. DITA offers powerful functions for typical management tasks—in particular for reuse, variables, and variant control.

Even more comprehensive management support is provided by content management systems. There is a whole range of XML-based content management systems already preconfigured for DITA. As a rule, however, you have to adapt and complete the configuration according to your own needs.

© The Author(s) 2016
S. Closs, *DITA – the Topic-Based XML Standard*,
SpringerBriefs in Applied Sciences and Technology,
DOI 10.1007/978-3-319-28349-4_11

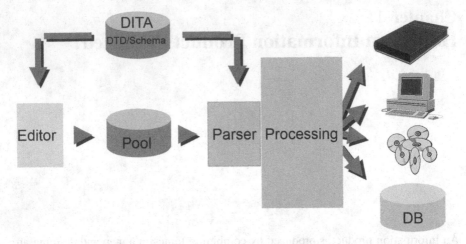

Fig. 11.1 XML production process

Further processing

For generating information products, further processing means formatting for the output medium. In the simplest case, you just need a Cascading Stylesheet (CSS) for formatting DITA sources. For a more ambitious design, you use XSL stylesheets (Fig. 11.2).

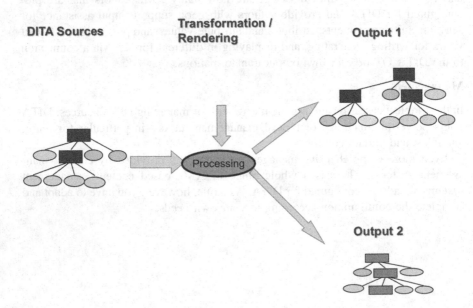

Fig. 11.2 Formatting with XSL stylesheets

Chapter 12
Production with DITA Open Toolkit

DITA Open Toolkit comes cost-free in different versions [6].

You can adapt the transformations that come with the toolkit to your own needs. But you need programming knowledge (CSS, XSL-FO, XSLT, ANT) according to the degree of your adaptation.

12.1 What's in DITA Open Toolkit?

Figure 12.1 shows the folders in the program directory of DITA Open Toolkit:

Folders in the DITA OT program directory

Table 12.1 lists the folder in the DITA OT program directory.

Folders in the plugins directory

The plugins directory contains the transformations as delivered for various output formats (Table 12.2). You adapt layouts in the respective subdirectories. If you have developed your own transformations or received any from third parties, install them in the plugins directory too.

12.2 Installing DITA Open Toolkit

The DITA Open Toolkit comes in several variants [6]:

- dita-ot-*version*.zip: contains the compiled DITA Open Toolkit for Windows
- dita-ot-*version*.tar.gz: contains the compiled DITA Open Toolkit for Linux and Mac
- Source Code: contains the Java source files of DITA Open Toolkit

© The Author(s) 2016
S. Closs, *DITA – the Topic-Based XML Standard*,
SpringerBriefs in Applied Sciences and Technology,
DOI 10.1007/978-3-319-28349-4_12

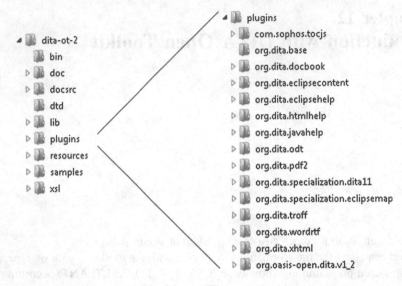

Fig. 12.1   Program directory of DITA Open Toolkit

Table 12.1   Program directory of DITA Open Toolkit

Folder	Contents
dita-ot-x	Base directory of DITA Open Toolkit, Version x
bin	Scripts for calling the DITA Open Toolkit
doc	Documentation of DITA Open Toolkit in HTML, PDF and CHM formats
docsrc	DITA sources of documentation of DITA Open Toolkit
dtd	DTDs
lib	Java program files
plugins	Folder for transformations, DTDs and own extensions
resources	Files for messages, etc.
samples	DITA sample documents and ANT sample scripts

Additional software

DITA Open Toolkit needs Java (JRE or JDK) in Version 7 or higher from Oracle
(Java [11]). For a transformation to HTML help (CHM), you need
Microsoft HTML Help Workshop [13].

Table 12.2 Plugins directory of DITA Open Toolkit

Folder	Contents
`com.sophos.tocjs`	Transformation to XHTML, with Javascript frameset
`org.dita.base`	Base files for all transformations
`org.dita.docbook`	Transformation to Docbook
`org.dita.eclipsecontent`	Transformation to normalized DITA with Eclipse project files
`org.dita.eclipsehelp`	Transformation to Eclipse help
`org.dita.htmlhelp`	Transformation to HTML help
`org.dita.javahelp`	Transformation to Java help
`org.dita.odt`	Transformation to Open Document Format (Open Office)
`org.dita.pdf2`	Transformation to PDF
`org.dita.specialization.dita11`	DITA 1.1 DTDs and schemas
`org.dita.specialization.eclipsemap`	EclipseMap DTDs and schemas
`org.dita.troff`	Transformation to Troff
`org.dita.wordrtf`	Transformation to Rich Text Format
`org.dita.xhtml`	Transformation to XHTML and HTML5, basis for all HTML-based transformations
`org.oasis-open.dita.v1_2`	DITA 1.2 DTDs and Schemas

Installing components

After downloading, unpack DITA Open Toolkit to a directory of your choice. For our examples, we use the following:

```
C:\dita-ot-2
```

Optional, but recommended for a simpler start of DITA Open Toolkit: add the pathname of the `bin` directory to the PATH system variable of your computer, here:

```
C:\dita-ot-2\bin
```

In addition to DITA Open Toolkit, Java must also be installed. DITA Open Toolkit is tested with Java Version 7.

If you want to create HTML help (CHM), install HTML Help Workshop.

12.3   Generating Output with DITA Open Toolkit

DITA Open Toolkit works with the command line. First, open the command line prompt (Windows) or a terminal window (Linux and Mac).

If you have added the pathname of the bin directory to the PATH system variable, you can start DITA Open Toolkit as follows:

```
dita
```

Otherwise, specify the path to DITA Open Toolkit, here:

```
C:\dita-ot-2\bin\dita
```

The command returns a brief overview of the call parameters.

12.4   The First Publication

To publish the DITA sample document ("garage sample") to XHTML, enter the following (if you have extended the PATH system variable):

```
dita  -f  xhtml  -i  C:\dita-ot-2\samples\hierarchy.
ditamap -o outdir
```

You will find the result in the outdir subdirectory of the current directory.

12.5   Parameters for Publication

To publish with the dita command, you have to specify at least the two parameters -f and -i. All other parameters are optional. Table 12.3 lists the parameters.

Table 12.3 DITA Open Toolkit parameters

Parameter	Meaning
-f < *output-format*>	Desired publication format
	DITA Open Toolkit includes the following output formats: docbook, eclipsecontent, eclipsehelp, html5, htmlhelp, javahelp, odt, pdf, pdf2, tocjs, troff, wordrtf, xhtml
-i < *input-file*>	Absolute or relative path to the map of your DITA document
-o < *output-directory*>	Absolute or relative pathname of the directory to which the publication is to be written
-filter < *input-file*>	Absolute or relative path to the ditaval file
-temp < *directory*>	Absolute or relative pathname of the directory to which temporary files are to be written
-v	Generates detailed logging output
-d	Generates detailed debugging output
-l	Absolute or relative pathname to the file to which logging output is to be written
- D < *parameter* >=<*value*>	Parameters for transformation
	Specify -D for each parameter
	Possible parameters for transformations are described in the DITA Open Toolkit documentation
-propertyfile < *file*>	Parameters for transformation
	You can define a whole parameter set in a file, particularly if you always want to use the same parameters

Chapter 13
DITA Specialization

With DITA, you can define new domains and types on the basis of predefined basic types. Such adaptations and extensions are called specialization. Using the inheritance principle, definitions for the output types are passed on to derived new types and can be specifically adapted and extended according to requirements. Figure 13.1 shows the principle using the example of specializations of the `reference` topic type:

You can make adaptations and extensions in several ways:

- You can extend topic types.
- For contents not covered by existing elements, you can introduce new domains.
- For special information structures, you can create your own domains.

DITA prescribes the rules for specialization. A new topic type must be built on an existing one and must further restrict the content.

> You should think carefully about a specialization. As a rule, the standard is sufficient for structuring contents meaningfully. And DITA is being continuously developed so that a complex specialization of your own could be a disadvantage if future standard versions also include these extensions. The `glossentry` and `troubleshooting` topic types are good examples of this.

© The Author(s) 2016
S. Closs, *DITA – the Topic-Based XML Standard*,
SpringerBriefs in Applied Sciences and Technology,
DOI 10.1007/978-3-319-28349-4_13

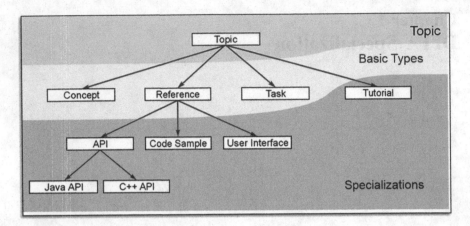

Fig. 13.1 Specializations of the `reference` topic type

Chapter 14
Why Use DITA?

DITA builds on topic-oriented structuring and therefore has all the benefits of this structuring technique if used correctly:

- Many different reuse options
- Efficient variant management
- Good ways of providing suitable access
- Support for collaboration

In combination with a suitable class concept, all documentation processes can be made more streamlined:

- For authors, it is easier to create and edit individual topics than a complete document.
- Authors with different expert knowledge can work simultaneously on topics. This makes revision faster without any loss of quality.
- You can test topics individually before the overall information product is finished.
- In the event of updates, only the new and revised topics have to be published.
- Translation of completed topics can start even if other topics still have to be edited.

DITA provides a formal framework for implementing tried and tested documentation techniques and practices that are suitable for meeting today's requirements. DITA is therefore excellently suited as a source format for equipping authors for fast and diverse developments.

Using DITA as a basis, you can efficiently set up a productive documentation environment for topic-based structuring, and develop it flexibly according to your needs.

The most notable advantages offered by DITA concepts and standardization are cost-savings and investment security. DITA provides a framework within which authors can start work immediately—without long-lasting structure-finding

© The Author(s) 2016
S. Closs, *DITA – the Topic-Based XML Standard*,
SpringerBriefs in Applied Sciences and Technology,
DOI 10.1007/978-3-319-28349-4_14

processes that cost a lot of money. It is to be hoped that growing proliferation and broad tool support will ensure a long-term available technical basis.

Moreover, the wide use of the same basic architecture across corporate boundaries promotes the continuous growth of structuring know-how for information products from which new standards can be created. New standards can arise not only for contents at the lowest level but also for the resulting information products, such as different types of manuals, online helps, and mobile contents. In turn, the wide use of such standards will make exchangeability and automation possible in far more dimensions from those of today. Furthermore, authors can save a lot of time and work that they can profitably invest in the content.

The more DITA is used, the more the authors are supported by tools. The output of DITA sources to EPUB format already exists. There are Wikis on Drupal basis that use DITA as their source format, and more and more tools are being created to generate contents automatically for mobile display from DITA-tagged data.

Summary

- DITA is an established standard suitable as a source format in any area and for all types of information products.
- Topic-based structuring is a good basis for meeting today's information management requirements flexibly and efficiently.
- Topic models (templates) of a class concept simplify the creation process and give authors more time for the actual contents.
- From the user's point of view, topics meet today's information expectations better than chapters or complete books. Good topics are short and easy to understand. They provide brief and targeted information. Equipped with metadata, they are suitable for just-in-time answers that can automatically be found on the basis of the current situation and the metadata.
- Standard environments can exchange contents with other standard environments and profit from the further developments of the standard and its tools.

Appendix
DITA 1.3 Overview

DITA 1.3 was released by OASIS on December 17, 2015 and is available on the OASIS web site [17].

This appendix summarizes the new DITA 1.3 features.

A.1 DITA 1.3 Editions

Version 1.3 is organized in three parts called editions:

- Base edition
- Technical content edition
- All-inclusive edition

Each edition is targeted at a different audience:

Figure A.1 from the DITA 1.3 overview, part 0, shows how the editions are related.

Base Edition

This edition contains the core DITA pieces for topic, map, and subject scheme map.

The base edition is designed for users who need only topics and maps and do not need the classic topic types of technical documentation such as concept, task, and reference.

This edition can be used just for authoring topics. It also can be used to develop specializations that do not need the classic topic types.

Technical Content Edition

This edition includes the base edition and the specializations for information typing: concept, task, and reference topics; machine industry task; troubleshooting topic; glossaries, bookmap, and classification map.

It is designed for authors who use topic types to modularize their content.

© The Author(s) 2016
S. Closs, *DITA – the Topic-Based XML Standard*,
SpringerBriefs in Applied Sciences and Technology,
DOI 10.1007/978-3-319-28349-4

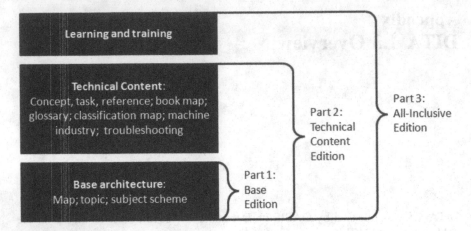

Fig. A.1 DITA 1.3 editions

All-Inclusive Edition

This edition includes the technical content edition and the specializations for learning and training.

It is designed for users who want all OASIS-approved specializations, as well as users who develop learning and training materials.

A.2   What's New in DITA 1.3

DITA 1.3 provides a variety of enhancements.

Table A.1 gives an overview of the DITA 1.3 features.

A.3   Enhanced Support for Troubleshooting Information

DITA 1.3 offers various ways of providing troubleshooting information:

- `note` element
- `task` topic type
- `troubleshooting` topic type

Troubleshooting enhancements for `note` element

With the new value `trouble` for the `type` attribute of the `note` element, a troubleshooting-related note can be explicitly identified.

Table A.1 DITA 1.3 features

For	DITA 1.3 enhancement
Troubleshooting	Troubleshooting support is described in more detail in the following section

User assistance	• New attributes for `resourceid`:
	  `appid`: to specify an ID for an application
	  `ux-context-string`: to specify a context id for the topic
	  `ux-source-priority`: to specify how to resolve conflicts between `resourceid` definitions that exist in both a map and a topic
	  `ux-windowref`: to reference a window
	• New element, `ux-window`, to specify a window or viewport in which an online help topic or Web page is displayed

Version management	• Further elements in topic prolog for version management • Release notes can be generated automatically
Variant management	• Branch filtering, which allows the application of different filters (`ditaval` conditions) to specific topic collections (branches) in a map • Scoped keys that support key definitions at different locations within a map structure. If a topic is re-used in several submaps, different key values can be specified, depending on the submap in which the topic appears • Expanded syntax for filtering attributes
Addressing	• New syntax for addressing an element within the same DITA topic (`#./label`) • Attribute `keyref` for `object` and `param` elements • Cross-deliverable linking
Sorting	• New `sort-as` element for use in sortable elements such as `title`, `searchtitle`, `navtitle`, `glossterm`, `dt`, `entry`, `stentry` where the base content is inadequate for sorting. The `sort-as` content is combined with the base content to construct the effective sort phrase
Deliverables	• New facility for key-based, cross-deliverable referencing • New conditional-processing attribute `deliveryTarget` replaces the `print` attribute and can have controlled values by using a subject scheme • Attribute `cascade` provides more control over cascading of metadata
Vocabulary	• Table extensions • `draft-comment`, `ph`, and `text` allowed in more places • More values for `style` attribute in a ditaval file • New `line-through` and `overline` elements • New `div` element for grouping elements in a topic
Learning and training	• New `learningObjectMap` • New `learningGroupMap` • New base domain and specialized domain for question-and-answer interactions
Integration of other standards	MathML equations and SVG figures can be directly integrated into DITA topics
Specialization	Parts of structural specializations can be reused by other structural specializations without requiring one to be specialized from the other

Troubleshooting enhancements for `task` topic type

The new element `steptroubleshooting` in the content model for `step` is intended to contain information that might assist users when a step does not complete successfully or does not produce the expected result.

The new section `tasktroubleshooting` in the `task` topic type is intended to contain information that might assist users when a task does not produce the expected result or complete successfully.

DITA troubleshooting topic

A troubleshooting topic describes how a problem can be solved.

It begins with a condition that describes the problem, followed by one or more cause-remedy pairs. Each cause-remedy pair describes a potential solution to the problem described in the condition.

Table A.2 shows specific `troubleshooting` elements in alphabetical order.

Example of a troubleshooting topic

For coffee which smells sour, the troubleshooting topic in Fig. A.2 describes how to solve the problem.

The DITA source for this troubleshooting topic is shown in Fig. A.3.

Table A.2 Selection of specific `troubleshooting` elements

Element	For marking
cause	One potential source of the problem described in the `condition` element
condition	Problem to be solved
remedy	Step-by-step procedure as a potential solution for one cause of the problem described in the `condition` element
responsibleParty	Party responsible for performing a remedy procedure
troublebody	Body of a troubleshooting topic
troubleshooting	Root of a troubleshooting topic
troublesolution	Container element for cause and remedy information

Coffee too sour

Condition

Your coffee smells hollow and sourish.

Cause: Coffee is too fresh

If the roasted coffee is too fresh, you will experience a lot of bubbles or quick dissipation.

Remedy

Allow the beans a couple of days to settle and degas.

Keep in mind that coffee will go stale 21 days past the roast date.

Cause: Not enough time to extract

Coffee that's too sour to the taste hasn't had enough time to extract.

Remedy

Let the coffee draw for 1-2 minutes longer.

By adapting the brewing time, you make sure that the coffee retains all important flavorings, and the bitter substances are not dissolved out of the coffee grounds.

Cause: Coffee grind too large

Particle size determines exactly how your coffee will extract, so a grind size that's large will require more extraction time than a grind size that's finer.

Remedy

1. Bring your grind down a bit.
2. Try again.

Fig. A.2 Coffee troubleshooting

```
<troubleshooting id="trblshtng">
    <title>Coffee too sour</title>
  - <troublebody>
    - <condition>
          <title>Condition</title>
          <p>Your coffee smells hollow and sourish.</p>
      </condition>
    + <troubleSolution>
    - <troubleSolution>
        - <cause>
              <title>Cause: Not enough time to extract</title>
              <p>Coffee that's too sour to the taste hasn't had enough time to
                  extract.</p>
          </cause>
        - <remedy>
              <title>Remedy</title>
            - <steps>
                - <step>
                      <cmd>Let the coffee draw for 1-2 minutes longer.</cmd>
                      <info>By adapting the brewing time, you make sure that the coffee
                          retains all important flavorings, and the bitter substances are
                          not dissolved out of the coffee grounds. </info>
                  </step>
              </steps>
          </remedy>
      </troubleSolution>
    + <troubleSolution>
  </troublebody>
</troubleshooting>
```

Fig. A.3 DITA troubleshooting topic

References

1. Bellamy L, et al (2012) DITA best practices: a roadmap for writing, editing, and architecting in DITA. IBM Press, Upper Saddle River
2. Closs S (2011) Single source publishing: modularer content für EPUB & Co., 2nd edn. entwickler.press, Frankfurt
3. Closs S (2014) Informationsarchitektur—Junge Disziplin mit großer Zukunft. Dok Magazin 3:69–72
4. Closs S (2014) Contextual content. tcworld e-magazine. http://www.tcworld.info/e-magazine/technical-communication/article/contextual-content/. Accessed 14 Aug 2015
5. Day D, Hargis G, Priestley M (2005) Frequently asked questions about the Darwin information typing architecture. http://www.ibm.com/developerworks/library/x-dita3/#N104. Accessed 14 Aug 2015
6. DITA OT: DITA open toolkit. http://www.dita-ot.org/download. Accessed 14 Aug 2015
7. DITA XML.org. http://dita.xml.org. Accessed 14 Aug 2015
8. Fritz M (2008) DITA in der Technischen Kommunikation—eine Entscheidungshilfe für den Einsatz, tekom, Stuttgart
9. Glushko RJ (2013) The discipline of organizing. MIT, Cambridge
10. Horn RE (1986) Engineering of documentation—the information mapping approach. Information Mapping Inc, Waltham
11. Java. https://www.java.com/en/download/. Accessed 24 Aug 2015
12. Muthig J, Schäflein-Armbruster R (2014) "Funktionsdesign®—methodische Entwicklung von Standards". In: Muthig J (Hrsg) Standardisierungsmethoden für die Technische Dokumentation 2nd Edition tcworld GmbH, Stuttgart, pp 41–73. See also http://www.tcworld.info/e-magazine/technical-communication/article/technical-documentation-needs-standardization/
13. MS Help WS: MS HTML help workshop. http://www.microsoft.com/en-us/download/details.aspx?id=21138. Accessed 14 Aug 2015
14. OASIS (2010) Darwin information typing architecture (DITA) Version 1.2. http://docs.oasis-open.org/dita/v1.2/spec/DITA1.2-spec.pdf. Accessed 14 Aug 2015
15. Rockley A (2003) Managing enterprise content, a unified content strategy. New Riders, Berkeley
16. DITA 1.3: Why three editions? http://docs.oasis-open.org/dita/dita-1.3-why-three-editions/v1.0/dita-1.3-why-three-editions-v1.0.html. Accessed 30 Jan 2016
17. OASIS package of the complete DITA 1.3 specifications and related files. http://docs.oasis-open.org/dita/dita/v1.3/os/dita-v1.3-os.zip. Accessed 30 Jan 2016

© The Author(s) 2016
S. Closs, *DITA – the Topic-Based XML Standard*,
SpringerBriefs in Applied Sciences and Technology,
DOI 10.1007/978-3-319-28349-4

Index

A
Access, 51
Attributes for filtering, 37
audience, 37

B
Benefits, 51
Book-oriented production, 1
Branch filtering, 38
Branches, 33

C
Cascading stylesheet, 42
Class concept, 13, 51
Class concept method, 11
Classic cross-references, 25
Classification, 9
Collaboration, 39, 51
Concept, 9
Conref mechanism, 27
Constraints, 39
Content reuse, 5
Context-independent, 8
Conventions, 39
Cost-savings, 51
Cross references, 23

D
Darwin information typing architecture, 5
Defining variable, 36
deliveryTarget, 37
Designing topic models, 13
Developing aâ£class concept, 11
Direct addressing, 31
DITA block elements, 15
DITA community, 4
DITA concept, 17

DITA glossentry, 19
DITA history, 5
DITA inline elements, 16
DITA map, 22
DITA open toolkit, 4, 43
DITA reference, 18
DITA specialization, 49
DITA task, 16
DITA topic, 8
DITA topic types, 9
ditaval, 37
Documentation of DITA standard, 4

E
Embedding, 27
Exchangeability, 52

F
Filter attributes, 37
Filter file ditaval, 37
Filtering, 37

G
Generating output withâ£DITA open toolkit, 46
Generic standard, 5
Glossentry, 9
Glossentry topic, 31

I
Indirect addressing, 31
Information product, 41
Information Typing, 5
Installing DITA open toolkit, 43

J
Just-in-time answers, 52

© The Author(s) 2016
S. Closs, *DITA – the Topic-Based XML Standard*,
SpringerBriefs in Applied Sciences and Technology,
DOI 10.1007/978-3-319-28349-4

Printed in the United States
By Bookmasters

Lecture Notes in Physics

Volume 957

The Lecture Notes in Physics

The series Lecture Notes in Physics (LNP), founded in 1969, reports new developments in physics research and teaching-quickly and informally, but with a high quality and the explicit aim to summarize and communicate current knowledge in an accessible way. Books published in this series are conceived as bridging material between advanced graduate textbooks and the forefront of research and to serve three purposes:

- to be a compact and modern up-to-date source of reference on a well-defined topic
- to serve as an accessible introduction to the field to postgraduate students and nonspecialist researchers from related areas
- to be a source of advanced teaching material for specialized seminars, courses and schools

Both monographs and multi-author volumes will be considered for publication. Edited volumes should, however, consist of a very limited number of contributions only. Proceedings will not be considered for LNP.

Volumes published in LNP are disseminated both in print and in electronic formats, the electronic archive being available at springerlink.com. The series content is indexed, abstracted and referenced by many abstracting and information services, bibliographic networks, subscription agencies, library networks, and consortia.

Proposals should be sent to a member of the Editorial Board, or directly to the managing editor at Springer:

Lisa Scalone
Springer Nature
Physics Editorial Department
Tiergartenstrasse 17
69121 Heidelberg, Germany
Lisa.Scalone@springernature.com

More information about this series at http://www.springer.com/series/5304

Timo A. Lähde • Ulf-G. Meißner

Nuclear Lattice Effective Field Theory

An Introduction

 Springer

Timo A. Lähde
Institute for Advanced Simulation
and Institute of Nuclear Physics
Forschungszentrum Jülich
Jülich, Germany

Ulf-G. Meißner
Helmholtz-Institut für Strahlen-und
Kernphysik (Theorie)
University of Bonn
Bonn, Germany

Institute for Advanced Simulation
and Institute of Nuclear Physics
Forschungszentrum Jülich
Jülich, Germany

ISSN 0075-8450   ISSN 1616-6361   (electronic)
Lecture Notes in Physics
ISBN 978-3-030-14187-5   ISBN 978-3-030-14189-9   (eBook)
https://doi.org/10.1007/978-3-030-14189-9

Library of Congress Control Number: 2019935140

This Springer imprint is published by the registered company Springer Nature Switzerland AG.
The registered company address is: Gewerbestrasse 11, 6330 Cham, Switzerland

Preface

Nuclear physics is a rich and fascinating field. Atomic nuclei are self-bound systems that make up a large chunk of the visible matter in the universe. Still, as they are governed by the strong interactions, the rich toolbox of theoretical nuclear physics can not yet cope with all the intricate phenomena observed, especially in some parts of the nuclear landscape that are of utmost importance for the element generation in stars. At present, accurate calculations can only be done if the neutrons and protons that make up the nucleus are not resolved into their quark-gluon substructure. This time-honored approach to the physics of atomic nuclei is pursued worldwide, and with the advent of radioactive beam facilities on various continents, much larger portions of the nuclear landscape can now be investigated both experimentally and theoretically.

The newest member of the nuclear theory toolbox is *nuclear lattice effective field theory*, whose foundations are laid out in this book. It combines the methodology of effective field theory for the forces between nucleons with simulation techniques such as Monte Carlo methods. Such stochastic methods allow for a numerically exact solution of the nuclear A-body problem, and combining these with more and more accurate nuclear forces enables one to go into directions not possible before. As will be stressed throughout this book, nuclear structure and reactions must be considered together and the appropriate methods to perform the pertinent calculations will be presented.

The aim of this book is to develop the underlying methods in detail and not so much reviewing all the interesting results obtained. We expect that this book will not only function as a useful reference for the theoretical and computational methods necessary to perform nuclear lattice effective field theory calculations but will also provide a solid understanding of the current status and future scope of this approach. We assume that the reader has some basic knowledge in quantum mechanics, many-body physics, group theory, and computational physics. In order to maintain a self-contained presentation, we have nevertheless included some of the basic formalism on chiral effective field theory, on the construction of lattice theories in terms of Grassmann fields and transfer matrices, as well as on the basics of scattering theory and Markov chains. We have, on purpose, made no effort to compactify the notation but rather decided to keep it as explicit as possible, which makes coding easier. The main body of the book deals with the nuclear Hamiltonian on the lattice at next-to-next-to-leading order, only in the last chapter we give a short summary of more recent developments. We expect that our book provides sufficient detail to enable the reader to easily comprehend such current and future developments. Finally, we hope that our book will inspire the reader to undertake their own state-of-the-art calculations in nuclear physics, or in other closely related fields of study.

The work presented here has been done as part of the "Nuclear Lattice Effective Field Theory Collaboration". We acknowledge support and insight of all the members who have contributed over the years. Special thanks go to Dean Lee, who has been developing and driving this method from the start and also to Serdar Elhatisari and Evgeny Epelbaum. We are also grateful to Véronique Bernard, Feng-Kun Guo, Thomas Luu, Bernard Metsch, and Carsten Urbach for carefully reading parts of this book. They can, however, not be held responsible for any error or unclear formulation. We further acknowledge the ongoing support by the Jülich Supercomputing Centre, which provided most of the computing resources required to perform the simulations described here. The financial support from the Deutsche Forschungsgemeinschaft, the VolkswagenStiftung, the Helmholtz Society, and the Chinese Academy of Sciences is greatfully acknowledged.

Finally, many thanks also go to Christian Caron who showed great patience and supported us throughout.

Jülich, Germany   Timo A. Lähde
Bonn, Germany   Ulf-G. Meißner

Contents

Chapter 1
Introduction to Effective Field Theory

In this chapter, the concepts underlying *Effective Field Theory* (from now on EFT, for short) will be discussed. This is essentially done as an introduction for the reader not familiar with these concepts. The EFT description of the nuclear forces in the continuum will be discussed in Chap. 2, so a more advanced reader might go to that chapter directly. Before embarking on the full machinery of Quantum Field Theory (QFT), it will be useful to study two simple examples from classical physics, namely the question why the sky is blue and how Newtonian gravity acts on human length scales. This will allow us to introduce some basic ingredients of EFT. The multipole expansion of classical electrostatics serves to further illustrate the separation into short- and long-distance physics, which is one of the cornerstones of any EFT. Two classical examples, namely Fermi's four-fermion approach to the weak interactions and the Euler-Heisenberg-Kockel effective Lagrangian for light-by-light scattering will help to familiarize the reader with some of the concepts underlying EFT. We then discuss the paradigm underlying effective field theory, and the so-called fundamental "theorem" due to Weinberg [1]. We put the word theorem in quotation marks since no general proof is given but the statement is so general that it is hard to imagine any counter-example, see also the discussion in Sect. 1.2. Then, the central issue of power counting, that is the organization of the various terms contributing to an observable under consideration, will be developed. For simplicity, we only consider a theory of interacting Goldstone bosons, more general formulations will be given in later chapters. Next, the meaning of the coupling constants appearing in the EFT expansion will be discussed and finally, we show how to deal with loop diagrams and the corresponding divergences. We also briefly discuss the complications with the power counting when including heavy degrees of freedom in the effective theory. We end this introduction with a summary of the concepts underlying EFTs. All this is more general than the EFT for nuclear physics discussed in Chap. 2, but we consider it important that the reader gets an idea of the broader perspective of these topics.

© Springer Nature Switzerland AG 2019
T. A. Lähde, U.-G. Meißner, *Nuclear Lattice Effective Field Theory*,
Lecture Notes in Physics 957, https://doi.org/10.1007/978-3-030-14189-9_1

1.1   Introducing the Concepts

Let us consider a few examples which are usually not analyzed in terms of EFT. Exact solutions to these problems exist, so that it is particularly simple to discuss the strengths and limitations of EFT. Our aim will be to introduce some concepts of EFT using these familiar examples. Note, however, that these do not exhibit all the intricacies of the full quantum theoretical treatment. We also briefly discuss two classical examples which actually predate the modern notion of EFT.

1.1.1   Why the Sky Is Blue

We consider the scattering of visible light on atoms in the atmosphere. Two very different length scales are involved. While the wavelength of visible light is a few thousand angstroms (Å), typical atomic sizes are a few a_0, that is a few Å, with $a_0 \simeq 0.53 \, \text{Å} = 0.53 \times 10^{-10} \, \text{m}$ the Bohr radius known from the quantum-mechanical treatment of the hydrogen atom. This entails a *separation of scales*,

$$\lambda_{\text{light}} \approx 5000 \, \text{Å} \gg \text{a few } a_0 \approx \text{a few Å},$$

or equivalently, a separation of energies since e.g. the photon energy is given by $\omega = c/\lambda$, and $\omega \ll E_{\text{atomic}} \sim 1/a_0$. Note that we have exceptionally displayed the speed of light in this formula. In general, we use rational units with $c = \hbar = 1$. The photons are thus insensitive to the atomic structure. It is now easy to write down an effective Hamiltonian that describes this process. To leading order, gauge invariance of the electromagnetic interactions together with the discrete symmetries (parity and time reversal invariance) requires

$$\mathcal{H}_{\text{eff}} = \chi^* \left[-\frac{1}{2} C_E \, \mathbf{E}^2 - \frac{1}{2} C_B \, \mathbf{B}^2 \right] \chi + \dots \qquad (1.1)$$

where χ denotes the atomic wave function and the ellipses stands for higher order terms in the electric ($\mathbf{E}$) and magnetic ($\mathbf{B}$) field strengths. The constants C_E and C_B are called *low-energy constants* as they parameterize the response of the low-energy photons to the atoms. Note that the numerical values of these constants are not fixed by any symmetry requirement, so in what follows we must find some means of extracting or estimating their values. The alert reader will have noticed that in Eq. (1.1) the Hamilton density has been written down, that is the energy of the electromagnetic field per volume. Consequently, the constants $C_{E,B}$ have dimension volume, so that we can write $C_E = k_E \, a_0^3$ and $C_B = k_B \, a_0^3$, as the Bohr radius is the only atomic length scale at our disposal. Here, $k_{E,B}$ are dimensionless constants. The determination of the $C_{E,B}$ has thus boiled down to pinning down the values of the $k_{E,B}$. In the absence of more detailed information, we assume *naturalness* of the low-energy constants $k_{E,B}$. This means that when one has normalized to the scale appropriate to the system, such coupling constants should be of order one, $k_{E,B} \sim \mathcal{O}(1)$. More precisely, the magnitude of $k_{E,B}$ is of order one since we do not even know their signs. Naturally, there are instances where such arguments are invalidated, but usually the physics behind unnaturally large or small couplings is understood. In fact, if one uses simple quantum mechanics for calculating Rayleigh scattering, one can determine these constants. We will not do this here but rather leave it as an exercise for the reader. Now both the magnitude of the electric and the magnetic field strengths are proportional to the photon energy, $|\mathbf{E}| \sim \omega$ and $|\mathbf{B}| \sim |\mathbf{k}| \sim \omega$, with $\mathbf{k}$ the photon momentum. Armed with this, we can deduce the cross section for light-atom scattering,

$$\frac{d\sigma}{d\Omega} = |\langle f | \mathcal{H}_{\text{eff}} | i \rangle|^2 \sim \omega^4 a_0^6 \left(1 + d \frac{\omega^2}{(\Delta E)^2} \right),$$ (1.2)

which is the desired result: The visible photons with higher energies (blue light) get scattered the strongest. We have also indicated in the round brackets corrections to the leading order result that stem from the excitation of atomic levels characterized by some energy scale ΔE and d is a numerical constant depending on the atomic structure. The above-mentioned assumption that the photons are insensitive to the structure of the atoms also means that atomic excitations are given by very different energy scales than the photon energy, so that $\omega \ll \Delta E$ and such corrections are small. The most naive estimate of ΔE is given by the ionization energy of the hydrogen atom, $\Delta E = 13.6$ eV. This duality between the excitation spectrum and the structure is very useful for exploring the properties of extended particles, but will not be considered in detail in what follows. For a more thorough discussion of light-atom scattering in EFT, see e.g. Ref. [2].

1.1.2 Newtonian Gravity: An Apple Falling from a Tree

Another instructive example from classical physics is given by an apple falling down from a tree. The height of the tree is denoted by r and the force acting on the apple is given by Newton's law, $\mathbf{F} = m\mathbf{g}$, with m the mass of the apple and $\mathbf{g}$ the gravitational acceleration pointing towards the center of the earth. An effective theory for this process can be obtained by expanding the gravitational interaction between the apple and the earth,

$$\begin{aligned} \frac{F}{m} &= \frac{G M}{(r + R)^2} \\ &= \frac{G M}{R^2} - \frac{2 G M}{R^3} r + \frac{3 G M}{R^4} r^2 + \dots \\ &= g \left(1 - 2x + 3x^2 + \dots \right), \end{aligned}$$ (1.3)

introducing the constant of acceleration g, expressed in terms of the gravitational constant G, the mass of the earth M and its radius R. Furthermore, x is a dimensionless *small parameter*, it measures the height of the tree in units of the earth radius, $x = r/R$. Numerically, $x = 3\,\text{m}/6800\,\text{km} \simeq 10^{-7}$, so that the expansion in Eq. (1.3) is rapidly converging for human length scales. Clearly, the radius of convergence of this expansion is given by the large scale R, which can therefore be considered the breakdown scale of the effective theory. More precisely, for Eq. (1.3) to be useful, we must require $x \ll 1$ (we ignore here the fact that the surface of the earth is a natural limit for an object falling off a tree). This expansion captures quite nicely the general philosophy of EFT. First, one must identify the relevant scales, here the height of the tree and the radius of the earth. It is then possible to write down an effective interaction (force) as a Taylor expansion in the small parameter x. As will be explained later, many interesting systems will not allow for simple Taylor expansions due to the absence of a mass gap. In writing down the expansion equation (1.3), one has to preserve the symmetries of the underlying theory, in this case e.g. rotational invariance. The strength of the EFT approach lies in the fact that a few constants capture the complicated dynamics,

$$F(x) = mg\,(1 + c_1 x + c_2 x^2 + \ldots)\,. \tag{1.4}$$

Here, g has to be determined from measurements (or can be deduced from the parameters of the underlying gravitational theory) and the higher order couplings c_i are integers, $c_1 = -2$ and $c_2 = 3$, due to the simple Taylor expansion performed. Note that these couplings are of natural size, that is of order one. Here we are faced with the situation that the underlying theory is known and thus the numerical values of the low-energy constants can be deduced. In more general cases, one is in a less fortunate situation and must determine these couplings from experiment or by performing some modeling. The expansion equation (1.3) also shows that one can work to a given accuracy by going to high enough orders in x. Furthermore, the theoretical uncertainty can be estimated, assuming c_3 to be natural, our calculation to second order in x should be modified by corrections of order $c_3 x^3 \sim 10^{-21}$. Of course, for larger values of x, one would need to keep more terms to reach such an accuracy. Thus, we have a well-defined expansion in a small parameter, which amounts to a *power counting*. The observable, here the gravitational force acting on the falling apple, is expressed as a power series in a small parameter with coefficients that can be calculated or must be determined from experiment. This concept will be worked out in more detail in the following paragraphs.

1.1.3   The Multipole Expansion

Consider a localized charge distribution ρ_1 which is centered at some point $\mathbf{r}'$ in space and whose extension is characterized by some scale R, see Fig. 1.1a. If we want to analyze its effects at some far away point $\mathbf{r}$, with $|\mathbf{r}' - \mathbf{r}| \gg R$, we can use the familiar expansion

$$\frac{1}{|\mathbf{r'}-\mathbf{r}|} = \sum_{l=0}^{\infty} \sum_{m=-l}^{l} \frac{4\pi}{2l+1} \frac{r_<^l}{r_>^{l+1}} Y_{lm}^*(\theta',\phi') Y_{lm}(\theta,\phi) \, ,$$

with $r_> = \max(r, r')$, $r_< = \min(r, r')$ and $r(r')$ the modulus of $\mathbf{r}(\mathbf{r'})$. This leads to the multipole expansion of the electrostatic potential

$$\Phi(\mathbf{r}) = \sum_{l,m} \frac{4\pi}{2l+1} Y_{lm}(\Omega) \int d^3r' Y_{lm}^*(\Omega') \frac{(r')^l}{r^{l+1}} = \sum_{l,m} \frac{4\pi}{2l+1} Y_{lm}(\Omega) \frac{Q_{lm}}{r^{l+1}} \, .$$

Here, $\Omega = (\theta, \phi)$ collects the angular variables and the Y_{lm} are the conventional spherical harmonics, as defined in Appendix A. The multipole moments Q_{lm} describe the charge distribution as seen from a large distance. In a Cartesian basis, we have

$$\Phi(r) = \frac{q}{r} + \frac{\mathbf{p}\cdot\mathbf{r}}{r^3} + \dots \, , \quad \mathbf{p} = \int d^3r' \mathbf{r'} \rho_1(\mathbf{r'}) \, ,$$

with q the total charge collected in ρ_1 and $\mathbf{p}$ the dipole moment. These terms correspond to short-distance operators, the short-range physics is expanded in the small parameter r'/r and the multipole moments correspond to the various coupling constants of the operators. In the case at hand, a measurement of the charge, the dipole moment and so on, allows one to reconstruct the charge distribution ρ_1. Note furthermore that the leading term in this expansion is only sensitive to global properties like the total charge and the finer structure of the charge distribution is only revealed if one can determine a certain number of the multipole moments. This is another general feature of EFT, low energies means that one has a low resolution. This might sound discouraging at first, but the multipole expansion also neatly demonstrates that with a certain amount of effort one is able to learn more about the system than just its charge. Note that for $r' \simeq r$, the multipole expansion ceases to be a useful tool to analyze the system. This is expected since then we no longer have the necessary scale separation.

Now let us complicate the situation somewhat and place a second charge distribution ρ_2 around the observation point $\mathbf{r}$, see Fig. 1.1b. If we now calculate the electrostatic potential, we have to explicitly account for this,

$$\Phi(\mathbf{r}) = \sum_{l,m} \frac{4\pi}{2l+1} Y_{lm}(\Omega) \frac{Q_{lm}}{r^{l+1}} + \int d^3r' \frac{\rho_2(\mathbf{r'})}{|\mathbf{r'}-\mathbf{r}|} \, .$$

The last term in this equation corresponds to the explicit inclusion of "long-range" physics. This is a feature of many physical systems that are characterized by different scales. The EFT amounts to the explicit treatment of the low-energy modes and a parametrization of the short-distance physics in terms of local operators

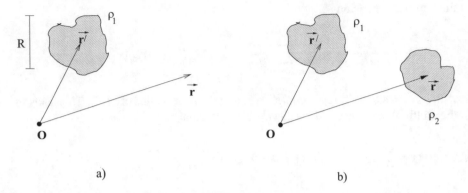

Fig. 1.1 (**a**) A charge distribution analyzed from far distances. (**b**) Two well separated charge distributions

with a priori unknown coupling constants. For this to make sense, the two charge distributions must be well separated and should not overlap.

Finally, we stress again that these examples are somewhat simple and do not cover all aspects of EFTs, in particular the appearance of loop corrections related to unitarity and the like. However, we have already seen a few of the underlying concepts at work. Before developing the full machinery, it is worth considering two classical examples that predated the notion of EFT by decades.

1.1.4 Two Classical Examples

Here, we briefly consider Fermi's weak interaction theory and the Euler-Heisenberg-Kockel Lagrangian for light-by-light scattering.

In 1933, Fermi proposed a theory of the weak interactions of the current-current form that included the massless neutrino postulated by Pauli a few years before [3]. In the electroweak Standard Model (SM), charged-current weak decays are mediated by W-bosons, which have a mass of about $80\,\mathrm{GeV}$. In typical low-energy processes like neutron β-decay, $n \rightarrow pe^-\bar{\nu}_e$, or in kaon decays like $K \rightarrow \pi \ell \nu$, where ℓ denotes a lepton, the energy release is about one MeV and a few $100\,\mathrm{MeV}$, respectively, which is much smaller than M_W. The fundamental process of the conversion of a down-quark into an up-quark through the emission of a W-boson that further decays into a lepton pair can be mapped onto a local four-fermion interaction of the Fermi type,

$$\frac{e^2}{8\sin\theta_W} \times \frac{1}{M_W^2 - q^2} \xrightarrow{q^2 \ll M_W^2} \frac{e^2}{8M_W^2 \sin\theta_W}\left\{1 + \frac{q^2}{M_W^2} + \dots\right\} = \frac{G_F}{\sqrt{2}} + \mathscr{O}\left(\frac{q^2}{M_W^2}\right),$$

Fig. 1.2 Light-by-light scattering. (**a**) It is given by the box graph in full QED. (**b**) Its low-energy approximation. Here, wiggly (solid) lines denote photons (electrons, positrons)

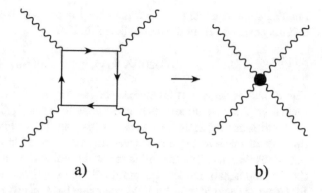

a)   b)

which allows one to express the phenomenological Fermi coupling constant G_F in terms of the SM parameters like the electric unit charge e, the weak mixing angle $\sin^2 \theta_W \simeq 1/4$ and the W mass. This is one of the cases where the high-energy theory is known and the parameters of the low-energy EFT can in fact be calculated. History has, of course, been quite different and the Fermi theory was abandoned as its cross section scales with energy squared, thus violating unitarity at $E \simeq 100$ GeV. Today we understand that this only means that we have pushed the EFT beyond its breakdown scale and need the full SM to perform systematic and precise calculations at such energies. In modern language, the electroweak SM is called an UV(ultraviolet)-completion of the Fermi theory.

In 1935/1936, Euler, Heisenberg and Kockel analyzed light-by-light scattering for photon energies ω well below the electron mass m_e, $\omega \ll m_e$ [4, 5]. Therefore, the photon-photon interaction must be representable by local interaction terms, see Fig. 1.2b. The only gauge invariant operators generated from the photon field four-vector A_μ are the field strength tensor $F_{\mu\nu} = \partial_\mu A_\nu - \partial_\nu A_\mu$ and its dual, $\tilde{F}_{\mu\nu} = (1/2)\epsilon_{\mu\nu\rho\sigma} F^{\rho\sigma}$, with $\epsilon_{\mu\nu\rho\sigma}$ the totally anti-symmetric Levi-Civita tensor. The latter is odd under parity and therefore can only appear quadratically, as the Lagrangian must be a Lorentz scalar and even under parity. This leads to the lowest order form

$$\mathscr{L}_{\text{eff}} = -\frac{1}{4} F_{\mu\nu} F^{\mu\nu} + \frac{\alpha^2}{360 m_e^4} \left[4(F_{\mu\nu} F^{\mu\nu})^2 + 7(F_{\mu\nu} \tilde{F}^{\mu\nu})^2 \right] + \dots , \qquad (1.5)$$

where the first term describes the free photon field and we have not considered gauge fixing terms, and the ellipses stands for terms with more derivatives. The resulting cross section for light-by-light scattering from the leading order four-photon interaction follows as

$$\sigma(\omega) = \frac{1}{(4\pi)^4} \frac{973}{10125\pi} \frac{e^8}{m_e^2} \left(\frac{\omega}{m_e} \right)^6 ,$$

which agrees with the small-ω limit of the full Quantum Electrodynamics (QED) cross section, that was calculated in 1950 by Karplus and Neuman [6]. In modern language, one has integrated out the heavy degrees of freedom, here the electrons and the positrons, from the theory (see Fig. 1.2),

$$\mathscr{L}_{\text{QED}}[\psi, \bar{\psi}, A_\mu] \to \mathscr{L}_{\text{eff}}[A_\mu] \, .$$

These massive degrees of freedom leave their imprint in the LECs accompanying the local four-photon operators, as these scale with $(e/m_e)^4$, since we have four vertices $\sim e$ and four propagators $\sim 1/m_e$ in diagram (a) of Fig. 1.2. We will encounter this role of massive degrees of freedom not included in the EFT on various other occasions. We note that the influence of the Euler-Heisenberg-Kockel approach on the development of modern quantum field theory is discussed in Ref. [7]. Finally, a first measurement of light-by-light scattering has recently be reported by the ATLAS collaboration at CERN [8].

1.2   Fundamental Theorem and Paradigma of EFT

Quantum field theories have led to some of the most astonishing and precise predictions in all of physics. An important pillar in the construction of abelian and non-abelian QFTs that combine to the so successful Standard Model of the strong and electroweak interactions, is the notion of renormalizability. This can be explained in a nutshell as follows: When calculating loop diagrams in QFT, divergences generated from the high-energy physics appear thus leading to infinities. To tame these, renormalization is used, i.e. the divergences can be shifted into a few parameters, whose physical and finite parts are determined from experiment. In such a renormalizable theory, more complicated loop graphs generate more divergences, but these too are made finite without invoking new parameters. A prime example is Quantum Electrodynamics, where a proper renormalization of the electron charge e and the electron mass m_e leads to a finite theory. The same program can be carried through, despite a few technical complications, for the non-abelian gauge theories underlying the electroweak and the strong interactions. However, while this approach has been tremendously successful, nowadays it has become clear that no theory is expected to work at *all* scales, e.g. the Standard Model must break down at the Planck scale (or even much earlier). From this point of view, all theories are essentially effective field theories, where unknown physics is lumped into a few coefficients of some local operators. The basic idea about renormalization today is that the influences of higher energy processes are localizable in a few structural properties which can be captured by an adjustment of parameters. With this in mind, we now are in the position to recall the steps to build an EFT, following closely the methods and ideas underlying QFT.

Let us first recall how a conventional, renormalizable quantum field theory is constructed. As an archetype, we use QED to illustrate the steps, without going into any detail. The rules for building and applying a QFT can be cast in the following form:

Step 1 Construct the action *invariant* under the desired *symmetries* in terms of the appropriate asymptotic fields. In the QED case, these are electrons, positrons and photons. The massive fields ($e^{\pm}$) are given by the spinors ψ with mass m_e and the massless gauge field by the four-vector A_μ. The corresponding Lagrangian density $\mathscr{L}_{QED}$ must be a Lorentz scalar, and the discrete symmetries (parity, charge conjugation and time reversal invariance) as well as local U(1) gauge invariance should be fulfilled. This leads to

$$\mathscr{L}_{QED} = -\frac{1}{4}F_{\mu\nu}F^{\mu\nu} + \bar{\psi}(i\slashed{\partial} - e\slashed{A})\psi - m_e\bar{\psi}\psi \,, \tag{1.6}$$

employing the standard notation $\slashed{a} = a_\mu\gamma^\mu$, with the γ_μ 4×4 matrices that generate a Clifford algebra, as we are dealing with anticommuting fields. Further, $F_{\mu\nu} = \partial_\mu A_\nu - \partial_\nu A_\mu$ is the field strength tensor, and e the electron charge. The theory is obviously invariant under local U(1) transformations, $\psi \to \psi' = U(x)\psi$, $\psi^\dagger \to (\psi')^\dagger = \psi^\dagger U^\dagger$ and $A_\mu \to A'_\mu = UA_\mu U^\dagger - (i/e)(\partial_\mu U)U^\dagger$, with $U(x) = \exp(-ie\lambda(x))$. We have omitted gauge fixing terms here. Note that gauge invariance forbids a photon mass term in four space-time dimensions. We also remark that the invariance of the action is not entirely sufficient to define the theory: Some symmetries do not survive quantization leading to the so-called anomalies. This fine but very important point will not be discussed any further here.

Step 2 Keep only *renormalizable* interactions. This restricts the possible terms in the Lagrangian to operators with canonical dimension $d_c \leq 4$. Since boson (fermion) fields have canonical dimension $d_c(\phi) = 1$ ($d_c(\psi) = 3/2$), we see that the first two terms in Eq. (1.6) have dimension four whereas the fermion mass term has dimension three. A Pauli-term $\bar{\psi}\sigma_{\mu\nu}F^{\mu\nu}\psi$, which would lead to an electron anomalous magnetic moment already at tree level, is thus excluded since it has $d_c = 5$. It is easy to construct the table of all potentially renormalizable interactions, which is left as an exercise for the reader. For our purposes, we will rather have to learn how to construct all terms for a given set of fields and interactions.

Step 3 *Quantize* the theory and calculate scattering processes (and bound states). In perturbation theory, one has to consider tree and loop diagrams. For example, in QED these contributions are accompanied by the fine structure constant $\alpha = e^2/4\pi = 1/137.06$, as depicted for the photon-fermion vertex in Fig. 1.3. The interactions dress the bare vertex and thus the electron acquires an anomalous magnetic moment a_e, which is expressed in terms of the bare parameters, i.e. the parameters appearing in the unrenormalized Lagrangian equation (1.6). Furthermore, most loop diagrams are not finite. This requires renormalization, see the next step.

Fig. 1.3 Tree, one-loop and two-loop topologies for the photon-electron vertex. These scale as e, e^3 and e^5, respectively. Only one representative example is shown at one- and two-loop order

Step 4   Determine the physical parameters from *experiment* and express the predictions in terms of these. This amounts to regularization and renormalization. In fact, in the QED case all appearing divergences at any order are made finite by the shifts

$$m_{bare} \to m_{phys} , \quad e_{bare} \to e_{phys}.$$

The corresponding one-loop expressions for these shifts can be found in any textbook on QFT. This renormalization program can be carried out to all orders systematically and the corresponding QFT can in principle be applied at all scales (modulo the appearance of Landau poles and alike which will not concern us here). Expressing e.g. the anomalous magnetic moment of the electron in terms of the physical (finite) parameters, one obtains a perfectly finite result:

$$\mu_e = g_e \frac{e}{2m_e} \frac{\sigma}{2} , \quad a_e = \frac{1}{2}(g_e - 2) ,$$

$$a_e = \sum_{n\geq 1} A_n \left(\frac{\alpha}{\pi}\right)^n ,$$

$$A_1 = \frac{1}{2} ,$$

$$A_2 = \frac{197}{144} + \left(\frac{1}{2} - 3\ln 2\right) \zeta(2) + \frac{3}{4}\zeta(3) = -0.328478\ldots , \qquad (1.7)$$

in terms of the Riemann ζ-function, $\zeta(p) = \sum_{n=1}^{\infty} 1/n^p$ and $\zeta(2) = \pi^2/6$. The 2×2 σ-matrices represent the spin of the fermions. Nowadays, this expansion has been worked out to tenth order and including also electroweak and hadronic corrections, see Ref. [9] for a review.

If one is not bothered by the appearance of divergences and their removal through a systematic regularization and renormalization procedure, the steps 1, 3, and 4 seem logically necessary. However, step 2 refers to the issue of renormalizability alluded to at the beginning of this section, namely that the theory should be operative at all scales. While it is true that for operators with a canonical dimension $d > 4$ one finds divergences in all amplitudes when utilizing perturbation theory, these can be removed through order-by-order renormalization. Still, this would be a rather useless approach as one would have to calculate an infinite number of diagrams. However, it has been realized that there exists a second consistent and predictive paradigm. It amounts to keeping steps 1, 3, and 4 but substituting step 2 by:

Step **2*** Work at *low energies* and expand the theory in *powers* of the energy.

Here, we take the word energy to be synonymous with momentum, mass, or any small parameter related to the scales in the problem. This will become more transparent in what follows. It is important to note that the term "small" refers to some scale in the problem that separates the low-energy ($E \ll \Lambda$) from the high-energy part ($E \gg \Lambda$) of the theory under consideration. The scale Λ will be referred to as the breakdown scale (sometimes referred to as the hard scale) in what follows. For such a scenario to work, we must have scale separation, that is we must be able to separate clearly low from high energies. Also, it is trivial to generalize such a scheme to a succession of well separated regimes. Clearly, with the new rule 2*, there is no more need to restrict the possible terms in the Lagrangian to renormalizable interactions, in fact, to be consistent with the symmetries one needs to consider *all* terms that can be constructed from the basic fields. Of course, the number of such operators is not limited; for example, any positive power of $F_{\mu\nu}$ in an abelian gauge field theory is invariant under gauge transformations. At first sight, one might be discouraged by this proliferation of operators, but all these are endowed with a dimension of energy (not to be confused with the canonical dimension discussed earlier). Also, each operator is accompanied by a coupling constant. Thus if we work to a given order in the energy expansion, we only have to consider a finite number of operators and must pin down a finite number of couplings to specify the theory. Predictions of EFTs therefore amount to relations among observables, parameterized by these coupling constants. Such a consistent QFT is called effective because the procedure obviously fails at high energies, that is at energies of the order of the breakdown scale. As we will see later, the word effective also has another meaning, since EFTs are a tool to efficiently perform accurate calculations. To summarize, EFT is a fully consistent QFT formulation based on the calculation of tree and loop diagrams and performing order-by-order renormalization.

Before presenting the so-called fundamental theorem, we briefly pause to discuss the issue of infinities. Loop diagrams are in general ultraviolet (UV) divergent, these divergences are generated by the high momenta of the particles running in the loops. This seems to contradict the notion of a low energy EFT, where all interactions (vertices) are constructed in the domain $E \ll \Lambda$. One might thus be tempted to question the validity of the approach altogether. First, we remark that the same happens in renormalizable theories like QED where a few basic vertices describe the theory at all scales. The issue here is of course regularization, to give the theory a definite meaning such divergences must be separated from the finite pieces of the loop integrals and be buried in some parameters, like e.g. the electron charge and mass in QED. How this is done in technical detail will be discussed later. Still, one might be worried that for an EFT that supposedly works only in some low-energy domain, such loop integrals do not make sense. In fact, any regularization will do since physics cannot depend on such a procedure. One could thus envision using a brute force cutoff in the loops, with its value supplied by the breakdown scale Λ. Such a method, when made Lorentz-invariant, will lead to polynomials in Λ, which could be handled by subtractions, provided the underlying symmetries (or Ward identities) are respected. If that is not the case, one must supply the theory with additional non-invariant counter terms to restore the symmetries (like e.g. gauge invariance or chiral symmetry). While that can be done, the procedure is in general very awkward. It is much more efficient to use a regularization scheme that preserves the symmetries in all intermediate steps, thus dimensional regularization is in most cases the method of choice, as discussed in more detail below. Note that in the discussion of the EFTs for few-nucleon systems, we will come back to this issue.

We are now in the position to display the fundamental theorem underlying all EFTs [1]:

Quantum Field Theory has no content besides unitarity, analyticity, cluster decomposition, and symmetries. Therefore, to calculate the S-matrix for any theory below some scale, simply use the most general effective Lagrangian consistent with these principles in terms of the appropriate asymptotic states.

While this sounds rather simple, it has deep consequences, as will be shown later in more detail. Let us direct the reader's attention especially to the last part of the second sentence—it is of utmost importance to formulate the EFT in terms of the degrees of freedom appropriate to the system and energy scales one considers. This is particularly important for the strong interactions. While these are formulated in terms of quark and gluon fields whose interaction can be compactly formulated in the terms of the QCD Lagrangian (see the next chapter), at low energies the corresponding EFT is indeed formulated in terms of pions and heavy matter fields (heavy as compared to the pions). Even more, for nuclear physics at very low energies one can even integrate out the pions ending with an EFT of nucleonic contact interactions, that exhibits some intricate features. To end this section, we note that using the chiral Ward identities of QCD the form of the chiral effective Lagrangian of pions coupled to external sources can indeed be derived [10].

1.2.1   Power Counting

We now must organize our energy expansion in a systematic way. We already know that we should consider tree and loop graphs, and there are in principle infinitely many. However, Weinberg's power counting theorem tells us how to organize all these terms. To be specific and make the arguments more transparent, we consider a

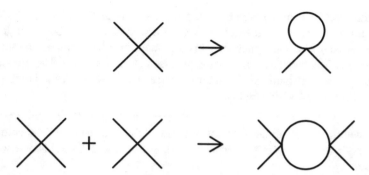

Fig. 1.4 How interactions generate loops. Upper line: Connecting two external lines leads to a tadpole (one-loop) graph. Lower line: Sewing together two tree-level four-boson vertices gives another typical one-loop graph, the so-called fish diagram

Lorentz-invariant theory of massless Goldstone bosons, so all vertices are made of two, four, ... derivatives, as invariance under parity transformations forbids terms with an odd number of derivatives. Note that Goldstone boson interactions are necessarily of derivative type, as explained in detail in Chap. 2. Now derivatives generate powers of four-momentum (loosely called energy) and thus an operator with n derivatives scales as $1/\Lambda^{n-4}$, where Λ is the scale limiting the low-energy domain. More precisely, an n derivative vertex scales with the small parameter $(E/\Lambda)^n$. Consequently, at small energies $E \ll \Lambda$ terms with many derivatives are highly suppressed. From these vertices we can construct all possible *tree* graphs, which in our theory scale as E^2, E^4, ..., ignoring for a moment factors stemming form the normalization of the external lines and other overall dimensionful factors. The leading term in e.g. $2 \to 2$ scattering is then $\sim E^2$, which is given by the four-boson vertex with the minimal number of derivatives. Furthermore, interactions also generate *loops*. This is depicted for two examples in Fig. 1.4. Let us consider the energy dimension of the $2 \to 2$ loop graph shown in this figure (the fish diagram). If the external four-momenta are called p_i ($i = 1, 2, 3, 4$) and we denote the internal (loop) momentum as p', we have for this diagram

$$\int d^4 p' \frac{P(p')}{(p'^2)\,(p' + p_2 - p_4)^2} \sim E^4$$

where $P(p')$ is a polynomial of fourth order in momenta (since each vertex supplies two powers of momentum). This argument is of course somewhat sloppy since such a diagram is divergent and needs to be regularized. If we use e.g. dimensional regularization, $\int d^4 p' \to \mu^{4-d} \int d^d p'$, one only produces factors of $\ln(E^2/\mu^2)$ through regularization. We remind the reader that in dimensional regularization, no power-law divergences are present and the appropriate mismatch in the dimension of the integral is compensated by factors of the regularization scale μ, see e.g. Ref. [11] for a detailed introduction to this method. Coming back to the fish diagram, we thus expect such a one-loop diagram to scale as the fourth power of the energy, which amounts to a suppression by two powers of energy compared to the leading tree graphs $\sim E^2$ (generated from the lowest dimensional four-boson vertex with two derivatives). One can generalize this argument to n-loop graphs and finds that contributions from such diagrams are suppressed by powers of E^{2n} as compared to the leading tree contribution $\sim E^2$. Putting pieces together, we are led to the *power counting theorem* of Weinberg [1]:

> The overall energy dimension (also called "chiral dimension" or counting index) of a diagram with N_d vertices of dimension d and L (Goldstone boson) loops is
>
> $$D = 2 + 2L + \sum_d N_d(d - 2) . \tag{1.8}$$

Proof The proof of this theorem goes as follows. Consider the effective Lagrangian for massless Goldstone bosons, $\mathscr{L}_{\text{eff}} = \sum_d \mathscr{L}^{(d)}$, where d is bounded from below. For interacting Goldstone bosons, $d \geq 2$ and the pion propagator has the form $D(q) = i/q^2$. With that, consider an L-loop diagram with I internal lines and V_d vertices of order d. The corresponding amplitude reads

$$Amp \propto \int (d^4q)^L \, \frac{1}{(q^2)^I} \prod_d (q^d)^{V_d}.$$

This can easily be understood. Each loop contributes a factor d^4q, the I propagators generate the factor q^{2I} (note that even for massive pions with mass M, the proof goes through as described as the mass does not play any role for the analysis of the UV behavior) and the last factor stems from the various vertices. Now let $Amp \sim q^D$, so that $D = 4L - 2I + \sum_d dV_d$. From topology we know that $L = I - \sum_d V_d + 1$, which allows us to eliminate I and we arrive at $D = 2 + 2L + \sum_d V_d(d - 2)$, which completes the proof.

Let us now analyze Eq. (1.8) in detail. The lowest value for the dimension D is given for $L = 0$ (no loops) and using entirely the lowest dimensional couplings with $d = 2$. For convenience, we will call this the leading order (LO) in what follows. The leading order result of energy (or chiral) dimension two is given by tree graphs. For the example of elastic Goldstone boson scattering depicted in Fig. 1.5, this leading term is a four-boson contact term with one dimension two insertion, as shown by the filled circle. We remark that the strength of this interaction can be related to the order parameter of the symmetry breaking (as will be discussed in more detail in later chapters). As will be shown later, for Goldstone boson interactions, the contributions from these diagrams can be calculated using the group structure of the conserved currents (charges). In the context of the strong interactions, this is the current algebra approach which has been developed long before the advent of QCD. For textbook discussions, see e.g. Refs. [12, 13]. However, the systematic

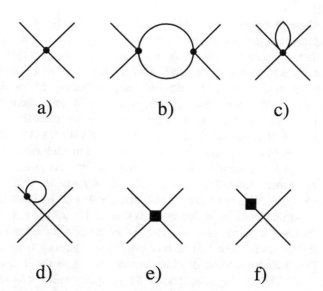

Fig. 1.5 $2 \rightarrow 2$ Goldstone boson scattering. The leading order $\sim E^2$ is given by the tree graph (**a**). At next-to-leading order $\sim E^4$, we have one-loop and tree corrections. The loop corrections are given by the fish graph (**b**), the tadpole (**c**) and wavefunction renormalization (**d**). The tree graphs with one dimension four insertion (as indicated by the filled square) fall into four-boson contact terms (**e**) and wavefunction renormalization terms (**f**). No exchange or crossed graphs are shown

expansion of any matrix element or transition current can only performed within the effective Lagrangian framework discussed here. The full beauty of this machinery is exposed in the ground-breaking papers of Gasser and Leutwyler [14, 15]. Here, we only discuss the structure of the EFT beyond leading order. At next-to-leading order (NLO), we have different contributions. From the lowest dimensional interactions, one can construct one-loop graphs ($L = 1$), which lead to $D = 4$ in the power counting formula. For the example of the scattering process, the various one-loop topologies are shown in Fig. 1.5b–d, respectively. These topologies are called the fish graph, the tadpole and wavefunction renormalization. However, there are other contributions at $D = 4$. Setting again $L = 0$, we get additional tree graphs with exactly one dimension four ($d = 4$) insertion at the four-boson vertex (**e**) or contributing to wavefunction renormalization (**f**). The coupling constants, often called low-energy constants (LECs), accompanying these terms are not fixed by symmetry. They must be determined from experiment. If the underlying field theory is known, one might be able to calculate these couplings. Often, models that capture some of the physics of the system under consideration allow one to estimate their values, for example in QCD, most LECs are well given in terms of the massive states that do not appear explicitly in the EFT. This will become more transparent in the chapters to follow. Here, we have been a bit sloppy. As already mentioned, the loops are in general UV infinite, so one has to renormalize the appearing divergences. Since all terms allowed by the symmetry are given by the effective

Lagrangian, these divergences can be absorbed in the low-energy constants (if we choose an appropriate regularization scheme). So only after this renormalization procedure, the finite pieces of the LECs are connected to observables and can be pinned down. At next-to-next-to-leading order (NNLO) ($D = 6$), we have three types of contributions. These are genuine two-loop diagrams with insertions of the $d = 2$ type exclusively, one-loop diagrams with exactly one insertion from the NLO terms ($d = 4$) and tree graphs with exactly one insertion from the NNLO terms with $d = 6$. It is obvious how to generalize this scheme to higher orders. A few remarks are in order. First, the number of LECs proliferates quickly as the order increases, but this should not be a concern if the energy scale is chosen small enough, or more precisely, if E/Λ is sufficiently smaller than one, with Λ the breakdown scale. We then have a small parameter and higher orders should be suppressed parametrically. A fine point is that so far we did not specify how we really calculate the matrix elements for Goldstone boson interactions, but the diagrammatic analysis (cf. Fig. 1.5) is most easily done starting from an effective Lagrangian that includes a tower of terms corresponding to the interactions with $d = 2, 4, 6, \ldots$. This is the most general Lagrangian which was discussed in the context of the "fundamental theorem". However, one can always add total derivatives to such an effective Lagrangian which in general leaves the physics of the effective field theory unchanged. To overcome this ambiguity, one should consider the generating functional instead. Here, we are only concerned with a general effective Lagrangian and make the implicit assumption that total derivatives can be ignored. So let us proceed but keep this in mind.

So far, we have considered strict Goldstone bosons, which are massless. More generally, if there is some small symmetry breaking, these particles will acquire a small mass. A generic boson mass term has the form

$$\mathscr{L}_{\mathrm{mass}} = -\frac{1}{2} M^2 \phi^2 , \tag{1.9}$$

where M is the boson mass and ϕ denotes the pseudoscalar field. As long as $M \ll \Lambda$, we can treat this mass as an additional small parameter. In such a case, one often speaks of pseudo-Goldstone bosons, because a true Goldstone boson must be massless. We will adopt a sloppy language and speak of Goldstone bosons throughout, leaving it as an exercise to the reader to supply the prefix "pseudo" whenever needed. It is obvious that if the explicit symmetry breaking cannot be treated perturbatively, the power counting breaks down. If the symmetry breaking is sufficiently weak, we have a dual expansion in the two small parameters E/Λ and M/Λ. To obtain a well-defined limit to the massless theory, we must keep E/M fixed as M goes to zero [14]. It is now straightforward to generalize our counting rules. Since the boson mass is a small energy, the leading order correction is of dimension two, so the boson mass term Eq. (1.9) comes together with the leading order two-derivative Goldstone boson interaction. Additional terms are generated at higher orders, at $D = 4$ we have e.g. terms quartic in M as well as mixed terms with two derivatives and two powers of M. It is often convenient to group the corresponding LECs into two classes. Class A collects the couplings that survive in

Fig. 1.6 Next-to-leading order corrections to the Goldstone boson self-energy. The filled box denotes a local NLO operator whereas the filled circle gives the lowest order four-boson interactions

the massless theory, whereas class B contains the symmetry-breaking LECs, i.e. the ones proportional to some power of M. Such a classification will be useful when we discuss the chiral properties of QCD as encoded in the pertinent low-energy effective field theory. To give an explicit example, the NLO corrections to the Goldstone boson self-energy are given by exactly two diagrams, the local NLO boson mass insertion and the tadpole, see Fig. 1.6. We will later calculate these terms explicitly.

So far, we have only considered Goldstone boson interactions. For calculating the response to external probes, as measured e.g. in the pion scalar and vector form factors, it is most convenient to employ the method of external fields (sources). This allows to construct building blocks that are invariant under the pertinent symmetries and build on that the most general effective Lagrangian to a given order. We refer the reader e.g. to Ref. [14] for a full exposition of the method.

1.2.2 Further Structural Aspects: Loops, Divergences and Low-Energy Constants

As stated before, loop graphs often generate infinities. This can most easily be seen by an explicit calculation of the tadpole graph in Fig. 1.6. Utilizing a mass-independent and symmetry-preserving regularization scheme, we have in d space-time dimensions (dimensional regularization):

$$
\begin{aligned}
\Delta_\pi(0) &= \frac{-i}{(2\pi)^d} \int d^d p \frac{1}{M^2 - p^2 - i\varepsilon} \\
&= (2\pi)^{-d} \int d^d k \frac{1}{M^2 + k^2} \quad \text{with } p_0 = ik_0, \quad -p^2 = k_0^2 + \mathbf{k}^2 \\
&= (2\pi)^{-d} \int d^d k \int_0^\infty d\lambda \exp(-\lambda(M^2 + k^2)) \\
&= (2\pi)^{-d} \int_0^\infty d\lambda \exp(-\lambda M^2) \underbrace{\int d^d k \exp(-\lambda k^2)}_{(\pi/\lambda)^{d/2}} \\
&= (4\pi)^{-d} M^{d-2} \Gamma\left(1 - \frac{d}{2}\right) .
\end{aligned}
\tag{1.10}
$$

In the second line, we have performed a Wick rotation to go from Minkowski to Euclidean space-time. The Γ function has a simple pole for $d = 4$, so that these infinities can be absorbed in the LECs,

$$g_i \rightarrow g_i^{\text{ren}} + \beta_i \frac{1}{d-4} , \tag{1.11}$$

where the renormalized and finite part g_i^{ren} can be obtained from a fit to data and β_i is the β-function of the corresponding operator. In fact, the appearing infinity can be absorbed in the LEC accompanying the operator corresponding to the tree graph in Fig. 1.6. This example nicely demonstrate the appearance of infinities and their elimination through the counter terms that appear at the same order in the chiral expansion. However, loop graphs serve another purpose, they generate the unitarity corrections to the strictly real tree graphs. This can be most easily understood by analyzing the fish graph of Fig. 1.5. To calculate its imaginary part, one makes use of the Cutkosky (cutting) rules [16], i.e. by putting the cut propagators on-shell. In any perturbative EFT, at a given order in the energy expansion, the real part of an observable is thus more precisely calculated than its imaginary part. There are cases, however, where the EFT needs a non-perturbative treatment, in that case unitarity can be fulfilled exactly. One obvious example is the generation of pion-pion resonances like the σ or the ρ meson, or the nuclear interactions to be discussed in the next chapter. One should, however, be aware that unitarization procedures are not unique and one therefore has to investigate possible dependences on the particular scheme.

There is one more intriguing structural aspect generated by (almost) massless particles in an EFT, here the Goldstone bosons of QCD, namely the so-called *chiral logarithms (logs)* or chiral singularities. Pions couple to themselves and matter fields like the nucleons, thus generating a cloud that is Yukawa-suppressed as $\exp(-Mr)/r$ for finite pion mass. In the chiral limit of vanishing quark and thus pion masses $M \rightarrow 0$, this Yukawa tail turns into a long-range Coulomb-like form. Consequently, S-matrix elements or transition currents can diverge in this limit. Famous examples are the pion vector radius or the nucleon isovector radius that scales as $\log(M)$ or the nucleon electromagnetic polarizabilities that scale as $1/M$. This is not a disaster but rather a natural consequence of having massless degrees of freedom, and it can serve as an important check for other calculational scheme, like e.g. the lattice formulation of QCD. Also, any model that is supposed to describe the pion or the nucleon structure at low energies should obey such constraints. Note also that such singularities are generated by loop graphs, so that in general there will also be pion-mass independent contributions at the same order of the calculation. A typical example is the isovector charge radius of the pion, or any other observable that displays a chiral log. Since the argument of a log must be a number, it really depends on M/μ, with μ some regularization scale. Similarly, the corresponding LEC that appears must also depend on μ, thus making the observable scale-independent. This, however, also means that it is in general not possible to assign a definite number to the pion cloud contribution of an observable since shifts in the regularization scale allow to shuffle strength from the long-range pion contribution to the shorter ranged contact term part. We refer the reader to Ref. [17] for a more detailed discussion of this point.

We end this section with a few remarks on the LECs. To be specific, consider a covariant and parity-invariant theory of Goldstone bosons, the latter being parameterized by some unitary, matrix-valued field U

$$\mathcal{L}_{\text{eff}} = g_2 \text{Tr}(\partial_\mu U \partial^\mu U^\dagger) + g_4^{(1)} \left[\text{Tr}(\partial_\mu U \partial^\mu U^\dagger) \right]^2$$
$$+ g_4^{(2)} \text{Tr}(\partial_\mu U \partial^\nu U^\dagger) \text{Tr}(\partial^\mu U \partial_\nu U^\dagger) + \dots , \tag{1.12}$$

with the $g_2, g_4^{(1)}, g_4^{(2)}, \dots$ the LECs. In such a theory, $g_2 \neq 0$, because it is a sufficient and necessary order parameter of spontaneous chiral symmetry breaking (in QCD, $g_2 \sim F_\pi^2$, with F_π the pion decay constant). In contradistinction, the LECs $g_4^{(1)}, g_4^{(2)}, \dots$ must be fixed from data (or be calculated from the underlying theory). Independent of their precise values, it is important to realize that the LECs encode information about the high mass states that are integrated out. This can most easily be understood by looking at the ρ meson contribution to pion-pion scattering, see Fig. 1.7. For momenta well below the ρ-mass, we can perform a Taylor-expansion of the vector-meson propagator (ignoring all tensor structures)

$$\frac{g^2}{M_\rho^2 - q^2} \xrightarrow{q^2 \ll M_\rho^2} \frac{g^2}{M_\rho^2} \left(1 + \frac{q^2}{M_\rho^2} + \dots \right) , \tag{1.13}$$

where g is the $\rho\pi\pi$ coupling constant. As we had already seen in the discussion of the Euler-Heisenberg-Kockel Lagrangian, integrating out the heavy field (here, the ρ meson) leads to a string of local contact interactions with increasing number of derivatives and the accompanying LECs are given in even powers of the coupling constant over the ρ-mass, thus leading to decoupling in the limit $M_\rho \to \infty$. In fact, one can analyze all couplings of the next-to-leading order chiral Lagrangian of pions coupled to external fields in this manner. The values of the LECs expressed in terms of resonance couplings and masses agree quite well with the numbers determined from experiment, see Refs. [18, 19]. Note, however, that such a modeling of the LECs only makes sense at a regularization scale close to the heavy meson mass and that it assumes that the remainder from other degrees of freedom is small.

Fig. 1.7 ρ meson contribution to pion-pion scattering at low energies

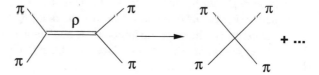

1.3   Inclusion of Heavy Fields

In this section, we briefly discuss the inclusion of heavy fields in the effective Lagrangian. To be definite, let us consider the chiral pion-nucleon Lagrangian. Its construction is straightforward, see the next chapter. Here, we want to discuss the issue of power counting in the presence of such a heavy scale. As the nucleon mass $m_N \sim 1\,\text{GeV}$ is comparable to the chiral symmetry breaking scale $\Lambda_\chi \simeq 4\pi F_\pi \sim 1.2\,\text{GeV}$, i.e. the breakdown scale of the chiral pion EFT, it is evident that only nucleon three-momenta can be soft. This complicates the power counting, as a naive application of the methods in loop diagrams leads to contributions at lower orders than expected from the power counting. There are various possibilities to overcome this problem. First, one can utilize the extreme non-relativistic limit in which the baryon mass term can be transformed systematically in a string of interaction terms proportional to inverse powers of the nucleon mass, similar to the Foldy-Wouthuysen transformation to understand the non-relativistic limit of the Dirac equation [20]. This so-called heavy baryon scheme has been proposed in Ref. [21] and further developed in Refs. [22, 23]. This scheme is also discussed in detail in the next chapter. Another option is to explicitly massage the loop integrals to get rid off the power counting violating contributions. In the so-called infrared regularization scheme [24], the integrals are analyzed in fractional dimensions which allows to uniquely separate the low- from the high-momentum components so that the latter can be explicitly absorbed in the LECs of the contact interactions. In the extended-on-mass-scheme [25], the power counting violating terms are subtracted from the loop diagrams directly, leaving one again with a consistent scheme that has a proper power counting. The detailed discussion and comparison of these schemes is given in the review [26]. Throughout this book, we will work within the heavy baryon framework.

1.4   Summary: A Short Recipe for the Construction of an EFT

The necessary and sufficient ingredients to construct an EFT can be compactly summarized as follows:

- *scale separation*—what is low, what is high?
- *active degrees of freedom*—what are the building blocks?
- *symmetries*—how are the interactions constrained by symmetries?
- *power counting*—how to organize the expansion in low over high?

With that, we now turn to the chiral EFT of nuclear physics, which will be at the heart of the physical systems considered in this book.

References

1. S. Weinberg, Phenomenological Lagrangians. Physica A **96**, 327 (1979)
2. B.R. Holstein, Effective interactions are effective interactions, in *Hadron Physics 2000*, Caraguatatuba (2000), pp. 57–107. arXiv:hep-ph/0010033
3. E. Fermi, An attempt of a theory of beta radiation. 1. Z. Phys. **88**, 161 (1934)
4. H. Euler, B. Kockel, Ueber die Streuung von Licht an Licht nach der Diracschen Theorie. Naturwiss. **23**, 246 (1935)
5. W. Heisenberg, H. Euler, Consequences of Dirac's theory of positrons. Z. Phys. **98**, 714 (1936)
6. R. Karplus, M. Neuman, The scattering of light by light. Phys. Rev. **83**, 776 (1951)
7. G.V. Dunne, The Heisenberg-Euler effective action: 75 years on. Int. J. Mod. Phys. A **27**, 1260004 (2012) [Int. J. Mod. Phys. Conf. Ser. **14**, 42 (2012)]
8. M. Aaboud et al., [ATLAS Collaboration], Evidence for light-by-light scattering in heavy-ion collisions with the ATLAS detector at the LHC. Nat. Phys. **13**(9), 852 (2017)
9. T. Aoyama, M. Hayakawa, T. Kinoshita, M. Nio, Quantum electrodynamics calculation of lepton anomalous magnetic moments: numerical approach to the perturbation theory of QED. PTEP **2012**, 01A107 (2012)
10. H. Leutwyler, On the foundations of chiral perturbation theory. Ann. Phys. **235**, 165 (1994)
11. J.C. Collins, *Renormalization* (Cambridge University Press, Cambridge, 1984)
12. S.L. Adler, R.F. Dashen, *Current Algebras and Applications to Particle Physics* (W.A. Benjamin, New York, 1968)
13. V. de Alfrao, S. Fubini, G. Furlan, C. Rossetti, *Currents in Hadron Physics* (North Holland, Amsterdam, 1973)
14. J. Gasser, H. Leutwyler, Chiral perturbation theory to one loop. Ann. Phys. **158**, 142 (1984)
15. J. Gasser, H. Leutwyler, Chiral perturbation theory: expansions in the mass of the strange quark. Nucl. Phys. B **250**, 465 (1985)
16. R.E. Cutkosky, Singularities and discontinuities of Feynman amplitudes. J. Math. Phys. **1**, 429 (1960)
17. V. Bernard, H.W. Fearing, T.R. Hemmert, U.-G. Meißner, The form-factors of the nucleon at small momentum transfer. Nucl. Phys. A **635**, 121 (1998) [Erratum-ibid. A **642**, 563 (1998)] [Nucl. Phys. A **642**, 563 (1998)]
18. G. Ecker, J. Gasser, A. Pich, E. de Rafael, The role of resonances in chiral perturbation theory. Nucl. Phys. B **321**, 311 (1989)
19. J.F. Donoghue, C. Ramirez, G. Valencia, The spectrum of QCD and chiral Lagrangians of the strong and weak interactions. Phys. Rev. D **39**, 1947 (1989)
20. L.L. Foldy, S.A. Wouthuysen, On the Dirac theory of spin 1/2 particle and its nonrelativistic limit. Phys. Rev. **78**, 29 (1950)
21. E.E. Jenkins, A.V. Manohar, Baryon chiral perturbation theory using a heavy fermion Lagrangian. Phys. Lett. B **255**, 558 (1991)
22. T. Mannel, W. Roberts, Z. Ryzak, A derivation of the heavy quark effective Lagrangian from QCD. Nucl. Phys. B **368**, 204 (1992)
23. V. Bernard, N. Kaiser, J. Kambor, U.-G. Meißner, Chiral structure of the nucleon. Nucl. Phys. B **388**, 315 (1992)
24. T. Becher, H. Leutwyler, Baryon chiral perturbation theory in manifestly Lorentz invariant form. Eur. Phys. J. C **9**, 643 (1999)
25. T. Fuchs, J. Gegelia, G. Japaridze, S. Scherer, Renormalization of relativistic baryon chiral perturbation theory and power counting. Phys. Rev. D **68**, 056005 (2003)
26. V. Bernard, Chiral perturbation theory and baryon properties. Prog. Part. Nucl. Phys. **60**, 82 (2008)

Chapter 2
Nuclear Forces in Chiral EFT

In this chapter, we discuss the formulation of the nuclear forces in chiral effective field theory (EFT) in the continuum. Again, we rather discuss the foundations than displaying very detailed calculations, as most of these are well documented in the literature. We begin with a discussion of the structure of the nuclear Hamilton operator and the scales pertinent to nuclear physics. The longest range part of the nuclear force is indeed given by pion exchange as predicted by Yukawa [1] long ago. This is nowadays firmly rooted in the spontaneously and explicitly broken chiral symmetry of QCD, see Sect. 2.3. Based on that, we give a short recipe to construct the chiral effective pion-nucleon Lagrangian in Sect. 2.4 and then discuss the extension to two or more nucleons, following the pioneering work of Weinberg [2, 3]. As nuclei are composed of non-relativistic nucleons and mesons, the underlying equation for the nuclear A-body system (where A is the atomic number) to be solved is the Schrödinger equation. So our aim is to construct the nuclear Hamiltonian, in which the various contributions to the potential are organized according to the power counting discussed below. Results for the two-nucleon system are displayed in Sect. 2.5, including a calculational scheme for the theoretical uncertainties and a short discussion of the approximate SU(4) symmetry of the nuclear forces.

2.1 The Nuclear Hamiltonian

The aim of this chapter is to outline the steps for developing a systematic and model independent theoretical framework capable to describe reactions involving nucleons up to center-of-mass (cms) three-momenta of (at least) the order of the pion mass M_π. Following the usual philosophy of EFT we aim at the most general parameterization of the amplitude consistent with the fundamental principles such as Lorentz invariance, unitarity, cluster separability and analyticity. Given that the

© Springer Nature Switzerland AG 2019
T. A. Lähde, U.-G. Meißner, *Nuclear Lattice Effective Field Theory*,
Lecture Notes in Physics 957, https://doi.org/10.1007/978-3-030-14189-9_2

energies of the nucleons we are interested in are well below the nucleon mass m_N, it is appropriate to make use of the non-relativistic expansion, which is an expansion in inverse powers of m_N. In the absence of external probes and below the pion production threshold, we are left with a potential theory in the framework of the quantum-mechanical A-body Schrödinger equation (in coordinate space representation)

$$
\left(H_0 + V\right)|\Psi\rangle = E|\Psi\rangle, \qquad H_0 = \sum_{i=1}^{A} \left(-\frac{\nabla_i^2}{2m_N}\right) + \mathcal{O}(m_N^{-3}), \tag{2.1}
$$

with $m_N = (m_p + m_n)/2 = 938.2\,\text{MeV}$ the nucleon mass. Here, m_p and m_n denote the proton and the neutron mass, respectively. The main task then reduces to the determination of the nuclear Hamiltonian $H_0 + V$, where the potential V itself can be decomposed into a string of terms as

$$
V = V_{2N} + V_{3N} + V_{4N} + \dots . \tag{2.2}
$$

Here, we made explicit the interactions between two, three and four nucleons, that are phenomenologically known to obey the hierarchy

$$
V_{2N} \gg V_{3N} \gg V_{4N} . \tag{2.3}
$$

Of course, the potential is not an observable and thus this statement needs to be understood in the following sense: Within a given framework, the two- and higher-body forces are determined and their contribution to a given observable, say a nuclear binding energy, can be calculated. Then, one finds that using two-nucleon forces alone turns out be a very good approximation. However, for doing precision calculations in nuclear physics, three-body forces are required, see e.g. [4]. We will come back to this issue in the following. Let us return to our task of determining the nuclear Hamiltonian based on the symmetries of QCD and allowing for a systematic and improvable representation. This can be accomplished using the framework of chiral perturbation theory. In what follows, we will outline the basic features of this approach, starting from the chiral symmetry of QCD and the construction of the corresponding chiral Lagrangian of pions and nucleons, with a particular emphasis on the basic features and the subtleties that appear in nuclear systems, as discussed in more detail in the following sections.

2.2   Scales in Nuclear Physics

To appreciate the complexity related to a theoretical description of the nuclear forces, it is most instructive to briefly discuss the pertinent scales arising in this problem. This can most easily be visualized by looking at the phenomenological

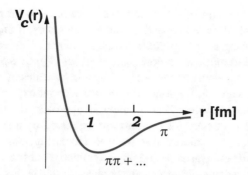

Fig. 2.1 Schematic plot of the central nucleon-nucleon potential. The longest range contribution is the one-pion-exchange, the intermediate range attraction is described by two-pion exchanges and other shorter ranged contributions. At even shorter distances, the nucleon-nucleon (NN) interaction is repulsive, that means that nucleons do not like to touch. The typical length scale associated with this so-called "hard core" is about 1 fm

central potential between two nucleons, as it appears e.g. in meson-theoretical approaches of the nuclear force, see Fig. 2.1. The longest range part of the interaction is the one-pion exchange (OPE) that is firmly rooted in QCD's chiral symmetry. Thus, the corresponding natural scale of the nuclear force problem is the Compton wavelength of the pion

$$\lambda_\pi = \frac{1}{M_\pi} \simeq 1.5 \, \text{fm} \,, \tag{2.4}$$

where $M_\pi = 139.57 \, \text{MeV}$ is the charged pion mass, making use of the conversion factor $1 \, \text{fm} = 1/(197.33 \, \text{MeV})$. The central intermediate range attraction is given by 2π exchange (and shorter ranged physics). Finally, the wavefunctions of two nucleons do not like to overlap, which is reflected in a short-range repulsion that can e.g. be modeled by vector meson exchange. From such considerations, one would naively expect to be able to describe nuclear binding in terms of energy scales of the order of the pion mass. However, the true binding energies of the nuclei are given by much smaller energy scales, between one and eight MeV per nucleon. Another measure for the shallow nuclear binding is the so called binding-momentum γ. In the deuteron, $\gamma = \sqrt{m_N B_D} \simeq 45 \, \text{MeV} \ll M_\pi$, with $B_D = 2.224 \, \text{MeV}$ the deuteron binding energy. The small value of γ signals the appearance of energy/momentum scales much below the pion mass. The most dramatic reflection of the complexity of the nuclear force problem is given by the values of the S-wave neutron-proton scattering lengths,

$$|a(^1S_0)| = 23.8 \, \text{fm} \gg 1/M_\pi \,, \quad a(^3S_1) = 5.4 \, \text{fm} \gg 1/M_\pi \,. \tag{2.5}$$

Here, we employ the standard spectroscopic notation for the NN partial waves, $^{2S+1}L_J$, with S the total spin of the two-nucleon system, L the angular momentum, $J = L + S$ the total angular momentum and $L = 0$ corresponds to the S-wave.

Thus, to properly set up an EFT for the forces between two (or more) nucleons, it is mandatory to deal with these very different energy scales. If one were to treat the large S-wave scattering lengths perturbatively, the range of the corresponding EFT would be restricted to momenta below $p_{\max} \sim 1/|a(^1S_0)| \simeq 8$ MeV. To overcome this barrier, one must generate the small binding energy scales by a non-perturbative resummation. This can e.g. be done in a theory without explicit pion degrees of freedom, the so-called pionless EFT. In such an approach, the limiting hard scale is the pion mass. The nuclear interactions are expressed by a string of local four-nucleon, six-nucleon, ... interactions with an increasing number of derivatives, constrained by Galilean invariance and other symmetries of the strong interactions. As pions are integrated out, chiral symmetry plays no role in this framework. While much can be learned from this approach, see e.g. Ref. [5], it is only appropriate to be used in few-nucleon systems and allows for precision calculations of very low energy processes. However, due to its simplicity, it can be used to introduce certain concepts most easily and we will make use of it when required. To go beyond few-nucleon systems, one must include the pions explicitly, as it is done in the pionfull or chiral nuclear EFT. There are essentially two methods of doing that. In an elegant approach suggested by Kaplan et al. [6], the pions are treated perturbatively and the nuclear bound states are generated by summing up the leading order four-nucleon interaction to infinite order. However, the strong tensor force so important for nuclear physics is not properly described in this approach. This can be explicitly shown by constructing NN phase shifts at next-to-next-to-leading order [7]. Here, we follow Weinberg's original suggestion that includes pions non-perturbatively and will be displayed in more detail below. However, the power counting is then applied to the irreducible potential rather than to the scattering amplitude. A different and more formal argument that shows the breakdown of a perturbative treatment of the EFT with two or more nucleons is related to the pinch singularities in the two-pion exchange diagram in the static limit as will be discussed later in the context of the explicit construction of the chiral nuclear EFT.

2.3   Chiral Symmetry of the Strong Interactions

First, we must discuss chiral symmetry in the context of QCD. Chromodynamics is a non-abelian SU(3)$_{\text{color}}$ gauge theory with $N_f = 6$ flavors of quarks, three of them being light (u, d, s) and the other three heavy (c, b, t). Here, light and heavy refers to the scale of QCD, $\Lambda_{\text{QCD}} \simeq 250$ MeV, determined from the running of the strong coupling constant. In what follows, we consider light quarks only (the heavy quarks are to be considered as decoupled). The QCD Lagrangian then reads

$$\mathscr{L}_{\text{QCD}} = -\frac{1}{2g^2}\text{Tr}\left(G_{\mu\nu}G^{\mu\nu}\right) + \bar{q}\,i\gamma^\mu D_\mu\,q - \bar{q}\mathscr{M}\,q$$

$$= \mathscr{L}_{\text{QCD}}^0 - \bar{q}\mathscr{M}\,q\,, \tag{2.6}$$

where we have absorbed the gauge coupling in the definition of the gluon field and color indices are suppressed. The three-component vector q collects the quark fields, $q^T(x) = (u(s), d(x), s(x))$. As far as the strong interactions are concerned, the different quarks u, d, s have identical properties, except for their masses. The quark masses are free parameters in QCD—the theory can be formulated for any value of the quark masses. In fact, light quark QCD can be well approximated by a fictitious world of massless quarks, denoted $\mathscr{L}_{QCD}^0$ in Eq. (2.6). Remarkably, this theory contains no adjustable parameter—the gauge coupling g merely sets the scale for the renormalization group invariant scale Λ_{QCD}. Furthermore, in the massless world left- and right-handed quarks are completely decoupled. The Lagrangian of massless QCD is invariant under separate unitary global transformations of the left- and right-hand quark fields, the so-called *chiral rotations*, $q_I \to V_I q_I$, $V_I \in U(3)$, $I = L, R$, leading to $3^2 = 9$ conserved left- and 9 conserved right-handed currents by virtue of Noether's theorem. These can be expressed in terms of vector $(V = L + R)$ and axial-vector $(A = L - R)$ currents

$$V_0^\mu = \bar{q} \gamma^\mu q , \quad V_a^\mu = \bar{q} \gamma^\mu \frac{\lambda_a}{2} q ,$$

$$A_0^\mu = \bar{q} \gamma^\mu \gamma_5 q , \quad A_a^\mu = \bar{q} \gamma^\mu \gamma_5 \frac{\lambda_a}{2} q . \tag{2.7}$$

Here, $a = 1, \ldots 8$, and the λ_a are Gell-Mann's SU(3) flavor matrices. The singlet axial current is anomalous, and thus not conserved. The actual symmetry group of massless QCD is generated by the charges of the conserved currents, it is $G_0 = SU(3)_R \times SU(3)_L \times U(1)_V$. The $U(1)_V$ subgroup of G_0 generates conserved baryon number since the isosinglet vector current counts the number of quarks minus antiquarks in a hadron. The remaining group $SU(3)_R \times SU(3)_L$ is often referred to as chiral SU(3). Note that one also considers the light u and d quarks only (with the strange quark mass fixed at its physical value), in that case, one speaks of chiral SU(2) and must replace the generators in Eq. (2.7) by the Pauli-matrices, $\tau_i, i = 1, 2, 3$. Let us mention that QCD is also invariant under the discrete symmetries of parity (P), charge conjugation (C) and time reversal (T). Although interesting in itself, we do not consider strong CP violation and the related θ-term in what follows, see e.g. [8].

The chiral symmetry is a symmetry of the Lagrangian of QCD but not of the ground state or the particle spectrum—to describe the strong interactions in nature, it is crucial that chiral symmetry is spontaneously broken. This can be most easily seen from the fact that hadrons do not appear in parity doublets. If chiral symmetry were exact, from any hadron one could generate by virtue of an axial transformation another state of exactly the same quantum numbers except of opposite parity. The spontaneous symmetry breaking leads to the formation of a quark condensate in the vacuum $\langle 0|\bar{q}q|0\rangle = \langle 0|\bar{q}_L q_R + \bar{q}_R q_L|0\rangle$, thus connecting the left- with the right-handed quarks. In the absence of quark masses this expectation value is flavor-independent: $\langle 0|\bar{u}u|0\rangle = \langle 0|\bar{d}d|0\rangle = \langle 0|\bar{s}s|0\rangle = \langle 0|\bar{q}q|0\rangle$. More precisely, the vacuum is only invariant under the subgroup of vector rotations times the baryon number current, $H_0 = SU(3)_V \times U(1)_V$. This is the generally accepted picture that is supported by general arguments [9] as well as lattice simulations of QCD (for a review, see [10]). In fact, the vacuum expectation value of the quark condensate is only one of the many possible order parameters characterizing the spontaneous symmetry violation—all operators that share the invariance properties of the vacuum qualify as order parameters. The quark condensate nevertheless enjoys a special role, it can be shown to be related to the density of small eigenvalues of the QCD Dirac operator (see [11] and more recent discussions in [12, 13]), $\lim_{\mathscr{M}\to 0}\langle 0|\bar{q}q|0\rangle = -\pi \rho(0)$. For free fields, $\rho(\lambda) \sim \lambda^3$ near $\lambda = 0$. Here, λ denotes an eigenvalue the QCD Dirac operator. Only if the eigenvalues accumulate near zero, one obtains a non-vanishing condensate. This scenario is indeed supported by lattice simulations and many model studies involving topological objects like instantons or monopoles.

Before discussing the implications of spontaneous symmetry breaking for QCD, we briefly remind the reader of Goldstone's theorem [14, 15]: to every generator of a spontaneously broken symmetry corresponds a massless excitation of the vacuum. These states are the *Goldstone bosons*, collectively denoted as pions $\pi(x)$ in what follows. Through the corresponding symmetry current the Goldstone bosons couple directly to the vacuum,

$$\langle 0|A^0(0)|\pi\rangle \neq 0 .\tag{2.8}$$

In fact, the non-vanishing of this matrix element is a *necessary and sufficient* condition for spontaneous symmetry breaking. In QCD, we have eight (three) Goldstone bosons for SU(3) (SU(2)) with spin zero and negative parity—the latter property is a consequence that these Goldstone bosons are generated by applying the axial charges on the vacuum. The dimensionful scale associated with the matrix element Eq. (2.8) is the pion decay constant F_π

$$\langle 0|A_\mu^a(0)|\pi^b(p)\rangle = i\delta^{ab} F_\pi p_\mu ,\tag{2.9}$$

which is a fundamental mass scale of low-energy QCD. In the world of massless quarks, the pion decay constant is F and its value differs from the physical one by terms proportional to the quark masses, to be introduced later, $F_\pi = F[1 + \mathscr{O}(\mathscr{M})]$. The physical value of F_π is 92.1 MeV, determined from pion decay $\pi \to \nu\mu$.

Of course, in QCD the quark masses are not exactly zero. The quark mass term leads to the so-called *explicit chiral symmetry breaking*. Consequently, the vector and axial-vector currents are no longer conserved (with the exception of the baryon number current)

$$\partial_\mu V_a^\mu = \frac{1}{2}i\bar{q}\,[\mathscr{M}, \lambda_a]\,q , \quad \partial_\mu A_a^\mu = \frac{1}{2}i\bar{q}\,\{\mathscr{M}, \lambda_a\}\,\gamma_5\,q .\tag{2.10}$$

However, the consequences of the spontaneous symmetry violation can still be analyzed systematically because the quark masses are *small*. QCD possesses what is called an approximate chiral symmetry. In that case, the mass spectrum of the unperturbed Hamiltonian and the one including the quark masses cannot be significantly different. Stated differently, the effects of the explicit symmetry breaking can be analyzed in perturbation theory. As a consequence, QCD has a remarkable mass gap —the pions (and, to a lesser extent, the kaons and the eta) are much lighter than all other hadrons. To be more specific, consider chiral SU(2). The second formula of Eq. (2.10) is nothing but a Ward-identity (WI) that relates the axial current $A^\mu = \bar{d}\gamma^\mu\gamma_5 u$ with the pseudoscalar density $P = \bar{d}i\gamma_5 u$,

$$\partial_\mu A^\mu = (m_u + m_d)\, P \,. \tag{2.11}$$

Taking on-shell pion matrix elements of this WI, one arrives at

$$M_\pi^2 = (m_u + m_d)\frac{G_\pi}{F_\pi}\,, \tag{2.12}$$

where the coupling G_π is given by $\langle 0|P(0)|\pi(p)\rangle = G_\pi$. This equation leads to some intriguing consequences: In the chiral limit, the pion mass is exactly zero—in accordance with Goldstone's theorem. More precisely, the ratio G_π/F_π is a constant in the chiral limit and the pion mass grows as $\sqrt{m_u + m_d}$ if the quark masses are turned on.

There is even further symmetry related to the quark mass term. It is observed that hadrons appear in isospin multiplets, characterized by very tiny splittings of the order of a few MeV. These are generated by the small quark mass difference $m_u - m_d$ and also by electromagnetic effects of the same size (with the notable exception of the charged to neutral pion mass difference that is almost entirely of electromagnetic origin). This can be made more precise: For $m_u = m_d$, QCD is invariant under SU(2) isospin transformations: $q \to q' = Uq$, with U a unitary matrix. In this limit, up and down quarks cannot be disentangled as far as the strong interactions are concerned. Rewriting of the QCD quark mass term allows to make the strong isospin violation explicit:

$$\begin{aligned}\mathscr{H}_{\mathrm{QCD}}^{\mathrm{SB}} &= m_u\,\bar{u}u + m_d\,\bar{d}d \\ &= \frac{m_u + m_d}{2}(\bar{u}u + \bar{d}d) + \frac{m_u - m_d}{2}(\bar{u}u - \bar{d}d)\,, \end{aligned} \tag{2.13}$$

where the first (second) term is an isoscalar (isovector). Extending these considerations to SU(3), one arrives at the eightfold way of Gell-Mann and Ne'eman [16] that played a decisive role in our understanding of the quark structure of the hadrons. The SU(3) flavor symmetry is also an approximate one, but the breaking is much stronger than it is the case for isospin. From this, one can directly infer that the quark mass difference $m_s - m_d$ must be much bigger than $m_d - m_u$.

The consequences of these broken symmetries can be analyzed systematically in a suitably tailored effective field theory (EFT), as discussed in more detail below. At this point, it is important to stress that the chiral symmetry of QCD plays a crucial role in determining the longest ranged parts of the nuclear force, which, as we will show, is given by Goldstone boson exchange between two and more nucleons. This was already stressed long ago, see e.g. [17] (and references therein) but only with the powerful machinery of chiral effective field theory this connection could be worked out model-independently, as we will show in what follows.

2.4 Chiral Lagrangian for Pions and Nucleons

Having discussed the fate of the chiral symmetry of QCD, we are now in the position to construct the corresponding effective chiral Lagrangian in terms of the asymptotically observed pion and nucleon fields. It shares the same symmetry properties as QCD and can be treated perturbatively for pion and pion-nucleon interactions (we restrict ourselves to the two-flavor case from now on). In this case, one speaks of chiral perturbation theory (CHPT). For the calculation of the nuclear forces, we also must include multi-nucleon operators, as will be discussed in more detail below. The chiral effective Lagrangian can be written as:

$$\mathscr{L}_{\text{eff}} = \mathscr{L}_{\pi\pi} + \mathscr{L}_{\pi N} + \mathscr{L}_{NN},$$

$$\mathscr{L}_{\pi\pi} = \mathscr{L}_{\pi\pi}^{(2)} + \mathscr{L}_{\pi\pi}^{(4)} + \mathscr{L}_{\pi\pi}^{(6)} + \dots,$$

$$\mathscr{L}_{\pi N} = \mathscr{L}_{\pi N}^{(1)} + \mathscr{L}_{\pi N}^{(2)} + \mathscr{L}_{\pi N}^{(3)} + \dots,$$

$$\mathscr{L}_{NN} = \mathscr{L}_{NN}^{(0)} + \mathscr{L}_{NN}^{(2)} + \dots,$$

$$(2.14)$$

where the superscript refers to the chiral dimension, that is number of external momentum or pion mass insertions, collectively denoted by a small (soft) scale $Q \ll 1\,\text{GeV}$ in what follows. We will explicitly derive the leading terms in this expansion, that is the terms in $\mathscr{L}_{\pi\pi}^{(2)}$ and $\mathscr{L}_{\pi N}^{(1)}$, respectively. For the higher order terms, we refer to the literature (see references given below). The multi-nucleon terms collected in $\mathscr{L}_{NN}$ will be discussed separately. We will also extend the power counting formula Eq. (1.8) in Sect. 1.2.1 to the case of the pion-nucleon Lagrangian and also for the inclusion of multi-nucleon contact interactions.

2.4.1 The Pion Lagrangian

First, it is a well established fact that the chiral symmetry in QCD is realized in a non-linear manner. The Lie algebra of $SU(2)_L \times SU(2)_R$ is isomorphic to that of SO(4). The smallest non-trivial representation of the four-dimensional rotation group SO(4) is four-dimensional. However, we have only the triplet of the pion fields at our disposal to parameterize the symmetry breaking $SU(2)_L \times SU(2)_R \to SU(2)_V$, thus the symmetry must be realized non-linearly. The general theory of the nonlinear realization of the chiral symmetry is due to Weinberg [18] and Callan, Coleman, Wess and Zumino [19, 20]. More precisely, the pion iso-triplet $\pi(x)$ is collected in a 2×2 matrix-valued field $U(\pi)$, with $\pi = \sum_{a=1}^{3} \pi^a(x)\tau^a \equiv \pi^a(x)\tau^a$ (throughout, we use the Einstein summation convention). U is unitary, $U^\dagger U = UU^\dagger = \mathbb{1}$, with $\mathbb{1} = \text{diag}(1, 1)$ the unit matrix in two dimensions. Popular choices of U are

$$U(\pi) = \exp(i\pi/F_\pi) \qquad \text{exponential parametrization}, \qquad (2.15)$$

$$U(\pi) = \sqrt{1 - \pi^2/F_\pi^2} \qquad \text{sigma model parametrization}. \qquad (2.16)$$

Expanding U in powers of π yields in general

$$U(\pi) = \mathbb{1} + i\frac{\pi}{F_\pi} - \frac{\pi^2}{2F_\pi^2} - i\alpha\frac{\pi^3}{F_\pi^3} + (8\alpha - 1)\frac{\pi^4}{8F_\pi^4} + \mathcal{O}(\pi^5). \qquad (2.17)$$

The constant α reflects the freedom in parameterizing U, e.g. $\alpha = 1/6$ for the exponential and $\alpha = 0$ for the sigma model representation. Of course, physical observables do not depend on the choice of α. Under $\text{SU}(2)_L \times \text{SU}(2)_R$, U transforms linearly,

$$U \to U' = LUR^\dagger, \qquad (2.18)$$

with $L, R \in \text{SU}(2)_{L,R}$. Since we know that Goldstone boson interactions must be of derivative nature, the first non-vanishing term in the effective Lagrangian takes the form

$$\mathcal{L}_{\pi\pi}^{(2)} = \frac{F_\pi^2}{4} \langle \partial_\mu U \partial^\mu U^\dagger \rangle, \qquad (2.19)$$

where $\langle \ldots \rangle$ denotes the trace in flavor space, and a derivative ∂_μ clearly counts as a small momentum Q. Terms with an odd number of derivatives vanish because of parity invariance. Therefore, in the Goldstone boson sector only terms with an even chiral dimension appear (we do not consider anomalous processes here). The invariance of $\mathcal{L}_{\pi\pi}^{(2)}$ under chiral transformations can easily be shown

$$\langle \partial_\mu U \partial^\mu U^\dagger \rangle \overset{\text{SU(2)}_L \times \text{SU(2)}_R}{\to} \langle \partial_\mu U' \partial^\mu (U')^\dagger \rangle = \langle L\partial_\mu U R^\dagger R \partial^\mu U L^\dagger \rangle$$

$$= \langle \underbrace{L^\dagger L}_{=\mathbb{1}} \partial_\mu U \underbrace{R^\dagger R}_{=\mathbb{1}} \partial^\mu U \rangle, \quad (2.20)$$

since we can cyclically permute under the trace and $L^\dagger L = R^\dagger R = \mathbb{1}$. Using Eq. (2.17), the factor F_π^2 in Eq. (2.19) follows from the requirement that the bosons have a properly normalized kinetic energy term, $\mathcal{L}_{\text{kin}} = (1/2)\partial_\mu \pi \partial^\mu \pi$, using $\langle \tau^a \tau^b \rangle = 2\delta^{ab}$. Carrying out the expansion in π further, the leading order term in Eq. (2.19) also generates interactions between 4, 6, 8, ... pions. All these are given in terms of one parameter, the LEC F_π. This is the power of chiral symmetry as it relates seemingly unrelated processes. Strictly speaking, the LEC that really appears in Eq. (2.19) is F, the pion decay constant in the chiral limit. For explicit calculations, this has to be taken into account, but we will only discuss this fine difference when required.

So far, we have considered massless pions. To account for the explicit symmetry breaking generated by the finite quark mass, it is most convenient to consider the quark mass term as generated from a scalar external Hermitian source $s(x)$, that transforms under chiral transformations as $s \to s' = LsR^\dagger = RsL^\dagger$. Setting $s(x) = \mathcal{M}$, one recovers the QCD quark mass term. As above, we can construct the lowest order term. It contains one insertion of the quark mass matrix and no derivatives,

$$\mathcal{L}^{(2)}_{\pi\pi,\text{SB}} = \frac{BF_\pi^2}{2} \langle \mathcal{M}U + \mathcal{M}^\dagger U^\dagger \rangle \,, \tag{2.21}$$

where B is a LEC of dimension [mass]. If CP is conserved, B must be real. Expanding again in powers of the pion field, we obtain

$$\mathcal{L}^{(2)}_{\pi\pi,\text{SB}} = (m_u + m_d)B \left[F_\pi^2 - \frac{\pi^2}{2} + \mathcal{O}(\pi^4) \right] \,. \tag{2.22}$$

The first term can be related to the QCD vacuum. Using $\partial \mathcal{L}_{\text{QCD}}/\partial m_q |_{m_q=0} = -\bar{q}q$, where q denotes the light flavors, we get for the vacuum expectation value of the scalar quark condensate

$$\langle 0|\bar{u}u|0 \rangle = \langle 0|\bar{d}d|0 \rangle = -BF_\pi^2[1 + \mathcal{O}(\mathcal{M})] \,. \tag{2.23}$$

Similarly, from the second term in Eq. (2.22) we can read off the charged pion mass,

$$\mathcal{L}_{\text{mass}} = -\frac{1}{2}M_\pi^2 \pi^2 \to M_\pi^2 = (m_u + m_d)B + \mathcal{O}(\mathcal{M}) \,. \tag{2.24}$$

Combining the last two equations gives the celebrated Gell-Mann–Oakes–Renner relation [21],

$$M_\pi^2 = -(m_u + m_d)\frac{\langle 0|\bar{u}u|0 \rangle}{F_\pi^2} + \mathcal{O}(\mathcal{M}^2) \,, \tag{2.25}$$

that is fulfilled in QCD to better than 94% [22]. It also explains why we count a quark mass insertion as two powers of small momentum since $M_\pi \sim Q$. Note that the leading order pion Lagrangian is isospin symmetric, as only the sum of the light quark masses appears. That is the reason why the charged to neutral pion mass splitting is almost entirely given by electromagnetic effects. Higher order terms for $\mathcal{L}^{(2)}_{\pi\pi}$ and the coupling to external fields (like e.g. photons) can be constructed along the same lines, see e.g. Refs. [23, 24]. The nowadays considered standard form of the fourth order chiral Lagrangian is given in Ref. [25] and the sixth order terms can be found e.g. in Ref. [26].

2.4.2 The Pion-Nucleon Lagrangian

We are now in the position to extend these considerations to include nucleons. More precisely, we are interested in describing reactions involving pions with external momenta of the order of M_π and (essentially) non-relativistic nucleons whose three-momenta are of the order of M_π. Similar to the triplet of pion fields, the isospin doublet of the nucleon fields

$$N = \begin{pmatrix} p \\ n \end{pmatrix} \tag{2.26}$$

should transform nonlinearly under the chiral $SU(2)_L \times SU(2)_R$ but linearly under the vector subgroup $SU(2)_V$. The unitary matrix U used so far is less useful when constructing the Lagrangian involving massive fields like the nucleons. It is more convenient to introduce its square root u, $U = u^2$. The transformation properties of u under chiral rotations can be read off from Eq. (2.18):

$$u \to u' = \sqrt{LUR^\dagger} \equiv Luh^{-1} = huR^\dagger, \tag{2.27}$$

where we have introduced the unitary matrix $h = h(L, R, U)$ that is given by $h = \sqrt{LUR^\dagger}^{-1} L\sqrt{U}$. It is sometimes referred to as the compensator field. The last equality in Eq. (2.27) follows from $U' = u'u' = Luh^{-1}u' = LuuR^\dagger$. Notice that since pions transform linearly under isospin rotations corresponding to $L = R = V$ with $U \to U' = VUV^\dagger$ and, accordingly, $u \to u' = VuV^\dagger$, the compensator field in this particular case becomes U-independent and coincides with V. It can be shown that $\{U, N\}$ defines a nonlinear realization of the chiral group if one demands that

$$N \to N' = hN. \tag{2.28}$$

as proven e.g. in Refs. [19, 20]. Moreover, this non-linear realization obviously fulfills the desired feature that pions and nucleons transform linearly under isospin rotations. Similar to the purely Goldstone boson case, one can show that all other possibilities to introduce the nucleon fields are identical with the above realization modulo nonlinear field redefinitions (this is nicely discussed in Georgi's book [27]). The most general chiral invariant Lagrangian for pions and nucleons can be constructed from covariantly transforming building blocks, i.e. $O_i \to O'_i = hO_ih^{-1}$, by writing down all possible terms of the form $\bar{N}O_1 \ldots O_nN$. The covariant derivative of the pion field is given by

$$u_\mu \equiv iu^\dagger(\partial_\mu U)u^\dagger = -\frac{\tau \cdot \partial_\mu \pi}{F} + \mathcal{O}(\pi^3) \rightarrow u'_\mu = hu_\mu h^{-1}, \qquad (2.29)$$

and is sometimes referred to as chiral vielbein. The derivative of the nucleon field, $\partial_\mu N$, does not transform covariantly, i.e. $\partial_\mu N \rightarrow (\partial_\mu N)' \neq h\partial_\mu N$ since the compensator field h does, in general, depend on space-time (through its dependence on U). The covariant derivative of the nucleon field $D_\mu N$, $D_\mu N \rightarrow (D_\mu N)' = hD_\mu N$, is given by

$$D_\mu N \equiv (\partial_\mu + \Gamma_\mu)N, \quad \text{with} \quad \Gamma_\mu \equiv \frac{1}{2}\left(u^\dagger\partial_\mu u + u\partial_\mu u^\dagger\right)$$

$$= \frac{i}{4F^2}\tau \cdot \pi \times \partial_\mu\pi + \mathcal{O}(\pi^4). \qquad (2.30)$$

We remark that due to the spinor nature of the nucleons, terms with odd numbers of derivatives are allowed in the pion-nucleon Lagrangian. A prominent example is the pseudo-vector πN coupling $\bar{N}\gamma_\mu\gamma_5\tau^a N\partial^\mu\pi^a$ that is in agreement with Goldstone's theorem. The so-called chiral connection Γ_μ can be used to construct higher covariant derivatives of the pion field, for example:

$$u_{\mu\nu} \equiv \partial_\mu u_\nu + [\Gamma_\mu, u_\nu]. \qquad (2.31)$$

To first order in derivatives, the most general pion-nucleon Lagrangian takes the form [28]

$$\mathscr{L}^{(1)}_{\pi N} = \bar{N}\left(i\gamma^\mu D_\mu - m_N + \frac{g_A}{2}\gamma^\mu\gamma_5 u_\mu\right)N, \qquad (2.32)$$

where m_N and g_A are the nucleon mass and the axial-vector coupling constant in the chiral limit, respectively. To keep the presentation simple, we have not denoted these by a different symbol, but the difference should be kept in mind. Contrary to the pion mass, the nucleon mass does not vanish in the chiral limit and introduces an additional hard scale in the problem. Consequently, terms proportional to D_0 and m_N in Eq. (2.32) are individually large. It can, however, be shown that $(i\gamma^\mu D_\mu - m_N)N \sim \mathcal{O}(Q)$ [28]. The appearance of the additional hard scale ($m_N \gg M_\pi$) associated with the nucleon mass invalidates the power counting for dimensionally regularized expressions since the contributions from loop integrals involving nucleon propagators are not automatically suppressed. To see this consider the correction to the nucleon mass m_N due to the pion loop shown in Fig. 2.2. Assuming that the nucleon and pion propagators scale as $1/Q$ and $1/Q^2$, respectively, and taking into account Q^4 from the loop integration and Q^2 from the derivatives stemming from the two pion-nucleon vertices $\sim g_A$, the pion loop contribution to the nucleon self-energy $\Sigma(p)$ is expected to be of the order $\sim Q^3$. Consequently, the corresponding nucleon mass shift $\delta m_N = \Sigma(m_N)$ is expected

Fig. 2.2 Pion loop contribution to the nucleon self energy. Solid (wiggly) lines represents the nucleon (the pion)

to be $\propto M_\pi^3$, since the pion mass is the only soft scale in the problem. An explicit calculation, however, shows that the resulting nucleon mass shift does not vanish in the chiral limit [28]:

$$\delta m_N \big|_{\text{one-loop}} \stackrel{\mathscr{M}\to 0}{=} -\frac{3g_A^2 m_N^3}{F_\pi^2}\left(L(\mu) + \frac{1}{32\pi^2}\ln\frac{m_N^2}{\mu^2}\right) + \mathscr{O}(d-4)\,, \qquad (2.33)$$

with

$$L(\mu) = \frac{\mu^{d-4}}{16\pi^2}\left[\frac{1}{d-4} - \frac{1}{2}(\ln(4\pi) + \Gamma'(1) + 1)\right]\,, \quad \Gamma'(1) = -0.577215\,, \qquad (2.34)$$

where μ is the scale of dimensional regularization and d the number of space-time dimensions. The result in Eq. (2.33) implies that the nucleon mass receives a contribution which is formally of the order $\sim m_N\,(m_N/4\pi F_\pi)^2$ and is not suppressed compared to m_N. The nucleon mass m_N that enters the lowest-order Lagrangian $\mathscr{L}_{\pi N}^{(1)}$ gets renormalized. This is in contrast to the purely mesonic sector where loop contributions are always suppressed by powers of the soft scale and the parameters F_π and B in the lowest-order Lagrangian $\mathscr{L}_{\pi\pi}^{(2)}$ remain unchanged by higher-order corrections (if a mass-independent regularization is used). There exist a variety of ways to ensure the correct power counting in the πN system, i.e. to make sure that only soft momenta $\sim Q$ flow through the diagrams. We refer to the review [29] for a detailed discussion of these methods.

The simplest way to ensure the proper power counting exploits the so-called heavy-baryon formalism [30, 31] which is closely related to the non-relativistic expansion due to Foldy and Wouthuysen [32]. It is also widely used in heavy-quark effective field theories, see e.g. [33]. The idea is to decompose the nucleon four-momentum p^μ according to

$$p_\mu = m_N v_\mu + k_\mu\,, \qquad (2.35)$$

with v_μ the four-velocity of the nucleon satisfying $v^2 = 1$ and k_μ is a small residual momentum, $v\cdot k \ll m_N$. One can thus decompose the nucleon field N into velocity eigenstates

$$H = e^{im_N v\cdot x}P_v^+ N\,, \qquad\qquad h = e^{im_N v\cdot x}P_v^- N\,, \qquad (2.36)$$

where $P_v^\pm = (1 \pm \gamma_\mu v^\mu)/2$ denote the corresponding projection operators, with $(P_v^\pm)^2 = P_v^\pm$, $P_v^+ + P_v^- = 1$, and $P_v^+ \cdot P_v^- = 0$. In the nucleon rest-frame with $v_\mu = (1, 0, 0, 0)$, the quantities N_v and h_v are related to the familiar large and small components of the free positive-energy Dirac field, respectively. One, therefore, usually refers to H and h as to the large and small components of N. The relativistic Lagrangian $\mathscr{L}_{\pi N}^{(1)}$ in Eq. (2.32) can be expressed in terms of H and h as:

$$\mathscr{L}_{\pi N}^{(1)} = \bar{H}\mathscr{A}H + \bar{h}\mathscr{B}H + \bar{H}\gamma_0\mathscr{B}^\dagger\gamma_0 h - \bar{h}\mathscr{C}h, \tag{2.37}$$

where

$$\mathscr{A} = i(v \cdot D) + g_A(S \cdot u),$$

$$\mathscr{B} = -\gamma_5\left[2i(S \cdot D) + \frac{g_A}{2}(v \cdot u)\right],$$

$$\mathscr{C} = 2m_N + i(v \cdot D) + g_A(S \cdot u), \tag{2.38}$$

with $S_\mu = i\gamma_5\sigma_{\mu\nu}v^\nu/2$ the nucleon (Pauli-Lubanski) spin operator, and $\sigma_{\mu\nu} = i[\gamma_\mu, \gamma_\nu]/2$. It obeys the following relations

$$\{S_\mu, S_\nu\} = \frac{1}{2}(v_\mu v_\nu - g_{\mu\nu}), [S_\mu, S_\nu] = i\varepsilon_{\mu\nu\alpha\beta}v^\alpha S^\beta, S \cdot v = 0, S^2 = \frac{1-d}{4}. \tag{2.39}$$

We use the convention $\varepsilon^{0123} = -1$. One can now use the equations of motion for the large and small component fields to completely eliminate h_v from the Lagrangian. Utilizing the more elegant path integral formulation [34], the heavy degrees of freedom can be integrated out performing a Gaussian integration over the (appropriately shifted) variables h, $\bar{h}$. This leads to the effective Lagrangian of the form [31]

$$\mathscr{L}_{\pi N}^{(1)} = \bar{H}\left[\mathscr{A} + (\gamma_0\mathscr{B}^\dagger\gamma_0)\mathscr{C}^{-1}\mathscr{B}\right]H = \bar{H}\left[i(v \cdot D) + g_A(S \cdot u)\right]H + \mathcal{O}(1/m_N). \tag{2.40}$$

Notice that the (large) nucleon mass term has disappeared from the Lagrangian, and the dependence on m_N in $\mathscr{L}_{\pi N}^{\text{eff}}$ resides entirely in a tower of new vertices suppressed by powers of $1/m_N$. In fact, all Dirac bilinears can now be expressed in terms of v_μ and S_μ, e.g. $\bar{H}\gamma_\mu H = v_\mu \bar{H}H$. As an exercise, the reader should work out the other bilinears or consult the detailed review [35]. The heavy-baryon propagator of the nucleon is simply

$$S_N(v \cdot k) = \frac{i}{v \cdot k + i\epsilon}, \qquad \epsilon \to 0^+, \tag{2.41}$$

It can also be obtained from the $1/m_N$ expansion of the Dirac propagator using Eq. (2.35) and assuming $v \cdot k \ll m_N$:

$$\frac{\slashed{p} + m}{p^2 - m^2 + i\epsilon} = \frac{\Lambda_+}{v \cdot k + i\epsilon} + \mathcal{O}\left(1/m_N\right) , \tag{2.42}$$

where $\Lambda_+ = (\slashed{p} + m_N)/(2m_N)$ is a projection operator on the states of positive energy. The lowest order pion-nucleon Lagrangian generates two vertices containing pions and nucleons that are of relevance to the construction of the nuclear forces. First, from the term $\sim g_A$, one gets the pseudo-vector pion-nucleon coupling with the corresponding Feynman rule

$$\frac{g_A}{F_\pi} S \cdot q\, \tau^a , \tag{2.43}$$

for an out-going pion with momentum q. In addition, the term $\sim D_\mu$ generates the celebrated Weinberg-Tomozawa term, where two pions are emitted from one point,

$$\frac{1}{4F_\pi^2} v \cdot (q_1 + q_2)\, \varepsilon^{abc}\, \tau^c , \tag{2.44}$$

with q_1 in-coming and q_2 out-going. As in the pion case, these terms generate a whole tower of multi-pion-nucleon interactions that are given in terms of g_A (odd powers) or $1/F_\pi$ (even powers). The heavy-baryon formulation obeys in contrast to the naive relativistic realistic formulation discussed above the power counting, i.e. the leading one-loop correction to the nucleon mass indeed scales as M_π^3,

$$\delta m_N \big|_{\text{one-loop}} = -\frac{3g_A^2 M_\pi^3}{32\pi F^2} . \tag{2.45}$$

Contrary to the relativistic CHPT result in Eq. (2.33), the loop correction in the heavy baryon approach is finite and vanishes in the chiral limit. The parameters in the lowest-order Lagrangian do not get renormalized due to higher-order corrections which are suppressed by powers of Q/Λ_χ. Notice further that Eq. (2.45) represents the leading contribution to the nucleon mass which is non-analytic in quark masses. It agrees with the time-honored result, see Ref. [36].

In the construction of the two- and three-nucleon forces, we will also need the chiral pion-nucleon Lagrangian at second order. The terms of relevance here are

$$\mathcal{L}_{\pi N}^{(2)} = \bar{H} \left\{ \frac{1}{2m_N}(v \cdot D)^2 - \frac{1}{2m_N} D^2 - \frac{ig_A}{2m_N}\{S \cdot D, v \cdot u\} + c_1 \langle \chi_+ \rangle + c_2 (v \cdot u)^2 \right.$$
$$\left. + c_3\, u \cdot u + c_4\, [S^\mu, S^\nu] u_\mu u_\nu + c_5\, \tilde{\chi}_+ + \dots \right\} H \tag{2.46}$$

with

$$\chi_\pm = u^\dagger \chi u^\dagger \pm u \chi^\dagger u , \quad \tilde{\chi}_+ = \chi_+ - \frac{1}{2}\langle \chi_+ \rangle \quad \chi = 2B\mathcal{M} . \tag{2.47}$$

The term $\sim \tilde{\chi}_+$ is sensitive to the quark mass difference $m_u - m_d$ and thus vanishes in the isospin limit. The LECs c_i have dimension [mass^{-1}], their natural size is $\sim g_A/\Lambda_\chi \simeq 1\,\text{GeV}^{-1}$. They can be determined from the analysis of pion-nucleon scattering, most precisely from the Roy-Steiner equation analysis [37]. Typical values are

$$c_1 = -1.10(3) , \quad c_2 = 3.57(4) , \quad c_3 = -5.54(6) , \quad c_4 = 4.17(4) , \quad c_5 = -0.09(1) , \tag{2.48}$$

all in units of GeV^{-1} (we use here the Weinberg counting of the nucleon mass as explained in Sect. 2.4.4). The value for c_1 is of natural size, whereas the unnaturally large values of c_2, c_3 and c_4 can be explained through the contribution form the $\Delta(1232)$ resonance, that leads to an enhancement $\Lambda_\chi/(m_\Delta - m_N) \simeq 3$. In case of c_4, there is also important contribution from t-channel vector meson exchange. For details, the reader should consult Ref. [38]. The small value of c_5 is related to the strong part of the neutron-proton mass difference that is about 2 MeV, i.e. much smaller than the nucleon mass. The complete pion-nucleon Lagrangian up-to-and-including the fourth order terms is given in Ref. [39].

Having constructed the chiral effective Lagrangian for pion-nucleon interactions, we must revisit the power counting discussed for the pion sector in Sect. 2.4.1. We simply spell out the differences to the pion case: (1) we have to account for the nucleon propagators that scale as $1/Q$, (2) we need to incorporate the pion-nucleon interactions that have one, two, three, ... derivatives, (3) in the heavy-baryon approach, no closed fermion loops appear so that exactly one nucleon line connecting the initial with the final state runs through each diagram in the single baryon sector. This also implies that the number of internal nucleon lines is equal to the sum of all pion-nucleon interactions minus one. Putting all this together gives for the chiral counting index

$$\nu = 1 + 2L + \sum_i V_i \Delta_i , \quad \text{with} \quad \Delta_i = -2 + \frac{1}{2}n_i + d_i , \tag{2.49}$$

with n_i the number of nucleon field operators at a vertex i with the chiral dimension Δ_i. Because of chiral symmetry, Δ_i is bounded from below, $\Delta_i \geq 0$.

As a typical process in the πN system, let us consider elastic pion-nucleon scattering. Tree graphs with $L = 0$ lead to contributions with $\nu = 1$ and $\nu = 2$, whereas the leading loop corrections appear at $\nu = 3$ and the one-loop amplitude is completed at $\nu = 4$. It is instructive to discuss the structure of the elastic pion-nucleon scattering amplitude as depicted in Fig. 2.3. At leading order $\sim Q$, we have tree graphs, the so-called Born diagrams (a) and the already mentioned Weinberg-

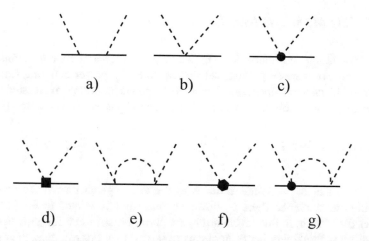

Fig. 2.3   Structure of the πN scattering amplitude as explained in the text. Only some typical diagrams at each order are shown and crossed graphs are not displayed. Solid (dashed) lines represents the nucleon (the pion). Filled circles/boxes/hexagons denote an insertion from $\mathscr{L}_{\pi N}^{(2)}$, $\mathscr{L}_{\pi N}^{(3)}$, and $\mathscr{L}_{\pi N}^{(4)}$, in order

Tomozawa contact term (b). First corrections appear at order Q^2. These are again tree graphs with exactly one insertion from $\mathscr{L}_{\pi N}^{(2)}$, (c), parameterized in terms of the LECs c_i, cf. Eq. (2.46). As stated before, these LECs contain information about the s-channel Δ and t-channel ρ-meson excitations. At order Q^3, we have further tree graphs with one insertion from $\mathscr{L}_{\pi N}^{(3)}$, (d), and the leading one-loop graphs (e). Contributions to order Q^4 are additional tree graphs with one insertion from $\mathscr{L}_{\pi N}^{(4)}$ (f) and one-loop graphs with the same topologies as at $\mathscr{O}(Q^3)$ but with exactly one insertion from $\mathscr{L}_{\pi N}^{(2)}$ as shown in (g). The number of LECs contributing to this reaction is, however, limited. If one writes down the most general polynomial contribution to the pion-nucleon amplitude that is quartic in momenta consistent with analyticity, crossing and unitarity, one has 13 independent terms. These can be mapped on various combinations of the LECs c_i, d_i and e_i. There is one additional third order LEC that parameterizes the deviation from the Goldberger-Treiman relation

$$g_{\pi N} = \frac{g_A m_N}{F_\pi}\left(1 - \frac{2M_\pi^2 d_{18}}{g_A}\right).\qquad(2.50)$$

Here, $g_{\pi N}$ is the pion-nucleon coupling constant, $g_{\pi N}^2/4\pi = 13.7 \pm 0.1$ [40]. In Nature, this relation is fulfilled within a few percent. The deviation from unity is parameterized by the LEC d_{18} from the third order pion-nucleon Lagrangian. Therefore, at one loop order we have to determine 14 LECs. For more details on this and the fourth order calculation of $\pi N \to \pi N$ in the heavy-baryon scheme, the reader is referred to Ref. [41] or to the more recent work in Ref. [42], that contains also uncertainty estimates based on the method discussed in Sect. 2.5.2.

2.4.3 The Multi-Nucleon Lagrangian

Before discussing the interactions between two, three and four nucleons, some
preparations are necessary. First, we must extend the power counting formula to
include multi-nucleon operators. Weinberg's power counting expression for N-
nucleon diagrams involving C separately connected pieces reads, see Ref. [43]:

$$v = 4 - N + 2(L - C) + \sum_i V_i \Delta_i , \qquad \Delta_i = d_i + \frac{1}{2} n_i - 2 . \qquad (2.51)$$

For $C = 0$ and $N = 1$, this reduces to the above given formula for processes
involving one nucleon. There is, however, one subtlety related to Eq. (2.51) that
requires discussion. In this formulation, the chiral dimension v for a given process
depends on the total number of nucleons in the system. For example, the one-pion
exchange in the two-nucleon system corresponds to $N = 2$, $L = 0$, $C = 1$ and
$\sum_i V_i \Delta_i = 0$ and, therefore, contributes at order $v = 0$. On the other hand, the same
process in the presence of a third (spectator) nucleon leads, according to Eq. (2.51),
to $v = -3$ since $N = 3$ and $C = 2$. See Fig. 2.4 for an example from the NN
interaction showing a connected and a disconnected diagram. The origin of this
apparent discrepancy stems from the different normalization of the 2N and 3N states
in momentum space:

$$2N : \quad \langle \mathbf{p}_1 \mathbf{p}_2 | \mathbf{p}_1' \mathbf{p}_2' \rangle = \delta^3(\mathbf{p}_1' - \mathbf{p}_1) \, \delta^3(\mathbf{p}_2' - \mathbf{p}_2) ,$$

$$3N : \quad \langle \mathbf{p}_1 \mathbf{p}_2 \mathbf{p}_3 | \mathbf{p}_1' \mathbf{p}_2' \mathbf{p}_3' \rangle = \delta^3(\mathbf{p}_1' - \mathbf{p}_1) \, \delta^3(\mathbf{p}_2' - \mathbf{p}_2) \, \delta^3(\mathbf{p}_3' - \mathbf{p}_3) .$$

$$(2.52)$$

It can be circumvented by assigning a chiral dimension to the transition operator
rather than to its matrix elements in the N-nucleon space. Adding the factor $3N - 6$
to the right-hand side of Eq. (2.51) in order to account for the normalization of the
N-nucleon states and to ensure that the LO contribution to the nuclear force appears

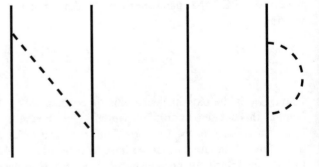

Fig. 2.4 Connected
contribution to the one-pion
exchange of the
nucleon-nucleon interaction
with $C = 1$ (left) and a
disconnected 2N contribution
(nucleon mass
renormalization) with $C = 2$
(right)

at order $\nu = 0$ we obtain

$$\nu = -2 + 2N + 2(L - C) + \sum_i V_i \Delta_i . \qquad (2.53)$$

In the heavy-baryon formulation outlined above, the leading terms for the NN and the 3N contact interactions including nucleons and pions take the form (we will explain later, why only the 3N interactions without derivatives is required):

$$\mathscr{L}_{NN}^{(0)} = -\frac{1}{2}C_S(\bar{N}N)(\bar{N}N) + 2C_T(\bar{N}SN) \cdot (\bar{N}SN) ,$$

$$\mathscr{L}_{NN}^{(1)} = \frac{D}{2}(\bar{N}N)(\bar{N}S \cdot uN) ,$$

$$\mathscr{L}_{NN}^{(2)} = -C_1\left[(\bar{N}DN) \cdot (\bar{N}DN) + ((D\bar{N})N) \cdot ((D\bar{N})N)\right]$$

$$-2(C_1 + C_2)(\bar{N}DN) \cdot ((D\bar{N})N) - C_2(\bar{N}N)$$

$$\cdot \left[(D^2\bar{N})N + \bar{N}D^2N\right] + \dots ,$$

$$\mathscr{L}_{NNN}^{(1)} = -\frac{E}{2}(\bar{N}N)(\bar{N}\tau N) \cdot (\bar{N}\tau N) . \qquad (2.54)$$

Here, the ellipses refer to terms which do not contribute to the nuclear forces up to next-to-next-to-leading order (N²LO) except for $\mathscr{L}_{NN}^{(2)}$ where only a few terms are shown in order to keep the presentation compact. The full list of terms will be given later together with the appropriate representation most suitable for a partial-wave analysis. The $C_{S,T}$, C_i, D and E denote the LECs related to multi-nucleon contact operators. For completeness, we rewrite the already discussed terms of the chiral pion-nucleon Lagrangian in the normalization corresponding to the so-called Weinberg power counting, which counts the nucleon mass differently from the standard CHPT approach, as discussed in detail in Sect. 2.4.4,

$$\mathscr{L}_{\pi N}^{(0)} = \bar{N}\left[i\,v \cdot D + g_A u \cdot S\right]N ,$$

$$\mathscr{L}_{\pi N}^{(1)} = \bar{N}\left[c_1\langle\chi_+\rangle + c_2(v \cdot u)^2 + c_3 u \cdot u\right.$$

$$\left. +c_4[S^\mu, S^\nu]u_\mu u_\nu + c_5\langle\hat{\chi}_+\rangle\right]N , \qquad (2.55)$$

$$\mathscr{L}_{\pi N}^{(2)} = \bar{N}\left[\frac{1}{2m_N}(v \cdot D)^2 - \frac{1}{2m_N}D \cdot D\right.$$

$$\left. +d_{16}S \cdot u\langle\chi_+\rangle + id_{18}S^\mu[D_\mu, \chi_-] + \dots\right]N .$$

It is instructive to construct the effective 4N Lagrangian without derivatives. In general, one could write down a contact term potential corresponding to the lowest order effective Lagrangian with four independent structures,

$$V_0 = a_0 + a_1(\boldsymbol{\sigma}_1 \cdot \boldsymbol{\sigma}_2) + a_2(\boldsymbol{\tau}_1 \cdot \boldsymbol{\tau}_2) + a_3(\boldsymbol{\sigma}_1 \cdot \boldsymbol{\sigma}_2)(\boldsymbol{\tau}_1 \cdot \boldsymbol{\tau}_2) \,, \tag{2.56}$$

where the a_i are functions of the nucleon three-momenta. Since we are dealing with fermions, the two-nucleon states must be antisymmetric and, correspondingly, the 2N potential must be antisymmetrized,

$$V_0^A = \frac{1}{2} \left(V_0 - \mathscr{A}[V_0] \right) \,, \tag{2.57}$$

in terms of the antisymmetrization operator

$$\mathscr{A}[V_0] = \mathscr{A}[\langle \alpha_1' \alpha_2' | V_0 | \alpha_1 \alpha_2 \rangle] = \langle \alpha_2' \alpha_1' | V_0 | \alpha_1 \alpha_2 \rangle \,, \tag{2.58}$$

where α_i denotes the spin, isospin and momentum quantum numbers of nucleon i. Following Ref. [44], in the 2N center-of-mass system (cms), antisymmetrization amounts to a left-multiplication with the spin-exchange operator $(1 + \boldsymbol{\sigma}_1 \cdot \boldsymbol{\sigma}_2)$, a left-multiplication with the isospin-exchange operator $(1 + \boldsymbol{\tau}_1 \cdot \boldsymbol{\tau}_2)$ and the substitution $\mathbf{p}' \rightarrow -\mathbf{p}'$, with $\mathbf{p}$ and $\mathbf{p}'$ the cms momenta in the initial and the final two-nucleon state, respectively. With that, the following identities are obtained:

$$\mathscr{A}[1] = \frac{1}{4} \left(1 + (\boldsymbol{\sigma}_1 \cdot \boldsymbol{\sigma}_2) + (\boldsymbol{\tau}_1 \cdot \boldsymbol{\tau}_2) + (\boldsymbol{\sigma}_1 \cdot \boldsymbol{\sigma}_2)(\boldsymbol{\tau}_1 \cdot \boldsymbol{\tau}_2) \right) \,,$$

$$\mathscr{A}[\boldsymbol{\sigma}_1 \cdot \boldsymbol{\sigma}_2] = \frac{1}{4} \left(3 - (\boldsymbol{\sigma}_1 \cdot \boldsymbol{\sigma}_2) + 3(\boldsymbol{\tau}_1 \cdot \boldsymbol{\tau}_2) - (\boldsymbol{\sigma}_1 \cdot \boldsymbol{\sigma}_2)(\boldsymbol{\tau}_1 \cdot \boldsymbol{\tau}_2) \right) \,,$$

$$\mathscr{A}[\boldsymbol{\tau}_1 \cdot \boldsymbol{\tau}_2] = \frac{1}{4} \left(3 + 3(\boldsymbol{\sigma}_1 \cdot \boldsymbol{\sigma}_2) - 3(\boldsymbol{\tau}_1 \cdot \boldsymbol{\tau}_2) - (\boldsymbol{\sigma}_1 \cdot \boldsymbol{\sigma}_2)(\boldsymbol{\tau}_1 \cdot \boldsymbol{\tau}_2) \right) \,,$$

$$\mathscr{A}[(\boldsymbol{\sigma}_1 \cdot \boldsymbol{\sigma}_2)(\boldsymbol{\tau}_1 \cdot \boldsymbol{\tau}_2)] = \frac{1}{4} \left(9 - 3(\boldsymbol{\sigma}_1 \cdot \boldsymbol{\sigma}_2) - 3(\boldsymbol{\tau}_1 \cdot \boldsymbol{\tau}_2) + (\boldsymbol{\sigma}_1 \cdot \boldsymbol{\sigma}_2)(\boldsymbol{\tau}_1 \cdot \boldsymbol{\tau}_2) \right) \,.$$

$$\tag{2.59}$$

It is now easy to perform the antisymmetrization of the potential (2.56),

$$V_0^A = \frac{1}{2} \left(3b_1 - (2b_1 + b_2)(\boldsymbol{\sigma}_1 \cdot \boldsymbol{\sigma}_2) + b_2(\boldsymbol{\tau}_1 \cdot \boldsymbol{\tau}_2) - b_1(\boldsymbol{\sigma}_1 \cdot \boldsymbol{\sigma}_2)(\boldsymbol{\tau}_1 \cdot \boldsymbol{\tau}_2) \right) \,, \tag{2.60}$$

in terms of two constants only,

$$b_1 = \frac{1}{4} \left(a_1 - a_2 - a_3 - 3a_4 \right) \,, \quad b_2 = \frac{1}{4} \left(-a_1 - 3a_2 + 5a_3 + 3a_4 \right) \,. \tag{2.61}$$

Note that only two independent constants appear in the antisymmetrized potential V_0^A, so that one has just two independent contact interactions at lowest order. These are commonly chosen as

$$V_0 = C_S + C_T \, \boldsymbol{\sigma}_1 \cdot \boldsymbol{\sigma}_2 \,, \tag{2.62}$$

with $C_S = a_1 - 2a_3 - 3a_4$ and $C_T = a_2 - a_4$. In a similar way, one can construct the minimal basis of higher order contact interactions. Note further that these results can also be obtained using Fierz transformations. The higher order terms as well the contact interactions between three nucleons can be worked out along the same lines.

2.4.4 Power Counting in the Two-Nucleon System

Naively, one could now use the so-constructed multi-nucleon Lagrangian to consider nucleon-nucleon scattering and other processes involving more than one nucleon. However, we know that there are shallow bound states like e.g. the deuteron that is located just 2.22 MeV below the neutron-proton continuum threshold. Such a bound state represents a non-perturbative phenomenon that cannot be described in perturbation theory. However, according to the counting rules displayed in Eq. (2.53), loops are suppressed by powers of the soft scale. This seems to suggest that the interaction is weak. Clearly, this conclusion is erroneous. The reason for this discrepancy can be traced back to the appearance of infrared (IR) divergences in the strict heavy baryon limit in diagrams that only contain two-nucleon intermediate states. The culprit is the two-pion exchange box diagram shown in the left panel of Fig. 2.5. For the static (heavy-baryon) propagator, its contribution is proportional to [3]

$$I_{\text{box}} = \int d^4q \, \frac{\tilde{P}(q)}{(q^0 + i\epsilon)(q^0 - i\epsilon)(q^2 + M_\pi^2)^2} \,, \tag{2.63}$$

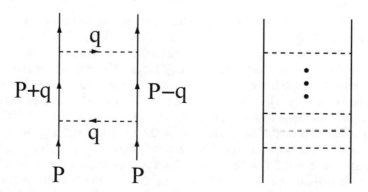

Fig. 2.5 Left panel: The two-pion exchange box diagram. Solid (dashed) lines denote nucleons (pions). P (q) denotes the nucleon (pion) four-momentum. Right panel: A typical ladder diagram. Such diagrams are reducible in the sense defined in the text

with $\tilde{P}(q)$ a polynomial in the pion momenta. As this polynomial is non-vanishing for $q^0 \to 0$, the integral is IR divergent, $I_{\text{box}} \sim \int dq_0 (q^0 + i\epsilon)^{-1} (q^0 - i\epsilon)^{-1}$. The integration contour is pinched between the two poles at $q^0 = \mp i\epsilon$, so this pole cannot be avoided by any contour deformation. This is completely different from the crossed box graph or any diagram corresponding to a single-nucleon process. Of course, this IR divergence is an artifact of the extreme non-relativistic approximation. If one repeats this calculation including the nucleon kinetic energy, the poles are shifted to $q^0 \simeq \pm [\mathbf{q}^2/2m_M - i\epsilon]$. This entails that the contribution from the box diagram is enhanced by $m_N/\mathbf{q} \sim m_N/Q$ compared to the expectation from the power counting for such a one-loop graph. These enhancement factors are responsible for the nuclear binding. However, this is not the whole story. If one counts, as usually done in the single-nucleon sector, the nucleon mass as $m_N \sim \Lambda_\chi$, loops are still suppressed by $Q m_N/\Lambda_\chi^2 \sim Q/\Lambda_\chi$. This can be understood from the Lippmann-Schwinger equation for the transition operator,

$$T = V + V G_0 T , \tag{2.64}$$

using a highly symbolic notation where the integration over momenta in the second term is not made explicit, G_0 is the two-nucleon propagator and V is the effective chiral potential. To leading order, we deal with the one-pion-exchange potential that carries the factor $\sim 1/F_\pi^2$. Further, the two-nucleon propagator scales as $G_0 \sim m_N/Q^2$, each momentum integration gives $Q^3/(4\pi^2)$ and $\Lambda = 4\pi F_\pi$. To deal with this, Weinberg proposed to treat the nucleon mass as a separate hard scale,

$$m_N \sim \Lambda_\chi^2/Q \gg \Lambda_\chi . \tag{2.65}$$

This together with Eq. (2.53) defines what is commonly called the *Weinberg power counting*. For example, in this counting, the first three terms given in Eq. (2.46) indeed only appear one order higher, cf. Eq. (2.55), and the values for the c_i given there in fact refer to the Weinberg counting (for a more detailed discussion, see the review [37]).

It is easy to convince oneself that this IR enhancement can be taken into account if one defines an *effective potential*, that includes all *irreducible* terms using the language of time-ordered perturbation theory (that is different from the Feynman diagram technique, for a discussion see e.g. [45]). The irreducible contributions are generated from the diagrams that include at least one pion in any intermediate state. The box diagram just discussed is clearly reducible in this sense, as is any higher order ladder type diagram, as depicted in the right panel of Fig. 2.5. Consider now an effective potential just consisting of the one-pion exchange type, $V_{\text{eff}} = V^{\text{OPE}}$, where the explicit form of V^{OPE} will be given below, and insert it into the Lippmann-Schwinger (LS) equation for the scattering operator, $T = V_{\text{eff}} + V_{\text{eff}} G T$, with G the two-nucleon propagator. Writing the LS equation in iterated form, $T = V_{\text{eff}} + V_{\text{eff}} G V_{\text{eff}} + V_{\text{eff}} G V_{\text{eff}} G V_{\text{eff}} + \ldots$, one sees immediately that the box diagram and all other higher order ladder graphs are generated. Following Weinberg, the effective chiral Lagrangian is therefore used to construct the effective potential between two, three, ... nucleons and this is used in a Schrödinger (or equivalent) equation to generate the scattering and the bound states. This potential will be discussed in more detail in the next section.

2.4.5 The Chiral Potential

The Weinberg power counting, Eqs. (2.53) and (2.65) does not only allow to organize the various contributions to the forces between two nucleons, but also gives a systematic expansion for the three- and four-nucleon forces. For practical purposes, one can rearrange the various contributions using some topological identities in a very compact form [46]

$$ \nu = -2 + \sum V_i \kappa_i , \quad \kappa_i = d_i = \frac{3}{2} n_i + p_i - 4 \qquad (2.66) $$

with p_i the number of pions at the given vertex i. The resulting nuclear potentials are schematically shown in Fig. 2.6, where some typical contributions to the potentials corresponding to the 2N, 3N and 4N forces from leading order (LO) up to next-to-next-to-next-to-leading order (N^{3}LO) are displayed. Here, LO means terms of order Q^0, NLO corresponds to $\mathscr{O}(Q^2)$, N^{2}LO (or NNLO) subsumes the terms of

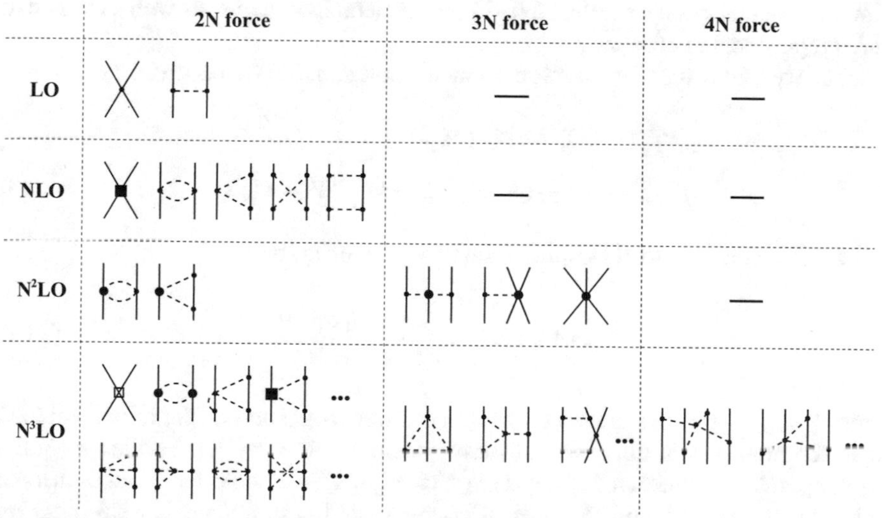

Fig. 2.6 Nuclear forces from chiral EFT based on Weinberg power counting. Beyond N^{2}LO, some typical diagrams are displayed. Solid and dashed lines denote nucleons and pions, respectively. Solid dots, filled circles and filled squares and crossed squares refer to vertices with $\Delta_i = 0, 1, 2$ and 4, respectively, cf. Eq. (2.53). Figure courtesy of Evgeny Epelbaum

order Q^3 and the N^{3}LO (or NNNLO) contributions are of $\mathcal{O}(Q^4)$, where Q is a small three-momentum or the pion mass, in terms of the breakdown scale Λ_{hard} that will be determined later. It is important to stress that the diagrams shown in Fig. 2.6 do actually not correspond to Feynman graphs but rather should be understood as a schematic visualization of the irreducible parts of the amplitude, namely those diagrams which do not correspond to iterations of the dynamical equation as discussed in the previous section.

Consider first the 2N force. Up to N^{3}LO, the NN potential involves contributions from one-, two- and three-pion exchanges and contact terms with up to four derivatives which parametrize the short-range interactions

$$V_{\text{NN}} = V_{1\pi} + V_{2\pi} + V_{3\pi} + V_{\text{cont}}, \tag{2.67}$$

The pattern for the chiral expansion displayed in Fig. 2.6 can be understood straight-forwardly. At LO, we have only the static pion exchange. Note that relativistic corrections to this are highly suppressed by factors of $\mathbf{p}^2/m_N^2$, with $\mathbf{p}$ a nucleon three-momentum in the cms frame. It is not possible to construct any term linear in small momenta, so NLO in fact is suppressed by two powers of Q^2 with respect to the leading order. At NLO, we therefore have the leading two-pion exchange diagrams, involving the leading order πNN and $\pi\pi NN$ couplings proportional to g_A and $1/F_\pi$, respectively. Then, at N^{2}LO, which is odd in momenta, one encounters the first corrections to the two-pion exchange (TPE) that are proportional to the LECs c_i and thus are somewhat enhanced. Three-pion exchange begins to contribute at N^{3}LO. At this order, we have further corrections to the TPE that involve one-insertion from the NNLO pion-nucleon Lagrangian as well as term that are proportional to c_i^2 or $c_i c_j$.

The well known static one-pion exchange potential (OPEP) is given by

$$V_{1\pi}^{pp} = V_{1\pi}^{nn} = V_{1\pi}(M_{\pi^0}),$$
$$V_{1\pi}^{np} = -V_{1\pi}(M_{\pi^0}) + 2(-1)^{I+1} V_{1\pi}(M_{\pi^\pm}), \tag{2.68}$$

where I denotes the total isospin of the two-nucleon system and

$$V_{1\pi}(M_\pi) = -\frac{g_A^2}{4F_\pi^2} \frac{\sigma_1 \cdot \mathbf{q}\, \sigma_2 \cdot \mathbf{q}}{q^2 + M_\pi^2}, \tag{2.69}$$

with $\mathbf{q} = \mathbf{p}' - \mathbf{p}$ refers to the momentum transfer with $\mathbf{p}$ and $\mathbf{p}'$ the initial and final nucleon momenta in the center-of-mass system, while $q \equiv |\mathbf{q}|$. Further, σ_i is the spin-operator of nucleon i ($i = 1, 2$) and $M_{\pi^0}/M_{\pi^\pm}$ denote the neutral/charged pion mass, respectively. We have accounted for the dominant isospin-breaking effect, namely the charged to neutral pion mass difference. Clearly, the OPEP has chiral dimension zero, as the two powers of Q in the numerator are cancelled by the terms of $\mathcal{O}(Q^2)$ in the denominator. The higher order potentials can be calculated using various methods, like Feynman diagram techniques, time-ordered perturbation theory or the method of unitary transformations. The relevant two-pion exchange potentials (TPEP) will be given later in lattice representation or can be found in the reviews [47, 48]. The three-pion exchange potential has been calculated in a series of papers by Kaiser, starting with Ref. [49]. It can, however, be shown explicitly that even the numerically dominant contribution proportional to the LEC c_4 plays only a negligible role in the 2N system and thus the three-pion exchange is usually neglected [50]. The contact interactions subsumed in V_{cont} come only in even powers of momenta due to parity conservation of the strong interactions. There are two terms at LO, see Eq. (2.62), and 7 (15) terms for on-shell scattering at NLO (N^3LO) that scale quadratically (quartically) in momenta. It has recently be shown that three of the 15 terms at N^3LO are redundant [51]. The corresponding LECs can be determined from a fit to neutron-proton scattering in a partial-wave representation. Clearly, the two constant terms at LO feed into the S-waves, whereas the two-derivative (four-derivative) terms at NLO (N^3LO) require P-waves (and D-waves) to be determined. The explicit partial-wave representation of the contact interactions will be discussed later, see Chaps. 4 and 5.

Clearly, the potential as defined here is UV divergent, e.g. the NLO contact terms scale quadratically in momenta, so when inserted into the Schrödinger (or LS) equation, some regularization is required. This has been much debated in the literature, but to keep the presentation simple, we stick to a modified version of the Weinberg approach, which amounts to introducing a cut-off in the LS equation. One advantage of such an approach, beyond its simplicity, is the ability to combine the resulting nuclear potentials with the available few- and many-body machinery which allows one to access observables beyond the NN system. A disadvantage is the appearance of cut-off artifacts, as the cut-off scale has to be chosen below the scale of chiral symmetry breaking, as otherwise unphysical, spurious bound states will be generated that drastically complicate applications to three- and more-nucleon systems. This means that the breakdown scale is below Λ_χ, one in fact finds $\Lambda_{hard} \simeq 500$ MeV. To reduce these cut-off artifacts, it is important to realize that chiral nuclear forces involve two distinct kinds of contributions: first, at large distances the potential is governed by contributions emerging from pion exchanges which are unambiguously determined by the chiral symmetry of QCD and experimental information on the pion-nucleon system is needed to pin down the relevant LECs. Second, the short-range part of the potential is parameterized by all possible contact interactions with increasing number of derivatives. It is desirable to introduce regularization in such a way that the long-range part of the interaction including especially the OPEP is not affected by the regulator. This can be achieved by cutting the pion-exchange(s) in coordinate space at a given separation $R = 0.8 \dots 1.2$ fm and regulating the contact interactions by a momentum-space Gaussian type of regulator with cut-off Λ, which can be related to each other by $\Lambda = 2R^{-1}$, for details the reader should consult Ref. [52] and references therein. A different semilocal regularization scheme was recently proposed in Ref. [51]. Furthermore, we note that the lattice considered later provides a natural UV cut-off as the maximum momentum for a given lattice spacing a at the edge of the Brillouin zone is $p_{max} = \pi/a$.

Next, we consider the three-nucleon potential. By close inspection of Fig. 2.6, the alert reader might wonder why there are no NLO contributions with leading order pion-nucleon vertices? It was shown by Weinberg [43] and van Kolck [53] that the leading order contributions cancel, partly on the diagrammatic level and partly by cancellations between reducible and irreducible diagrams up to suppressed corrections from the nucleon kinetic energy. The leading three-nucleon forces (3NFs) are therefore given by the three topologies at N^2LO. The two-pion exchange topology, which is a reflection of the time-honored Fujita-Miyazawa 3NF [54], has the longest range and is determined entirely in terms of the pion-nucleons LECs $c_{1,3,4}$ (for a detailed discussion of the Fujita-Miyazawa force in view of chiral EFT, see Ref. [55]). Next, there is a term of one-pion range, given in terms of the LEC D, cf. Eq. (2.54), that can be determined from a variety of reactions like e.g. $NN \rightarrow NN\pi$ or $pp \rightarrow de^+\nu_e$. This in fact is a beautiful manifestation of the power of chiral symmetry that connects seemingly unrelated processes. In addition, we have the six-nucleon contact term which is proportional to the LEC E, see again Eq. (2.54). It can only be determined in three-nucleon (or heavier) systems, like e.g. from the triton binding energy. Explicit expressions of all these contributions will be given in Chap. 4.

Four-nucleon forces only appear at N^3LO and have so far not been included in explicit calculations. At this order, they are given parameter-free, see Ref. [46]. In what follows, we will not further consider them explicitly.

2.5   The Two-Nucleon System

In this section, we discuss some pertinent issues related to the two-nucleon (2N) system. A compilation of the relevant formulae is given in Appendix B. First, we display a few state-of-the-art results, however, without going into any details. Then, we show how one can obtain error estimates at a given level of the chiral expansion. We end this section with a brief discussion of Wigner's time-honored observation that the NN interaction shows an approximate spin-isospin SU(4) symmetry, which will play an important role in the later discussion of the so-called "sign problem" in the nuclear Monte Carlo simulations, that are at the heart of this monograph.

2.5.1   Results in the Two-Nucleon System

The chiral expansion of the NN interaction in the continuum has been carried out to fifth order in the chiral expansion, that amounts to N^4LO [50, 56]. Before showing some results, let us dissect the physics behind the expansion displayed in Fig. 2.6. First, one has to determine the method to fix the 2N LECs. This can be done by either fitting to the existing partial wave analyses or to the data in the two-nucleon system directly. The former method is easier to implement and will be considered in what

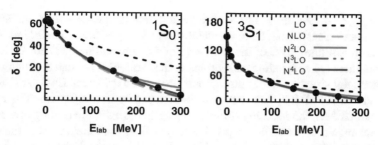

Fig. 2.7 Chiral expansion of the np S-waves up to fifth order. The $^1S_0$ ($^3S_1$) wave is shown in the left (right) panel as function of the nucleon laboratory energy E_{lab}. The various orders are given in the legend of the right panel. The filled circles refer to the Nijmegen PWA. Figure courtesy of Evgeny Epelbaum

follows. More precisely, the Nijmegen partial wave analysis (PWA) [57] will be taken as a benchmark. Second, one has to realize that it is important to differentiate between the partial waves with low angular momentum (S, P, D) and the so-called peripheral waves (G, H, I, ...), which are highly suppressed because of the angular momentum barrier and are well described by chiral one- and two-pion exchanges already at LO and NLO [44]. Therefore, let us concentrate on the lower partial waves, especially the S-waves. At LO, there are just two LECs that are usually fitted to the large scattering lengths in both the $^1S_0$ and the $^3S_1$ partial wave, respectively, cf. Eq. (2.5). As can be seen from Fig. 2.7, while for the $^3S_1$ wave, the description is quite decent, one also sees that the prediction for the $^1S_0$ is quickly deviates from the data. This can be understood from the fact that with one LEC, one can describe the rise due to the large scattering length $\sim 1/(8\,\text{MeV})$ but not the fall-off. This is different in the triplet wave due to the deuteron pole. At NLO, the description of the S-waves notably improves, but also some P- and D-waves are described well. At this order, we have seven LECs that feed into the S- and P-waves as well as the $^3S_1$-$^3D_1$ mixing angle ϵ_1. This is further improved at N^{2}LO, largely due to the subleading TPE graphs. Then at N^{3}LO, most partial waves are well described, one has 12 LECs that feed into the S-, P- and D-waves and the corresponding triplet mixing angles ϵ_1 and ϵ_2. One also has to account for the isospin-breaking between the np, nn and pp systems, which can be accounted for by adding two LO isospin-breaking contact interactions. Note that formally, these terms only appear at NLO, but for a better comparison to the Nijmegen PWA, they are already considered at LO. Further improvements especially to higher pion momenta are obtained at N^{4}LO, up to the pion production threshold.

Let us now consider a few assorted results from the fifth order study of Ref. [50]. At this order in the chiral expansion, one has no new contact interactions as compared to N^{3}LO (odd number of derivatives), but new TPEP contributions. These are given by graphs with one insertion from the fourth order pion-nucleon Lagrangian of order Q^4 proportional to the LECs e_i as well as correlated TPE graphs with one insertion from $\mathscr{L}_{\pi N}^{(2)} \sim c_i$ and TPE graphs with one insertion from $\mathcal{L}_{\pi N}^{(3)} \sim d_i$. All these LECs have been determined earlier from the analysis of pion-nucleon scattering, here, the values from Ref. [58] were used. Thus, such a calculation is a fine probe of the relevance of the sub-leading TPE corrections. In addition, there are subleading corrections to the three-pion exchange potential that are potentially sizeable as they are proportional to the enhanced LEC c_4. However, it was shown in Ref. [50] that such contribution can be well represented by contact terms, with the remainder being completely negligible, as is the leading order three-pion exchange at N^3LO. For the precise values of the parameters used, the reader is referred to Refs. [50, 52]. The OPEP and TPEP are regularized in r-space by multiplying with the function

$$f\left(\frac{r}{R}\right) = \left[1 - \exp\left(-\frac{r^2}{R^2}\right)\right]^6, \qquad (2.70)$$

where the cutoff R is chosen in the range of $R = 0.8 \dots 1.2$ fm. The corresponding hard (breakdown) scale is estimated as $\Lambda_{hard} = 600$ MeV for the cutoffs $R = 0.8 - 1.0$ fm, $\Lambda_{hard} = 500$ MeV for $R = 1.1$ fm and $\Lambda_{hard} = 400$ MeV for $R = 1.2$ fm to account for the increasing cut-off artifacts. For the contact interactions, a nonlocal Gaussian regulator in momentum space,

$$V_{cont}^{reg}(\mathbf{p}', \mathbf{p}) = \exp\left(-\frac{p'^2}{\Lambda^2}\right) V(\mathbf{p}', \mathbf{p}) \exp\left(-\frac{p^2}{\Lambda^2}\right) \qquad (2.71)$$

with the cutoff $\Lambda = 2R^{-1}$ is used. The inclusion of the fifth order TPE leads to a marked improvement in the description of the neutron-proton (np) and proton-proton (pp) scattering data taken from the Nijmegen partial wave analysis [57], see Table 2.1. The resulting chiral expansion of the two S-wave partial wave amplitudes $^1S_0$ and $^3S_1$ is displayed in Fig. 2.7, as a function of the energy in the laboratory system E_{lab} that is related to the nucleon center-of-mass three-momentum squared via $\mathbf{p}^2 = p^2 = m_p^2 E_{lab}(E_{lab} + 2m_n)/((m_p + m_n)^2 + 2E_{lab}m_p)$.

The deuteron properties are also well described at this order. Using the binding energy as input, one obtains for the asymptotic S-state normalization $A_S = 0.8844[0.8846(9)]$ fm$^{-1/2}$, for the asymptotic D/S state ratio $\eta = 0.0255[0.0256(4)]$, for the deuteron matter radius $r_d = 1.972[1.97535(85)]$ fm,

Table 2.1 χ^2/datum for the description of the Nijmegen np and pp phase shifts [57] at different orders in the chiral expansion for the cutoff $R = 0.9$ fm

E_{lab} bin [MeV]	LO	NLO	N^2LO	N^3LO	N^4LO
0–100	360 (5750)	31 (102)	4.5 (15)	0.7 (0.8)	0.3 (0.3)
0–200	480 (9150)	63 (560)	21 (130)	0.7 (0.7)	0.3 (0.6)

Only those channels are included which have been used in the N^3LO/N^4LO fits

and for the quadrupole moment $Q_d = 0.271\,[0.2859(3)]\,\mathrm{fm}^2$, where the values in the square brackets are the empirical numbers. The discrepancy in the quadrupole moment comes from the neglect of relativistic and meson-exchange corrections, that still need to be worked out consistently. In models, such corrections are typically $\Delta Q_d \simeq 0.01\,\mathrm{fm}^2$, closing the gap between theory and experiment. From the above and the more detailed analyses available in the literature, it becomes obvious that the chiral expansion converges in the 2N system and that at this order of the expansion, a precise description of the data is achieved. Similar studies in the three- and four-nucleon systems are on-going. It goes beyond the scope of this monograph to discuss these in detail here.

2.5.2 Error Estimates

Having established that the chiral expansion of the two-nucleon potential converges and leads to an accurate description of the np and pp partial waves, it is mandatory to work out the theoretical uncertainty with which a considered observable at a given order is afflicted. More precisely, we want to quantify the neglect of the orders not considered. Of course, there are other possible sources of uncertainty, like from the knowledge of the pion-nucleon LECs that determine the long-range part of the potential or the ones inherent to the data or partial wave analysis used to pin down the NN LECs. For the applications considered here, there are many indications that the truncation of the chiral series indeed generates the dominant uncertainty, so we will only discuss this in the following.

In a strictly perturbative expansion, working to Nth order $\sim Q^N$, the corresponding theoretical error would simply be given by $\pm c(Q/\Lambda_{\mathrm{hard}})^{N+1}$, where c is a number of order one and Λ_{hard} the pertinent breakdown scale of the system under consideration. In a non-perturbative setting as we encounter it here, matters are a bit more complicated. In what follows, we present a method that was first advocated in Ref. [52] in the study of the nucleon-nucleon interaction as it can easily be extended to other observables. Here, we are dealing with a double expansion, and one thus needs to define the small parameter in a way that includes both small external momenta p as well as the pion mass M_π. This can be achieved setting

$$Q = \max \left(\frac{p}{\Lambda_{\mathrm{hard}}}, \frac{M_\pi}{\Lambda_{\mathrm{hard}}} \right). \tag{2.72}$$

Clearly, at low momenta, that is momenta below the pion mass, the error is dominated by the pion mass corrections. A very conservative way of estimating the uncertainty is to take the maximum of all the differences of the lower orders one

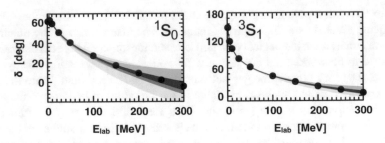

Fig. 2.8 Results for the neutron-proton S-waves up to fifth order for the cutoff $R = 0.9$ fm in comparison with the Nijmegen PWA. The bands of increasing width show estimated theoretical uncertainty at N^4LO (red), N^3LO (blue), N^2LO (green), and NLO (yellow). Figure courtesy of Evgeny Epelbaum

has considered for a given observable $X(p)$ at order Q^N, i.e.

$$\Delta X^N(p) = \max \left(Q^{N+1} \cdot |X^{LO}(p)|, \, Q^{N-1} \cdot |X^{NLO}(p) -^{LO}(p)|, \right.$$

$$\left. Q^{N-2} \cdot |X^{N^2LO}(p) -^{NLO}(p)|, \ldots, Q \cdot |X^{N^NLO}(p) -^{N^{N-1}LO}(p)| \right) .$$

$$(2.73)$$

Note that the particular jump in the prefactor of the second term with respect to the first term is due to the fact that the NLO corrections are suppressed by two powers in the small momenta compared to the LO terms. As stated, this method is very easy to implement, but it does not provide a statistical interpretation. However, as shown in Ref. [59] in the framework of a Bayesian analysis, such a procedure emerges from one class of naturalness priors considered, and that all such priors result in consistent quantitative predictions for 68% degree-of-believe intervals, see also the further study in Ref. [60].

In Fig. 2.8, the two S-waves in neutron-proton scattering are shown with the corresponding error bands at NLO, N^2LO, N^3LO and N^4LO. One sees the expected pattern, with increasing order in the chiral expansion, the uncertainty becomes smaller. This shows that the underlying power counting is indeed working. This is also reflected in similar plots for the observables in the np system, like the differential cross section, the vector analyzing power, the polarization transfer coefficients as well as the spin correlation parameters. We will leave this issue here but return to it when we consider such and other few- and many-nucleon systems on the lattice.

2.5.3 Wigner SU(4) Symmetry in the Two-Nucleon System

The last issue we want to discuss in this chapter is the approximate SU(4) spin-isospin symmetry of the nuclear forces. This can be understood from the observation that the nuclear forces are almost equal between protons and neutrons, which led to the introduction of isotopic spin (isospin). Further, to a first rough approximation, one can also ignore the spin-dependent terms of the nuclear interaction. This lead Wigner in a series of papers in the 1930s to make a number of predictions for nuclear multiplets and decays based on this SU(4) symmetry, and considering its breaking through various terms in the nuclear forces like spin-spin interactions [61–63]. This Wigner super-multiplet theory enjoyed considerable phenomenological success for some time but then was essentially forgotten. Renewed interest was triggered by the work in Ref. [64], which established the connection to the nuclear EFT.

If the nuclear Hamiltonian does not depend on spin and isospin, then it can easily be shown that it is invariant under SU(4) transformations $N \to UN$, $U \in SU(4)$, with $N = (p, n)^T$ the nucleon field (we only display the infinitesimal form of this unitary operator here),

$$N \to N + \delta N, \qquad \delta N = i\epsilon_{\mu\nu}\sigma^\mu \tau^\nu N. \tag{2.74}$$

Here, the $\epsilon_{\mu\nu}$, $\mu, \nu = 0, 1, 2, 3$, are real infinitesimal group parameters, $\sigma^\mu = (1, \sigma_i)$ and $\tau^\mu = (1, \tau_i)$ with $i = 1, 2, 3$. The σ matrices act in spin space and the τ matrices in isospin space, respectively. We have set $\epsilon_{00} = 0$. In case that $\epsilon_{00} \neq 0$, we actually deal with an SU(4)×U(1) transformation, where the U(1) corresponds to conserved baryon number, which we will not consider further.

The basic concepts underlying the SU(4) spin-isospin symmetry can most easily be understood in the pionless EFT, which to leading order is given by the Lagrangian

$$\mathcal{L}_{\not{\pi}} = \mathcal{L}_{\text{kin}} + \mathcal{L}_{\text{int}}$$

$$= N^\dagger \left(i\partial_t + \frac{\nabla^2}{2m_N} \right) N - \frac{1}{2} \left(C_s (N^\dagger N)^2 + C_T (N^\dagger \sigma N)^2 \right) + \dots \tag{2.75}$$

where the ellipsis denotes terms with more derivatives. The corresponding LECs for the partial waves $^1S_0$ and $^3S_1$ are related to the ones in Eq. (2.75) by $C(^1S_0) = C_S - 3C_T$ and $C(^3S_1) = C_S + C_T$. It is obvious that while the interaction term $\sim C_S$ is invariant under Wigner SU(4), the term $\sim C_T$ is not. Let us make the relation to the Wigner symmetry in the NN system more precise. In fact, in the limit that the NN singlet and triplet scattering lengths become infinite, one does not only encounter a fascinating scale invariance but also invariance under spin-isospin transformations. Working in a partial wave basis and to leading order in pionless EFT, the LECs $C(^1S_0)$ and $C(^3S_1)$ can be fixed from the corresponding scattering lengths or equivalently, from the binding momentum of the virtual/bound state pole in the singlet/triplet channel, respectively,

$$C(^1S_0) = \frac{4\pi}{m_N} \frac{1}{\gamma_s - \mu}, \qquad C(^3S_1) = \frac{4\pi}{m_N} \frac{1}{\gamma_t - \mu}, \tag{2.76}$$

with μ the scale of dimensional regularization. More precisely, it is advantageous to use a scheme in which not only the poles in $D = 4$ but also the ones in lower dimensions are subtracted [6]. Wigner SU(4) symmetry is recovered for $\gamma_t = \gamma_s$. In nature, we have $\gamma_s = -7.9\,\text{MeV}$ and $\gamma_t = 45.7\,\text{MeV}$. At this order in pionless EFT, $\gamma_s = 1/a_{np}^{S=0}$ and $\gamma_t = 1/a_{np}^{S=1}$. Thus, for $\mu \gg \gamma_s, \gamma_t$, the Wigner SU(4) symmetry is approximate in the NN system. Further, the exact symmetry limit corresponds to a scale invariant non-relativistic system, $x \to \lambda x$, $t \to \lambda^2 t$ and $N \to \lambda^{-3/2} N$, with λ the scale parameter. The corresponding partial wave cross section in this limit read $\sigma_s(p) = \sigma_t(p) = 4\pi/p^2$, with p the scattering momentum. For further discussions on the SU(4) symmetry in the continuum, we refer to Refs. [64, 65]. Its consequences for the lattice formulation of chiral EFT will be considered in Chap. 8.

References

1. H. Yukawa, On the interaction of elementary particles. Proc. Phys. Math. Soc. Jpn. **17**, 48 (1935)
2. S. Weinberg, Nuclear forces from chiral Lagrangians. Phys. Lett. B **251**, 288 (1990)
3. S. Weinberg, Effective chiral Lagrangians for nucleon – pion interactions and nuclear forces. Nucl. Phys. B **363**, 3 (1991)
4. N. Kalantar-Nayestanaki, E. Epelbaum, J.G. Messchendorp, A. Nogga, Signatures of three-nucleon interactions in few-nucleon systems. Rep. Prog. Phys. **75**, 016301 (2012)
5. E. Braaten, H.-W. Hammer, Universality in few-body systems with large scattering length. Phys. Rep. **428**, 259 (2006)
6. D.B. Kaplan, M.J. Savage, M.B. Wise, A new expansion for nucleon-nucleon interactions. Phys. Lett. B **424**, 390 (1998)
7. S. Fleming, T. Mehen, I.W. Stewart, NNLO corrections to nucleon-nucleon scattering and perturbative pions. Nucl. Phys. A **677**, 313 (2000)
8. Peccei, R.D.: The Strong CP Problem and Axions. Lect. Notes Phys. 741, 3 (2008)
9. C. Vafa, E. Witten, Restrictions on symmetry breaking in vector – like gauge theories. Nucl. Phys. B **234**, 173 (1984)
10. A.S. Kronfeld, Twenty-first century lattice gauge theory: results from the QCD Lagrangian. Ann. Rev. Nucl. Part. Sci. **62**, 265 (2012)
11. T. Banks, A. Casher, Chiral symmetry breaking in confining theories. Nucl. Phys. B **169**, 103 (1980)
12. H. Leutwyler, A. Smilga, Spectrum of Dirac operator and role of winding number in QCD. Phys. Rev. D **46**, 5607 (1992)
13. J. Stern, Two alternatives of spontaneous chiral symmetry breaking in QCD (1998). arXiv:hep-ph/9801282
14. J. Goldstone, Field theories with 'superconductor' solutions. Nuovo Cimento **19**, 154 (1961)
15. J. Goldstone, A. Salam, S. Weinberg, Broken symmetries. Phys. Rev. **127**, 965 (1962)
16. M. Gell-Mann, Y. Ne'eman (eds.), *The Eightfold Way* (W. A. Benjamin, New York, 1964)
17. G.E. Brown, Isn't it time to calculate the nucleon-nucleon force? Comments Nucl. Part. Phys. **4**, 140 (1970)
18. S. Weinberg, Nonlinear realizations of chiral symmetry. Phys. Rev. **166**, 1568 (1968)
19. S.R. Coleman, J. Wess, B. Zumino, Structure of phenomenological Lagrangians. 1. Phys. Rev. **177**, 2239 (1969)
20. C.G. Callan Jr., S.R. Coleman, J. Wess, B. Zumino, Structure of phenomenological Lagrangians. 2. Phys. Rev. **177**, 2247 (1969)

21. M. Gell-Mann, R.J. Oakes, B. Renner, Behavior of current divergences under SU(3) x SU(3). Phys. Rev. **175**, 2195 (1968)
22. G. Colangelo, J. Gasser, H. Leutwyler, The Quark condensate from K(e4) decays. Phys. Rev. Lett. **86**, 5008 (2001)
23. J. Gasser, H. Leutwyler, Chiral perturbation theory to one loop. Ann. Phys. **158**, 142 (1984)
24. J. Gasser, H. Leutwyler, Chiral perturbation theory: expansions in the mass of the strange quark. Nucl. Phys. B **250**, 465 (1985)
25. S. Bellucci, J. Gasser, M.E. Sainio, Low-energy photon-photon collisions to two loop order. Nucl. Phys. B **423**, 80 (1994) [Erratum-ibid. B **431**, 413 (1994)]
26. J. Bijnens, G. Colangelo, G. Ecker, The mesonic chiral Lagrangian of order p^6. J. High Energy Phys. **9902**, 020 (1999)
27. H. Georgi, *Weak Interactions and Modern Particle Theory* (Benjamin/Cummings, Menlo Park, 1984)
28. J. Gasser, M.E. Sainio, A. Svarc, Nucleons with chiral loops. Nucl. Phys. B **307**, 779 (1988)
29. V. Bernard, Chiral perturbation theory and baryon properties. Prog. Part. Nucl. Phys. **60**, 82 (2008)
30. E.E. Jenkins, A.V. Manohar, Baryon chiral perturbation theory using a heavy fermion Lagrangian. Phys. Lett. B **255**, 558 (1991)
31. V. Bernard, N. Kaiser, J. Kambor, U.-G. Meißner, Chiral structure of the nucleon. Nucl. Phys. B **388**, 315 (1992)
32. L.L. Foldy, S.A. Wouthuysen, On the Dirac theory of spin 1/2 particle and its nonrelativistic limit. Phys. Rev. **78**, 29 (1950)
33. A.V. Manohar, M.B. Wise, *Heavy Quark Physics* (Cambridge University Press, Cambridge, 2000)
34. T. Mannel, W. Roberts, Z. Ryzak, A derivation of the heavy quark effective Lagrangian from QCD. Nucl. Phys. B **368**, 204 (1992)
35. V. Bernard, N. Kaiser, U.-G. Meißner, Chiral dynamics in nucleons and nuclei. Int. J. Mod. Phys. E **4**, 193 (1995)
36. P. Langacker, H. Pagels, Applications of chiral perturbation theory: mass formulas and the decay eta to 3 pi. Phys. Rev. D **10**, 2904 (1974)
37. M. Hoferichter, J. Ruiz de Elvira, B. Kubis, U.-G. Meißner, Roy-Steiner-equation analysis of pion-nucleon scattering. Phys. Rep. **625**, 1 (2016)
38. V. Bernard, N. Kaiser, U.-G. Meißner, Aspects of chiral pion – nucleon physics. Nucl. Phys. A **615**, 483 (1997)
39. N. Fettes, U.-G. Meißner, M. Mojzis, S. Steininger, The chiral effective pion nucleon Lagrangian of order p^4. Ann. Phys. **283**, 273 (2000) [Erratum-ibid. **288**, 249 (2001)]
40. V. Baru, C. Hanhart, M. Hoferichter, B. Kubis, A. Nogga, D.R. Phillips, Precision calculation of threshold pi- d scattering, pi N scattering lengths, and the GMO sum rule. Nucl. Phys. A **872**, 69 (2011)
41. N. Fettes, U.-G. Meißner, Pion nucleon scattering in chiral perturbation theory. 2.: fourth order calculation. Nucl. Phys. A **676**, 311 (2000)
42. D. Siemens, V. Bernard, E. Epelbaum, A. Gasparyan, H. Krebs, U.-G. Meißner, Elastic pion-nucleon scattering in chiral perturbation theory: a fresh look. Phys. Rev. C **94**, 014620 (2016)
43. S. Weinberg, Three body interactions among nucleons and pions. Phys. Lett. B **295**, 114 (1992)
44. N. Kaiser, R. Brockmann, W. Weise, Peripheral nucleon-nucleon phase shifts and chiral symmetry. Nucl. Phys. A **625**, 758 (1997)
45. E. Epelbaum, Few-nucleon forces and systems in chiral effective field theory. Prog. Part. Nucl. Phys. **57**, 654 (2006)
46. E. Epelbaum, Four-nucleon force using the method of unitary transformation. Eur. Phys. J. A **34**, 197 (2007)
47. E. Epelbaum, H.W. Hammer, U.-G. Meißner, Modern theory of nuclear forces. Rev. Mod. Phys. **81**, 1773 (2009)
48. R. Machleidt, D.R. Entem, Chiral effective field theory and nuclear forces. Phys. Rep. **503**, 1 (2011)

49. N. Kaiser, Chiral 3 pi exchange N N potentials: results for representation invariant classes of diagrams. Phys. Rev. C **61**, 014003 (2000)
50. E. Epelbaum, H. Krebs, U.-G. Meißner, Precision nucleon-nucleon potential at fifth order in the chiral expansion. Phys. Rev. Lett. **115**, 122301 (2015)
51. P. Reinert, H. Krebs, E. Epelbaum, Semilocal momentum-space regularized chiral two-nucleon potentials up to fifth order. Eur. Phys. J. A **54**, 86 (2018)
52. E. Epelbaum, H. Krebs, U.-G. Meißner, Improved chiral nucleon-nucleon potential up to next-to-next-to-next-to-leading order. Eur. Phys. J. A **51**, 53 (2015)
53. U. van Kolck, Few nucleon forces from chiral Lagrangians. Phys. Rev. C **49**, 2932 (1994)
54. J. Fujita, H. Miyazawa, Pion theory of three-body forces. Prog. Theor. Phys. **17**, 360 (1957)
55. U.-G. Meißner, The Fujita-Miyazawa force in the light of effective field theory. AIP Conf. Proc. **1011**, 49 (2008)
56. D.R. Entem, R. Machleidt, Y. Nosyk, High-quality two-nucleon potentials up to fifth order of the chiral expansion. Phys. Rev. C **96**, 024004 (2017)
57. V.G.J. Stoks, R.A.M. Kompl, M.C.M. Rentmeester, J.J. de Swart, Partial wave analysis of all nucleon-nucleon scattering data below 350-MeV. Phys. Rev. C **48**, 792 (1993)
58. H. Krebs, A. Gasparyan, E. Epelbaum, Chiral three-nucleon force at N4LO I: longest-range contributions. Phys. Rev. C **85**, 054006 (2012)
59. R.J. Furnstahl, N. Klco, D.R. Phillips, S. Wesolowski, Quantifying truncation errors in effective field theory. Phys. Rev. C **92**, 024005 (2015)
60. J.A. Melendez, S. Wesolowski, R.J. Furnstahl, Bayesian truncation errors in chiral effective field theory: nucleon-nucleon observables. Phys. Rev. C **96**, 024003 (2017)
61. E. Wigner, On the consequences of the symmetry of the nuclear Hamiltonian on the spectroscopy of nuclei. Phys. Rev. **51**, 106 (1937)
62. E. Wigner, On the structure of nuclei beyond oxygen. Phys. Rev. **51**, 947 (1937)
63. E.P. Wigner, On coupling conditions in light nuclei and the lifetimes of beta-radioactivities. Phys. Rev. **56**, 519 (1939)
64. T. Mehen, I.W. Stewart, M.B. Wise, Wigner symmetry in the limit of large scattering lengths. Phys. Rev. Lett. **83**, 931 (1999)
65. J. Vanasse, D.R. Phillips, Three-nucleon bound states and the Wigner-SU(4) limit. Few Body Syst. **58**, 26 (2017)

Chapter 3
Lattice Formulations

In Chap. 1, we introduced the concept of EFT, and in Chap. 2 we applied the EFT method to the low-energy sector of QCD. This led us to the framework of chiral EFT and provided us with a description of the interactions between nucleons, which is valid at low energies, as encountered by nucleons in nuclei. In order to use chiral EFT to predict the properties of nuclei and make contact with phenomena of current interest, we need the ability to compute spectra, transitions and other properties of many-body systems (such as nuclei or nuclear matter) starting from the chiral EFT Hamiltonian. Not surprisingly, there are several different choices one can make for the calculational method with which to describe interacting low-energy nucleons. More precisely, there are essentially two pathways to proceed if one wants to base the many-body calculation on the forces discussed in the preceding chapter. One much studied type of approach is to use these chiral continuum forces in combination with a standard and well-developed few- or many-body method, such as the Faddeev-Yakubovsky integral equations for few-nucleon systems or the (no-core) shell model, coupled cluster theory, etc. for larger systems. Such approaches are vigorously pursued by many researchers world-wide. However, in our context, we require a method that gives access to light and medium-mass nuclei, while simultaneously preserving the ab initio nature of the theory in the following sense: We will use the lattice, which is a representation of the world in a finite volume with discrete space-time coordinates, as the computational tool for the solving the A-nucleon Schrödinger equation, with properly adjusted chiral EFT forces as input. Here, A is the atomic number, that is the sum of the number of neutrons and protons in the nucleus under consideration. Clearly, for a given lattice set-up, the appearing LECs have to be refitted. Through most parts of this book, we will use few-body systems with $A \leq 4$ to pin down these parameters, which allows us to make truly parameter-free predictions for larger nuclei. This approach is called *nuclear lattice EFT* (NLEFT), or *nuclear lattice simulations*, for short, and will be worked out in detail in this and the following chapters. Another important consideration is the computational scaling of the method with A, as well as the suitability to large-scale

© Springer Nature Switzerland AG 2019
T. A. Lähde, U.-G. Meißner, *Nuclear Lattice Effective Field Theory*,
Lecture Notes in Physics 957, https://doi.org/10.1007/978-3-030-14189-9_3

high-performance computing. In this context, we refer by "light nuclei" to systems up to carbon ($A = 12$) and by "medium-mass nuclei" to systems with $A > 12$ up to the calcium region. Because of these considerations, our treatment of current chiral EFT work will focus on the Auxiliary-Field Quantum Monte Carlo (AFQMC) method. We shall first describe the theoretical aspects and turn to the details of the Monte Carlo calculation in Chap. 6. We end this chapter with a discussion of the symmetries of the lattice Hamiltonian.

3.1   Some Basic Considerations

To apply Monte Carlo methods (or other ways of solving the A-body problem) considered in what follows, the chiral EFT Hamiltonian is formulated on a cubic spatial lattice of volume L^3 with lattice spacing a, with $L = Na$ and N an integer. This is depicted in Fig. 3.1. Often, one speaks of the lattice size giving just N or the lattice volume is given as N^3, recognizing that a is the fundamental length scale on the lattice. Further, the time coordinate is Wick rotated to Euclidean time, $t \to -it$, which is a prerequisite to perform simulations in a finite volume, as will become clear later. The lattice spacing a also serves as the ultraviolet (UV) cutoff scale

$$\Lambda \equiv \frac{\pi}{a}, \tag{3.1}$$

of the EFT, in terms of natural units such that $\hbar = k_B = c = 1$, as is customary in nuclear and high-energy physics. Here, $\hbar$, k_B and c denote the Planck's constant divided by 2π, the Boltzmann constant and the speed of light in vacuum, respectively. The lattice therefore provides a natural regularization of the UV divergent EFT, which is a welcome feature of this approach. We recall that in nuclear chiral EFT, the cutoff Λ should be larger than M_π, but smaller than the

Fig. 3.1 Graphical representation of a cubic spatial lattice with spatial volume L^3 and lattice spacing a. The discrete Euclidean time dimension with temporal spacing δ is not shown

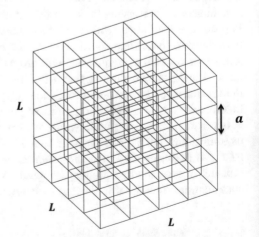

hard scale $\Lambda \simeq 600\,\mathrm{MeV}$. In practice, this translates to lattice spacings between $a \simeq 1.0\,\mathrm{fm}$ and $a \simeq 2.0\,\mathrm{fm}$. The optimal choice of the cutoff is currently dictated by computational considerations, as the full chiral EFT Monte Carlo probability measure is, in general, not positive definite. We note that such a "sign problem" commonly occurs in AFQMC calculations due to repulsive interactions, as will be discussed in more detail in Chap. 8. In the case of chiral EFT, decreasing the lattice spacing renders sign problem more severe, as the short-range repulsion in the LO interaction is then more pronounced. Additionally, the higher-order contributions at NLO and NNLO in chiral EFT may also have a detrimental impact on the sign problem. Because of these considerations, most of the lattice chiral EFT calculations described here have been performed with $a = 1.97\,\mathrm{fm}$. However, we remark that by now calculations employ also smaller lattice spacings, but these have not yet been systematically used to explore the structure of nuclei.

It should be carefully noted that although the lattice spacings used in chiral EFT are very large compared to those of Lattice QCD, the purpose of chiral EFT is not to resolve the substructure of quarks and gluons within individual nucleons. Also, in lattice chiral EFT the purpose is not to examine the behavior of the theory in the continuum limit $a \rightarrow 0$, but rather for a range of lattice spacings within the window of applicability of the EFT. To some extent, the computational situation of lattice chiral EFT is then intermediate between those encountered in condensed matter physics and in Lattice QCD. In the former field, the lattice spacing a (and often also the system size L) is a physical quantity which determines the scales appropriate to the problem at hand. Finally, we remark that it is possible to consider a more complicated approach, in which the EFT cut-off scale Λ is decoupled from the lattice spacing a. In such a set-up, one could investigate the continuum limit $a \rightarrow 0$. For a first step in this direction, see Ref. [1], but we will not consider this any further here.

3.2 Grassmann Fields and Transfer Matrices

In order to access the many-body properties of the chiral EFT Hamiltonian, we shall perform Euclidean time projection (to be described in Chap. 6) using the "transfer matrix" operator. The transfer matrix essentially amounts to the exponential of the lattice Hamiltonian over a slice of Euclidean time of length δ. In order to introduce the transfer matrix, we shall first consider the partition function $\mathscr{Z}$ and formulate this in terms of a Grassmann path integral. We then establish the correspondence between the Grassmann path integral and transfer matrix formulations, which allows us to introduce the "Euclidean time projection amplitude" which is closely related to $\mathscr{Z}$ as formulated in terms of transfer matrices. Effectively, this amounts to repeatedly applying the transfer matrix operator to a "trial state" of A nucleons, such that after a sufficient number of Euclidean time steps, the ground state (and the first few excited states) will form the dominant contributions to the projection amplitude.

Another noteworthy issue is that many operators in the two-nucleon sector of the chiral EFT Hamiltonian are four-fermion contact interactions, which are furthermore dependent on spin and isospin. In our treatment of lattice chiral EFT, we shall treat such terms by means of a Hubbard-Stratonovich (HS) transformation [2, 3] that decouples the nucleons at the price of introducing "auxiliary fields" which parameterize the interactions between nucleons. In essence, in the auxiliary-field formulation the individual nucleons are independent particles which only interact with the background auxiliary field that encodes the interactions between nucleons. It should be pointed out that the HS transformation is an exact mathematical identity, in fact, it is nothing but the well-known Gaussian quadrature. As the HS transformation reduces the transfer matrix operator to a bilinear form in the fermion operators, the latter can be integrated out, thus yielding a "fermion determinant" with a path integral over the auxiliary fields.

The significant advantage of the transfer matrix operator formulated in terms of auxiliary fields is that it lends itself readily to Monte Carlo computation via the AFQMC framework. This method provides the needed favorable scalability with nucleon number A, while providing an unapproximated (though stochastical) solution to the problem of finding the properties of the lattice Hamiltonian for a few-nucleon system. Here, we shall concentrate on the development of the Grassmann, transfer matrix and auxiliary-field formalisms, such that the aspects particular to lattice chiral EFT can be straightforwardly treated in Chap. 4.

3.2.1   Grassmann Variables

We recall that anti-commuting variables (so-called Grassmann variables), are required for a consistent formulation of fermionic fields. In order to introduce the path-integral formalism for the fermionic partition function, we first discuss some of the basic properties of Grassmann variables, functions of Grassmann variables, as well as their differentiation and integration.

We define η_k as a set of Grassmann variables that satisfy the anticommutation relations

$$\eta_k \eta_l + \eta_l \eta_k \equiv \{\eta_k, \eta_l\} = 0, \tag{3.2}$$

for all k and l. We note that this implies that $\eta_k^2 = 0$. Hence, if we assume that functions of Grassmann variables can be expanded in a Taylor series, the most general function of a Grassmann variable has the form

$$f(\eta) \equiv a + b\eta, \tag{3.3}$$

where a and b are ordinary numbers. We may also define differentiation of Grassmann variables according to

$$\frac{\overrightarrow{\partial}}{\partial \eta_k} \eta_l = -\eta_l \frac{\overleftarrow{\partial}}{\partial \eta_k} = \delta_{kl}, \tag{3.4}$$

where the direction of the arrow denotes "left" and "right" differentiation, respectively. This gives

$$\frac{\overrightarrow{\partial}}{\partial \eta_k} \eta_l \eta_k = -\eta_l = \eta_l \eta_k \frac{\overleftarrow{\partial}}{\partial \eta_k}, \tag{3.5}$$

for a product of Grassmann variables. For the integration of Grassmann variables, we note the properties

$$\int d\eta_k = 0, \qquad \int d\eta_k \eta_l = -\int \eta_l d\eta_k = \delta_{kl}, \tag{3.6}$$

such that the operations of differentiation and integration are identical for Grassmann variables.

We can also introduce complex Grassmann variables $\xi \equiv \eta_k + i\eta_l$, the properties of which are given in terms of those established for real Grassmann variables. This gives

$$\xi^2 = \xi^{*2} = 0, \qquad \xi^* \xi = i[\eta_k, \eta_l], \tag{3.7}$$

where $\xi^* \equiv \eta_k - i\eta_l$. Using ξ and ξ^*, we may define a Grassmann algebra by associating a generator ξ_α with each annihilation operator a_α, and a generator ξ_α^* with each creation operator $a_\alpha^\dagger$.

3.2.2 Partition Function

We are now in a position to consider the grand canonical partition function $\mathscr{Z}$, which is the basic object of interest in the development of lattice Monte Carlo formulations, as all desired observables can be computed using appropriately chosen derivatives of $\mathscr{Z}$, as will be shown in detail in later chapters. The partition function is given by

$$\mathscr{Z} \equiv \text{Tr} \exp\left(-H(a_x^\dagger, a_x)/T\right), \tag{3.8}$$

where T is the temperature and the Hamiltonian H is a function of the creation operators $a_x^\dagger$ and the annihilation operators a_x. The index x refers to the relevant degrees of freedom which may include spin, isospin and (spatial) lattice sites.

By the trace of an operator A, we refer to the operation

$$\text{Tr}\, A = \sum_n \langle n|A|n \rangle, \tag{3.9}$$

where $\{|n\rangle\}$ denotes a complete set of states in the Fock space. We proceed by introducing a fermionic "coherent state"

$$|\xi\rangle \equiv \exp\left(-\sum_\alpha \xi_\alpha a_\alpha^\dagger\right)|0\rangle = \prod_\alpha (1 - \xi_\alpha a_\alpha^\dagger)|0\rangle, \tag{3.10}$$

which satisfies

$$a_\alpha|\xi\rangle = \xi_\alpha|\xi\rangle, \quad \langle \xi|a_\alpha^\dagger = \langle \xi|\xi_\alpha^*, \tag{3.11}$$

and the "closure relation"

$$\int \prod_\alpha d\xi_\alpha^* d\xi_\alpha \exp\left(-\sum_\beta \xi_\beta^* \xi_\beta\right) |\xi\rangle\langle\xi| = \mathbb{1}, \tag{3.12}$$

where $\mathbb{1}$ denotes the unit operator in the fermionic Fock space. In most of what follows, we shall for simplicity write

$$\int \prod_\alpha d\xi_\alpha^* d\xi_\alpha \to \int d\xi^* d\xi, \tag{3.13}$$

such that the indices denoting the degrees of freedom corresponding to lattice sites, spin etc., collectively denoted α, are suppressed.

The closure relation allows us to write

$$\text{Tr}\, A = \int d\xi^* d\xi \exp(-\xi^*\xi) \sum_n \langle n|\xi\rangle\langle\xi|A|n\rangle, \tag{3.14}$$

where the anticommutation relations give

$$\langle n|\xi\rangle\langle\xi|A|n\rangle = \langle -\xi|A|n\rangle\langle n|\xi\rangle, \tag{3.15}$$

such that

$$\text{Tr}\, A = \int d\xi^* d\xi \exp(-\xi^*\xi)\langle -\xi|A \sum_n |n\rangle\langle n|\xi\rangle, \tag{3.16}$$

by which we arrive at the result

$$\text{Tr } A = \int d\xi^* d\xi \, \exp(-\xi^*\xi)\langle-\xi|A|\xi\rangle, \tag{3.17}$$

where it should be carefully noted that the minus sign in the matrix element is a direct consequence of the anticommuting property of the Grassmann variables. As we shall see, this minus sign reappears in the form of antiperiodic boundary conditions in the temporal lattice dimension. On the other hand, the boundary conditions in the spatial dimensions may be freely chosen, although periodic boundary conditions are the most common choice.

So far, we have not assumed any special property for the operator A. However, if we assume normal ordering where creation operators are on the left and annihilation operators on the right, we can make use of the relation

$$\langle\xi|A(a_\alpha^\dagger, a_\alpha)|\xi'\rangle = \exp\left(\sum_\beta \xi_\beta^* \xi_\beta'\right) A(\xi_\alpha^*, \xi_\alpha'), \tag{3.18}$$

where $A(a_\alpha^\dagger, a_\alpha)$ is an arbitrary, normal ordered function of all the $a_\alpha^\dagger$ and a_α. For normal-ordered operators A, we can then recast the trace in the form

$$\text{Tr} (: A :) = \int d\xi^* d\xi \, \exp(-2\xi^*\xi) A(-\xi^*, \xi), \tag{3.19}$$

where we have used the double colon notation to indicate normal ordering. Let us evaluate this expression explicitly, using the parameterization

$$A(\xi^*, \xi) = a_0 + a_1\xi + \overline{a_1}\xi^* + a_{12}\xi^*\xi, \tag{3.20}$$

which is valid for arbitrary functions of Grassmann variables. We then find

$$\text{Tr} (: A :) = \int d\xi^* d\xi (1 - 2\xi^*\xi) a_0 + a_1\xi - \overline{a_1}\xi^* - a_{12}\xi^*\xi,$$

$$= 2a_0 + a_{12}, \tag{3.21}$$

using the rules for Grassmann integration and by commuting the Grassmann variables where necessary. Using this result, we may determine that

$$\text{Tr} (: A :) = \int d\xi^* d\xi \, \exp(-2\xi^*\xi) A(-\xi^*, \xi) = \int d\xi d\xi^* \, \exp(2\xi^*\xi) A(\xi^*, \xi), \tag{3.22}$$

and we note that by commuting $d\xi$ and $d\xi^*$, we can absorb the minus signs in the right-hand side and obtain the result given in Ref. [4]. We shall next consider how this relation can be generalized to products of normal-ordered functions of several creation and annihilation operators, as encountered in lattice field theories.

3.2.3 Euclidean Time Evolution

We shall now return to the problem of the grand canonical partition function $\mathscr{Z}$, which we recall is given by

$$\mathscr{Z} \equiv \operatorname{Tr} \exp\left(-\beta H(a_x^\dagger, a_x)\right), \tag{3.23}$$

where $\beta \equiv 1/T$ is the inverse temperature, and x labels spin, isospin and position (spatial lattice sites). Using Eq. (3.17), we can write

$$\mathscr{Z} = \int \left[\prod_x d\xi_x^* d\xi_x\right] \exp\left(-\sum_x \xi_x^* \xi_x\right) \langle -\xi | \exp(-\beta H(a_x^\dagger, a_x)) | \xi \rangle, \tag{3.24}$$

in the Grassmann formalism. However, even if H is normal ordered, the exponential of H is in general not, which means that we cannot use Eq. (3.18) to eliminate the creation and annihilation operators from the matrix element. However, we note that by means of the closure relation, we can write

$$\langle -\xi | \exp(-\beta H(a_x^\dagger, a_x)) | \xi \rangle = \int \left[\prod_x d\xi_{x,1}^* d\xi_{x,1}\right] \exp\left(-\sum_x \xi_{x,1}^* \xi_{x,1}\right) \tag{3.25}$$

$$\times \langle -\xi | \exp(-\beta H(a_x^\dagger, a_x)/2) | \xi_1 \rangle$$

$$\times \langle \xi_1 | \exp(-\beta H(a_x^\dagger, a_x)/2) | \xi \rangle,$$

where we may then repeat this procedure and insert additional complete sets of states until the exponential has been subdivided into N_t factors, formally

$$\exp(-\beta H) \rightarrow \exp(-\delta H) \times \exp(-\delta H) \times \ldots \times \exp(-\delta H), \tag{3.26}$$

where $\delta \equiv \beta/N_t$ denotes the step size in the Euclidean time evolution. If we then include the Euclidean time dimension in our definition of the lattice, we may express the partition function as

$$\mathscr{Z} = \int \prod_{t=0}^{N_t-1} \left[\prod_x d\xi_{x,t}^* d\xi_{x,t}\right] \exp\left(-\sum_x \xi_{x,t+1}^* \xi_{x,t+1}\right)$$

$$\times \langle \xi_{t+1} | \exp(-\delta H(a_x^\dagger, a_x)) | \xi_t \rangle, \tag{3.27}$$

where we note that the fermionic minus sign in Eq. (3.24) is accounted for by means of the antiperiodic boundary conditions $\xi_{x,N_t} = -\xi_{x,0}$ and $\xi^*_{x,N_t} = -\xi^*_{x,0}$. The advantage of using a formalism with a discretized Euclidean time evolution is that for sufficiently small δ, we may regard the exponential of H to also be normal ordered if H is. If we now apply Eq. (3.18), we find

$$\langle \xi_{t+1} | \exp(-\delta H(a^\dagger_x, a_x)) | \xi_t \rangle = \exp\left(\sum_x \xi^*_{x,t+1} \xi_{x,t} \right)$$

$$\times \exp\left(-\delta H(\xi^*_{x,t+1}, \xi_{x,t}) \right) + \mathcal{O}(\delta^2), \quad (3.28)$$

where the terms of $\mathcal{O}(\delta^2)$ represent contributions in an expansion of $\exp(-\delta H)$ that are not normal ordered. The discretization error associated with neglecting such terms is of $\mathcal{O}(\delta)$ and vanishes as $N_t \to \infty$, hence δ can be viewed as the lattice spacing in the Euclidean time direction.

3.2.4 From Path Integrals to Transfer Matrices

If we combine the results from the previous subsections, we may express the partition function as

$$\mathcal{Z} \simeq \int \prod_{t=0}^{N_t-1} \left[\prod_x d\xi^*_{x,t} d\xi_{x,t} \right] \exp\left(\sum_x \xi^*_{x,t+1} \left(\xi_{x,t} - \xi_{x,t+1} \right) \right)$$

$$\times \exp\left(-\delta H(\xi^*_{x,t+1}, \xi_{x,t}) \right) + \mathcal{O}(\delta^2), \quad (3.29)$$

such that this expression becomes exact in the "Hamiltonian limit" $\delta \to 0$. In what follows, we shall assume that this condition is satisfied and not write out the $\mathcal{O}(\delta^2)$ any more. We shall now switch to a more explicit notation where the abstract index x is written out in terms of spatial lattice sites $\mathbf{n}$ and spin and isospin degrees of freedom, denoted by i and j, respectively. This gives

$$\mathcal{Z} = \int \mathcal{D}\xi^* \mathcal{D}\xi \, \exp\left(\sum_{t=0}^{N_t-1} \sum_{\mathbf{n},i,j} \xi^*_{i,j}(\mathbf{n}, t+1) \left[\xi_{i,j}(\mathbf{n}, t) - \xi_{i,j}(\mathbf{n}, t+1) \right] \right)$$

$$\times \prod_{t=0}^{N_t-1} \exp\left(-\delta H \left(\xi^*_{i',j'}(\mathbf{n}', t+1), \xi_{i,j}(\mathbf{n}, t) \right) \right), \quad (3.30)$$

where we have introduced the path integral notation

$$\int \left[\prod_{t=0}^{N_t-1} \left\{ \prod_{\mathbf{n},i,j} d\xi_{i,j}^*(\mathbf{n},t)\, d\xi_{i,j}(\mathbf{n},t) \right\} \right] \rightarrow \int \left[\prod_{t=0}^{N_t-1} d\xi^*(t)\, d\xi(t) \right] \rightarrow \int \mathscr{D}\xi^* \mathscr{D}\xi,$$

(3.31)

and we note that in Eq. (3.30), the indices $\mathbf{n}', i', j'$ indicate that H in general contains not only on-site interactions but also "hopping" terms which connect sites separated by one or more spatial lattice spacings. Similarly, H may also contain interactions which couple different spin and isospin quantum numbers. We also recall the antiperiodic boundary conditions

$$\xi_{i,j}(\mathbf{n}, N_t) = -\xi_{i,j}(\mathbf{n}, 0), \quad \xi_{i,j}^*(\mathbf{n}, N_t) = -\xi_{i,j}^*(\mathbf{n}, 0),$$

(3.32)

where translational invariance implies

$$\xi_{i,j}(\mathbf{n}, t + N_t) = -\xi_{i,j}(\mathbf{n}, t), \quad \xi_{i,j}^*(\mathbf{n}, t + N_t) = -\xi_{i,j}^*(\mathbf{n}, t),$$

(3.33)

while in the spatial lattice directions we take

$$\xi_{i,j}(\mathbf{n} + L\hat{e}_l, t) = \xi_{i,j}(\mathbf{n}, t), \quad \xi_{i,j}^*(\mathbf{n} + L\hat{e}_l, t) = \xi_{i,j}^*(\mathbf{n}, t),$$

(3.34)

such that periodic boundary conditions are satisfied. Here, $\hat{e}_l$ denotes a unit vector in (spatial) lattice direction $l = 1, 2, 3$.

Finally, we shall introduce a new set of Grassmann fields ζ^*, ζ according to

$$\zeta_{i,j}(\mathbf{n}, t) \equiv \xi_{i,j}(\mathbf{n}, t), \quad \zeta_{i,j}^*(\mathbf{n}, t) \equiv \xi_{i,j}^*(\mathbf{n}, t+1),$$

(3.35)

where the ζ^*, ζ also satisfy antiperiodic boundary conditions in the temporal and periodic boundary conditions in the spatial directions. In terms of the new Grassmann fields, we find

$$\mathscr{Z} = \int \mathscr{D}\zeta\, \mathscr{D}\zeta^* \exp \left(\sum_{t=0}^{N_t-1} \sum_{\mathbf{n},i,j} \zeta_{i,j}^*(\mathbf{n}, t) \Big[\zeta_{i,j}(\mathbf{n}, t) - \zeta_{i,j}(\mathbf{n}, t+1) \Big] \right)$$

$$\times \prod_{t=0}^{N_t-1} \exp \left(-\delta H \left(\zeta_{i',j'}^*(\mathbf{n}', t), \zeta_{i,j}(\mathbf{n}, t) \right) \right),$$

(3.36)

where it should be noted that the ordering of $\mathscr{D}\zeta\, \mathscr{D}\zeta^*$ versus $\mathscr{D}\xi^* \mathscr{D}\xi$ is significant, and arises through reordering and anticommutation of the time slices in Eq. (3.31). Based on Eqs. (3.30) and (3.36), we may also write

$$\mathscr{Z} \equiv \int \mathscr{D}\xi^* \mathscr{D}\xi \, \exp(-S(\xi^*, \xi)) = \int \mathscr{D}\zeta\, \mathscr{D}\zeta^* \exp(-S(\zeta^*, \zeta)),$$

(3.37)

where S is referred to as the action.

So far, we have considered the Grassmann path integral formulation of the partition function. However, we may also simply perform the decomposition of $\exp(-\beta H)$ into N_t Euclidean time slices without introducing Grassmann fields, in which case we obtain

$$\mathscr{Z} \simeq \mathrm{Tr}\{M^{N_t}\} + \mathcal{O}(\delta^2), \tag{3.38}$$

where we have defined the normal ordered transfer matrix operator

$$M \equiv \; : \exp\left(-\delta H(a^\dagger_{i',j'}(\mathbf{n}'), a_{i,j}(\mathbf{n}))\right) :, \tag{3.39}$$

which represents the exponential of the Hamiltonian over one Euclidean lattice time step δ. This "Trotter-Suzuki" decomposition, $\exp[\delta(A + B)] = \lim_{\delta \to 0}[\exp(\delta A) \times \exp(\delta B)] + \mathcal{O}(\delta^2)$, of the Euclidean time evolution operator circumvents the problem of having to diagonalize H in order to evaluate $\exp(-\beta H)$. In many cases H can be decomposed into components that are straightforward to diagonalize, such as the kinetic energy term T and potential operator V. For example, T is usually either a sparse matrix consisting of hopping terms connecting the first few nearest-neighbor lattice sites, or can alternatively be treated as diagonal in momentum space. On the other hand, V is often an on-site contact interaction and as such already diagonal in coordinate space. Such a structure suggests that algorithms based on the Fast Fourier Transform (FFT) technique are optimal for the numerical evaluation of Trotter-Suzuki decomposed evolution operators. It should be noted that the Trotter-Suzuki decomposition is not unique, and it can be carried out to higher orders with the aim of reducing the dependence on the temporal lattice spacing δ.

The choice of the Grassmann versus the transfer matrix representation of the partition function is a matter of calculational convenience. The Grassmann formalism provides a basis for a large variety of highly successful lattice quantum field theories with strongly interacting fermions, such as Lattice QCD, strongly interacting QED in $2+1$ (QED$_3$) and $3+1$ (QED$_4$) dimensions, and theories with contact interactions such as the Thirring and Gross-Neveu models. However, it should be noted that in such Lorentz-invariant theories a symmetric time derivative is required. The asymmetric time derivative used in Eq. (3.36) is suitable for non-relativistic theories such as lattice chiral EFT and variants of the Hubbard model commonly employed in the areas of condensed matter and atomic physics. An intermediate case of recent interest is given by the hexagonal tight-binding description of graphene where long-range electron-electron interactions are incorporated using the Grassmann formalism. In the case of graphene, the existence of two distinct triangular sublattices makes the (non-relativistic) problem suitable for a symmetric time derivative, with greatly reduced errors due to the finite δ.

3.3   Cold Atoms: A First Example

As a first example illustrating the transfer matrix and Grassmann formalisms, we shall consider the low-energy EFT for the "unitarity limit", which can be derived from any theory of two-component fermions with infinite scattering length and negligible higher-order scattering effects (vanishing effective range and shape parameters) at the low-momentum scale in question. These two-fermion components may e.g. correspond to hyperfine states of ^{40}K, with the interaction given either by a full multi-channel Hamiltonian or by a simplified two-channel model [5–7]. While alkali atoms such as ^{6}Li and ^{40}K are convenient for evaporative cooling experiments due to their predominantly elastic collisions, the starting point does not matter as long as the S-wave scattering length is tuned to infinity in order to produce a zero-energy resonance.

We shall here denote the atomic mass by m and label the two hyperfine levels as spin-up and spin-down states, although such a connection to actual intrinsic spin is not necessary. The EFT Hamiltonian is then given by

$$H \equiv H_{\text{free}} + V, \tag{3.40}$$

with the non-relativistic free Hamiltonian (or kinetic energy term)

$$H_{\text{free}} = \frac{1}{2m} \sum_{i=\uparrow,\downarrow} \int d^3 r \, \nabla a_i^\dagger(\mathbf{r}) \cdot \nabla a_i(\mathbf{r}), \tag{3.41}$$

and we may use integration by parts to obtain

$$H_{\text{free}} = -\frac{1}{2m} \sum_{i=\uparrow,\downarrow} \int d^3 r \, a_i^\dagger(\mathbf{r}) \nabla^2 a_i(\mathbf{r}), \tag{3.42}$$

where the Laplacian can be discretized by means of a finite difference formula. The simplest choice is the three-point formula for the Laplacian, which gives the (dimensionless) free lattice Hamiltonian

$$\tilde{H}_{\text{free}} = \frac{3}{\tilde{m}} \sum_{\mathbf{n},i=\uparrow,\downarrow} a_i^\dagger(\mathbf{n}) a_i(\mathbf{n})$$

$$- \frac{1}{2\tilde{m}} \sum_{\mathbf{n},i=\uparrow,\downarrow} \sum_{l=1}^{3} \left[a_i^\dagger(\mathbf{n}) a_i(\mathbf{n} + \hat{e}_l) + a_i^\dagger(\mathbf{n}) a_i(\mathbf{n} - \hat{e}_l) \right], \tag{3.43}$$

where $\hat{e}_l$ denotes a unit vector in (spatial) lattice direction l, and a is the (spatial) lattice spacing. In lattice chiral EFT, we shall consider "improved" higher-order Laplacian operators that minimize the effects of the finite lattice spacing. Here, we have introduced the notation $\tilde{m} \equiv ma$, where m is the particle mass, such that $\tilde{m}$ is expressed in units of the lattice spacing. We also define $\tilde{H}_{\text{free}} \equiv a H_{\text{free}}$.

The contact interaction between spin-up and spin-down particles is included through

$$V \equiv C \int d^3r \, \rho_\uparrow^{a^\dagger, a}(\mathbf{r}) \rho_\downarrow^{a^\dagger, a}(\mathbf{r}), \tag{3.44}$$

where we have defined the density operators

$$\rho_\uparrow^{a^\dagger, a}(\mathbf{r}) \equiv a_\uparrow^\dagger(\mathbf{r}) a_\uparrow(\mathbf{r}), \quad \rho_\downarrow^{a^\dagger, a}(\mathbf{r}) \equiv a_\downarrow^\dagger(\mathbf{r}) a_\downarrow(\mathbf{r}), \tag{3.45}$$

for spin-up and spin-down particles, respectively, and C is the coupling constant. We also define

$$\rho^{a^\dagger, a}(\mathbf{r}) \equiv \rho_\uparrow^{a^\dagger, a}(\mathbf{r}) + \rho_\downarrow^{a^\dagger, a}(\mathbf{r}), \tag{3.46}$$

for the total particle density. On the lattice, we have

$$\tilde{V} = \tilde{C} \sum_{\mathbf{n}} \rho_\uparrow^{a^\dagger, a}(\mathbf{n}) \rho_\downarrow^{a^\dagger, a}(\mathbf{n}), \tag{3.47}$$

where $\tilde{V} \equiv aV$ and $\tilde{C} \equiv C/a^2$, such that $\tilde{C}$ corresponds to the coupling in units of the lattice spacing. Note also that we have introduced the lattice density operators $\rho(\mathbf{n})$ according to the definition in Eq. (3.45). The transfer matrix M is then given by

$$M = \, : \exp\left(-\tilde{H}_{\text{free}} \alpha_t - \tilde{C} \alpha_t \sum_{\mathbf{n}} \rho_\uparrow^{a^\dagger, a}(\mathbf{n}) \rho_\downarrow^{a^\dagger, a}(\mathbf{n})\right) : \, , \tag{3.48}$$

where $\alpha_t \equiv \delta/a$ is the ratio of temporal and spatial lattice spacings. We note that this expression is identical to the three-dimensional Hubbard model, where an attractive interaction corresponds to $\tilde{C} < 0$.

Similarly, in the Grassmann formalism we have

$$S(\zeta^*, \zeta) \equiv S_{\text{free}}(\zeta^*, \zeta) + \tilde{C} \alpha_t \sum_{\mathbf{n}, t} \rho_\uparrow(\mathbf{n}, t) \rho_\downarrow(\mathbf{n}, t), \tag{3.49}$$

for the lattice action. In analogy with the transfer matrix formalism, we define the densities

$$\rho_\uparrow(\mathbf{n}, t) \equiv \zeta_\uparrow^*(\mathbf{n}, t) \zeta_\uparrow(\mathbf{n}, t), \quad \rho_\downarrow(\mathbf{n}, t) \equiv \zeta_\downarrow^*(\mathbf{n}, t) \zeta_\downarrow(\mathbf{n}, t), \tag{3.50}$$

for spin-up and spin-down particles. In terms of these, we again find

$$\rho(\mathbf{n}, t) \equiv \rho_\uparrow(\mathbf{n}, t) + \rho_\downarrow(\mathbf{n}, t), \tag{3.51}$$

for the total density. In Eq. (3.49), the free fermion action is given by

$$S_{\text{free}}(\zeta^*, \zeta) = \sum_{\mathbf{n},t,i=\uparrow,\downarrow} \zeta_i^*(\mathbf{n}, t)\left[\zeta_i(\mathbf{n}, t+1) - \zeta_i(\mathbf{n}, t)\right]$$

$$+ 6h \sum_{\mathbf{n},t,i=\uparrow,\downarrow} \zeta_i^*(\mathbf{n}, t)\zeta_i(\mathbf{n}, t)$$

$$- h \sum_{\mathbf{n},t,i=\uparrow,\downarrow} \sum_{l=1}^{3} \zeta_i^*(\mathbf{n}, t)\left[\zeta_i(\mathbf{n} + \hat{e}_l, t) + \zeta_i(\mathbf{n} - \hat{e}_l, t)\right], \quad (3.52)$$

with

$$h \equiv \alpha_t/(2\tilde{m}), \tag{3.53}$$

as the "hopping amplitude".

Given Eq. (3.52), it is useful to note some similarities and differences to applications in the field of condensed matter physics, e.g. the "tight-binding" descriptions of carbon nanomaterials, such as graphene. For two spatial dimensions, provided that the central term proportional to $6h$ is dropped, Eq. (3.52) effectively describes the hopping of electrons between the π-orbitals of carbon ions. Then, $h = \kappa\delta$ can be identified with the hopping amplitude κ, which is given by the degree of overlap between the orbitals of the neighboring ions. Empirically, one finds that $\kappa \simeq 2.8$ eV in actual graphene monolayers, while the next-to-nearest-neighbor hopping amplitude is suppressed by an order of magnitude with respect to κ. Hence, in condensed matter systems, the "lattice effects" arising from the use of "truncated" expressions such as Eq. (3.52) are physical and form the basis of their electronic properties, provided that the temporal extent of the Euclidean lattice satisfies $\kappa\delta \ll 1$. In contrast, we shall find that "improved" lattice derivative operators are essential for NLEFT.

We have so far not considered the case of a non-zero chemical potential μ coupled to the total particle density, in which case the lattice Hamiltonian is

$$\tilde{H}(\mu) \equiv \tilde{H}_{\text{free}} + \tilde{C} \sum_{\mathbf{n}} \rho_\uparrow^{a^\dagger,a}(\mathbf{n})\rho_\downarrow^{a^\dagger,a}(\mathbf{n}) - \tilde{\mu} \sum_{\mathbf{n}} \rho^{a^\dagger,a}(\mathbf{n}), \tag{3.54}$$

where $\tilde{\mu}$ denotes the chemical potential in units of the lattice spacing. This gives

$$M(\mu) = M \exp\left\{\tilde{\mu}\alpha_t \sum_{\mathbf{n}} \rho^{a^\dagger,a}(\mathbf{n})\right\}, \tag{3.55}$$

for the transfer matrix at finite $\tilde{\mu}$, where it should be noted that while M is normal-ordered, such a prescription has not been applied to the exponential introduced by

$\tilde{\mu} \neq 0$. In the Grassmann formulation, we find

$$\mathscr{Z}(\mu) = \int \mathscr{D}\zeta \, \mathscr{D}\zeta^* \exp(-S(\zeta^*, \zeta, \mu)), \qquad (3.56)$$

with

$$S(\zeta^*, \zeta, \mu) = S(\zeta^*, e^{\tilde{\mu}\alpha_t}\zeta) + \sum_{\mathbf{n},t,i=\uparrow,\downarrow} (1 - e^{\tilde{\mu}\alpha_t})\zeta_i^*(\mathbf{n}, t)\zeta_i(\mathbf{n}, t + 1), \qquad (3.57)$$

where the first term on the l.h.s. equals Eq. (3.49) with $\zeta \to \exp(\tilde{\mu}\alpha_t)\zeta$. So far, we have taken a common chemical potential for the spin-up and spin-down particles. We may also introduce separate chemical potentials $\mu_\uparrow$ and $\mu_\downarrow$ for the two spin states, however when $\mu_\uparrow \neq \mu_\downarrow$ (a situation referred to as the asymmetric Fermi gas) the corresponding probability measure for AFQMC calculations is no longer positive definite.

Having established the Grassmann path integral and transfer matrix formulations, we are presented with the problem of developing these into viable and efficient algorithms for Monte Carlo calculations. It should be noted that if the action or transfer matrix were entirely bilinear in the Grassmann fields or the creation and annihilation operators, then these could be "integrated out" in order to obtain a formalism exclusively based on complex-valued quantities. However, this property of non-interacting nucleons (or in general: fermions) is apparently spoiled by the contact interactions we have considered so far, as these are quartic in the fields and operators. Nevertheless, we may recover a similar situation by introducing an auxiliary-field transformation that decouples the fermions. In the auxiliary-field formulation, the individual fermions have no mutual interactions, and merely interact with a fluctuating background field which encodes the relevant fermion–fermion interactions.

It should also be noted that many equivalent approaches exist which work directly with the Grassmann fields without an auxiliary-field transformation. A recent example of such a study is the application of the "fermion bag" method to the Thirring model in Ref. [8] (and references therein).

3.4 Auxiliary Fields

We shall next recast the Grassmann path integral and transfer matrix formulations into a form which is bilinear in the Grassmann fields and creation and annihilation operators. Specifically, we shall simplify the treatment of the four-fermion contact interaction by re-expressing the Grassmann path integral using an "auxiliary field" coupled to the particle density. We note that such a formulation has been used in several lattice studies at non-zero temperature [9–15].

We introduce the auxiliary field s by means of the transformation

$$\mathscr{Z} = \prod_{\mathbf{n},t} \left[\int d_A s(\mathbf{n}, t) \right] \int \mathscr{D}\zeta \, \mathscr{D}\zeta^* \exp(-S_A(\zeta^*, \zeta, s)), \tag{3.58}$$

where

$$S_A(\zeta^*, \zeta) \equiv S_{\text{free}}(\zeta^*, \zeta) - \sum_{\mathbf{n},t} A[s(\mathbf{n}, t)] \, \rho(\mathbf{n}, t), \tag{3.59}$$

and we note that due to the simple structure of the contact interaction and the anti-commutation properties of the Grassmann variables, there exists a large choice of auxiliary-field transformations which reproduce the same lattice action. This freedom in the choice of auxiliary-field transformation can often be exploited in order optimize algorithmic factors, such as numerical stability and severity of the sign problem. The numerical performance of four different auxiliary-field transformations has been compared in Ref. [16] for the problem of the unitary Fermi gas.

As a concrete example, we consider the Gaussian auxiliary-field transformation

$$\int d_A s(\mathbf{n}, t) = \frac{1}{\sqrt{2\pi}} \int_{-\infty}^{+\infty} ds(\mathbf{n}, t) \, \exp(-s^2(\mathbf{n}, t)/2), \tag{3.60}$$

with

$$A[s(\mathbf{n}, t)] = \sqrt{-\tilde{C}\alpha_t} \, s(\mathbf{n}, t), \tag{3.61}$$

which is similar to the original Hubbard-Stratonovich (HS) transformation [2, 3]. This is an example of a continuous HS transformation, which can be used as a basis for a Monte Carlo algorithm with global lattice updates such as Hybrid Monte Carlo, Chap. 6. Although here the auxiliary field s assumes an unbounded integration range, compact HS transformations where the integration range is finite, are also commonly used. Another category of auxiliary fields is provided by discrete HS transformations, for example

$$\int d_A s(\mathbf{n}, t) = \int_{-1/2}^{+1/2} ds(\mathbf{n}, t), \tag{3.62}$$

with

$$A[s(\mathbf{n}, t)] = \sqrt{-\tilde{C}\alpha_t} \, \text{sgn}[s(\mathbf{n}, t)], \tag{3.63}$$

which is similar to that used in Ref. [17]. Here, we define $\text{sgn}(x)$ as $+1$ for positive values of x, and -1 for negative values of x. Discrete HS transformations are commonly used in algorithms where local lattice updates (such as random Metropolis moves) are sufficient. This is often the case in low-dimensional problems, such as those encountered in condensed matter physics.

Without referring to any specific HS transformation, we shall now establish how the original Grassmann path integral is recovered when the auxiliary field s is integrated out. For this purpose, we leave $d_A s(\mathbf{n}, t)$ and $A[s(\mathbf{n}, t)]$ unspecified and consider the integral

$$\int d_A s(\mathbf{n}, t) \, \exp\left\{ A[s(\mathbf{n}, t)] \left(\rho_\uparrow(\mathbf{n}, t) + \rho_\downarrow(\mathbf{n}, t) \right) \right\}, \tag{3.64}$$

where we will temporarily suppress the arguments $(\mathbf{n}, t)$ in order to shorten the notation. This gives

$$\int d_A s \, \exp\left[A(s)(\rho_\uparrow + \rho_\downarrow) \right] = \int d_A s \left[1 + A(s)(\rho_\uparrow + \rho_\downarrow) + A^2(s)\rho_\uparrow \rho_\downarrow \right], \tag{3.65}$$

where we first require that

$$\int d_A s(\mathbf{n}, t) = 1, \tag{3.66}$$

and

$$\int d_A s(\mathbf{n}, t) \, A[s(\mathbf{n}, t)] = 0, \tag{3.67}$$

such that

$$\int d_A s \, \exp\left[A(s)(\rho_\uparrow + \rho_\downarrow) \right] = 1 + \int d_A s \, A^2(s)\rho_\uparrow \rho_\downarrow = \exp\left[\int d_A s \, A^2(s)\rho_\uparrow \rho_\downarrow \right], \tag{3.68}$$

and the remaining condition then reads

$$\int d_A s(\mathbf{n}, t) \, A^2[s(\mathbf{n}, t)] = -\tilde{C}\alpha_t, \tag{3.69}$$

which allows us to reproduce the original Grassmann path integral. For a general treatment of higher-order density operators $\sim \rho^n$ ($n \geq 3$) within the HS method, see Ref. [18].

We may now also consider how the expressions for the partition function and the action with auxiliary fields are modified in the presence of a non-zero chemical potential μ. As for the Grassmann path integral without auxiliary fields, we find

$$\mathscr{Z}(\mu) = \prod_{\mathbf{n},t} \left[\int d_A s(\mathbf{n}, t) \right] \int \mathscr{D}\zeta \, \mathscr{D}\zeta^* \exp(-S_A(\zeta^*, \zeta, s, \mu)), \tag{3.70}$$

with

$$S_A(\zeta^*, \zeta, s, \mu) = S_A(\zeta^*, e^{\tilde{\mu}\alpha_t}\zeta, s) + \sum_{\mathbf{n},t,i=\uparrow,\downarrow} (1 - e^{\tilde{\mu}\alpha_t})\zeta_i^*(\mathbf{n}, t)\zeta_i(\mathbf{n}, t+1). \tag{3.71}$$

In the transfer matrix formulation, we first note that the partition function in the presence of the auxiliary field s is of the form

$$\mathscr{Z} = \prod_{\mathbf{n},t} \left[\int d_A s(\mathbf{n}, t) \right] \mathrm{Tr} \Big\{ M_A(s, N_t - 1) \cdots M_A(s, 0) \Big\}, \tag{3.72}$$

where the auxiliary-field transfer matrix $M_A(s, t)$ or "timeslice" is given by

$$M_A(s, t) = : \exp\left(-\tilde{H}_{\text{free}}\alpha_t + \sum_{\mathbf{n}} A[s(\mathbf{n}, t)] \rho^{a^\dagger, a}(\mathbf{n}) \right) : . \tag{3.73}$$

Finally, at non-zero chemical potential μ, we find

$$\mathscr{Z}(\mu) = \prod_{\mathbf{n},t} \left[\int d_A s(\mathbf{n}, t) \right] \mathrm{Tr} \Big\{ M_A(s, N_t - 1, \mu) \cdots M_A(s, 0, \mu) \Big\}, \tag{3.74}$$

for the partition function, where the auxiliary-field transfer matrix $M_A(s, t, \mu)$ is

$$M_A(s, t, \mu) = M_A(s, t) \exp\left\{ \tilde{\mu}\alpha_t \sum_{\mathbf{n}} \rho^{a^\dagger, a}(\mathbf{n}) \right\}. \tag{3.75}$$

As we have now established the Grassmann path integral and transfer matrix lattice formulations both with and without auxiliary fields, we need to also take into account the fact that often the object of interest is the properties of the ground state of the system. In that case, one must first tune the chemical potential μ to obtain the desired (average) particle number and thereafter take the low-temperature limit $T \to 0$. As in particular the latter limit is often numerically troublesome, a significant advantage can be obtained if the calculation is performed directly for the ground state at fixed particle number. As will be shown in Chap. 6, such a framework is provided by the Projection Monte Carlo (PMC) method, which can be straightforwardly combined with the AFQMC algorithm in the transfer matrix formulation.

3.5 Pionless EFT for Nucleons: A Second Example

So far, we have introduced the concepts of transfer matrices and auxiliary fields using as an example a contact interaction which is appropriate to describe systems composed of cold atoms, where the details of the interaction are not of major significance. Before introducing the lattice formulation of chiral EFT, we shall therefore consider a system of intermediate complexity, namely nucleons at momenta much smaller than the pion mass M_π. For such low momenta, all interactions between nucleons generated by the strong nuclear force can be treated as local. In that case, the Hamiltonian (3.40) also describes the interactions of low-energy neutrons at leading order in the EFT expansion. For systems composed of protons as well as neutrons, we label the nucleon annihilation operators with two indices,

$$a_{0,0} \equiv a_{\uparrow,p}, \quad a_{1,0} \equiv a_{\downarrow,p}, \quad a_{0,1} \equiv a_{\uparrow,n}, \quad a_{1,1} \equiv a_{\downarrow,n}, \tag{3.76}$$

where the first index ($\uparrow$, $\downarrow$) denotes the spin projection, and the second index (p, n) denotes the isospin projection. Similarly, we define Pauli matrices σ_S with $S = 1, 2, 3$ acting in spin space, and σ_I with $I = 1, 2, 3$ acting in isospin space. The isospin matrices are sometimes also called τ_i, $i = 1, 2, 3$. We note that the symbols S and I are also occasionally used to denote total spin and isospin quantum numbers. Thus, if isospin-breaking and electromagnetic effects are neglected, the "pionless" EFT possesses exact SU(2) spin and SU(2) isospin symmetry.

At LO in pionless EFT, the Hamiltonian is

$$H_{\mathrm{LO}} \equiv H_{\mathrm{free}} + V_{\mathrm{LO}}, \tag{3.77}$$

where the free Hamiltonian is again given by

$$H_{\mathrm{free}} = \frac{1}{2m_N} \sum_{i,j=0,1} \int d^3r \, \nabla a_{i,j}^\dagger(\mathbf{r}) \cdot \nabla a_{i,j}(\mathbf{r}), \tag{3.78}$$

with m_N the mass of the nucleon. The LO interaction is

$$V_{\mathrm{LO}} \equiv V + V_{I^2} + V^{(3N)}, \tag{3.79}$$

with the SU(4)-symmetric two-nucleon contact term (see also the discussion in Chap. 2)

$$V \equiv \frac{C}{2} \int d^3r \, : \left[\rho^{a^\dagger,a}(\mathbf{r}) \right]^2 :, \tag{3.80}$$

with

$$\rho^{a^\dagger,a}(\mathbf{r}) \equiv \sum_{i,j=0,1} a_{i,j}^\dagger(\mathbf{r}) a_{i,j}(\mathbf{r}), \tag{3.81}$$

which is otherwise similar to the interaction term in the previously considered EFT for cold atoms, except for the "symmetry factor" of $1/2$ due to the indistinguishable nucleons. In the pionless EFT, we also have the isospin-dependent contact interaction

$$V_{I2} \equiv \frac{C_{I2}}{2} \sum_{I=1,2,3} \int d^3r : \left[\rho_I^{a^\dagger,a}(\mathbf{r})\right]^2 :, \tag{3.82}$$

where we have defined the isospin-density operators

$$\rho_I^{a^\dagger,a}(\mathbf{r}) \equiv \sum_{i,j,j'=0,1} a_{i,j}^\dagger(\mathbf{r}) \left[\tau_I\right]_{jj'} a_{i,j'}(\mathbf{r}). \tag{3.83}$$

Unlike full chiral EFT, in the pionless EFT an SU(4)-symmetric three-nucleon term also appears at LO, and is given by

$$V^{(3N)} \equiv \frac{D}{6} \int d^3r : \left[\rho^{a^\dagger,a}(\mathbf{r})\right]^3 :, \tag{3.84}$$

where we note that the three-nucleon term is required to account for an instability in the limit of zero-range interactions and for a consistent renormalization in the extreme non-relativistic limit, see Refs. [19–22].

In the lattice formulation of pionless EFT, the LO transfer matrix M is given by

$$M = : \exp\left(-\tilde{H}_{\text{free}}\alpha_t - \frac{\tilde{C}\alpha_t}{2} \sum_{\mathbf{n}} \left[\rho^{a^\dagger,a}(\mathbf{n})\right]^2 - \frac{\tilde{C}_{I2}\alpha_t}{2} \sum_{\mathbf{n}} \left[\rho_I^{a^\dagger,a}(\mathbf{n})\right]^2 \right.$$
$$\left. - \frac{\tilde{D}\alpha_t}{6} \sum_{\mathbf{n}} \left[\rho^{a^\dagger,a}(\mathbf{n})\right]^3 \right) :, \tag{3.85}$$

which leads to the complication that we need to account for multiple contact interactions when introducing the auxiliary-field formalism. This feature will be prominent in lattice chiral EFT, where (depending on the level of improvement of the LO action) we require as many as sixteen auxiliary-field components. In the case of pionless EFT as LO, we require four auxiliary-field components in order to account for the three components of the isospin-dependent contact interaction and the single SU(4)-symmetric interaction. We thus find

$$\mathscr{Z} = \prod_{\mathbf{n},t} \left[\int d_A s(\mathbf{n}, t)\right] \prod_{\mathbf{n},t,I} \left[\frac{1}{\sqrt{2\pi}} \int_{-\infty}^{\infty} ds_I(\mathbf{n}, t) \exp\left(-\frac{1}{2}s_I^2(\mathbf{n}, t)\right)\right]$$
$$\times \text{Tr}\left\{M_A(s, s_I, N_t - 1) \cdots M_A(s, s_I, 0)\right\}, \tag{3.86}$$

with

$$M_A(s, s_I, t) = \; :\exp\left(-\tilde{H}_{\text{free}}\alpha_t + \sum_{\mathbf{n}} A[s(\mathbf{n}, t)]\,\rho^{a^\dagger, a}(\mathbf{n})\right.$$
$$\left. + i\sqrt{\tilde{C}_{I2}\alpha_t}\sum_{\mathbf{n}, I} s_I(\mathbf{n}, t)\,\rho_I^{a^\dagger, a}(\mathbf{n})\right):, \qquad (3.87)$$

where we have introduced the auxiliary-field components s_I using a continuous, Gaussian HS transformation and left the SU(4)-symmetric part $A[s(\mathbf{n}, t)]$ unspecified. As for the EFT of cold atoms, we may now determine the conditions for reproducing the original LO interactions. We require

$$\int d_A s(\mathbf{n}, t) = 1, \qquad (3.88)$$

and

$$\int d_A s(\mathbf{n}, t)\, A[s(\mathbf{n}, t)] = 0, \qquad (3.89)$$

followed by

$$\int d_A s(\mathbf{n}, t)\, A^2[s(\mathbf{n}, t)] = -\tilde{C}\alpha_t, \qquad (3.90)$$

and

$$\int d_A s(\mathbf{n}, t)\, A^3[s(\mathbf{n}, t)] = -\tilde{D}\alpha_t, \qquad (3.91)$$

with

$$\int d_A s(\mathbf{n}, t)\, A^4[s(\mathbf{n}, t)] = 3\tilde{C}^2\alpha_t^2, \qquad (3.92)$$

as the final consistency condition.

In the case of the EFT for cold atoms, we found that a positive definite probability measure suitable for Monte Carlo calculations could be obtained as long as the interaction remains attractive. However, for pionless EFT at LO, the situation turns out to be more intricate. If we introduce the notation

$$\langle A^k \rangle \equiv \int d_A s(\mathbf{n}, t)\, A^k[s(\mathbf{n}, t)], \quad k = 0, 1, 2, \ldots \qquad (3.93)$$

we can define

$$Q \equiv \begin{pmatrix} \langle A^0 \rangle & \langle A^1 \rangle & \langle A^2 \rangle \\ \langle A^1 \rangle & \langle A^2 \rangle & \langle A^3 \rangle \\ \langle A^2 \rangle & \langle A^3 \rangle & \langle A^4 \rangle \end{pmatrix} = \begin{pmatrix} 1 & 0 & -\tilde{C}\alpha_t \\ 0 & -\tilde{C}\alpha_t & -\tilde{D}\alpha_t \\ -\tilde{C}\alpha_t & -\tilde{D}\alpha_t & 3\tilde{C}^2\alpha_t^2 \end{pmatrix}, \tag{3.94}$$

and it can be shown that the positivity of the Monte Carlo probability measure is equivalent to the condition that the matrix Q be positive semi-definite [23]. In other words, we require that

$$\det Q = -2\tilde{C}^3\alpha_t^3 - \tilde{D}^2\alpha_t^2 \geq 0, \tag{3.95}$$

where for attractive interactions $\tilde{C} < 0$. Thus, there exists a delicate interplay between the two-nucleon and three-nucleon interactions, and furthermore the positivity condition is more easily violated in the Hamiltonian limit $\alpha_t \to 0$.

3.6 Symmetries on the Lattice

Here, we will elaborate on the issue of symmetries and their representation on the lattice. As is well known, symmetries play an important role in physics, and we have already discussed the pertinent symmetries of the strong interactions in Chaps. 1 and 2. Clearly, we expect the discrete representation of space-time in a lattice formulation (more precisely: a hyper-cubic box) to have consequences for the various symmetries under consideration. These will be discussed here.

To be specific, let us consider a non-relativistic system with Hamiltonian

$$H \equiv T + V, \tag{3.96}$$

in terms of the kinetic energy T and the potential energy V. This is also the appropriate framework for our upcoming discussion of NLEFT. Without loss of generality, let us assume that V is spherically symmetric (we will return to this issue in Chap. 5). In the continuum, this entails rotational invariance, i.e. the conservation of the squared angular momentum $\hat{L}^2$, and of one of its components,

$$[H, \hat{L}^2] = [H, \hat{L}_z] = 0, \tag{3.97}$$

usually taken to be $\hat{L}_z$. Stated differently, the continuum eigenstates of H can be labeled according to the eigenvalues $l(l + 1)$ of $\hat{L}^2$ (note that we reserve the symbol L for the physical dimension of the lattice), the quadratic Casimir operator of the group SO(3), and by those of its third component $\hat{L}_z$ with eigenvalue m (the Casimir operator of the rotations in the plane),

$$\text{SO}(3) \supset \text{SO}(2)$$
$$\downarrow \qquad \downarrow \qquad\qquad (3.98)$$
$$l \qquad m$$

i.e. as a basis of the $(2l + 1)$-dimensional irreducible representation of SO(3) and eigenstates of rotations about the z-axis.

The lattice regularization of H reduces the SO(3) rotational symmetry of continuous space to the cubic rotational (or octahedral) group SO(3, Z), which consists of 24 rotations of $\pi/2$ (or multiples thereof) about the x, y and z axes, see e.g. Refs. [24–26]. The elements of SO(3, Z) can be expressed in terms of rotation operators $\hat{R}_j(\phi)$ of SO(3), where $j \in x, y, z$ denotes a fixed axis, and $\phi = n\pi/2$ (with n an integer). Then, the angular momentum operators of SO(3, Z) can be defined by

$$\hat{R}_j(\phi) \equiv \exp(-i\hat{L}_j\phi), \qquad\qquad (3.99)$$

from which it follows that the eigenvalues L_j of $\hat{L}_j$ are integers modulo 4. The set of rotations by $\pi/2$ forms an abelian group, which is isomorphic to the cyclic group of order four, denoted $\mathscr{C}_4$. The $2l + 1$ elements of angular momentum transform according to the irreducible representations or "irreps" of SO(3). However, under SO(3, Z) these are in most cases reducible, and can be decomposed into the five irreps of SO(3, Z), which are denoted as $\Gamma^{(i)} = A_1^{\pm}, A_2^{\pm}, E^{\pm}, T_1^{\pm}$, and $T_2^{\pm}$ (for positive and negative parity), and have dimension 1, 1, 2, 3, and 3, respectively. Examples of the representation of SO(Z, 3) irreps in terms of the spherical harmonics $Y_{l,m}(\theta, \phi)$, along with some of the properties of these irreps are given in Table 3.1. Furthermore, the decomposition of SO(3) states with $l \leq 9$ in terms of the $\Gamma^{(i)}$ are shown in Table 3.2. It should be noted that each of the

Table 3.1 Examples of irreducible SO(3, Z) representations

Irrep $\Gamma^{(i)}$	$L_z(\text{mod } 4)$	$Y_{l,m}(\theta, \phi)$
A_1^+	0	$Y_{0,0}$
T_1^-	0, 1, 3	$Y_{1,0}, Y_{1,1}, Y_{1,-1}$
E^+	0, 2	$Y_{2,0}, \frac{1}{\sqrt{2}}(Y_{2,-2} + Y_{2,2})$
T_2^+	1, 2, 3	$Y_{2,1}, \frac{1}{\sqrt{2}}(Y_{2,-2} - Y_{2,2}), Y_{2,-1}$
A_2^-	2	$\frac{1}{\sqrt{2}}(Y_{3,2} - Y_{3,-2})$

The quantum number L_z indicates that a phase factor of $\exp(i\hat{L}_z\phi)$ arises from a cubic rotation $\phi = n\pi/2$ about the z-axis (with n an integer). Note that $L_z = 3$ is equivalent to $L_z = -1$, and $L_z = 2$ to $L_z = -2$

Table 3.2 Decompositions of orbital angular momentum eigenstates $|l, m\rangle$ into irreps of SO(3, Z), for angular momenta $l \leq 9$

Angular mom. l	Decomposition in terms of $\Gamma^{(i)}$
0	A_1^+
1	T_1^-
2	$E^+ \oplus T_2^+$
3	$A_2^- \oplus T_1^- \oplus T_2^-$
4	$A_1^+ \oplus E^+ \oplus T_1^+ \oplus T_2^+$
5	$E^- \oplus T_1^{-(1)} \oplus T_1^{-(2)} \oplus T_2^-$
6	$A_1^+ \oplus A_2^+ \oplus E^+ \oplus T_1^+ \oplus T_2^{+(1)} \oplus T_2^{+(2)}$
7	$A_2^- \oplus E^- \oplus T_1^{-(1)} \oplus T_1^{-(2)} \oplus T_2^{-(1)} \oplus T_2^{-(2)}$
8	$A_1^+ \oplus E^{+(1)} \oplus E^{+(2)} \oplus T_1^{+(1)} \oplus T_1^{+(2)} \oplus T_2^{+(1)} \oplus T_2^{+(2)}$
9	$A_1^- \oplus A_2^- \oplus E^- \oplus T_1^{-(1)} \oplus T_1^{-(2)} \oplus T_1^{-(3)} \oplus T_2^{-(1)} \oplus T_2^{-(2)}$

Note that the dimensionality $2l + 1$ of an SO(3) irrep must equal the sum of dimensionalities of the SO(3, Z) irreps in each decomposition. In general, SO(3, Z) irreps may appear multiple times in a given decomposition, as is the case for $l \geq 5$. Conversely, states that belong to the A_1^+ representation on the lattice correspond to continuum states with $l = 0, 4, 6, 8, \ldots$

irreps of SO(3, Z) have a non-zero overlap with infinitely many irreps of SO(3), which means that the two-nucleon states that transform according to a given irrep of SO(3, Z) receive contributions from the phase shifts in infinitely many partial waves. When the lattice Hamiltonian is diagonalized together with $\hat{R}_j(\phi)$, the simultaneous eigenstates can be labeled according to the irreducible representations of SO(3, Z) and $\mathscr{C}_4$,

$$
\begin{array}{ccc}
\mathcal{O} & \supset & \mathscr{C}_4 \\
\downarrow & & \downarrow \\
\Gamma & & \hat{L}_j
\end{array}
\tag{3.100}
$$

where $\Gamma \in A_1, A_2, E, T_1, T_2$. We shall return to the issue of the reduced rotational symmetry at various places, notably for rotational symmetry breaking operators in Chaps. 4 and 5. Also, in Appendix C we discuss methods to overcome (or at least minimize) the effects of rotational symmetry breaking.

Let us now consider additional symmetries on the lattice, starting with translational invariance. As the spatial coordinates assume discrete values on the lattice sites (subject to periodic boundary conditions), the momenta are also quantized as $p_i = (2\pi/L)n_i$ for the lattice dimensions $i = 1, 2, 3$, where n_i an integer that assumes values within the first Brillouin zone (see Chap. 4). Translations then correspond to discrete shifts along the lattice directions, and this lattice translational invariance can be used e.g. to formulate lattice derivatives in terms of forward differences only, as discussed in Chap. 4. Let us also discuss the discrete symmetries of parity $\mathscr{P}$, charge conjugation, and time reversal. The latter two play essentially no role in non-relativistic systems, as discussed here. The strong interactions are invariant under parity, such that

$$[H, \mathscr{P}] = 0, \tag{3.101}$$

which is also reflected in our continuum Hamiltonian H. This feature is preserved by the formulation on a cubic lattice. The corresponding invariance allows for the construction of projectors onto the two irreducible representations of the parity group ($\approx \mathscr{C}_2$, the cyclic group of order two), labelled by $+$ and $-$,

$$P_{\pm} = \mathbb{1} \pm \mathscr{P}, \tag{3.102}$$

which act on the lattice eigenfunctions of H. We have already used this fact in the above discussions on the lattice representation of angular momentum eigenstates.

Finally, let us mention that internal symmetries, such as isospin, are not affected by the lattice discretization. For the non-relativistic systems considered here, it is always possible to find a Hamiltonian which maintains exact parity invariance and all internal symmetries. Therefore, the symmetry or antisymmetry of intrinsic spin on the lattice is unambiguous, and the same as in continuous space. Consider e.g. the case of two spin-1/2 particles where the total intrinsic spin is $S = 0$ or $S = 1$. This is sufficient to specify the intrinsic spin representation completely. If the intrinsic spin is symmetric, then the SO(3, Z) representation is A_1. Similarly, if the intrinsic spin is antisymmetric, then the SO(3, Z) representation is T_1. These are in one-to-one correspondence with the representations $S = 0$ and $S = 1$ in continuous space, respectively. It is thus possible to borrow the continuous space notation for the two cases (as in Chap. 5). Note that for particles with higher intrinsic spin, there is in general some unphysical mixing on the lattice among even values of S and among odd values of S. The same mixing on the lattice occurs among even and odd values of orbital angular momentum l, regardless of the intrinsic spin.

References

1. I. Montvay, C. Urbach, Exploratory investigation of nucleon-nucleon interactions using Euclidean Monte Carlo simulations. Eur. Phys. J. A **48**, 38 (2012)
2. R.L. Stratonovich, On a method of calculating quantum distribution functions. Sov. Phys. Dokl. **2**, 416 (1958)
3. J. Hubbard, Calculation of partition functions. Phys. Rev. Lett. **3**, 77 (1959)
4. M. Creutz, Transfer matrices and lattice fermions at finite density. Found. Phys. **30**, 487 (2000)
5. K. Góral, T. Köhler, S.A. Gardiner, E. Tiesinga, P.S. Julienne, Adiabatic association of ultracold molecules via magnetic field tunable interactions. J. Phys. B **37**, 3457 (2004)
6. M.H. Szymanska, K. Góral, T. Köhler, K. Burnett, Conventional character of the BCS-BEC cross-over in ultra-cold gases of 40K. Phys. Rev. A **72**, 013610 (2005)
7. N. Nygaard, R. Piil, K. Mølmer, Feshbach molecules in a one-dimensional optical lattice. Phys. Rev. A **77**, 021601(R) (2008)

8. S. Chandrasekharan, Fermion bag approach to fermion sign problems. Eur. Phys. J. A **49**, 90 (2013)
9. J.W. Chen, D.B. Kaplan, A lattice theory for low-energy fermions at finite chemical potential. Phys. Rev. Lett. **92**, 257002 (2004)
10. D. Lee, B. Borasoy, T. Schäfer, Nuclear lattice simulations with chiral effective field theory. Phys. Rev. C **70**, 014007 (2004)
11. D. Lee, T. Schäfer, Neutron matter on the lattice with pionless effective field theory. Phys. Rev. C **72**, 024006 (2005)
12. D. Lee, T. Schäfer, Cold dilute neutron matter on the lattice. I. Lattice virial coefficients and large scattering lengths. Phys. Rev. C **73**, 015201 (2006)
13. D. Lee, T. Schäfer, Cold dilute neutron matter on the lattice. II. Results in the unitary limit. Phys. Rev. C **73**, 015202 (2006)
14. T. Abe, R. Seki, Lattice calculation of thermal properties of low-density neutron matter with NN effective field theory. Phys. Rev. C **79**, 054002 (2009)
15. T. Abe, R. Seki, From low-density neutron matter to the unitary limit. Phys. Rev. C **79**, 054003 (2009)
16. D. Lee, The ground state energy at unitarity. Phys. Rev. C **78**, 024001 (2008)
17. J.E. Hirsch, Discrete Hubbard-Stratonovich transformation for fermion lattice models. Phys. Rev. B **28**, 4059 (1983)
18. C. Körber, E. Berkowitz, T. Luu, Sampling general N-body interactions with auxiliary fields. Europhys. Lett. **119**(6), 60006 (2017)
19. P.F. Bedaque, H.W. Hammer, U. van Kolck, Renormalization of the three-body system with short range interactions. Phys. Rev. Lett. **82**, 463 (1999)
20. P.F. Bedaque, H.W. Hammer, U. van Kolck, The three boson system with short range interactions. Nucl. Phys. A **646**, 444 (1999)
21. P.F. Bedaque, H.W. Hammer, U. van Kolck, Effective theory of the triton. Nucl. Phys. A **676**, 357 (2000)
22. E. Epelbaum, J. Gegelia, U.-G. Meißner, D.L. Yao, Renormalization of the three-boson system with short-range interactions revisited. Eur. Phys. J. A **53**, 98 (2017)
23. J.W. Chen, D. Lee, T. Schäfer, Inequalities for light nuclei in the Wigner symmetry limit. Phys. Rev. Lett. **93**, 242302 (2004)
24. R.C. Johnson, Angular momentum on a lattice. Phys. Lett. **114B**, 147 (1982)
25. B. Berg, A. Billoire, Glueball spectroscopy in four-dimensional SU(3) lattice gauge theory. 1. Nucl. Phys. B **221**, 109 (1983)
26. J.E. Mandula, G. Zweig, J. Govaerts, Representations of the rotation reflection symmetry group of the four-dimensional cubic lattice. Nucl. Phys. B **228**, 91 (1983)

Chapter 4
Lattice Chiral Effective Field Theory

We have introduced the concepts of Grassmann path integrals, auxiliary fields and transfer matrix operators in Chap. 3, and illustrated these with a number of examples relevant to cold atomic gases and low-energy nucleons. Here, we apply these methods to formulate lattice chiral EFT (or NLEFT), based on the continuum discussion in Chap. 2. We shall mainly use the term NLEFT when referring to the combination of lattice chiral EFT with Monte Carlo (MC) and Euclidean time projection methods (to be discussed in Chaps. 5–7). In addition to presenting the current "state of the art", we shall also outline some of the historical developments of lattice chiral EFT for the benefit of the reader. In this chapter, we will consider lattice chiral EFT up-to-and-including NNLO contributions. Perhaps the most significant new feature we encounter here is that LO effects are treated non-perturbatively whereas NLO and NNLO effects are incorporated as perturbations. When formulating the LO lattice action for the two-nucleon force (2NF) of chiral EFT, we encounter the one-pion exchange (OPE) potential as well as momentum-dependent "smeared" contact interactions. More precisely, the LO contact terms are, of course, momentum-independent, but to overcome instabilities when four nucleons occupy the same lattice site, these contact interactions are smeared, which makes them momentum-dependent. At NLO, we find further contact terms with two powers of momenta, along with the two-pion exchange (TPE) contributions, which we can effectively absorb into the NLO contact terms at the relatively low cutoff momenta mostly used in lattice chiral EFT. At NNLO, we need to account for the three-nucleon force (3NF), and we also discuss the treatment of electromagnetic and strong isospin-symmetry breaking contributions. We conclude this chapter with a discussion of the expected significance of neglected higher-order (N3LO and above) contributions. It should be noted that the main focus here is to formulate the NLEFT transfer matrix operator, whereas the numerical treatment is mainly given for the two- and three-nucleon problems in Chap. 5, and for the few-nucleon systems and larger nuclei in Chaps. 6 and 7.

© Springer Nature Switzerland AG 2019
T. A. Lähde, U.-G. Meißner, *Nuclear Lattice Effective Field Theory*,
Lecture Notes in Physics 957, https://doi.org/10.1007/978-3-030-14189-9_4

4.1 Lattice Formulation and Notation

In our formulation of lattice chiral EFT, we shall make use of the notational conventions developed for the pionless EFT in Chap. 3 and complement it when necessary. We begin by recalling a number of basic aspects. We define integer-valued lattice position vectors

$$\mathbf{n} \equiv (n_1, n_2, n_3), \tag{4.1}$$

on a three-dimensional spatial lattice, such that $\mathbf{r} \equiv \mathbf{n}a$. We also define lattice momentum vectors

$$\mathbf{k} \equiv (k_1, k_2, k_3) = \left(\frac{2\pi}{N} \hat{k}_1, \frac{2\pi}{N} \hat{k}_2, \frac{2\pi}{N} \hat{k}_3 \right), \tag{4.2}$$

such that $\mathbf{k}_{\text{phys}} \equiv \mathbf{k}/a$, where the $\hat{k}_i$ are integer-valued components. The length (in physical units) of the spatial and temporal dimensions are given by $L \equiv Na$ and $L_t \equiv N_t \delta$, respectively. Here, N_t indicates the number of Euclidean time steps, and the ratio of temporal to spatial lattice spacings is denoted $\alpha_t \equiv \delta/a$.

We apply periodic boundary conditions to the cubic box of length L. Hence, the momentum modes are quantized by the condition

$$\exp(ik_{\text{phys},i}L) = \exp(ik_i N) = 1, \tag{4.3}$$

and restricted to an interval of length $2\pi/a$, referred to as the first Brillouin zone (B_Z). A standard choice of the B_Z is

$$B_Z \equiv \left[-\frac{\pi}{a} < k_{\text{phys},i} \le \frac{\pi}{a} \right], \tag{4.4}$$

which contains the quantized values of $\mathbf{k}_{\text{phys}}$, and the domain of momentum integration. In the B_Z, the components $\hat{k}_i$ of Eq. (4.2) assume the values

$$\hat{k}_i = -\frac{N}{2} + 1, \ldots, -1, 0, 1, \ldots, \frac{N}{2} - 1, \frac{N}{2}, \tag{4.5}$$

for even N, such that the mode π/a on the zone boundary is identified with the mode $-\pi/a$. Similarly, we have

$$\hat{k}_i = -\frac{(N-1)}{2}, \ldots, -1, 0, 1, \ldots, \frac{N-1}{2}, \tag{4.6}$$

for odd N. The momenta become continuous in the infinite-volume limit, but remain restricted to the B_Z. A more flexible representation of the lattice momenta can be achieved using so-called twisted boundary conditions, as worked out for NLEFT in

Ref. [1]. For creation and annihilation operators along with Pauli matrices acting in spin and isospin space, we use the notation established in Eq. (3.76).

For NLEFT, we shall define different levels of improvement [2, 3] for the derivatives that appear in the nucleon kinetic energy operators, in the pion-nucleon coupling, in the momentum-dependent NLO contact terms and elsewhere. By improvement, we mean the following: Any derivative or operator has a simplest representation in terms of the lattice variables, e.g. a finite difference which involves just neighboring lattice points. However, this simplest representation is afflicted with the largest discretization errors, and thus it is expected to be improved by invoking further lattice sites. Depending on the operator considered, the corresponding improved representation involves various powers of the lattice spacing a. Thus, at higher levels of improvement, we expect the discretization effects due to the lattice to be increasingly suppressed, at the price of greater computational complexity. We introduce the improvement coefficients $\theta_{v,j}$ through

$$k_l \equiv \sum_{j=1}^{v+1} (-1)^{j+1} \theta_{v,j} \sin(jk_l) + \mathcal{O}(k^{2v+2}), \tag{4.7}$$

and the $\omega_{v,j}$ according to

$$\frac{k_l^2}{2} \equiv \sum_{j=0}^{v+1} (-1)^j \omega_{v,j} \cos(jk_l) + \mathcal{O}(k^{2v+2}), \tag{4.8}$$

where v denotes the level of improvement. For instance, $v = 0$ (no improvement) corresponds to

$$\theta_{0,1} = 1, \quad \omega_{0,0} = 1, \quad \omega_{0,1} = 1, \tag{4.9}$$

while for improved lattice actions, we use the terminology "$\mathcal{O}(a^{2v})$ improvement". Thus, for $v = 1$ or $\mathcal{O}(a^2)$ improvement, we have

$$\theta_{1,1} = \frac{4}{3}, \quad \theta_{1,2} = \frac{1}{6}, \tag{4.10}$$

and

$$\omega_{1,0} = \frac{5}{4}, \quad \omega_{1,1} = \frac{4}{3}, \quad \omega_{1,2} = \frac{1}{12}, \tag{4.11}$$

while for $v = 2$ or $\mathcal{O}(a^4)$ improvement, we find

$$\theta_{2,1} = \frac{3}{2}, \quad \theta_{2,2} = \frac{3}{10}, \quad \theta_{2,3} = \frac{1}{30}, \tag{4.12}$$

and

$$\omega_{2,0} = \frac{49}{36}, \quad \omega_{2,1} = \frac{3}{2}, \quad \omega_{2,2} = \frac{3}{20}, \quad \omega_{2,3} = \frac{1}{90}, \tag{4.13}$$

and so on.

Due to the derivative coupling of pions to nucleons, and the appearance of momentum-dependent operators at NLO and higher orders in NLEFT, we need to define corresponding derivative operators on the lattice. For instance, we define the spatial directional derivatives

$$\nabla_{l,(\nu)} f(\mathbf{n}) \equiv \frac{1}{2} \sum_{j=1}^{\nu+1} (-1)^{j+1} \theta_{\nu,j} \left[f(\mathbf{n} + j\hat{e}_l) - f(\mathbf{n} - j\hat{e}_l) \right], \tag{4.14}$$

where $f(\mathbf{n})$ is an arbitrary lattice function and the $\hat{e}_l$ denote unit vectors in lattice directions $l = 1, 2, 3$. As several different values of the improvement index ν are used for the nucleons, pions and auxiliary fields, ν will not be directly specified in most equations that involve derivatives. Higher derivatives may be defined by repeated application of Eq. (4.14), for instance

$$\nabla^2_{l,(\nu)} f(\mathbf{n}) \equiv \nabla_{l,(\nu)} \left[\nabla_{l,(\nu)} f(\mathbf{n}) \right], \tag{4.15}$$

where we note that such a definition of the second derivative would be undesirable for the nucleon kinetic energy term, as it has two zeros per Brillouin zone in each spatial lattice dimension. Hence, a fermion doubling problem would be introduced akin to that encountered in Lattice QCD and other lattice theories of relativistic chiral fermions. In order to avoid the doubling problem, we use the alternative definition

$$\tilde{\nabla}^2_{l,(\nu)} f(\mathbf{n}) \equiv - \sum_{j=0}^{\nu+1} (-1)^j \omega_{\nu,j} \left[f(\mathbf{n} + j\hat{e}_l) + f(\mathbf{n} - j\hat{e}_l) \right], \tag{4.16}$$

for second derivatives in the pion and nucleon kinetic energy terms. Nevertheless, the fact that $\nabla^2_{l,(\nu)}$ has two zeros in per Brillouin zone turns out to be a beneficial feature for the operators in the NLO action, which are treated perturbatively. Since the NLO coefficients are tuned at momenta less than half the cutoff momentum, this feature prevents the interactions from becoming too strong at momenta close to the cutoff scale.

As an alternative to Eq. (4.14), we can use the eight vertices of a unit cube on the lattice, which gives

$$\nabla_{l,(c)} f(\mathbf{n}) \equiv \frac{1}{4} \sum_{\mathbf{d}} (-1)^{d_l+1} f(\mathbf{n} + \mathbf{d}), \tag{4.17}$$

where we have used the notation

$$\mathbf{d} \equiv d_1\hat{e}_1 + d_2\hat{e}_2 + d_3\hat{e}_3, \qquad \sum_{\mathbf{d}} \equiv \sum_{d_1=0}^{1}\sum_{d_2=0}^{1}\sum_{d_3=0}^{1}. \qquad (4.18)$$

As we shall see in Chap. 5, the mixing angles in neutron-proton scattering exhibit significant lattice artifacts that can be traced to the breaking of rotational invariance by the lattice. While the "cubic derivative" of Eq. (4.17) originally supplanted the symmetric difference of Eq. (4.14), the former was ultimately found to confer only marginally improved discretization artifacts at the price of significantly complicating the formalism. The most recent NLEFT calculations therefore exclusively use Eq. (4.14).

4.1.1 Local Densities and Currents

We shall next define the local density operators needed for lattice chiral EFT. While these are similar to those encountered in the pionless EFT, the number of structures needed here is larger both due to the more involved expressions at LO, as well as the fact that we shall here consider operators up to NNLO in the EFT expansion.

In the Grassmann and transfer matrix formulations respectively, we have the local SU(4)-invariant densities (the role of the SU(4) symmetry was already discussed in Chap. 2)

$$\rho(\mathbf{n}, t) \equiv \sum_{i,j=0,1} \zeta_{i,j}^*(\mathbf{n}, t)\zeta_{i,j}(\mathbf{n}, t), \qquad (4.19)$$

or

$$\rho^{a^\dagger,a}(\mathbf{n}) \equiv \sum_{i,j=0,1} a_{i,j}^\dagger(\mathbf{n})a_{i,j}(\mathbf{n}), \qquad (4.20)$$

and we also define the local spin densities for $S = 1, 2, 3$ according to

$$\rho_S(\mathbf{n}, t) \equiv \sum_{i,j,i'=0,1} \zeta_{i,j}^*(\mathbf{n}, t)\left[\sigma_S\right]_{ii'}\zeta_{i,j}(\mathbf{n}, t), \qquad (4.21)$$

or

$$\rho_S^{a^\dagger,a}(\mathbf{n}) \equiv \sum_{i,j,i'=0,1} a_{i,j}^\dagger(\mathbf{n})\left[\sigma_S\right]_{ii'}a_{i',j}(\mathbf{n}), \qquad (4.22)$$

followed by the isospin densities for $I = 1, 2, 3$ given by

$$\rho_I(\mathbf{n}, t) \equiv \sum_{i,j,j'=0,1} \zeta_{i,j}^*(\mathbf{n}, t) \left[\tau_I\right]_{jj'} \zeta_{i,j}(\mathbf{n}, t), \tag{4.23}$$

or

$$\rho_I^{a^\dagger,a}(\mathbf{n}) \equiv \sum_{i,j,j'=0,1} a_{i,j}^\dagger(\mathbf{n}) \left[\tau_I\right]_{jj'} a_{i',j}(\mathbf{n}), \tag{4.24}$$

and finally the spin-isospin densities

$$\rho_{S,I}(\mathbf{n}, t) \equiv \sum_{i,j,i',j'=0,1} \zeta_{i,j}^*(\mathbf{n}, t) \left[\sigma_S\right]_{ii'} \left[\tau_I\right]_{jj'} \zeta_{i,j}(\mathbf{n}, t), \tag{4.25}$$

and

$$\rho_{S,I}^{a^\dagger,a}(\mathbf{n}) \equiv \sum_{i,j,i',j'=0,1} a_{i,j}^\dagger(\mathbf{n}) \left[\sigma_S\right]_{ii'} \left[\tau_I\right]_{jj'} a_{i',j'}(\mathbf{n}). \tag{4.26}$$

For each static density, we also introduce an associated current density operator. These first appear at NLO when we construct all lattice chiral EFT operators with two powers of momenta. As the current operators involve lattice derivatives, we need to specify the level of improvement in each case. Here, we shall merely introduce the operator structures and leave this detail unspecified. We first define the lth component of the SU(4)-invariant current density as

$$\Pi_l^{a^\dagger,a}(\mathbf{n}) \equiv \sum_{i,j=0,1} a_{i,j}^\dagger(\mathbf{n}) \nabla_{l,(v)} a_{i,j}(\mathbf{n}) - \sum_{i,j=0,1} \left[\nabla_{l,(v)} a_{i,j}^\dagger(\mathbf{n})\right] a_{i,j}(\mathbf{n}), \tag{4.27}$$

or

$$\Pi_l^{a^\dagger,a}(\mathbf{n}) \equiv \frac{1}{4} \sum_{\mathbf{d}} \sum_{i,j=0,1} (-1)^{d_l+1} a_{i,j}^\dagger(\mathbf{n} + \mathbf{d}(-l)) a_{i,j}(\mathbf{n} + \mathbf{d}), \tag{4.28}$$

where we have defined the result of reflecting the lth component of $\mathbf{d}$ about the center of the unit lattice cube in terms of

$$\mathbf{d}(-l) \equiv \mathbf{d} + (1 - 2d_l)\hat{e}_l. \tag{4.29}$$

We then also need the lth component of the spin current density

$$\Pi_{l,S}^{a^\dagger,a}(\mathbf{n}) \equiv \sum_{i,j,i'=0,1} a_{i,j}^\dagger(\mathbf{n}) \left[\sigma_S\right]_{ii'} \nabla_{l,(v)} a_{i',j}(\mathbf{n})$$

$$- \sum_{i,j,i'=0,1} \left[\nabla_{l,(v)} a_{i,j}^\dagger(\mathbf{n})\right] \left[\sigma_S\right]_{ii'} a_{i',j}(\mathbf{n}), \tag{4.30}$$

or

$$\Pi_{l,S}^{a^\dagger,a}(\mathbf{n}) \equiv \frac{1}{4} \sum_{\mathbf{d}} \sum_{i,j,i'=0,1} (-1)^{d_l+1} a_{i,j}^\dagger(\mathbf{n}+\mathbf{d}(-l)) \left[\sigma_S\right]_{ii'} a_{i',j}(\mathbf{n}+\mathbf{d}), \quad (4.31)$$

and the lth component of the isospin current density

$$\Pi_{l,I}^{a^\dagger,a}(\mathbf{n}) \equiv \sum_{i,j,j'=0,1} a_{i,j}^\dagger(\mathbf{n}) \left[\tau_I\right]_{jj'} \nabla_{l,(v)} a_{i,j'}(\mathbf{n})$$

$$- \sum_{i,j,j'=0,1} \left[\nabla_{l,(v)} a_{i,j}^\dagger(\mathbf{n})\right] \left[\tau_I\right]_{jj'} a_{i,j'}(\mathbf{n}), \quad (4.32)$$

and finally the lth component of the spin-isospin current density

$$\Pi_{l,S,I}^{a^\dagger,a}(\mathbf{n}) \equiv \sum_{i,j,i',j'=0,1} a_{i,j}^\dagger(\mathbf{n}) \left[\sigma_S\right]_{ii'} \left[\tau_I\right]_{jj'} \nabla_{l,(v)} a_{i',j'}(\mathbf{n})$$

$$- \sum_{i,j,i',j'=0,1} \left[\nabla_{l,(v)} a_{i,j}^\dagger(\mathbf{n})\right] \left[\sigma_S\right]_{ii'} \left[\tau_I\right]_{jj'} a_{i',j'}(\mathbf{n}), \quad (4.33)$$

were the equivalent expressions using $\nabla_{l,(c)}$ can be straightforwardly inferred from that of the spin-current density. We recall that the "cubic derivative" was ultimately abandoned in favor of improved directional derivatives, and is included here for the sake of completeness.

Armed with these notational building blocks, we are now in a position to formulate lattice chiral EFT up to NNLO. For the LO results, we shall use both the Grassmann and transfer matrix formulations. However, the NLO and NNLO contributions will be treated in the transfer matrix formulation only, as that formulation is the one which is ultimately used in practical calculations.

4.1.2 Free Nucleon Action

We are now in a position to discuss the various contributions to the NLEFT action, and we shall begin with the free (or kinetic energy) component. As for the pionless

EFT, the simplest choice for the kinetic energy term of the nucleons is the action

$$S_{\tilde{N}N}(\zeta^*, \zeta) = \sum_{\mathbf{n},t,i,j} \zeta_{i,j}^*(\mathbf{n}, t)\left[\zeta_{i,j}(\mathbf{n}, t+1) - \zeta_{i,j}(\mathbf{n}, t)\right]$$

$$+6h \sum_{\mathbf{n},t,i,j} \zeta_{i,j}^*(\mathbf{n}, t)\zeta_{i,j}(\mathbf{n}, t) \tag{4.34}$$

$$-h \sum_{\mathbf{n},t,i,j} \sum_{l=1}^{3} \zeta_{i,j}^*(\mathbf{n}, t)\left[\zeta_{i,j}(\mathbf{n}+\hat{e}_l, t) + \zeta_{i,j}(\mathbf{n}-\hat{e}_l, t)\right],$$

with

$$h \equiv \frac{\alpha_t}{2\tilde{m}_N}, \tag{4.35}$$

where $\tilde{m}_N \equiv m_N a$ is the nucleon mass, expressed in units of the (spatial) lattice spacing. We also consider the $\mathcal{O}(a^4)$ improved action

$$S_{\tilde{N}N}(\zeta^*, \zeta) = \sum_{\mathbf{n},t,i,j} \zeta_{i,j}^*(\mathbf{n}, t)\left[\zeta_{i,j}(\mathbf{n}, t+1) - \zeta_{i,j}(\mathbf{n}, t)\right]$$

$$+\frac{49h}{6} \sum_{\mathbf{n},t,i,j} \zeta_{i,j}^*(\mathbf{n}, t)\zeta_{i,j}(\mathbf{n}, t)$$

$$-\frac{3h}{2} \sum_{\mathbf{n},t,i,j} \sum_{l=1}^{3} \zeta_{i,j}^*(\mathbf{n}, t)\left[\zeta_{i,j}(\mathbf{n}+\hat{e}_l, t) + \zeta_{i,j}(\mathbf{n}-\hat{e}_l, t)\right]$$

$$+\frac{3h}{20} \sum_{\mathbf{n},t,i,j} \sum_{l=1}^{3} \zeta_{i,j}^*(\mathbf{n}, t)\left[\zeta_{i,j}(\mathbf{n}+2\hat{e}_l, t) + \zeta_{i,j}(\mathbf{n}-2\hat{e}_l, t)\right]$$

$$-\frac{h}{90} \sum_{\mathbf{n},t,i,j} \sum_{l=1}^{3} \zeta_{i,j}^*(\mathbf{n}, t)\left[\zeta_{i,j}(\mathbf{n}+3\hat{e}_l, t) + \zeta_{i,j}(\mathbf{n}-3\hat{e}_l, t)\right],$$

$$\tag{4.36}$$

which includes hopping terms up to third-nearest-neighbors in the spatial dimensions of the lattice. We expect such an improved lattice dispersion relation to be beneficial when scattering phase shifts are determined on the lattice, which we shall discuss in Chap. 5. In the transfer matrix formulation, we find that $\mathcal{O}(a^4)$ improvement of the dispersion relation corresponds to a lattice Hamiltonian with the kinetic energy term given by

$$\tilde{H}_{\text{free}} = \frac{49}{12\tilde{m}_N} \sum_{n,i,j} a_{i,j}^\dagger(\mathbf{n}) a_{i,j}(\mathbf{n})$$

$$- \frac{3}{4\tilde{m}_N} \sum_{\mathbf{n},i,j} \sum_{l=1}^{3} \left[a_{i,j}^\dagger(\mathbf{n}) a_{i,j}(\mathbf{n}+\hat{e}_l) + a_{i,j}^\dagger(\mathbf{n}) a_{i,j}(\mathbf{n}-\hat{e}_l) \right]$$

$$+ \frac{3}{40\tilde{m}_N} \sum_{\mathbf{n},i,j} \sum_{l=1}^{3} \left[a_{i,j}^\dagger(\mathbf{n}) a_{i,j}(\mathbf{n}+2\hat{e}_l) + a_{i,j}^\dagger(\mathbf{n}) a_{i,j}(\mathbf{n}-2\hat{e}_l) \right]$$

$$- \frac{1}{180\tilde{m}_N} \sum_{\mathbf{n},i,j} \sum_{l=1}^{3} \left[a_{i,j}^\dagger(\mathbf{n}) a_{i,j}(\mathbf{n}+3\hat{e}_l) + a_{i,j}^\dagger(\mathbf{n}) a_{i,j}(\mathbf{n}-3\hat{e}_l) \right]. \quad (4.37)$$

The standard lattice dispersion relation, which corresponds to the lowest-order free nucleon lattice Hamiltonian, is given by

$$\omega(\mathbf{p}) = \frac{1}{\tilde{m}_N} \sum_{l=1}^{3} \left(1 - \cos p_l \right), \quad (4.38)$$

and in general terms, we define

$$\omega^{(\nu)}(\mathbf{p}) \equiv \frac{1}{\tilde{m}_N} \sum_{j=0}^{\nu+1} \sum_{l=1}^{3} (-1)^j \omega_{\nu,j} \cos(j p_l), \quad (4.39)$$

such that Eq. (4.38) is recovered for $\nu = 0$ with the coefficients $\omega_{0,0} = 1$ and $\omega_{0,1} = 1$. Similarly, the dispersion relation of Eq. (4.37) is found for $\omega_{2,0} = 49/36$, $\omega_{2,1} = 3/2$, $\omega_{2,2} = 3/20$ and $\omega_{2,3} = 1/90$.

While such improved actions are clearly superior to the standard (unimproved) action, the effects of lattice artifacts can be further suppressed without introducing additional non-locality. Instead of requiring the lattice dispersion relation to be accurate to a given order, we may tune the dispersion relation for a given ν to optimally match the continuum one in the first Brillouin zone. One possibility is then the "well-tempered" lattice action [4], defined by

$$\omega_{\text{wt}}^{(\nu)}(\mathbf{p}) = \omega^{(\nu-1)}(\mathbf{p}) + s_w \left[\omega^{(\nu)}(\mathbf{p}) - \omega^{(\nu-1)}(\mathbf{p}) \right], \quad (4.40)$$

where the coefficient s_w is fixed using the integral relation

$$\prod_{l=1}^{3} \int_{-\pi/a}^{+\pi/a} dp_l \, \omega_{\text{wt}}^{(\nu)}(\mathbf{p}) = \prod_{l=1}^{3} \int_{-\pi/a}^{+\pi/a} dp_l \, \frac{\mathbf{p}^2}{2\tilde{m}_N}, \quad (4.41)$$

from which we find that the well-tempered action in general provides better agreement with the continuum form at large momenta. In practice, well-tempered actions exhibit a smaller decrease at high momenta, at the price of slightly overshooting the continuum dispersion relation at low momenta. Another class of improved actions is given by the "stretched" actions, defined in similar terms by

$$\omega_{\mathrm{st}}^{(\nu)}(\mathbf{p}) \equiv \omega^{(\nu)}(\mathbf{p}) + s_t \left[\omega^{(\nu)}(\mathbf{p}) - \omega^{(\nu-1)}(\mathbf{p}) \right], \tag{4.42}$$

where the coefficient s_t can be tuned to minimize the effect of lattice artifacts on the two-nucleon phase shifts and mixing angles. While most NLEFT calculations use the stretched $\mathcal{O}(a^4)$-improved action with $s_t = 10$, it should be noted that the choice of improvement is usually based on trial and error, with the aim of optimizing agreement with experiment at a low order in the EFT.

4.1.3 Inclusion of Pions

A major complication in lattice chiral EFT as compared to the cases of pionless EFT and the EFT for cold atoms is the appearance of OPE at LO. As this contribution amounts to an instantaneous (static) potential [5–9], we take this into account by omitting the time derivatives in the free pion lattice action. Thus, the pion field does not propagate in time and does not couple to physical pions. Rather, our "pion field" is essentially an auxiliary field which can be integrated out in order to reproduce the original OPE potential. A distinct advantage of this formalism is that the nucleon self-energy vanishes and that the nucleon mass is not renormalized by interaction effects. Note that to include interactions with physical low-energy pions, one simply inserts the corresponding operators with external pion fields.

The action for free pions in NLEFT is thus

$$S_{\pi\pi}(\pi_I) \equiv \frac{\alpha_t M_\pi^2}{2} \sum_{I=1}^{3} \sum_{\mathbf{n},t} \pi_I(\mathbf{n},t)\pi_I(\mathbf{n},t) - \frac{\alpha_t}{2} \sum_{I=1}^{3} \sum_{\mathbf{n},t} \sum_{l=1}^{3} \pi_I(\mathbf{n},t)\tilde{\nabla}_{l,(\nu)}^2 \pi_I(\mathbf{n},t),$$

$$\tag{4.43}$$

where π_I is the pion field with isospin index $I = 1, 2, 3$. We note that the pion fields at different Euclidean time slices are decoupled, which is a consequence of the absence of time derivatives in the instantaneous formulation of $S_{\pi\pi}$. This has the effect of generating instantaneous propagation at each time slice when OPE contributions are computed. By means of the improvement coefficients established previously, we shall write this explicitly for $\nu = 2$, giving

$$S_{\pi\pi}(\pi_I) = \alpha_t \left(\frac{M_\pi^2}{2} + 3\omega_{2,0}\right) \sum_{I=1}^{3} \sum_{\mathbf{n},t} \pi_I(\mathbf{n},t)\pi_I(\mathbf{n},t)$$

$$- \alpha_t\omega_{2,1} \sum_{I=1}^{3} \sum_{\mathbf{n},t} \sum_{l=1}^{3} \pi_I(\mathbf{n},t)\pi_I(\mathbf{n}+\hat{e}_l,t)$$

$$+ \alpha_t\omega_{2,2} \sum_{I=1}^{3} \sum_{\mathbf{n},t} \sum_{l=1}^{3} \pi_I(\mathbf{n},t)\pi_I(\mathbf{n}+2\hat{e}_l,t)$$

$$- \alpha_t\omega_{2,3} \sum_{I=1}^{3} \sum_{\mathbf{n},t} \sum_{l=1}^{3} \pi_I(\mathbf{n},t)\pi_I(\mathbf{n}+3\hat{e}_l,t), \qquad (4.44)$$

where we have exploited translational invariance in order to write everything in terms of forward differences. If we Fourier transform Eq. (4.44) into momentum space, we find

$$S_{\pi\pi}(\pi_I) = \frac{1}{N^3} \sum_{I=1}^{3} \sum_{\mathbf{q},t} \pi_I(-\mathbf{q},t)\pi_I(\mathbf{q},t) \qquad (4.45)$$

$$\times \left[\frac{\alpha_t M_\pi^2}{2} + \alpha_t \sum_{l=1}^{3} \left(\omega_{2,0} - \omega_{2,1}\cos q_l + \omega_{2,2}\cos(2q_l)\right.\right.$$

$$\left.\left. -\omega_{2,3}\cos(3q_l)\right)\right],$$

for the improved free pion action.

Before we proceed further, we shall introduce a slight modification of Eq. (4.44) by expressing the action $S_{\pi\pi}$ in terms of a rescaled pion field π_I', according to

$$\pi_I'(\mathbf{n},t) \equiv \sqrt{q_\pi}\,\pi_I(\mathbf{n},t), \qquad (4.46)$$

$$q_\pi \equiv \alpha_t(M_\pi^2 + 6\omega_{2,0}), \qquad (4.47)$$

where the exact expression for q_π depends on the order of improvement. In terms of these new variables

$$S_{\pi\pi}(\pi_I') = \frac{1}{2} \sum_{I=1}^{3} \sum_{\mathbf{n},t} \pi_I'(\mathbf{n},t)\pi_I'(\mathbf{n},t)$$

$$- \frac{\alpha_t\omega_{2,1}}{q_\pi} \sum_{I=1}^{3} \sum_{\mathbf{n},t} \sum_{l=1}^{3} \pi_I'(\mathbf{n},t)\pi_I'(\mathbf{n}+\hat{e}_l,t)$$

$$+ \frac{\alpha_t \omega_{2,2}}{q_\pi} \sum_{I=1}^{3} \sum_{\mathbf{n},t} \sum_{l=1}^{3} \pi_I'(\mathbf{n}, t) \pi_I'(\mathbf{n} + 2\hat{e}_l, t)$$

$$- \frac{\alpha_t \omega_{2,3}}{q_\pi} \sum_{I=1}^{3} \sum_{\mathbf{n},t} \sum_{l=1}^{3} \pi_I'(\mathbf{n}, t) \pi_I'(\mathbf{n} + 3\hat{e}_l, t), \qquad (4.48)$$

and if we Fourier transform Eq. (4.48) to momentum space, we find

$$S_{\pi\pi}(\pi_I') = \frac{1}{2N^3} \sum_{I=1}^{3} \sum_{\mathbf{q},t} \pi_I'(-\mathbf{q}, t) \pi_I'(\mathbf{q}, t) \qquad (4.49)$$

$$\times \left[1 + \frac{2\alpha_t}{q_\pi} \sum_{l=1}^{3} \left(-\omega_{2,1} \cos q_l + \omega_{2,2} \cos(2q_l) - \omega_{2,3} \cos(3q_l) \right) \right],$$

$$\equiv \frac{1}{2N^3} \sum_{I=1}^{3} \sum_{\mathbf{q},t} \pi_I'(-\mathbf{q}, t) \, D_\pi^{-1}(\mathbf{q}) \, \pi_I'(\mathbf{q}, t), \qquad (4.50)$$

where

$$D_\pi(\mathbf{q}) \equiv \left[1 + \frac{2\alpha_t}{q_\pi} \sum_{l=1}^{3} \left(-\omega_{2,1} \cos q_l + \omega_{2,2} \cos(2q_l) - \omega_{2,3} \cos(3q_l) \right) \right]^{-1},$$

$$(4.51)$$

is the propagator for (rescaled) pions. It should be understood that the sole purpose of rescaling the pion field is to obtain shorter and more convenient expressions for the construction of a MC algorithm. Using Eq. (4.51), we immediately find the instantaneous pion two-point function at spatial separation $\mathbf{n}$,

$$\langle \pi_I'(\mathbf{n}, t) \pi_I'(\mathbf{0}, t) \rangle = \frac{\int \mathcal{D}\pi_I' \, \pi_I'(\mathbf{n}, t) \pi_I'(\mathbf{0}, t) \, \exp\left(-S_{\pi\pi}\right)}{\int \mathcal{D}\pi_I' \, \exp\left(-S_{\pi\pi}\right)}$$

$$= \frac{1}{N^3} \sum_{\mathbf{q}} \exp\left(-i\,\mathbf{q} \cdot \mathbf{n}\right) D_\pi(\mathbf{q}), \qquad (4.52)$$

where no summation over the isospin index I is implied. Due to the gradient coupling of pions to the nucleons, we anticipate the need for a "two-derivative" pion correlator, which we define as

$$\left\langle \left[\nabla_{S_1,(v)} \pi_I'(\mathbf{n}, t) \right] \nabla_{S_2,(v)} \pi_I'(\mathbf{0}, t) \right\rangle \equiv G_{S_1 S_2}^{(v)}(\mathbf{n}, M_\pi), \qquad (4.53)$$

where again no summation over the isospin index I is implied. The level of improvement of the gradient operators should also be specified, and need in general not coincide with that of Eq. (4.51). We shall elaborate further on this point when we discuss the OPE potential and the perturbative improvement of the derivative couplings. We may also express the free pion action in the form

$$S_{\pi\pi}(\pi'_I) = \frac{1}{2} \sum_{I=1}^{3} \sum_{\mathbf{n},t} \pi'_I(\mathbf{n}, t)\, D_\pi^{-1}(\mathbf{n} - \mathbf{n}')\, \pi'_I(\mathbf{n}', t), \qquad (4.54)$$

where

$$D_\pi^{-1}(\mathbf{n} - \mathbf{n}') \equiv \frac{1}{N^3} \sum_{\mathbf{q}} \frac{\exp(-i\mathbf{q}\cdot\mathbf{n})\exp(i\mathbf{q}\cdot\mathbf{n}')}{D_\pi(\mathbf{q})}, \qquad (4.55)$$

using our expression for the pion propagator.

So far, we have considered all the pions to have a common mass M_π, and we shall now briefly outline the modifications that result from having separate masses $M_{\pi\pm}$ and M_{π^0} for the charged and neutral pions, respectively. We again define rescaled pion fields according to

$$\pi'_{1,2}(\mathbf{n}, t) \equiv \sqrt{q_{\pi\pm}}\, \pi_{1,2}(\mathbf{n}, t), \quad \pi'_3(\mathbf{n}, t) \equiv \sqrt{q_{\pi^0}}\, \pi_3(\mathbf{n}, t), \qquad (4.56)$$

and

$$q_{\pi\pm} \equiv \alpha_t(M_{\pi\pm}^2 + 6\omega_{2,0}), \quad q_{\pi^0} \equiv \alpha_t(M_{\pi^0}^2 + 6\omega_{2,0}), \qquad (4.57)$$

in terms of which we find

$$D_{\pi\pm}(\mathbf{q}) \equiv \left[1 + \frac{2\alpha_t}{q_{\pi\pm}} \sum_{l=1}^{3} \left(-\omega_{2,1}\cos q_l + \omega_{2,2}\cos(2q_l) - \omega_{2,3}\cos(3q_l) \right) \right]^{-1}, \tag{4.58}$$

and

$$D_{\pi^0}(\mathbf{q}) \equiv \left[1 + \frac{2\alpha_t}{q_{\pi^0}} \sum_{l=1}^{3} \left(-\omega_{2,1}\cos q_l + \omega_{2,2}\cos(2q_l) - \omega_{2,3}\cos(3q_l) \right) \right]^{-1}, \tag{4.59}$$

for the propagators of the charged and neutral pions, respectively.

We shall next consider the treatment of the pion-nucleon coupling in lattice chiral EFT. In the Grassmann formulation, we define the pion-nucleon lattice action according to

$$S_{\pi \bar{N} N}(\pi_I', \zeta^*, \zeta) \equiv \frac{g_A \alpha_t}{2 F_\pi \sqrt{q_\pi}} \sum_{I,S=1}^{3} \sum_{\mathbf{n},t} \left[\nabla_{S,(v)} \pi_I'(\mathbf{n}, t) \right] \rho_{S,I}(\mathbf{n}, t), \qquad (4.60)$$

where the spin-isospin density is defined in Eq. (4.25), and the derivative acts on the pion field. Also, $F_\pi \simeq 92\,\mathrm{MeV}$ is the weak pion decay constant and $g_A \simeq 1.27$ the axial coupling of the nucleon. Again, we ignore the fact that these quantities are actually given in the chiral limit. The equivalent expression for the pion-nucleon lattice action in the transfer matrix formulation can be obtained by substituting

$$\sum_{\mathbf{n},t} \left[\nabla_{S,(v)} \pi_I'(\mathbf{n}, t) \right] \rho_{S,I}(\mathbf{n}, t) \to \sum_{\mathbf{n}} \left[\nabla_{S,(v)} \pi_I'(\mathbf{n}, t) \right] \rho_{S,I}^{a^\dagger, a}(\mathbf{n}). \qquad (4.61)$$

Before we move on to the description of the nucleon-nucleon contact interactions at LO in NLEFT, we note that we shall only consider transfer matrices with explicit pion fields in this chapter. The case where the pion field is integrated out is of significance for the treatment of the two- and three-nucleon problems, and will be presented in Chap. 5. We also note that in present NLEFT calculations, the free pion action is treated similarly to the free nucleon action, i.e. with $v = 2$ and a "stretching factor" $s_t = 10$. On the other hand, the gradient coupling of the pions is treated at LO with $v = 0$, with $\mathcal{O}(a^2)$ improvement included perturbatively at NLO. It should also be noted that the "cubic derivative" has been used for the gradient coupling of the pions in earlier lattice chiral EFT work, and details of that formalism can found in the corresponding publications.

4.2 The LO Action and Transfer Matrix

At LO in NLEFT, we have two contact interactions similar to those of pionless EFT. These are

$$S_{\bar{N} N \bar{N} N}(\zeta^*, \zeta) \equiv \frac{\tilde{C} \alpha_t}{2} \sum_{\mathbf{n},t} \left[\rho(\mathbf{n}, t) \right]^2 + \frac{\tilde{C}_{I^2} \alpha_t}{2} \sum_{I=1}^{3} \sum_{\mathbf{n},t} \left[\rho_I(\mathbf{n}, t) \right]^2, \qquad (4.62)$$

where $\rho(\mathbf{n}, t)$ and $\rho_I(\mathbf{n}, t)$ are the SU(4)-symmetric and isospin densities defined in Eqs. (4.19) and (4.23), respectively. As in the case of pionless EFT, we introduce auxiliary fields using the HS transformations

$$\exp\left(-\frac{\tilde{C} \alpha_t}{2} \left[\rho(\mathbf{n}, t) \right]^2 \right) = \frac{1}{\sqrt{2\pi}} \int_{-\infty}^{+\infty} ds \, \exp\left(-\frac{s^2}{2} + \sqrt{-\tilde{C} \alpha_t} \, \rho(\mathbf{n}, t) \, s \right),$$

$$\qquad (4.63)$$

and

$$\exp\left(-\frac{\tilde{C}_{I^2}\alpha_t}{2}\sum_{I=1}^{3}[\rho_I(\mathbf{n}, t)]^2\right) = \int\left(\prod_{I=1}^{3}\frac{ds_I}{\sqrt{2\pi}}\right) \tag{4.64}$$

$$\times \exp\left(-\sum_{I=1}^{3}\frac{s_I^2}{2} + i\sqrt{\tilde{C}_{I^2}\alpha_t}\sum_{I=1}^{3}\rho_I(\mathbf{n}, t)s_I\right),$$

at each point on the lattice. Note that the physical values of the LO operator coefficients are such that the factors inside the square roots are positive, see also Sect. 3.4. This allows us to define the auxiliary field actions

$$S_{ss}(s, s_I) \equiv \frac{1}{2}\sum_{\mathbf{n},t}s^2(\mathbf{n}, t) + \frac{1}{2}\sum_{I=1}^{3}\sum_{\mathbf{n},t}s_I^2(\mathbf{n}, t), \tag{4.65}$$

and

$$S_{s\tilde{N}N}(s, s_I, \zeta^*, \zeta) \equiv -\sqrt{-\tilde{C}\alpha_t}\sum_{\mathbf{n},t}\rho(\mathbf{n}, t)s(\mathbf{n}, t)$$

$$-i\sqrt{\tilde{C}_{I^2}\alpha_t}\sum_{I=1}^{3}\sum_{\mathbf{n},t}\rho_I(\mathbf{n}, t)s_I(\mathbf{n}, t), \tag{4.66}$$

in terms of which the original contact interaction terms without auxiliary fields are given by

$$\exp\left(-S_{\tilde{N}N\tilde{N}N}(\zeta^*, \zeta)\right) = \int \mathcal{D}s\,\mathcal{D}s_I\,\exp\left(-S_{ss}(s, s_I) - S_{s\tilde{N}N}(s, s_I, \zeta^*, \zeta)\right), \tag{4.67}$$

where

$$\int \mathcal{D}s\mathcal{D}s_I \equiv \int \prod_{\mathbf{n},t}\left[\frac{ds(\mathbf{n}, t)}{\sqrt{2\pi}}\right]\prod_{I=1}^{3}\prod_{\mathbf{n},t}\left[\frac{ds_I(\mathbf{n}, t)}{\sqrt{2\pi}}\right], \tag{4.68}$$

is the functional integration measure. If we combine all the components, we find

$$\mathcal{Z}_{LO} = \int \mathcal{D}\zeta\,\mathcal{D}\zeta^*\,\mathcal{D}\pi_I'\,\mathcal{D}s\,\mathcal{D}s_I\,\exp\left(-S_{LO}(\pi_I', s, s_I, \zeta^*, \zeta)\right), \tag{4.69}$$

where

$$\int \mathcal{D}\pi_I'\mathcal{D}s\mathcal{D}s_I \equiv \int \prod_{\mathbf{n},t}\left[\frac{ds(\mathbf{n}, t)}{\sqrt{2\pi}}\right]\prod_{I=1}^{3}\prod_{\mathbf{n},t}\left[\frac{ds_I(\mathbf{n}, t)d\pi_I'(\mathbf{n}, t)}{2\pi}\right], \tag{4.70}$$

such that

$$S_{LO}(\pi'_I, s, s_I, \zeta^*, \zeta) = S_{\bar{N}N}(\zeta^*, \zeta) + S_{\pi\pi}(\pi'_I) + S_{\pi\bar{N}N}(\pi'_I, \zeta^*, \zeta)$$
$$+ S_{ss}(s, s_I) + S_{s\bar{N}N}(s, s_I, \zeta^*, \zeta), \qquad (4.71)$$

summarizes the LO action. Here, we have explicitly displayed for each component of the action the dependence on the pion fields, the auxiliary fields and the Grassmann fields.

In the transfer matrix formulation with auxiliary fields, we find for the partition function

$$\mathscr{Z}_{LO_1} = \int \mathscr{D}\pi'_I \mathscr{D}s \, \mathscr{D}s_I \, \exp\left(-S_{\pi\pi}(\pi'_I) - S_{ss}(s, s_I)\right)$$

$$\times \mathrm{Tr}\left\{ M_{LO_1}(\pi'_I, s, s_I, N_t - 1) \cdots M_{LO_1}(\pi'_I, s, s_I, 0) \right\}, \qquad (4.72)$$

where the transfer matrix operator is

$$M_{LO_1}(\pi'_I, s, s_I, t) = : \exp\left(-\tilde{H}_{\text{free}}\alpha_t + \sqrt{-\tilde{C}\alpha_t} \sum_{\mathbf{n}} s(\mathbf{n}, t)\, \rho^{a^\dagger, a}(\mathbf{n}) \right.$$

$$+ i\sqrt{\tilde{C}_{I^2}\alpha_t} \sum_{I=1}^{3} \sum_{\mathbf{n}} s_I(\mathbf{n}, t)\, \rho_I^{a^\dagger, a}(\mathbf{n})$$

$$\left. - \frac{g_A \alpha_t}{2F_\pi \sqrt{q_\pi}} \sum_{I,S=1}^{3} \sum_{\mathbf{n}} \left[\nabla_{S,(v)} \pi'_I(\mathbf{n}, t) \right] \rho_{S,I}^{a^\dagger, a}(\mathbf{n}) \right) :,$$

$$(4.73)$$

and the nucleon kinetic energy $\tilde{H}_{\text{free}}$ is given, for example, by the $\mathcal{O}(a^4)$-improved expression with $v = 2$.

4.2.1 Zero-Range Clustering Instability

Before we conclude our treatment of NLEFT at LO, we shall discuss the level of accuracy that can be expected from what has been presented so far. This will involve a discussion of results obtained for the two-nucleon system and light nuclei using the methods that will be presented in Chaps. 5 and 6, but since these will serve as justification for the further refinement of the LO action, we shall present the main conclusions here. As will be discussed in Chap. 5, we already encounter some problems when studying neutron-proton scattering at LO in chiral EFT. While these problems are likely to be rectified at higher orders in the EFT expansion, we may

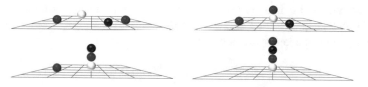

Fig. 4.1 Sketch of the possible configurations of nucleons on the lattice. The different colors of the particles represent spin and isospin degrees of freedom. Figure courtesy of Dean Lee

press on and also consider systems with more than two nucleons. This is possible since the analysis of the two-nucleon system allows us to fit the coefficients $\tilde{C}$ and $\tilde{C}_{I2}$ of the contact terms in the LO action, once the lattice spatial and temporal lattice spacings a and δ have been specified.

It is important to take a closer look at the possible configuration that nucleons can take on the lattice. This is exemplified in Fig. 4.1 for the case of four nucleons for a two-dimensional spatial lattice. All nucleons can be on different lattice sites, but one can also have doublets, triplets and even quartets (one proton with spin up, one proton with spin down, one neutron with spin up and one neutron with spin down). These are all types of configurations in harmony with the Pauli principle. This will be of importance when we discuss the clustering in nuclear physics, that can nicely be investigated with the methods of NLEFT. Let us return to the discussion of the LO action. In fact, one finds that the momentum-independent contact interactions in NLEFT at LO, when fitted to the nucleon-nucleon S-wave scattering lengths, do not provide a very realistic description of the ^{4}He system, the alpha particle. In particular, we find that the ground state is significantly overbound and consists almost exclusively of the quantum state with all four nucleons occupying the same lattice site. Such a "clustering instability" can be understood as the result of two contributing factors. Firstly, NLEFT at LO gives a rather poor description of S-wave scattering above a center-of-mass momentum of $\simeq$50 MeV. Effectively, the momentum-independent LO contact interactions are too strong at high momenta. Secondly, a combinatorial enhancement of the contact interactions occurs when more than two nucleons occupy the same lattice site. An analogous effect has been noted and studied in two-dimensional large-N droplets with zero-range attraction [10], and in systems of higher-spin fermions in optical traps and lattices [11, 12].

In order to illustrate how the clustering instability arises, we first consider the situation at LO in the SU(4)-symmetric pionless EFT using a Hamiltonian lattice formalism. Let $E_1^{\text{localized}}$ denote the expectation value of the kinetic energy of a single nucleon localized on a single lattice site. Further, let $V_2 < 0$ denote the potential energy between two nucleons occupying the same lattice site. If the two-nucleon scattering length has been fixed, then both $E_1^{\text{localized}}$ and V_2 scale quadratically with the momentum cutoff $\Lambda \equiv \pi/a$, giving

$$E_1^{\text{localized}} \sim -V_2 \sim \frac{\Lambda^2}{2m_N}, \tag{4.74}$$

where we note that a detailed calculation of V_2 for infinite scattering length can be found in Ref. [13]. The total energies associated with two, three, or four nucleons occupying the same lattice site are then

$$E_2^{localized} = 2E_1^{localized} + V_2, \tag{4.75}$$

$$E_3^{localized} = 3E_1^{localized} + 3V_2, \tag{4.76}$$

$$E_4^{localized} = 4E_1^{localized} + 6V_2, \tag{4.77}$$

from which we note that the kinetic energy scales as the number of nucleons A, while the potential energy scales as $\binom{A}{2}$. At LO in the pionless EFT, it can be shown [13] that $V_2 < -E_1^{localized}$. Therefore, $E_3^{localized}$ is negative and scales with Λ^2. This produces an instability in the three-nucleon system in the absence of three-body forces or other stabilizing effects. Clearly, this problem becomes more severe as A is increased. As the Pauli exclusion principle prevents more than four nucleons from occupying the same lattice site, we expect the problem to be most pronounced in the four-nucleon system.

The instability of the triton for zero-range forces was first noted by Thomas [14] as early as 1935. There have been a number of more recent studies of the triton in the pionless EFT, as well as of more general three-body systems with short-range interactions and large scattering lengths, see Refs. [15–22]. It has also been shown that when the cutoff-dependence in the three-nucleon system is removed using a three-nucleon contact interaction, then the binding energy of the four-nucleon system is also independent of the cutoff [23]. This is related to the so-called Tjon line (or band) to be discussed later [24–26]. In our lattice Hamiltonian notation, we denote by V_3 the potential energy associated with the three-nucleon contact interaction. The new localized energies are then

$$E_2^{localized} = 2E_1^{localized} + V_2, \tag{4.78}$$

$$E_3^{localized} = 3E_1^{localized} + 3V_2 + V_3, \tag{4.79}$$

$$E_4^{localized} = 4E_1^{localized} + 6V_2 + 4V_3, \tag{4.80}$$

such that $E_4^{localized}$ would clearly be stabilized by the $4V_3$ term, provided that $V_3 > 0$ is sufficiently large. However, for realistic nuclear binding energies, it was found that the desired cutoff-independence in ^{4}He does not manifest itself until $\Lambda \geq 8\,\text{fm}^{-1} \simeq 1.6\,\text{GeV}$ [26]. Unfortunately, such a large value of the momentum cutoff provides a very difficult starting point for lattice MC simulations of realistic light nuclei. This value of the cutoff equals a lattice spacing of $a = \pi/\Lambda \simeq 0.4\,\text{fm}$. Such a small lattice spacing is clearly outside the range of applicability of chiral EFT, as one would start to resolve the inner structure of the nucleons. Also, from a purely computational point of view, such a combination of small lattice spacing and strong repulsive forces would make lattice MC simulations nearly impossible due to severe sign and phase oscillations (see Chaps. 6 and 8 for more details). We shall therefore adopt a different approach suitable for a larger lattice spacing of $a \simeq 2\,\text{fm}$, which equals $\Lambda \simeq 2.3\,M_\pi$. While we shall again restrict ourselves to chiral EFT at LO, we shall introduce higher-derivative operators which improve the S-wave scattering amplitude at higher momenta. We expect that this should remove the clustering instability in the four-nucleon system.

4.2.2 Momentum-Dependent LO Contact Interactions

As discussed above, the momentum-independent contact interactions of NLEFT at LO become too strong at large center-of-mass momenta, and hence lead to a large overbinding of the light nuclei, as well as a rather poor description of S-wave nucleon-nucleon scattering. While we expect that these contact interactions are appropriately weakened at NLO and higher orders in the EFT expansion, we also require that the NLO component should remain a (relatively) small perturbation. This is because a strongly repulsive NLO interaction would lead to an unacceptable level of sign oscillations if treated non-perturbatively.

We shall therefore seek to improve the description of nucleon-nucleon scattering at LO by introducing momentum-dependent contact interactions. This also allows us to avoid the clustering instability associated with point-like contact interactions. Specifically, we shall introduce a momentum-dependent "smearing" of the LO contact interactions, which we shall tune to the empirical average effective range in the S-wave channels of the two-nucleon system. By doing so we are effectively absorbing a significant part of the higher-order contributions into an "improved" LO lattice action. This procedure should not be confused with operator smearing in Lattice QCD, which is used to enhance the ground-state signal in correlator calculations.

In order to implement the momentum-dependent smearing, we begin by Fourier transforming the LO contact interactions in the transfer matrix, giving

$$
-\frac{\tilde{C}\alpha_t}{2}\sum_{\mathbf{n}}\left[\rho^{a^\dagger,a}(\mathbf{n})\right]^2 - \frac{\tilde{C}_{I^2}\alpha_t}{2}\sum_{I=1}^{3}\sum_{\mathbf{n}}\left[\rho_I^{a^\dagger,a}(\mathbf{n})\right]^2
$$

$$
= -\frac{\alpha_t}{2N^3}\sum_{\mathbf{q}}\left[\tilde{C}\,\rho^{a^\dagger,a}(\mathbf{q})\rho^{a^\dagger,a}(-\mathbf{q}) + \tilde{C}_{I^2}\sum_{I=1}^{3}\rho_I^{a^\dagger,a}(\mathbf{q})\rho_I^{a^\dagger,a}(-\mathbf{q})\right],
$$

$$(4.81)$$

which we can, for instance, replace by the momentum-dependent interaction

$$
-\frac{\alpha_t}{2N^3}\sum_{\mathbf{q}} D_s(\mathbf{q})\left[\tilde{C}\,\rho^{a^\dagger,a}(\mathbf{q})\rho^{a^\dagger,a}(-\mathbf{q}) + \tilde{C}_{I^2}\sum_{I=1}^{3}\rho_I^{a^\dagger,a}(\mathbf{q})\rho_I^{a^\dagger,a}(-\mathbf{q})\right],
$$

$$(4.82)$$

where the auxiliary field "propagator" $D_s(\mathbf{q})$ is taken to be

$$D_s(\mathbf{q}) \equiv D_{s,0}^{-1} \exp\left(-\frac{b_2}{2}\mathbf{q}^2 - \frac{b_4}{4}\mathbf{q}^4 - \frac{b_6}{8}\mathbf{q}^6 - \frac{b_8}{16}\mathbf{q}^8 - \cdots\right), \qquad (4.83)$$

with the normalization factor

$$D_{s,0} \equiv \frac{1}{N^3}\sum_{\mathbf{q}} \exp\left(-\frac{b_2}{2}\mathbf{q}^2 - \frac{b_4}{4}\mathbf{q}^4 - \frac{b_6}{8}\mathbf{q}^6 - \frac{b_8}{16}\mathbf{q}^8 - \cdots\right), \qquad (4.84)$$

where in practice, we set either $b_2 \neq 0$ or $b_4 \neq 0$, and all the other $b_i = 0$. The original momentum-independent contact interactions are recovered for $b_i \to 0$.

On the lattice, we also need to choose a representation of $\mathbf{q}^2$, and we note that the simplest choice is

$$\frac{\mathbf{q}^2}{2} = \sum_{l=1}^{3}\left(1 - \cos q_l\right), \qquad (4.85)$$

which equals Eq. (4.8) for the unimproved case of $\nu = 0$. In order to fully accommodate the smeared contact interactions, we also need to modify the auxiliary-field component of the lattice action according to

$$S_{ss} \equiv \frac{1}{2}\sum_{\mathbf{n},\mathbf{n}',t} s(\mathbf{n},t)\, D_s^{-1}(\mathbf{n}-\mathbf{n}')\, s(\mathbf{n}',t)$$

$$+ \frac{1}{2}\sum_{I=1}^{3}\sum_{\mathbf{n},\mathbf{n}',t} s_I(\mathbf{n},t)\, D_s^{-1}(\mathbf{n}-\mathbf{n}')\, s_I(\mathbf{n}',t), \qquad (4.86)$$

with

$$D_s^{-1}(\mathbf{n}-\mathbf{n}') \equiv \frac{1}{N^3}\sum_{\mathbf{q}} \frac{\exp(-i\mathbf{q}\cdot\mathbf{n})\exp(i\mathbf{q}\cdot\mathbf{n}')}{D_s(\mathbf{q})}, \qquad (4.87)$$

at which point the remaining task is to determine the coefficients b_i, which can be done for instance using the empirical S-wave average effective range for nucleon-nucleon scattering. The treatment of the two-nucleon problem will be covered in Chap. 5. The improved LO action is treated non-perturbatively within the lattice MC calculation, while higher-order interactions are included as a perturbative expansion in powers of Q/Λ. This is sketched in Fig. 4.2.

What we have so far illustrated is arguably the most straightforward approach to improving the LO contact interactions by introducing a momentum-dependent smearing. Indeed, as will be shown in Chap. 5, the "smearing coefficient" b_4 is

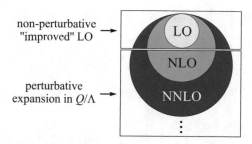

Fig. 4.2 Illustration of the lattice chiral EFT (or NLEFT) formalism. The calculation consists of a non-perturbative and a perturbative component. The "improved" LO action with momentum-dependent contact terms is iterated non-perturbatively, and therefore a significant part of the NLO and higher-order contributions are absorbed into the LO component. As a result, the magnitude of the higher-order corrections is greatly reduced. These are treated in perturbation theory, as an expansion in powers of Q/Λ

adjusted to account for the average effective ranges in the singlet and triplet two-nucleon S-wave scattering channels. The description of the S-wave phase shifts is then dramatically improved at momenta up to the cutoff scale determined by the lattice spacing. On the other hand, the P-wave phase shifts turn out to already be satisfactorily described by the unimproved LO action, as the P-waves are dominated by OPE interaction effects. Unfortunately, the improved LO contact terms are found to induce significant attractive forces in the P-wave channels, which in turn need to be cancelled by a relatively large NLO contribution.

In order to remove the tendency to worsen the level of agreement for the P-waves while the S-waves are improved upon, a more sophisticated version of the smeared LO contact interactions with a more general operator structure has been put forward. This will be referred to as the LO_3 transfer matrix, which we define as

$$
M_{LO_3} =: \exp\Bigg(-\tilde{H}_{\text{free}}\alpha_t \tag{4.88}
$$

$$
- \frac{\alpha_t}{2N^3} \sum_{\mathbf{q}} D_s(\mathbf{q}) \Big[\tilde{C}_{S=0,I=1} \tilde{V}_{S=0,I=1}(\mathbf{q})
$$

$$
+ \tilde{C}_{S=1,I=0} \tilde{V}_{S=1,I=0}(\mathbf{q}) \Big]
$$

$$
- \frac{g_A \alpha_t}{2 F_\pi \sqrt{q_\pi}} \sum_{I,S=1}^{3} \sum_{\mathbf{n}} \Big[\nabla_{S,(\nu)} \pi'_I(\mathbf{n},t) \Big] \rho_{S,I}^{a^\dagger,a}(\mathbf{n}) \Bigg) \,,
$$

where we have introduced separate coupling constants for the isospin-singlet and isospin-triplet channels. Here, the OPE contribution is given in terms of the (rescaled) pion fields. Equivalently, these may be integrated out, giving

$$M_{\text{LO}_3} = : \exp\left(-\tilde{H}_{\text{free}}\alpha_t - \tilde{V}_\pi \alpha_t \right. \tag{4.89}$$

$$-\frac{\alpha_t}{2N^3} \sum_{\mathbf{q}} D_s(\mathbf{q}) \left[\tilde{C}_{S=0,I=1} \tilde{V}_{S=0,I=1}(\mathbf{q}) \right.$$

$$\left. \left. + \tilde{C}_{S=1,I=0} \tilde{V}_{S=1,I=0}(\mathbf{q}) \right] \right) :,$$

which is useful in particular for $A \leq 3$, for which MC methods are not needed. In Eq. (4.89), the OPE potential is given by

$$\tilde{V}_\pi \equiv -\frac{\alpha_t}{2} \left(\frac{g_A}{2F_\pi}\right)^2 \sum_{I,S,S'=1}^3 \sum_{\mathbf{n},\mathbf{n'}} \rho_{S,I}^{a^\dagger,a}(\mathbf{n}) \, \rho_{S',I}^{a^\dagger,a}(\mathbf{n'}) \, \frac{G_{SS'}^{(v)}(\mathbf{n}-\mathbf{n'}, M_\pi)}{q_\pi}, \tag{4.90}$$

where the two-derivative pion correlator of Eq. (4.53) is given by

$$G_{SS'}^{(v)}(\mathbf{n}-\mathbf{n'}, M_\pi) \equiv \frac{1}{N^3} \sum_{\mathbf{q}} \exp(-i\mathbf{q} \cdot \mathbf{n}) \exp(i\mathbf{q} \cdot \mathbf{n'}) G_{SS'}^{(v)}(\mathbf{q}, M_\pi), \tag{4.91}$$

where

$$G_{SS'}^{(v)}(\mathbf{q}, M_\pi) = q_{S,(v)} \, q_{S',(v)} \, D_\pi(\mathbf{q}), \tag{4.92}$$

with the propagator $D_\pi(\mathbf{q})$ of Eq. (4.51). Note that the explicit appearance of factors of α_t and q_π in Eq. (4.90) is due to the definition of the correlator (4.53) and the pion propagator in terms of rescaled pion fields. Similar factors will appear whenever the OPE correlator is used, such as for the various pion-exchange contributions to the three-nucleon force at NNLO. For MC simulations of NLEFT at LO, we take $v = 0$ for the factors of $q_{S,(v)}$ according to Eq. (4.7). This choice is made in order to minimize sign oscillations in the MC simulations. The perturbative inclusion of effects due to $v = 1$ improvement (treated as NLO corrections) is discussed later in this chapter, in the context of operators that remove rotational symmetry breaking lattice artifacts.

The smeared contact interactions in Eq. (4.89) are most conveniently expressed in momentum space. The required density-density correlations are

$$\tilde{V}_{S=0,I=1}(\mathbf{q}) = \frac{3}{16} : \rho^{a^\dagger,a}(\mathbf{q})\rho^{a^\dagger,a}(-\mathbf{q}) : -\frac{3}{16} : \sum_{S=1}^3 \rho_S^{a^\dagger,a}(\mathbf{q})\rho_S^{a^\dagger,a}(-\mathbf{q}) : \tag{4.93}$$

$$+\frac{1}{16} : \sum_{I=1}^3 \rho_I^{a^\dagger,a}(\mathbf{q})\rho_I^{a^\dagger,a}(-\mathbf{q}) : -\frac{1}{16} : \sum_{I,S=1}^3 \rho_{S,I}^{a^\dagger,a}(\mathbf{q})\rho_{S,I}^{a^\dagger,a}(-\mathbf{q}) :,$$

for the spin-singlet isospin-triplet channel, and

$$\tilde{V}_{S=1,I=0}(\mathbf{q}) = \frac{3}{16} : \rho^{a^\dagger,a}(\mathbf{q})\rho^{a^\dagger,a}(-\mathbf{q}) : + \frac{1}{16} : \sum_{S=1}^{3} \rho_S^{a^\dagger,a}(\mathbf{q})\rho_S^{a^\dagger,a}(-\mathbf{q}) : \quad (4.94)$$

$$- \frac{3}{16} : \sum_{I=1}^{3} \rho_I^{a^\dagger,a}(\mathbf{q})\rho_I^{a^\dagger,a}(-\mathbf{q}) : - \frac{1}{16} : \sum_{I,S=1}^{3} \rho_{S,I}^{a^\dagger,a}(\mathbf{q})\rho_{S,I}^{a^\dagger,a}(-\mathbf{q}) : ,$$

for the spin-triplet isospin-singlet channel. Here, the coupling constants of the density operators have been chosen such that the LO$_3$ interaction corresponds to the nucleon-nucleon scattering amplitude

$$\mathscr{A}\left[V_{\text{LO}_3}\right] = C_{S=0,I=1} D_s(\mathbf{q}) P_\sigma^{S=0} P_\tau^{I=1} + C_{S=1,I=0} D_s(\mathbf{q}) P_\sigma^{S=1} P_\tau^{I=0}$$

$$- \left(\frac{g_A}{2F_\pi}\right)^2 \frac{(\tau_A \cdot \tau_B)(\mathbf{q} \cdot \sigma_A)(\mathbf{q} \cdot \sigma_B)}{\mathbf{q}^2 + M_\pi^2}, \quad (4.95)$$

where we have defined the projection operators

$$P_\sigma^{S=0} \equiv \frac{1}{4} - \frac{\sigma_A \cdot \sigma_B}{4}, \quad P_\sigma^{S=1} \equiv \frac{3}{4} + \frac{\sigma_A \cdot \sigma_B}{4}, \quad (4.96)$$

and

$$P_\tau^{I=0} \equiv \frac{1}{4} - \frac{\tau_A \cdot \tau_B}{4}, \quad P_\tau^{I=1} \equiv \frac{3}{4} + \frac{\tau_A \cdot \tau_B}{4}, \quad (4.97)$$

treating the nucleons A and B as distinguishable particles. Equation (4.95) should be compared with the amplitude

$$\mathscr{A}\left[V_{\text{LO}_1}\right] = C + C_{I^2}\tau_A \cdot \tau_B - \left(\frac{g_A}{2F_\pi}\right)^2 \frac{(\tau_A \cdot \tau_B)(\mathbf{q} \cdot \sigma_A)(\mathbf{q} \cdot \sigma_B)}{\mathbf{q}^2 + M_\pi^2}, \quad (4.98)$$

for the unimproved, momentum-independent contact interactions. The major effect of the spin- and isospin-dependent projection operators in Eq. (4.95) is to negate the effect of the momentum-dependent smearing on the P-wave nucleon-nucleon phase shifts, while necessitating a more general operator structure.

Hence, the transfer matrix for the LO$_3$ action in the auxiliary-field formulation requires sixteen auxiliary field components (in addition to the pion fields). One of these components is associated with the total nucleon density ρ, three with the spin density ρ_S, another three with the isospin density ρ_I, with an additional nine components for the spin-isospin density $\rho_{S,I}$. In terms of the coupling constants

$$\tilde{C} \equiv \frac{1}{16}(3\tilde{C}_{S=0,I=1} + 3\tilde{C}_{S=1,I=0}), \tag{4.99}$$

$$\tilde{C}_{S^2} \equiv \frac{1}{16}(-3\tilde{C}_{S=0,I=1} + \tilde{C}_{S=1,I=0}), \tag{4.100}$$

$$\tilde{C}_{I^2} \equiv \frac{1}{16}(\tilde{C}_{S=0,I=1} - 3\tilde{C}_{S=1,I=0}), \tag{4.101}$$

$$\tilde{C}_{S,I} \equiv \frac{1}{16}(-\tilde{C}_{S=0,I=1} - \tilde{C}_{S=1,I=0}), \tag{4.102}$$

and the same notation as in Eq. (4.73), we find

$$
\begin{aligned}
M_{\mathrm{LO}_3}(\pi'_I, s, s_S, s_I, s_{S,I}, t) =\ :\exp\Bigg(& -\tilde{H}_{\mathrm{free}}\alpha_t + \sqrt{-\tilde{C}\alpha_t}\ \sum_{\mathbf{n}} s(\mathbf{n}, t)\,\rho^{a^\dagger,a}(\mathbf{n}) \\
& + i\sqrt{\tilde{C}_{S^2}\alpha_t}\ \sum_{S=1}^{3}\sum_{\mathbf{n}} s_S(\mathbf{n}, t)\rho_S^{a^\dagger,a}(\mathbf{n}) \\
& + i\sqrt{\tilde{C}_{I^2}\alpha_t}\ \sum_{I=1}^{3}\sum_{\mathbf{n}} s_I(\mathbf{n}, t)\rho_I^{a^\dagger,a}(\mathbf{n}) \\
& + i\sqrt{\tilde{C}_{S,I}\alpha_t}\ \sum_{I,S=1}^{3}\sum_{\mathbf{n}} s_{S,I}(\mathbf{n}, t)\rho_{S,I}^{a^\dagger,a}(\mathbf{n}) \\
& - \frac{g_A\alpha_t}{2F_\pi\sqrt{q_\pi}}\sum_{I,S=1}^{3}\sum_{\mathbf{n}}\Big[\nabla_{S,(\nu)}\pi'_I(\mathbf{n}, t)\Big] \\
& \times \rho_{S,I}^{a^\dagger,a}(\mathbf{n})\Bigg)\ :,
\end{aligned}
\tag{4.103}
$$

for the transfer matrix operator. The physical parameters are taken to be $C_{S=0,I=1}$ and $C_{S=1,I=0}$, from which the coupling constants of the individual density operators in the transfer matrix can be computed.

We also need to account for the more general operator structure in the auxiliary field action, which for the case of the LO_3 transfer matrix is modified to

$$S_{ss} \equiv \frac{1}{2}\sum_{\mathbf{n},\mathbf{n}',t} s(\mathbf{n}, t)\, D_s^{-1}(\mathbf{n} - \mathbf{n}')\, s(\mathbf{n}', t)$$

$$+ \frac{1}{2}\sum_{S=1}^{3}\sum_{\mathbf{n},\mathbf{n}',t} s_S(\mathbf{n}, t)\, D_s^{-1}(\mathbf{n} - \mathbf{n}')\, s_S(\mathbf{n}', t)$$

$$+ \frac{1}{2} \sum_{l=1}^{3} \sum_{\mathbf{n},\mathbf{n}',t} s_l(\mathbf{n},t) \, D_s^{-1}(\mathbf{n}-\mathbf{n}') \, s_l(\mathbf{n}',t)$$

$$+ \frac{1}{2} \sum_{l,S=1}^{3} \sum_{\mathbf{n},\mathbf{n}',t} s_{S,l}(\mathbf{n},t) \, D_s^{-1}(\mathbf{n}-\mathbf{n}') \, s_{S,l}(\mathbf{n}',t), \qquad (4.104)$$

where the function D_s^{-1} is given by Eq. (4.87). Finally, we note that the LO_3 transfer matrix uses an $\mathcal{O}(a^4)$-improved expression (equivalent to $\nu = 2$) for the exponent of the auxiliary field propagator $D_s(\mathbf{q})$, such that

$$D_s(\mathbf{q}) \equiv D_{s,0}^{-1} \exp\left(-b_4 \frac{\mathbf{q}^4}{4}\right), \quad D_{s,0} \equiv \frac{1}{N^3} \sum_{\mathbf{q}} \exp\left(-b_4 \frac{\mathbf{q}^4}{4}\right), \qquad (4.105)$$

where

$$\frac{\mathbf{q}^2}{2} = \sum_{l=1}^{3} \left(\omega_{2,0} - \omega_{2,1} \cos q_l + \omega_{2,2} \cos(2q_l) - \omega_{2,3} \cos(3q_l)\right), \qquad (4.106)$$

and we note that the "stretching" of the $\nu = 2$ improvement coefficients applied to the free nucleon and pion dispersion relations has not been applied in the case of the auxiliary field propagator.

We conclude our discussion of NLEFT at LO by noting that the LO_3 action allows for a simultaneously good description of both S-wave and P-wave nucleon-nucleon scattering phase shifts without resorting to large NLO corrections. The spherical wall method used for extracting the scattering phase shifts as well as examples of results obtained from the different actions at LO and NLO are given in Chap. 5.

4.3   Lattice Chiral EFT at Higher Orders

We shall now continue our treatment of lattice chiral EFT by considering effects beyond the LO amplitude. These include contact interactions with two powers of momenta at NLO, isospin-breaking and Coulomb effects, two-pion exchange interactions at NLO and NNLO, as well as the three-nucleon interaction which contributes at NNLO. We note that all of these effects are treated perturbatively in most implementations of lattice chiral EFT, though there have been studies in the two-nucleon systems, where these effects were included non-perturbatively. As we have already noted for the pionless EFT, the three-nucleon interaction should in that case be included at LO in order to stabilize the triton in the limit of short-range interactions. In the same spirit, should the need arise, some components of the two- and three-nucleon interactions in lattice chiral EFT could also be treated non-perturbatively in the future. However, it should be kept in mind that the inclusion of such repulsive interaction components non-perturbatively could significantly worsen the complex sign and phase oscillations that already occur at LO. For this reason, the calculational framework has been carefully tuned to keep the size of higher-order contributions small. This includes the use of a highly improved LO action in order to minimize the impact of NLO corrections, and the tuning of the temporal lattice spacing δ in order to minimize the size of the three-nucleon force contributions at NNLO.

4.3.1 Contact Interactions at NLO

At NLO, we have seven independent contact interactions with two derivatives. Before discussing these, we note that we also must include corrections to the two LO contact interactions, according to

$$\Delta\tilde{V} \equiv \frac{1}{2}\,\Delta\tilde{C} : \sum_{\mathbf{n}} \rho^{a^{\dagger},a}(\mathbf{n})\rho^{a^{\dagger},a}(\mathbf{n}) : , \qquad (4.107)$$

$$\Delta\tilde{V}_{I^2} \equiv \frac{1}{2}\,\Delta\tilde{C}_{I^2} : \sum_{I=1}^{3}\sum_{\mathbf{n}} \rho_I^{a^{\dagger},a}(\mathbf{n})\rho_I^{a^{\dagger},a}(\mathbf{n}) : , \qquad (4.108)$$

which are also treated perturbatively. There are several reasons for the inclusion of such terms in the NLO lattice action. As we shall see, at $a \simeq 2\,\text{fm}$ the TPE terms at NLO and NNLO can be expanded in powers of momenta. A part of the TPE expansion is then absorbed into $\Delta\tilde{C}$ and $\Delta\tilde{C}_{I^2}$, along with two of the four possible electromagnetic contact interactions that regularize the (singular) Coulomb interaction between protons. Another reason for the inclusion of $\Delta\tilde{C}$ and $\Delta\tilde{C}_{I^2}$ is to absorb the "back-reaction" on the LO coupling constants when the NLO contribution to the two-nucleon phase shifts is added to a previously optimized LO analysis (as detailed in Chap. 5). This $\simeq 10\%$ shift in the LO constants is treated perturbatively as part of the NLO contribution, as the light nuclei would otherwise be significantly overbound at LO.

The seven independent NLO contact interactions with two derivatives are

$$\tilde{V}_{q^2} \equiv -\frac{1}{2}\,\tilde{C}_{q^2} : \sum_{\mathbf{n}} \rho^{a^{\dagger},a}(\mathbf{n})\sum_{l=1}^{3}\nabla_{l,(v)}^2\rho^{a^{\dagger},a}(\mathbf{n}) : , \qquad (4.109)$$

$$\tilde{V}_{I^2,q^2} \equiv -\frac{1}{2}\,\tilde{C}_{I^2,q^2} : \sum_{I=1}^{3}\sum_{\mathbf{n}} \rho_I^{a^{\dagger},a}(\mathbf{n})\sum_{l=1}^{3}\nabla_{l,(v)}^2\rho_I^{a^{\dagger},a}(\mathbf{n}) : , \qquad (4.110)$$

$$\tilde{V}_{S^2,q^2} \equiv -\frac{1}{2}\tilde{C}_{S^2,q^2} : \sum_{S=1}^{3}\sum_{\mathbf{n}} \rho_S^{a^\dagger,a}(\mathbf{n}) \sum_{l=1}^{3} \nabla_{l,(v)}^2 \rho_S^{a^\dagger,a}(\mathbf{n}) :, \qquad (4.111)$$

$$\tilde{V}_{S^2,I^2,q^2} \equiv -\frac{1}{2}\tilde{C}_{S^2,I^2,q^2} : \sum_{I,S=1}^{3}\sum_{\mathbf{n}} \rho_{S,I}^{a^\dagger,a}(\mathbf{n}) \sum_{l=1}^{3} \nabla_{l,(v)}^2 \rho_{S,I}^{a^\dagger,a}(\mathbf{n}) :, \qquad (4.112)$$

$$\tilde{V}_{(q\cdot S)^2} \equiv \frac{1}{2}\tilde{C}_{(q\cdot S)^2} : \sum_{\mathbf{n}} \left[\sum_{S=1}^{3} \nabla_{S,(v)} \rho_S^{a^\dagger,a}(\mathbf{n})\right] \sum_{S'=1}^{3} \nabla_{S',(v)} \rho_{S'}^{a^\dagger,a}(\mathbf{n}) :, \qquad (4.113)$$

$$\tilde{V}_{I^2,(q\cdot S)^2} \equiv \frac{1}{2}\tilde{C}_{I^2,(q\cdot S)^2} : \sum_{I=1}^{3}\sum_{\mathbf{n}} \left[\sum_{S=1}^{3} \nabla_{S,(v)} \rho_{S,I}^{a^\dagger,a}(\mathbf{n})\right] \sum_{S'=1}^{3} \nabla_{S',(v)} \rho_{S',I}^{a^\dagger,a}(\mathbf{n}) :, \qquad (4.114)$$

and

$$\tilde{V}_{(iq\times S)\cdot k} \equiv -\frac{i}{2}\tilde{C}_{(iq\times S)\cdot k} : \sum_{S=1}^{3}\sum_{\mathbf{n}}\sum_{l,l'=1}^{3} \varepsilon_{l,S,l'} \qquad (4.115)$$

$$\times \left(\Pi_l^{a^\dagger,a}(\mathbf{n})\nabla_{l',(v)}\rho_S^{a^\dagger,a}(\mathbf{n}) + \Pi_{l,S}^{a^\dagger,a}(\mathbf{n})\nabla_{l',(v)}\rho^{a^\dagger,a}(\mathbf{n})\right) :,$$

with $\varepsilon_{a,b,c}$ the three-dimensional Levi-Civita tensor. We note that the "spin-orbit" interaction of Eq. (4.115) vanishes unless $S = 1$, and that it is antisymmetric under the exchange of $\mathbf{q}$ and $\mathbf{k}$ in the continuum limit. Therefore, this interaction should contribute only in odd-parity channels with $S = 1$. However, at non-zero lattice spacing, the exact t-u channel antisymmetry of Eq. (4.115) is violated. Thus, we expect lattice artifacts to appear in even-parity channels with $S = 1$. We note that such artifacts may be eliminated by projecting onto $I = 1$, and for this purpose we include the operator

$$\tilde{V}_{I^2,(iq\times S)\cdot k} \equiv -\frac{i}{2}\tilde{C}_{I^2,(iq\times S)\cdot k} : \sum_{I,S=1}^{3}\sum_{\mathbf{n}}\sum_{l,l'=1}^{3} \varepsilon_{l,S,l'} \qquad (4.116)$$

$$\times \left(\Pi_{l,I}^{a^\dagger,a}(\mathbf{n})\nabla_{l',(v)}\rho_{S,I}^{a^\dagger,a}(\mathbf{n}) + \Pi_{l,S,I}^{a^\dagger,a}(\mathbf{n})\nabla_{l',(v)}\rho_I^{a^\dagger,a}(\mathbf{n})\right) :,$$

with

$$\tilde{C}_{(iq\times S)\cdot k} \equiv \frac{1}{4}\left(3\tilde{C}_{(iq\times S)\cdot k}^{I=1} + \tilde{C}_{(iq\times S)\cdot k}^{I=0}\right), \quad \tilde{C}_{I^2,(iq\times S)\cdot k} \equiv \frac{1}{4}\left(\tilde{C}_{(iq\times S)\cdot k}^{I=1} - \tilde{C}_{(iq\times S)\cdot k}^{I=0}\right), \qquad (4.117)$$

where we have introduced separate coupling constants for the $I = 0$ and $I = 1$ channels. If we define

$$\tilde{V}^{I=1}_{(iq \times S) \cdot k} \equiv \tilde{V}_{(iq \times S) \cdot k} + \tilde{V}_{I^2, (iq \times S) \cdot k}, \tag{4.118}$$

and set

$$\tilde{C}_{(iq \times S) \cdot k} \equiv \frac{3}{4} \tilde{C}^{I=1}_{(iq \times S) \cdot k}, \quad \tilde{C}_{I^2, (iq \times S) \cdot k} \equiv \frac{1}{4} \tilde{C}^{I=1}_{(iq \times S) \cdot k}, \tag{4.119}$$

then the lattice artifacts in the spin-orbit interaction for odd-parity channels with $S = 1$ are eliminated.

We shall next describe how the NLO (and other higher-order contributions) are treated in perturbation theory. Such a treatment is justified if the higher-order terms constitute a small correction to the LO result. As discussed earlier, a major motivation for the introduction of momentum-dependent contact interactions at LO was to accomplish such a situation. As the perturbative treatment involves the calculation of operator expectation values, we introduce the isospin-independent source terms

$$U(\varepsilon, t) \equiv \sum_{\mathbf{n}} \varepsilon_1(\mathbf{n}, t) \, \rho^{a^\dagger, a}(\mathbf{n}) + \sum_{S=1}^{3} \sum_{\mathbf{n}} \varepsilon_2(\mathbf{n}, t, S) \, \rho_S^{a^\dagger, a}(\mathbf{n})$$

$$+ \sum_{S=1}^{3} \sum_{\mathbf{n}} \varepsilon_3(\mathbf{n}, t, S) \, \nabla_{S,(v)} \, \rho^{a^\dagger, a}(\mathbf{n})$$

$$+ \sum_{S,S'=1}^{3} \sum_{\mathbf{n}} \varepsilon_4(\mathbf{n}, t, S, S') \, \nabla_{S,(v)} \, \rho_{S'}^{a^\dagger, a}(\mathbf{n})$$

$$+ \sum_{\mathbf{n}} \sum_{l=1}^{3} \varepsilon_5(\mathbf{n}, t, l) \, \nabla^2_{l,(v)} \, \rho^{a^\dagger, a}(\mathbf{n})$$

$$+ \sum_{S=1}^{3} \sum_{\mathbf{n}} \sum_{l=1}^{3} \varepsilon_6(\mathbf{n}, t, l, S) \, \nabla^2_{l,(v)} \, \rho_S^{a^\dagger, a}(\mathbf{n})$$

$$+ \sum_{\mathbf{n}} \sum_{l=1}^{3} \varepsilon_7(\mathbf{n}, t, l) \, \Pi_l^{a^\dagger, a}(\mathbf{n}) + \sum_{S=1}^{3} \sum_{\mathbf{n}} \sum_{l=1}^{3} \varepsilon_8(\mathbf{n}, t, l, S) \, \Pi_{l,S}^{a^\dagger, a}(\mathbf{n}), \tag{4.120}$$

and the isospin-dependent source terms

$$
U_{I^2}^{(\varepsilon,t)} \equiv \sum_{I=1}^{3} \sum_{\mathbf{n}} \varepsilon_9(\mathbf{n}, t, I)\, \rho_I^{a^\dagger,a}(\mathbf{n}) + \sum_{I,S=1}^{3} \sum_{\mathbf{n}} \varepsilon_{10}(\mathbf{n}, t, S, I)\, \rho_{S,I}^{a^\dagger,a}(\mathbf{n})
$$

$$
+ \sum_{I,S=1}^{3} \sum_{\mathbf{n}} \varepsilon_{11}(\mathbf{n}, t, S, I)\, \nabla_{S,(v)}\, \rho_I^{a^\dagger,a}(\mathbf{n})
$$

$$
+ \sum_{I,S,S'=1}^{3} \sum_{\mathbf{n}} \varepsilon_{12}(\mathbf{n}, t, S, S', I)\, \nabla_{S,(v)}\, \rho_{S',I}^{a^\dagger,a}(\mathbf{n})
$$

$$
+ \sum_{I=1}^{3} \sum_{\mathbf{n}} \sum_{l=1}^{3} \varepsilon_{13}(\mathbf{n}, t, l, I)\, \nabla_{l,(v)}^2\, \rho_I^{a^\dagger,a}(\mathbf{n})
$$

$$
+ \sum_{I,S=1}^{3} \sum_{\mathbf{n}} \sum_{l=1}^{3} \varepsilon_{14}(\mathbf{n}, t, l, S, I)\, \nabla_{l,(v)}^2\, \rho_{S,I}^{a^\dagger,a}(\mathbf{n})
$$

$$
+ \sum_{I=1}^{3} \sum_{\mathbf{n}} \sum_{l=1}^{3} \varepsilon_{15}(\mathbf{n}, t, l, I)\, \Pi_{l,I}^{a^\dagger,a}(\mathbf{n})
$$

$$
+ \sum_{I,S=1}^{3} \sum_{\mathbf{n}} \sum_{l=1}^{3} \varepsilon_{16}(\mathbf{n}, t, l, S, I)\, \Pi_{l,S,I}^{a^\dagger,a}(\mathbf{n}), \tag{4.121}
$$

which are sufficient to account for all operators up-to-and-including NNLO in the EFT expansion. With these additional fields and linear functionals, we define

$$
M(\pi_I', s, s_S, s_I, s_{S,I}, \varepsilon_i, t) \equiv : M_{LO_3}(\pi_I', s, s_S, s_I, s_{S,I}, t)\, \exp\big(U(\varepsilon_i, t)
$$

$$
+ U_{I^2}(\varepsilon_i, t)\big) :, \tag{4.122}
$$

where ε_i collectively denotes the source fields $\varepsilon_1 \ldots \varepsilon_{16}$. While we have explicitly used the M_{LO_3} transfer matrix in Eq. (4.122), a similar expression can be straight-forwardly obtained for transfer matrices with fewer auxiliary field components. We therefore use the notation

$$
\int \mathscr{D}s\, \mathscr{D}s_S\, \mathscr{D}s_I\, \mathscr{D}s_{S,I} \to \int \mathscr{D}\sigma, \tag{4.123}
$$

and

$$
M(\pi_I', s, s_S, s_I, s_{S,I}, \varepsilon_i, t) \to M(\pi_I', \sigma, \varepsilon_i, t), \tag{4.124}
$$

when the exact structure of the auxiliary-field sector of the theory is not essential. With the help of Eq. (4.122), we define $M(\varepsilon_i, t)$ according to

$$M(\varepsilon_i, t) \equiv \frac{\int \mathscr{D}\pi_I' \mathscr{D}\sigma \, \exp\left(-S_{\pi\pi}(t) - S_{ss}(t)\right) M(\pi_I', \sigma, \varepsilon_i, t)}{\int \mathscr{D}\pi_I' \mathscr{D}\sigma \, \exp\left(-S_{\pi\pi}(t) - S_{ss}(t)\right)}, \qquad (4.125)$$

from which we recover

$$\lim_{\varepsilon_i \to 0} M(\varepsilon_i, t) = M_{\mathrm{LO}}, \qquad (4.126)$$

in the limit $\varepsilon_i \to 0$. Here, M_{LO} equals the LO transfer matrix with the auxiliary and pion fields integrated out. More details on the transfer matrix formalism without auxiliary fields will be given in Chap. 5 in the context of the two- and three-nucleon problems.

To first order in perturbation theory, we need the expectation values of the NLO operators. We define the NLO transfer matrix as

$$M_{\mathrm{NLO}} \equiv M_{\mathrm{LO}} + \frac{\int \mathscr{D}\pi_I' \mathscr{D}\sigma \, \exp\left(-S_{\pi\pi}(t) - S_{ss}(t)\right) \Delta M_{\mathrm{NLO}}(\pi_I', \sigma, t)}{\int \mathscr{D}\pi_I' \mathscr{D}\sigma \, \exp\left(-S_{\pi\pi}(t) - S_{ss}(t)\right)}, \qquad (4.127)$$

where ΔM_{NLO} consists of a sum of bilinear derivatives of $M(\varepsilon_i, t)$ with respect to the source fields ε_i in the limit $\varepsilon_i \to 0$, according to

$$\begin{aligned}
\Delta M_{\mathrm{NLO}}(\pi_I', \sigma, t) \equiv &-\frac{1}{2}\Delta\tilde{C}\alpha_t \sum_{\mathbf{n}} \left(\frac{\delta}{\delta\varepsilon_1(\mathbf{n}, t)}\right)^2 M(\pi_I', \sigma, \varepsilon_i, t)\Big|_{\varepsilon_i \to 0} \\
&-\frac{1}{2}\Delta\tilde{C}_{I^2}\alpha_t \sum_{I=1}^{3}\sum_{\mathbf{n}} \left(\frac{\delta}{\delta\varepsilon_9(\mathbf{n}, t, I)}\right)^2 M(\pi_I', \sigma, \varepsilon_i, t)\Big|_{\varepsilon_i \to 0} \\
&+\frac{1}{2}\tilde{C}_{q^2}\alpha_t \sum_{\mathbf{n}}\sum_{l=1}^{3} \frac{\delta}{\delta\varepsilon_1(\mathbf{n}, t)} \frac{\delta}{\delta\varepsilon_5(\mathbf{n}, t, l)} M(\pi_I', \sigma, \varepsilon_i, t)\Big|_{\varepsilon_i \to 0} \\
&+\frac{1}{2}\tilde{C}_{I^2,(q\cdot S)^2}\alpha_t \sum_{I,S,S'=1}^{3}\sum_{\mathbf{n}} \\
&\times \frac{\delta}{\delta\varepsilon_{12}(\mathbf{n}, t, S, S, I)} \frac{\delta}{\delta\varepsilon_{12}(\mathbf{n}, t, S', S', I)} M(\pi_I', \sigma, \varepsilon_i, t)\Big|_{\varepsilon_i \to 0} \\
&+ \dots,
\end{aligned} \qquad (4.128)$$

where we have only included a few terms explicitly in order to illustrate the principle. By substitution of Eq. (4.128) into Eq. (4.127), we find

$$
\begin{aligned}
M_{\mathrm{NLO}} = M_{\mathrm{LO}} - \alpha_t \Bigg(\frac{1}{2} \Delta \tilde{C} \alpha_t \sum_{\mathbf{n}} \left(\frac{\delta}{\delta \varepsilon_1(\mathbf{n}, t)} \right)^2 M(\varepsilon_i, t) \Bigg|_{\varepsilon_i \to 0} \\
+ \frac{1}{2} \Delta \tilde{C}_{I^2} \alpha_t \sum_{I=1}^{3} \sum_{\mathbf{n}} \left(\frac{\delta}{\delta \varepsilon_9(\mathbf{n}, t, I)} \right)^2 M(\varepsilon_i, t) \Bigg|_{\varepsilon_i \to 0} \\
- \frac{1}{2} \tilde{C}_{q^2} \alpha_t \sum_{\mathbf{n}} \sum_{l=1}^{3} \frac{\delta}{\delta \varepsilon_1(\mathbf{n}, t)} \frac{\delta}{\delta \varepsilon_5(\mathbf{n}, t, l)} M(\varepsilon_i, t) \Bigg|_{\varepsilon_i \to 0} \\
- \frac{1}{2} \tilde{C}_{I^2,(q \cdot S)^2} \alpha_t \sum_{I,S,S'=1}^{3} \sum_{\mathbf{n}} \\
\times \frac{\delta}{\delta \varepsilon_{12}(\mathbf{n}, t, S, S, I)} \frac{\delta}{\delta \varepsilon_{12}(\mathbf{n}, t, S', S', I)} M(\varepsilon_i, t) \Bigg|_{\varepsilon_i \to 0} \\
+ \dots \Bigg),
\end{aligned}
\tag{4.129}
$$

for the NLO transfer matrix. It should be noted that an alternative convention has been used in the literature, whereby the $U(\varepsilon_i, t)$ and $U_{I^2}(\varepsilon_i, t)$ were multiplied by factors of $\sqrt{\alpha_t}$. While such a convention appears to simplify the treatment of the NLO contribution, it would introduce unnecessary complications at NNLO and higher orders. Hence, such factors are absent from Eq. (4.122). Schematically, we may write the full NLO expression as

$$
M_{\mathrm{NLO}} = M_{\mathrm{LO}} - \alpha_t : \left(\Delta V^{(0)} + V^{(2)} \right) M_{\mathrm{LO}} :,
\tag{4.130}
$$

where

$$
\Delta V^{(0)} \equiv \Delta \tilde{V} + \Delta \tilde{V}_{I^2},
\tag{4.131}
$$

due to the shifts in the two LO contact operators, and

$$
V^{(2)} \equiv \tilde{V}_{q^2} + \tilde{V}_{I^2,q^2} + \tilde{V}_{S^2,q^2} + \tilde{V}_{S^2,I^2,q^2} + \tilde{V}_{(q \cdot S)^2} + \tilde{V}_{I^2,(q \cdot S)^2} + \tilde{V}_{(iq \times S) \cdot k}^{I=1},
\tag{4.132}
$$

for the seven independent NLO contact operators. This amounts to a total of nine operators with unknown coefficients. As described in Chap. 5, we determine the coefficients of the NLO operators from the phase shifts and mixing angles of neutron-proton (np) scattering. The 2NF can then be used for predictive calculations of heavier nuclei.

Before we move on to the three-nucleon operators at $\mathscr{O}(Q^3)$ or NNLO, we shall complete our discussion of operators that appear at $\mathscr{O}(Q^2)$ by considering the TPE potential, operators that address issues related to rotational symmetry breaking on the lattice, electromagnetic effects, and the effects of strong isospin breaking due to the unequal neutral and charged pion masses. We emphasize that the focus in this chapter has been on the development of the formalism rather than on the calculational methods. More detail on the implementation and numerical evaluation of NLO and higher-order contributions will be given in Chap. 5 for systems with $A \leq 3$, and for MC calculations of heavier nuclei in Chaps. 6 and 7.

4.3.2 Two-Pion Exchange Contributions

So far, we have treated the two-nucleon contact interactions with two powers of momentum which appear at NLO. For completeness, we should also include the TPE contributions to the two-nucleon force which appear at NLO and NNLO. For low values of the momentum cutoff (or large lattice spacing a), the TPE contributions become indistinguishable from the contact terms at NLO, and can be absorbed into the latter. We shall return to this point after presenting the explicit expressions. In any case, we need the full TPE lattice expressions, since we are interested in studying a range of lattice spacings 1 fm $\leq a \leq$ 2 fm, for which the TPE contribution should be accounted for explicitly. The TPE amplitude at NLO is

$$
\mathscr{A}\left[V_{\pi\pi}^{(2)}\right] \equiv L(\mathbf{q}) \left[C_1^{\pi\pi} M_\pi^2 + C_2^{\pi\pi} \mathbf{q}^2 + C_3^{\pi\pi} M_\pi^2 \, D_{\pi\pi}(\mathbf{q}) \right] \tau_A \cdot \tau_B
$$
$$
+ L(\mathbf{q}) \left[C_4^{\pi\pi} (\mathbf{q} \cdot \boldsymbol{\sigma}_A)(\mathbf{q} \cdot \boldsymbol{\sigma}_B) + C_5^{\pi\pi} \mathbf{q}^2 \boldsymbol{\sigma}_A \cdot \boldsymbol{\sigma}_B \right], \quad (4.133)
$$

with the coupling constants

$$
C_1^{\pi\pi} = -\frac{5g_A^4 - 4g_A^2 - 1}{96\pi^2 F_\pi^4}, \quad C_2^{\pi\pi} = -\frac{23g_A^4 - 10g_A^2 - 1}{384\pi^2 F_\pi^4}, \quad (4.134)
$$

$$
C_3^{\pi\pi} = -\frac{g_A^4}{32\pi^2 F_\pi^4}, \quad C_4^{\pi\pi} = -C_5^{\pi\pi} = -\frac{3g_A^4}{64\pi^2 F_\pi^4}, \quad (4.135)
$$

the loop function

$$
L(\mathbf{q}) \equiv \frac{\sqrt{4M_\pi^2 + \mathbf{q}^2}}{2|\mathbf{q}|} \ln \left(\frac{\sqrt{4M_\pi^2 + \mathbf{q}^2} + |\mathbf{q}|}{\sqrt{4M_\pi^2 + \mathbf{q}^2} - |\mathbf{q}|} \right), \quad (4.136)
$$

and

$$D_{\pi\pi}(\mathbf{q}) \equiv \frac{4M_\pi^2}{4M_\pi^2 + \mathbf{q}^2}, \tag{4.137}$$

which resembles the OPE propagator for an object with mass $2M_\pi$. For a derivation of the TPE expressions at NLO, see e.g. Refs. [6, 27].

On the lattice, we need to account for the central, tensor and spin-spin components of the NLO TPE potential. This gives

$$V_{\text{TPE}}^{(2)} = \sum_{I=1}^{3} \sum_{\mathbf{n},\mathbf{n}'} W_C^{(2)}(\mathbf{n} - \mathbf{n}')\rho_I(\mathbf{n})\rho_I(\mathbf{n}') + \sum_{S,S'=1}^{3} \sum_{\mathbf{n},\mathbf{n}'} T_{S,S'}^{(2)}(\mathbf{n} - \mathbf{n}')\rho_S(\mathbf{n})\rho_{S'}(\mathbf{n}')$$

$$+ \sum_{S=1}^{3} \sum_{\mathbf{n},\mathbf{n}'} V_S^{(2)}(\mathbf{n} - \mathbf{n}')\rho_S(\mathbf{n})\rho_S(\mathbf{n}'), \tag{4.138}$$

where the explicit expressions are

$$W_C^{(2)}(\mathbf{n} - \mathbf{n}') \equiv \frac{1}{N^3} \sum_{\mathbf{q}} \exp(-i\mathbf{q} \cdot \mathbf{n}) \exp(i\mathbf{q} \cdot \mathbf{n}') W_C^{(2)}(\mathbf{q}), \tag{4.139}$$

with

$$W_C^{(2)}(\mathbf{q}) = \frac{1}{2} L(\mathbf{q}) \left[C_1^{\pi\pi} M_\pi^2 + C_2^{\pi\pi} \mathbf{q}^2 + C_3^{\pi\pi} M_\pi^2 D_{\pi\pi}(\mathbf{q}) \right], \tag{4.140}$$

for the central term,

$$T_{S,S'}^{(2)}(\mathbf{n} - \mathbf{n}') \equiv \frac{1}{N^3} \sum_{\mathbf{q}} \exp(-i\mathbf{q} \cdot \mathbf{n}) \exp(i\mathbf{q} \cdot \mathbf{n}') T_{S,S'}^{(2)}(\mathbf{q}), \tag{4.141}$$

with

$$T_{S,S'}^{(2)}(\mathbf{q}) = \frac{1}{2} L(\mathbf{q}) C_4^{\pi\pi} q_{S,(\nu=0)} q_{S',(\nu=0)}, \tag{4.142}$$

for the tensor term, and

$$V_S^{(2)}(\mathbf{n} - \mathbf{n}') \equiv \frac{1}{N^3} \sum_{\mathbf{q}}' \exp(-i\mathbf{q} \cdot \mathbf{n}) \exp(i\mathbf{q} \cdot \mathbf{n}') V_S^{(2)}(\mathbf{q}), \tag{4.143}$$

with

$$V_S^{(2)}(\mathbf{q}) = \frac{1}{2} L(\mathbf{q}) C_5^{\pi\pi} \mathbf{q}^2, \tag{4.144}$$

for the spin-spin term. For consistency with the treatment of derivative operators in the NLO contact terms, we take on the lattice

$$\mathbf{q}^2 = \sum_{l=1}^{3} \sin^2 q_l, \quad |\mathbf{q}| = \sqrt{\mathbf{q}^2}, \tag{4.145}$$

for all the components of Eq. (4.138).

At NNLO in the Weinberg power counting scheme, no additional two-nucleon contact terms appear. However, the TPE amplitude receives a sub-leading contribution, given by

$$\mathscr{A}\left[V_{\pi\pi}^{(3)}\right] \equiv A(\mathbf{q})\,(2M_\pi^2 + \mathbf{q}^2)\left[D_1^{\pi\pi}\,M_\pi^2 + D_2^{\pi\pi}\,\mathbf{q}^2\right] \tag{4.146}$$

$$+ A(\mathbf{q})\,(4M_\pi^2 + \mathbf{q}^2)\left[D_4^{\pi\pi}\,(\mathbf{q}\cdot\boldsymbol{\sigma}_A)(\mathbf{q}\cdot\boldsymbol{\sigma}_B)\right.$$

$$\left. + D_5^{\pi\pi}\,\mathbf{q}^2\boldsymbol{\sigma}_A\cdot\boldsymbol{\sigma}_B\right]\tau_A\cdot\tau_B,$$

with the coupling constants

$$D_1^{\pi\pi} = -\frac{3g_A^2\,(2c_1 - c_3)}{8\pi\,F_\pi^4}, \quad D_2^{\pi\pi} = \frac{3g_A^2\,c_3}{16\pi\,F_\pi^4}, \tag{4.147}$$

$$D_4^{\pi\pi} = -D_5^{\pi\pi} = -\frac{g_A^2\,c_4}{32\pi\,F_\pi^4}, \tag{4.148}$$

and the loop function

$$A(\mathbf{q}) \equiv \frac{1}{2|\mathbf{q}|}\arctan\left(\frac{|\mathbf{q}|}{2M_\pi}\right). \tag{4.149}$$

We parameterize the NNLO TPE lattice potential as for NLO TPE, giving

$$V_{\text{TPE}}^{(3)} = \sum_{\mathbf{n},\mathbf{n}'} V_C^{(3)}(\mathbf{n}-\mathbf{n}')\rho(\mathbf{n})\rho(\mathbf{n}') + \sum_{I,S,S'=1}^{3}\sum_{\mathbf{n},\mathbf{n}'} T_{S,S'}^{(3)}(\mathbf{n}-\mathbf{n}')\rho_{S,I}(\mathbf{n})\rho_{S',I}(\mathbf{n}')$$

$$+ \sum_{I,S=1}^{3}\sum_{\mathbf{n},\mathbf{n}'} W_S^{(3)}(\mathbf{n}-\mathbf{n}')\rho_{S,I}(\mathbf{n})\rho_{S,I}(\mathbf{n}'), \tag{4.150}$$

where the explicit expressions are

$$V_C^{(3)}(\mathbf{n}-\mathbf{n}') \equiv \frac{1}{N^3}\sum_{\mathbf{q}}\exp(-i\mathbf{q}\cdot\mathbf{n})\exp(i\mathbf{q}\cdot\mathbf{n}')\,V_C^{(3)}(\mathbf{q}), \tag{4.151}$$

with

$$V_C^{(3)}(\mathbf{q}) = \frac{1}{2} A(\mathbf{q}) \, (2M_\pi^2 + \mathbf{q}^2) \left[D_1^{\pi\pi} M_\pi^2 + D_2^{\pi\pi} \mathbf{q}^2 \right], \tag{4.152}$$

for the central term,

$$T_{S,S'}^{(3)}(\mathbf{n} - \mathbf{n}') \equiv \frac{1}{N^3} \sum_{\mathbf{q}} \exp(-i\mathbf{q} \cdot \mathbf{n}) \exp(i\mathbf{q} \cdot \mathbf{n}') \, T_{S,S'}^{(3)}(\mathbf{q}), \tag{4.153}$$

with

$$T_{S,S'}^{(3)}(\mathbf{q}) = \frac{1}{2} A(\mathbf{q}) \, (4M_\pi^2 + \mathbf{q}^2) \, D_4^{\pi\pi} \, q_{S,(\nu=0)} q_{S',(\nu=0)}, \tag{4.154}$$

for the tensor term, and

$$W_S^{(3)}(\mathbf{n} - \mathbf{n}') \equiv \frac{1}{N^3} \sum_{\mathbf{q}} \exp(-i\mathbf{q} \cdot \mathbf{n}) \exp(i\mathbf{q} \cdot \mathbf{n}') \, W_S^{(3)}(\mathbf{q}), \tag{4.155}$$

with

$$W_S^{(3)}(\mathbf{q}) = \frac{1}{2} A(\mathbf{q}) \, (4M_\pi^2 + \mathbf{q}^2) \, D_5^{\pi\pi} \, \mathbf{q}^2, \tag{4.156}$$

for the spin-spin term. Again, one uses the definition (4.145) for the lattice momenta in the NNLO TPE potential.

Before we conclude our discussion of the NLO and NNLO operators in lattice chiral EFT, let us consider in which cases the explicit inclusion of the TPE components is called for. With the notable exception of the most recent calculations of np scattering presented in Chap. 5, most NLEFT work uses a spatial lattice spacing of $a = 1.97$ fm, which corresponds to a (relatively low) cutoff momentum $\Lambda \equiv \pi/a = 314$ MeV $\approx 2.3 \, M_\pi$. In such cases, the treatment of the TPE has followed the Weinberg power counting scheme with some additional simplifications made possible by the low value of Λ. Specifically, for nearly all $|\mathbf{q}| < \Lambda$, the NLO TPE potential can be expanded in powers of $\mathbf{q}^2/(4M_\pi^2)$, using the Taylor expansions

$$L(\mathbf{q}) \simeq 1 + \frac{1}{3} \left(\frac{\mathbf{q}^2}{4M_\pi^2} \right) + \mathcal{O} \left(\frac{\mathbf{q}^4}{16M_\pi^4} \right), \tag{4.157}$$

and

$$L(\mathbf{q}) \, D_{\pi\pi}(\mathbf{q}) \simeq 1 - \frac{2}{3} \left(\frac{\mathbf{q}^2}{4M_\pi^2} \right) + \mathcal{O} \left(\frac{\mathbf{q}^4}{16M_\pi^4} \right), \tag{4.158}$$

which are found to deviate significantly from the full forms only for momenta near $\pi/a \simeq 2.3\,M_\pi$, where the EFT is already problematic due to large cutoff effects. For the NNLO TPE potential, we may similarly expand Eq. (4.146) in powers of $\mathbf{q}^2/(4M_\pi^2)$, using the Taylor expansion

$$4M_\pi\,A(\mathbf{q}) \simeq 1 - \frac{1}{3}\left(\frac{\mathbf{q}^2}{4M_\pi^2}\right) + \mathcal{O}\left(\frac{\mathbf{q}^4}{16M_\pi^4}\right), \tag{4.159}$$

for Eq. (4.149). In practice, there is little advantage in retaining the full non-local TPE potential at the (relatively large) lattice spacing $a \simeq 2\,\text{fm}$. Instead, we retain only the simplified structure

$$\tilde{V}_{\text{NLO}} \equiv \tilde{V}_{\text{LO}} + \Delta\tilde{V}^{(0)} + \tilde{V}^{(2)}, \tag{4.160}$$

where

$$\tilde{V}_{\text{LO}} \equiv \tilde{V}^{(0)} + \tilde{V}_\pi, \tag{4.161}$$

such that the terms in the expansion of the NLO and NNLO TPE amplitudes (4.133) and (4.146) with up to two powers of $\mathbf{q}$ are absorbed as a redefinition of the coefficients of $\Delta\tilde{V}^{(0)}$ and $\tilde{V}^{(2)}$. Hence, in the simplified Weinberg power counting at low cutoff, no additional terms appear in the two-nucleon potential at NNLO, such that the only new NNLO contributions are due to the three-nucleon interactions. However, for $a \simeq 1\,\text{fm}$, such a Taylor expansion of the TPE potential is no longer justified, and the full structure should be kept (as shown for np scattering in Chap. 5). Finally, it should be noted that a number of alternatives to the Weinberg power counting have been discussed in the literature, see e.g. Refs. [28–31]. However, at low values of Λ, the possible advantages of such alternative schemes appear numerically small [32].

4.3.3   Rotational Symmetry Breaking Operators

The lattice regularization reduces the full three-dimensional rotational group SO(3) to the cubic subgroup SO(3, Z). We shall next discuss the relative sizes of lattice artifacts induced by such rotational symmetry breaking, and to what extent these have been incorporated in the lattice action. A more detailed treatment of the rotational group is given in Chap. 3. Local two-nucleon operators that break rotational invariance first appear at $\mathcal{O}(Q^2)$. These include two-nucleon operators with amplitudes proportional to

$$\sum_{l=1}^{3} q_l^2\,(\sigma_A)_l\,(\sigma_B)_l, \tag{4.162}$$

and

$$(\tau_A \cdot \tau_B) \sum_{l=1}^{3} q_l^2 \, (\sigma_A)_l \, (\sigma_B)_l \,, \tag{4.163}$$

and we note that such operators are clearly very similar to $\tilde{V}_{S^2,q^2}$ and $\tilde{V}_{S^2,I^2,q^2}$ which do not break rotational invariance and are already included in our set of local $\mathcal{O}(Q^2)$ operators at NLO. However, the operators (4.162) and (4.163) also generate transition matrix elements, which introduce unphysical mixing effects. For instance, such mixing is generated between the tensor-coupled $^3S(D)_1$ and $^3D(G)_3$ channels. Here, we use the following notation: The ordering of the partial waves, e.g. $S(D)$ rather than $D(S)$, indicates that this state becomes a pure S-wave when the strength of the tensor component of the interaction continuously approaches zero. Another commonly used notation for the partial-wave mixing is $^3S_1 - {}^3D_1$ and $^3D_3 - {}^3G_3$. In what follows, we will adhere to the round bracket notation.

On the lattice, we introduce

$$\tilde{V}_{SSqq} \equiv \frac{1}{2} \, \tilde{C}_{SSqq} : \sum_{S=1}^{3} \sum_{\mathbf{n}} \left[\nabla_{S,(v)} \rho_S^{a^\dagger,a}(\mathbf{n}) \right] \nabla_{S,(v)} \rho_S^{a^\dagger,a}(\mathbf{n}) :\,, \tag{4.164}$$

and

$$\tilde{V}_{I^2,SSqq} \equiv \frac{1}{2} \, \tilde{C}_{I^2,SSqq} : \sum_{I,S=1}^{3} \sum_{\mathbf{n}} \left[\nabla_{S,(v)} \rho_{S,I}^{a^\dagger,a}(\mathbf{n}) \right] \nabla_{S,(v)} \rho_{S,I}^{a^\dagger,a}(\mathbf{n}) :\,, \tag{4.165}$$

which are invariant under the cubic rotational group SO(3, Z) but not under general SO(3) transformations. We now adjust the $I = 0$ combination of these operators to eliminate the unphysical mixing of the $^3D_3$ partial wave in the $^3S(D)_1$ channel. For this purpose, we take

$$\tilde{C}_{SSqq} \equiv \frac{1}{4} \left(3\tilde{C}_{SSqq}^{I=1} + \tilde{C}_{SSqq}^{I=0} \right), \quad \tilde{C}_{I^2,SSqq} \equiv \frac{1}{4} \left(\tilde{C}_{SSqq}^{I=1} - \tilde{C}_{SSqq}^{I=0} \right), \tag{4.166}$$

as the coupling constants for the $I = 0$ and $I = 1$ channels. If we define

$$\tilde{V}_{SSqq}^{I=0} \equiv \tilde{V}_{SSqq} + \tilde{V}_{I^2,SSqq}, \tag{4.167}$$

and set

$$\tilde{C}_{SSqq} \equiv \frac{1}{4} \, \tilde{C}_{SSqq}^{I=0}, \quad \tilde{C}_{I^2,SSqq} \equiv -\frac{1}{4} \, \tilde{C}_{SSqq}^{I=0}, \tag{4.168}$$

we may tune $\tilde{C}_{SSqq}^{I=0}$ to eliminate the unphysical mixing of the $^3D_3$ partial wave. In principle, we may also include the $I = 1$ combination in order to eliminate the unphysical mixing of the $^3F_4$ partial wave in the $^3P(F)_2$ channel. However, as this is found to be a very small effect, the $I = 1$ coefficient is set to zero. The effects of rotational symmetry breaking on the mixing angles in np scattering are discussed in more detail in Chap. 5.

In addition to the local rotational symmetry-breaking terms discussed above, non-local lattice artifacts associated with the OPE potential also appear. These include $\mathcal{O}(Q^2)$ terms proportional to

$$\frac{\tau_A \cdot \tau_B}{\mathbf{q}^2 + M_\pi^2} \left[(\mathbf{q} \cdot \boldsymbol{\sigma}_A) \sum_{l=1}^{3} q_l^3 (\sigma_B)_l + (\mathbf{q} \cdot \boldsymbol{\sigma}_B) \sum_{l=1}^{3} q_l^3 (\sigma_A)_l \right], \qquad (4.169)$$

from the gradient coupling of the pion, and

$$\tau_A \cdot \tau_B \frac{(\mathbf{q} \cdot \boldsymbol{\sigma}_A)(\mathbf{q} \cdot \boldsymbol{\sigma}_B)}{(\mathbf{q}^2 + M_\pi^2)^2} \sum_{l=1}^{3} q_l^4, \qquad (4.170)$$

from the pion propagator. For $|\mathbf{q}| < M_\pi$, the components of Eqs. (4.169) and (4.170) with $S = 0$ coincide with the local $\mathcal{O}(Q^2)$ terms discussed above. However, for $M_\pi < |\mathbf{q}| < \Lambda$, the non-locality of these lattice artifacts becomes apparent. As we shall discuss in more detail in Chap. 5, such lattice artifacts may manifest themselves in the $^3S(D)_1$ mixing angle ε_1. In order to remove such $\mathcal{O}(Q^2)$ effects, we employ a perturbative improvement of the pion-nucleon gradient coupling. This is accomplished using the operator

$$\tilde{V}_\pi^{DX} \equiv -\frac{\alpha_t}{2} \left(\frac{g_A}{2F_\pi} \right)^2 : \sum_{I,S,S'=1}^{3} \sum_{\mathbf{n},\mathbf{n}'} \rho_{S,I}^{a^\dagger,a}(\mathbf{n}) \, \rho_{S',I}^{a^\dagger,a}(\mathbf{n}') \qquad (4.171)$$

$$\times \left(\frac{G_{SS'}^{(\nu=1)}(\mathbf{n} - \mathbf{n}', M_\pi)}{q_\pi} - \frac{G_{SS'}^{(\nu=0)}(\mathbf{n} - \mathbf{n}', M_\pi)}{q_\pi} \right) :,$$

where we have followed the conventions of the OPE potential given in Eq. (4.90). We recall that the OPE potential in the (non-perturbative) LO treatment in itself makes use of a (stretched) improved pion propagator, while the gradient couplings correspond to the unimproved expression with $\nu = 0$. In Eq. (4.171), we account perturbatively for the improvement of the gradient couplings from $\nu = 0$ to $\nu = 1$.

When the rotational symmetry-breaking terms are added to the NLO expressions, we obtain

$$V^{(2)} \rightarrow V^{(2)} + \tilde{V}_{SSqq} + \tilde{V}_{I^2,SSqq} + \tilde{V}_\pi^{DX}, \qquad (4.172)$$

which completes our treatment of the NLO contributions in NLEFT. When we present results labeled "NLO" in Chaps. 5 and 7, we are in most cases referring to a combination of NLO contact terms, shifts of the LO contact interactions, rotational symmetry-breaking operators, and any other contributions that may be counted as NLO. Unless otherwise stated, the momentum operators in the NLO contact terms are unimproved and correspond, unlike the nucleon and pion kinetic energy terms, to the operator $\nabla^2_{l,(v)}$ which has two zeros per Brillouin zone in each spatial dimension. Such a "doubling problem" is a useful feature for the operators in the NLO action, since it prevents interactions that are tuned at low momenta from becoming too strong in the vicinity of the cutoff momentum.

Finally, we note that non-local corrections to the OPE potential are also generated at $\mathcal{O}(\alpha_t Q^2/\tilde{m}_N)$ by the finite temporal lattice spacing δ. However, one finds that such discretization errors remain negligible due to the relatively small value of $\delta = (150\,\text{MeV})^{-1}$. This has been explicitly checked by comparison of nucleon-nucleon scattering observables for different choices of δ. Our choice of δ will be discussed further in the context of the NNLO three-nucleon interaction in Chap. 5, and for issues related to the Tjon correlation of the binding energies for $A = 3$ and $A = 4$, to be discussed in Chap. 8.

4.3.4 Isospin Breaking and Electromagnetism

We shall now consider the effects of strong isospin breaking and the Coulomb interaction between protons on NLEFT. These effects will be collectively labeled "EMIB" for brevity. The treatment of EMIB in EFT has been addressed in the literature, see Refs. [33–40]. In the power counting scheme proposed by Ref. [40], the isospin-breaking OPE and Coulomb interactions are taken to be of the same magnitude as the $\mathcal{O}(Q^2)$ or NLO contributions to the 2NF. For the isospin-symmetric interactions, the convention $M_\pi \equiv M_\pi^0$ has been used for the pion masses. The isospin-violating OPE amplitude due to pion mass differences is

$$\mathcal{A}\left[V_\pi^{\text{IB}}\right] = -\left(\frac{g_A}{2F_\pi}\right)^2 \left[(\tau_1)_A (\tau_1)_B + (\tau_2)_A (\tau_2)_B\right]$$

$$\times (\mathbf{q} \cdot \boldsymbol{\sigma}_A)(\mathbf{q} \cdot \boldsymbol{\sigma}_B)\left(\frac{1}{\mathbf{q}^2 + M_{\pi^\pm}^2} - \frac{1}{\mathbf{q}^2 + M_{\pi^0}^2}\right), \quad (4.173)$$

and on the lattice, the isospin-breaking OPE operator is

$$\tilde{V}_\pi^{\text{IB}} \equiv -\frac{\alpha_t}{2}\left(\frac{g_A}{2F_\pi}\right)^2 : \sum_{I,S,S'=1}^{3} \sum_{\mathbf{n},\mathbf{n}'} \rho_{S,I}^{a^\dagger,a}(\mathbf{n})\, \rho_{S',I}^{a^\dagger,a}(\mathbf{n}') \qquad (4.174)$$

$$\times \left(\frac{G_{SS'}^{(v=0)}(\mathbf{n} - \mathbf{n}', M_{\pi^\pm})}{q_{\pi^\pm}} - \frac{G_{SS'}^{(v=0)}(\mathbf{n} - \mathbf{n}', M_{\pi^0})}{q_{\pi^0}}\right) : ,$$

where the pion propagators are again improved similarly to the OPE expression in Eq. (4.90), and the gradient couplings in the pion correlation functions use unimproved lattice momenta with $\nu = 0$.

Electromagnetic effects in pp scattering can be straightforwardly accounted for by the long-range Coulomb repulsion between protons, which in position space is described by the amplitude

$$\mathscr{A}[V_{\rm EM}] = \frac{\alpha_{\rm EM}}{r} \left(\frac{1+\tau_3}{2}\right)_A \left(\frac{1+\tau_3}{2}\right)_B, \tag{4.175}$$

with $\alpha_{\rm EM} \simeq 1/137$ the electromagnetic fine-structure constant. The corresponding lattice operator is

$$\tilde{V}_{\rm EM} \equiv \frac{1}{2} : \sum_{\mathbf{n},\mathbf{n}'} \frac{\alpha_{\rm EM}}{R(\mathbf{n}-\mathbf{n}')} \frac{1}{4} \left[\rho^{a^\dagger,a}(\mathbf{n}) + \rho^{a^\dagger,a}_{I=3}(\mathbf{n})\right] \left[\rho^{a^\dagger,a}(\mathbf{n}') + \rho^{a^\dagger,a}_{I=3}(\mathbf{n}')\right] :, \tag{4.176}$$

where R denotes the separation on the lattice, given by

$$R(\mathbf{n}) \equiv \max(1/2, |\mathbf{n}|), \tag{4.177}$$

where the factor $1/2$ removes the singularity associated with the $\sim 1/r$ Coulomb potential. Such a convention is necessary, as $R(\mathbf{n}) = |\mathbf{n}|$ vanishes when two protons occupy the same lattice site. However, any other value besides $1/2$ could also be used without any observable effect, as an explicit pp contact interaction is included.

In addition to the long-range Coulomb potential, we include all possible two-nucleon contact interactions (nn, pp, $S = 0$ np and $S = 1$ np). As we shall fit the isospin-symmetric interaction coefficients using np scattering data (for details, see Chap. 5), the additional np contact interactions are merely linear combinations of the ΔV and ΔV_{I^2} terms in the NLO amplitude. This leaves the amplitudes

$$\mathscr{A}[V_{nn}] = C_{nn} \left(\frac{1-\tau_3}{2}\right)_A \left(\frac{1-\tau_3}{2}\right)_B, \tag{4.178}$$

and

$$\mathscr{A}[V_{pp}] = C_{pp} \left(\frac{1+\tau_3}{2}\right)_A \left(\frac{1+\tau_3}{2}\right)_B, \tag{4.179}$$

for the contribution from isospin-breaking contact interactions. On the lattice, the nn contact operator is

$$\tilde{V}_{nn} \equiv \frac{1}{2}\tilde{C}_{nn} : \sum_{\mathbf{n}} \frac{1}{4} \left[\rho^{a^\dagger,a}(\mathbf{n}) - \rho^{a^\dagger,a}_{I=3}(\mathbf{n})\right] \left[\rho^{a^\dagger,a}(\mathbf{n}') - \rho^{a^\dagger,a}_{I=3}(\mathbf{n}')\right] :, \tag{4.180}$$

and the pp operator is

$$\tilde{V}_{pp} \equiv \frac{1}{2}\tilde{C}_{pp} : \sum_{\mathbf{n}} \frac{1}{4}\left[\rho^{a^{\dagger},a}(\mathbf{n}) + \rho^{a^{\dagger},a}_{I=3}(\mathbf{n})\right]\left[\rho^{a^{\dagger},a}(\mathbf{n}') + \rho^{a^{\dagger},a}_{I=3}(\mathbf{n}')\right] :, \quad (4.181)$$

for which the coefficients can determined from the observed inequality of the np, nn and pp scattering lengths. We add these isospin-breaking terms perturbatively to the NLO transfer matrix, giving the electromagnetic and isospin-breaking NLO transfer matrix

$$M_{\text{NLO+EMIB}} = M_{\text{NLO}} - \alpha_t : \left(\tilde{V}^{\text{IB}}_\pi + \tilde{V}_{\text{EM}} + \tilde{V}_{nn} + \tilde{V}_{pp}\right) M_{\text{LO}} :, \quad (4.182)$$

where we use the term "EMIB contribution" to denote collectively the strong and electromagnetic isospin-breaking effects.

Before we move on to a description of the three-nucleon interaction which first contributes at NNLO, we note that the isospin-breaking contributions in our treatment of NLEFT are (in general) perturbative. Hence, the energy splitting between "mirror nuclei" such as ^{3}H and ^{3}He can be obtained from the same, non-perturbative, LO calculation. In addition to calculating the isospin-flipped $\tilde{V}_{\text{EM}}$, the other necessary modification is $C_{nn} \leftrightarrow C_{pp}$ for $\tilde{V}_{nn}$ and $\tilde{V}_{pp}$.

4.3.5 Three-Nucleon Interactions

The three-nucleon interaction or three-nucleon force (3NF) first appears at NNLO or $\mathcal{O}(Q^3)$ in the chiral EFT expansion. This 3NF consists of a pure contact interaction $V^{(3N)}_{\text{contact}}$, an OPE contribution $V^{(3N)}_{\text{OPE}}$, and a TPE contribution $V^{(3N)}_{\text{TPE}}$ (see Figs. 2.6 and 4.3). The TPE contribution to the 3NF is closely related to the pion-nucleon scattering amplitude, as noted long ago [41]. As for the 2NF at NLO, we shall introduce the lattice operators and relate them to continuum amplitudes, along with a discussion of various different conventions for their coupling constants.

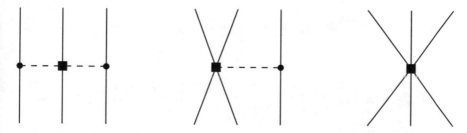

Fig. 4.3 Two-pion exchange (TPE), one-pion exchange (OPE) and contact contributions to the three-nucleon force (3NF) in NLEFT. The TPE contribution is conventionally further subdivided into three parts, into contributions proportional to the pion-nucleon coupling constants c_i

Due to the constraints of isospin symmetry, spin symmetry and Fermi statistics, there is only one independent 3NF contact interaction [19, 42, 43]. On the lattice, we choose to write this contact interaction as a product of total nucleon densities, giving

$$\tilde{V}_{\text{contact}}^{(3N)} \equiv \frac{1}{6} \tilde{D}_{\text{contact}} : \sum_{\mathbf{n}} \left[\rho^{a^\dagger, a}(\mathbf{n}) \right]^3 : . \tag{4.183}$$

Further, the OPE potential is written as

$$\tilde{V}_{\text{OPE}}^{(3N)} \equiv -\tilde{D}_{\text{OPE}} \frac{g_A}{2F_\pi} \frac{\alpha_t}{q_\pi} \sum_{I,S,S'=1}^{3} \sum_{\mathbf{n},\mathbf{n}'} \left\langle \left[\nabla_{S',(\nu)} \pi'_I(\mathbf{n}', t) \right] \nabla_{S,(\nu)} \pi'_I(\mathbf{n}, t) \right\rangle$$

$$\times : \rho_{S',I}^{a^\dagger, a}(\mathbf{n}') \, \rho_{S,I}^{a^\dagger, a}(\mathbf{n}) \, \rho^{a^\dagger, a}(\mathbf{n}) : \tag{4.184}$$

$$= -\tilde{D}_{\text{OPE}} \frac{g_A}{2F_\pi} \alpha_t \sum_{I,S,S'=1}^{3} \sum_{\mathbf{n},\mathbf{n}'} \frac{G_{SS'}^{(\nu)}(\mathbf{n} - \mathbf{n}', M_\pi)}{q_\pi}$$

$$\times : \rho_{S',I}^{a^\dagger, a}(\mathbf{n}') \, \rho_{S,I}^{a^\dagger, a}(\mathbf{n}) \, \rho^{a^\dagger, a}(\mathbf{n}) :, \tag{4.185}$$

where the angled brackets denote the pion correlation function defined in Eq. (4.52), which for the 3NF contributions may appear with either one or two derivatives. The TPE contribution to the 3NF is conventionally split into three parts, as the corresponding pion-nucleon scattering amplitude contains three distinct contributions. These are defined as

$$\tilde{V}_{\text{TPE1}}^{(3N)} \equiv \tilde{D}_{\text{TPE1}} \left(\frac{g_A}{2F_\pi} \right)^2 \frac{\alpha_t^2}{q_\pi^2} \sum_{I,S,S'=1}^{3} \sum_{\mathbf{n},\mathbf{n}'} \left\langle \left[\nabla_{S',(\nu)} \pi'_I(\mathbf{n}', t) \right] \nabla_{S,(\nu)} \pi'_I(\mathbf{n}, t) \right\rangle$$

$$\tag{4.186}$$

$$\times \sum_{S''=1}^{3} \sum_{\mathbf{n}''} \left\langle \left[\nabla_{S'',(\nu)} \pi'_I(\mathbf{n}'', t) \right] \nabla_{S,(\nu)} \pi'_I(\mathbf{n}, t) \right\rangle$$

$$: \rho_{S',I}^{a^\dagger, a}(\mathbf{n}') \, \rho_{S'',I}^{a^\dagger, a}(\mathbf{n}'') \, \rho^{a^\dagger, a}(\mathbf{n}) :,$$

or

$$\tilde{V}_{\text{TPE1}}^{(3N)} = \tilde{D}_{\text{TPE1}} \left(\frac{g_A}{2F_\pi} \right)^2 \alpha_t^2 \sum_{I,S,S'=1}^{3} \sum_{\mathbf{n},\mathbf{n}'} \frac{G_{SS'}^{(\nu)}(\mathbf{n} - \mathbf{n}', M_\pi)}{q_\pi} \tag{4.187}$$

$$\times \sum_{S''=1}^{3} \sum_{\mathbf{n}''} \frac{G_{SS''}^{(\nu)}(\mathbf{n} - \mathbf{n}'', M_\pi)}{q_\pi} : \rho_{S',I}^{a^\dagger, a}(\mathbf{n}') \, \rho_{S'',I}^{a^\dagger, a}(\mathbf{n}'') \, \rho^{a^\dagger, a}(\mathbf{n}) :,$$

for the contribution proportional to c_3,

$$\tilde{V}_{\text{TPE2}}^{(3N)} \equiv \tilde{D}_{\text{TPE2}} \left(\frac{g_A}{2F_\pi}\right)^2 M_\pi^2 \frac{\alpha_t^2}{q_\pi^2} \sum_{I,S'=1}^{3} \sum_{\mathbf{n},\mathbf{n}'} \left\langle \left[\nabla_{S',(v)} \pi'_I(\mathbf{n}', t)\right] \pi'_I(\mathbf{n}, t)\right\rangle$$

$$(4.188)$$

$$\times \sum_{S''=1}^{3} \sum_{\mathbf{n}''} \left\langle \left[\nabla_{S'',(v)} \pi'_I(\mathbf{n}'', t)\right] \pi'_I(\mathbf{n}, t)\right\rangle$$

$$: \rho_{S',I}^{a^\dagger,a}(\mathbf{n}') \, \rho_{S'',I}^{a^\dagger,a}(\mathbf{n}'') \, \rho^{a^\dagger,a}(\mathbf{n}) : ,$$

or

$$\tilde{V}_{\text{TPE2}}^{(3N)} = \tilde{D}_{\text{TPE2}} \left(\frac{g_A}{2F_\pi}\right)^2 M_\pi^2 \alpha_t^2 \sum_{I,S,S'=1}^{3} \sum_{\mathbf{n},\mathbf{n}'} \frac{G_{S'}^{(v)}(\mathbf{n} - \mathbf{n}', M_\pi)}{q_\pi}$$

$$(4.189)$$

$$\times \sum_{S''=1}^{3} \sum_{\mathbf{n}''} \frac{G_{S''}^{(v)}(\mathbf{n} - \mathbf{n}'', M_\pi)}{q_\pi} : \rho_{S',I}^{a^\dagger,a}(\mathbf{n}') \, \rho_{S'',I}^{a^\dagger,a}(\mathbf{n}'') \, \rho^{a^\dagger,a}(\mathbf{n}) : ,$$

for the contribution proportional to c_1. Here, we note that the pion correlation functions with one derivative are obtained from those with two derivatives by simply dropping one of the momentum factors in the numerator of Eq. (4.52). Finally, we have

$$\tilde{V}_{\text{TPE3}}^{(3N)} \equiv \tilde{D}_{\text{TPE3}} \left(\frac{g_A}{2F_\pi}\right)^2 \frac{\alpha_t^2}{q_\pi^2} \sum_{I_1,S_1,S'=1}^{3} \sum_{\mathbf{n},\mathbf{n}'} \left\langle \left[\nabla_{S',(v)} \pi'_{I_1}(\mathbf{n}', t)\right] \nabla_{S_1,(v)} \pi'_{I_1}(\mathbf{n}, t)\right\rangle$$

$$\times \sum_{I_2,S_2,S''=1}^{3} \sum_{\mathbf{n}''} \left\langle \left[\nabla_{S'',(v)} \pi'_{I_2}(\mathbf{n}'', t)\right] \nabla_{S_2,(v)} \pi'_{I_2}(\mathbf{n}, t)\right\rangle$$

$$\times \sum_{I_3,S_3=1}^{3} \varepsilon_{S_1,S_2,S_3} \varepsilon_{I_1,I_2,I_3} : \rho_{S',I_1}^{a^\dagger,a}(\mathbf{n}') \, \rho_{S'',I_2}^{a^\dagger,a}(\mathbf{n}'') \, \rho_{S_3,I_3}^{a^\dagger,a}(\mathbf{n}) : ,$$

$$(4.190)$$

or

$$\tilde{V}_{\text{TPE3}}^{(3N)} \equiv \tilde{D}_{\text{TPE3}} \left(\frac{g_A}{2F_\pi}\right)^2 \alpha_t^2 \sum_{I_1,S_1,S'=1}^{3} \sum_{\mathbf{n},\mathbf{n}'} \frac{G_{S'S_1}^{(v)}(\mathbf{n} - \mathbf{n}', M_\pi)}{q_\pi}$$

$$\times \sum_{I_2,S_2,S''=1}^{3} \sum_{\mathbf{n}''} \frac{G_{S''S_2}^{(v)}(\mathbf{n} - \mathbf{n}'', M_\pi)}{q_\pi}$$

$$\times \sum_{I_3,S_3=1}^{3} \varepsilon_{S_1,S_2,S_3} \varepsilon_{I_1,I_2,I_3} : \rho_{S',I_1}^{a^\dagger,a}(\mathbf{n}') \, \rho_{S'',I_2}^{a^\dagger,a}(\mathbf{n}'') \, \rho_{S_3,I_3}^{a^\dagger,a}(\mathbf{n}) : .$$

$$(4.191)$$

for the contribution proportional to c_4, where the $\varepsilon_{i,j,k}$ are Levi-Civita tensors for spin and isospin.

We are now in a position to consider the continuum limits of the 3NF expressions and relate the coupling constants to physical observables. As for the two-nucleon interactions considered so far, we take the nucleons to be distinguishable and use the notation

$$\sum_{P(A,B,C)} f(A, B, C) \equiv f(A, B, C) + f(A, C, B) + f(B, A, C)$$

$$+ f(C, A, B) + f(B, C, A) + f(C, B, A), \qquad (4.192)$$

and

$$\sum_{P(A,B,C)} f(A, B) \equiv f(A, B) + f(A, C) + f(B, A)$$

$$+ f(C, A) + f(B, C) + f(C, B), \qquad (4.193)$$

to denote summation over all permutations of the (distinguishable) nucleons A, B and C. In the continuum limit, the tree-level scattering amplitudes are

$$\mathscr{A}\left[V^{(3N)}_{\text{contact}}\right] = D_{\text{contact}} = \frac{1}{2}E \sum_{P(A,B,C)} \tau_A \cdot \tau_B, \qquad (4.194)$$

where the latter expression corresponds to writing the 3NF contact term as a $\tau_A \cdot \tau_B$ interaction with summation over all nucleon labels. In the continuum limit, the 3NF OPE scattering amplitude is given by

$$\mathscr{A}\left[V^{(3N)}_{\text{OPE}}\right] = -D_{\text{OPE}} \frac{g_A}{2F_\pi} \sum_{P(A,B,C)} \frac{\mathbf{q}_A \cdot \boldsymbol{\sigma}_A}{\mathbf{q}_A^2 + M_\pi^2} (\mathbf{q}_A \cdot \boldsymbol{\sigma}_B)(\tau_A \cdot \tau_B), \qquad (4.195)$$

where $\mathbf{q}_A$ denotes the difference between the final and initial momenta of nucleon A. Several sets of related coupling constants for the 3NF are encountered in the literature. The relationship between our lattice coupling constants for the 3NF contact and 3NF OPE terms, the alternative constants D, E introduced in Chap. 2, and the sometimes used couplings c_D, c_E can be summarized as in Epelbaum et al. [43]

$$D_{\text{contact}} = -3E, \quad E \equiv \frac{c_E}{F_\pi^4 \Lambda_{\text{hard}}}, \qquad (4.196)$$

and

$$D_{\text{OPE}} = \frac{D}{4F_\pi}, \quad D \equiv \frac{c_D}{F_\pi^2 \Lambda_{\text{hard}}}, \qquad (4.197)$$

where $\Lambda_{\text{hard}} \simeq 600\,\text{MeV}$. Just as the coupling constants in the two-nucleon sector are determined by fitting two-nucleon scattering phase shifts, we determine c_D and c_E by matching to observables for $A = 3$. For example, we may use the physical triton binding energy at infinite volume to determine one of these constants. Given c_D or c_E, the other one may then be constrained using a second observable, such as the nucleon-deuteron (nd) scattering phase shifts. More details on this topic will be given in Chap. 5.

For the TPE contributions to the 3NF, we find the scattering amplitudes

$$\mathscr{A}\left[V_{\text{TPE1}}^{(3N)}\right] = D_{\text{TPE1}} \left(\frac{g_A}{2F_\pi}\right)^2 \sum_{P(A,B,C)} \frac{(\mathbf{q}_A \cdot \boldsymbol{\sigma}_A)(\mathbf{q}_B \cdot \boldsymbol{\sigma}_B)}{(\mathbf{q}_A^2 + M_\pi^2)(\mathbf{q}_B^2 + M_\pi^2)} (\mathbf{q}_A \cdot \mathbf{q}_B)(\tau_A \cdot \tau_B),$$

(4.198)

$$\mathscr{A}\left[V_{\text{TPE2}}^{(3N)}\right] = D_{\text{TPE2}} \left(\frac{g_A}{2F_\pi}\right)^2 M_\pi^2 \sum_{P(A,B,C)} \frac{(\mathbf{q}_A \cdot \boldsymbol{\sigma}_A)(\mathbf{q}_B \cdot \boldsymbol{\sigma}_B)}{(\mathbf{q}_A^2 + M_\pi^2)(\mathbf{q}_B^2 + M_\pi^2)} (\tau_A \cdot \tau_B),$$

(4.199)

and

$$\mathscr{A}\left[V_{\text{TPE3}}^{(3N)}\right] = D_{\text{TPE3}} \left(\frac{g_A}{2F_\pi}\right)^2 \sum_{P(A,B,C)} \frac{(\mathbf{q}_A \cdot \boldsymbol{\sigma}_A)(\mathbf{q}_B \cdot \boldsymbol{\sigma}_B)}{(\mathbf{q}_A^2 + M_\pi^2)(\mathbf{q}_B^2 + M_\pi^2)}$$
$$\times \left[(\mathbf{q}_A \times \mathbf{q}_B) \cdot \boldsymbol{\sigma}_C\right]\left[(\tau_A \times \tau_B) \cdot \tau_C\right],$$

(4.200)

where

$$D_{\text{TPE1}} = \frac{c_3}{F_\pi^2}, \quad D_{\text{TPE2}} = -\frac{2c_1}{F_\pi^2}, \quad D_{\text{TPE3}} = \frac{c_4}{2F_\pi^2},$$

(4.201)

in terms of the couplings c_1, c_3 and c_4 introduced in Chap. 2. While the most precise values have been obtained from the Roy-Steiner equation analysis of pion-nucleon scattering, most NLEFT calculations used older values extracted from the chiral perturbation theory analysis of low-energy pion-nucleon scattering data. These are $c_1 = -0.81\,\text{GeV}^{-1}$, $c_3 = -4.7\,\text{GeV}^{-1}$ and $c_4 = 3.4\,\text{GeV}^{-1}$, which have been determined in Refs. [44, 45]. For future applications, the values extracted from the Roy-Steiner equation analysis listed in Chap. 2 should be used. Note further that the aforementioned relation between the elastic pion-nucleon scattering amplitude and the TPE contribution to the 3NF becomes apparent.

We also apply the perturbative analysis introduced for the NLO contact operators to the NNLO interactions. We therefore define the NNLO transfer matrix

$$M_{\text{NNLO}} \equiv M_{\text{NLO}} + \frac{\int \mathscr{D}\pi_I' \mathscr{D}\sigma \, \exp\left(-S_{\pi\pi}(t) - S_{ss}(t)\right) \Delta M_{\text{NNLO}}(\pi_I', \sigma, t)}{\int \mathscr{D}\pi_I' \mathscr{D}\sigma \, \exp\left(-S_{\pi\pi}(t) - S_{ss}(t)\right)},$$

(4.202)

where

$$\Delta M_{\text{NNLO}}(\pi'_I, \sigma, t) \equiv \Delta M_{\text{contact}}^{(3N)}(\pi'_I, \sigma, t) + \Delta M_{\text{OPE}}^{(3N)}(\pi'_I, \sigma, t)$$
$$+ \Delta M_{\text{TPE1}}^{(3N)}(\pi'_I, \sigma, t) + \Delta M_{\text{TPE2}}^{(3N)}(\pi'_I, \sigma, t)$$
$$+ \Delta M_{\text{TPE3}}^{(3N)}(\pi'_I, \sigma, t), \tag{4.203}$$

summarizes the contact, OPE and TPE contributions to the NNLO 3NF. As for the NLO terms, we now obtain the contributions to the NNLO transfer matrix in terms of functional derivatives with respect to the source fields ε_i in Eqs. (4.120) and (4.121). The three-nucleon contact term is given by

$$\Delta M_{\text{contact}}^{(3N)}(\pi'_I, \sigma, t) = -\frac{1}{6} \tilde{D}_{\text{contact}} \alpha_t \sum_{\mathbf{n}} \left(\frac{\delta}{\delta \varepsilon_1(\mathbf{n}, t)} \right)^3 M(\pi'_I, \sigma, \varepsilon_i, t) \Bigg|_{\varepsilon_i \to 0}, \tag{4.204}$$

the OPE contribution by

$$\Delta M_{\text{OPE}}^{(3N)}(\pi'_I, \sigma, t) = -\tilde{D}_{\text{OPE}} \frac{\alpha_t}{\sqrt{q_\pi}} \sum_{I,S=1}^{3} \sum_{\mathbf{n}} \nabla_{S,(v)} \pi'_I(\mathbf{n}, t)$$
$$\times \frac{\delta}{\delta \varepsilon_{10}(\mathbf{n}, t, S, I)} \frac{\delta}{\delta \varepsilon_1(\mathbf{n}, t)} M(\pi'_I, \sigma, \varepsilon_i, t) \Bigg|_{\varepsilon_i \to 0}, \tag{4.205}$$

and the TPE terms by

$$\Delta M_{\text{TPE1}}^{(3N)}(\pi'_I, \sigma, t) = -\tilde{D}_{\text{TPE1}} \frac{\alpha_t}{q_\pi} \sum_{I,S=1}^{3} \sum_{\mathbf{n}}$$
$$\times \left\{ \left[\nabla_{S,(v)} \pi'_I(\mathbf{n}, t) \right] \nabla_{S,(v)} \pi'_I(\mathbf{n}, t) \right.$$
$$\left. - \left\langle \left[\nabla_{S,(v)} \pi'_I(\mathbf{n}, t) \right] \nabla_{S,(v)} \pi'_I(\mathbf{n}, t) \right\rangle \right\}$$
$$\times \frac{\delta}{\delta \varepsilon_1(\mathbf{n}, t)} M(\pi'_I, \sigma, \varepsilon_i, t) \Bigg|_{\varepsilon_i \to 0}, \tag{4.206}$$

$$\Delta M_{\text{TPE2}}^{(3N)}(\pi_I', \sigma, t) = -\tilde{D}_{\text{TPE2}} \, M_\pi^2 \frac{\alpha_t}{q_\pi} \sum_{I=1}^{3} \sum_{\mathbf{n}}$$

$$\times \left\{ \pi_I'(\mathbf{n}, t)\pi_I'(\mathbf{n}, t) - \langle \pi_I'(\mathbf{n}, t)\pi_I'(\mathbf{n}, t) \rangle \right\}$$

$$\times \left. \frac{\delta}{\delta \varepsilon_1(\mathbf{n}, t)} \, M(\pi_I', \sigma, \varepsilon_i, t) \right|_{\varepsilon_i \to 0}, \tag{4.207}$$

and

$$\Delta M_{\text{TPE3}}^{(3N)}(\pi_I', \sigma, t) = -\tilde{D}_{\text{TPE3}} \, \frac{\alpha_t}{q_\pi} \sum_{I_1, I_2, I_3, S_1, S_2, S_3 = 1}^{3} \sum_{\mathbf{n}}$$

$$\times \varepsilon_{S_1, S_2, S_3} \varepsilon_{I_1, I_2, I_3} \left[\nabla_{S_1, (v)} \pi_{I_1}'(\mathbf{n}, t) \right] \nabla_{S_2, (v)} \pi_{I_2}'(\mathbf{n}, t)$$

$$\times \left. \frac{\delta}{\delta \varepsilon_{10}(\mathbf{n}, t, S_3, I_3)} \, M(\pi_I', \sigma, \varepsilon_i, t) \right|_{\varepsilon_i \to 0}. \tag{4.208}$$

Schematically, we may then write the full NNLO expression as

$$M_{\text{NNLO}} = M_{\text{NLO}} - \alpha_t : V^{(3N)} M_{\text{LO}} :, \tag{4.209}$$

where

$$V^{(3N)} \equiv \tilde{V}_{\text{contact}}^{(3N)} + \tilde{V}_{\text{OPE}}^{(3N)} + \tilde{V}_{\text{TPE1}}^{(3N)} + \tilde{V}_{\text{TPE2}}^{(3N)} + \tilde{V}_{\text{TPE3}}^{(3N)}, \tag{4.210}$$

summarizes the contributions from the 3NF. It should be noted that in pionless EFT, the three-nucleon contact interaction is included at LO [17–19]. This is required in order to stabilize the three-nucleon system in the limit of zero-range interactions [14]. In NLEFT, the LO interactions have non-zero range, which allows us to circumvent this problem. Nevertheless, for a relatively large lattice spacing of $a \simeq 2$ fm, the three-nucleon contact interaction may turn out to be numerically large and to require a non-perturbative treatment. Such a treatment of three-body contact interactions on the lattice has been discussed in Refs. [21, 46, 47].

4.3.6   Contributions Beyond NNLO

The "state-of-the-art" lattice NLEFT calculations contain all operators up-to-and-including NNLO. Some residual error is expected to appear due to omitted higher-order interactions, starting at $\mathcal{O}(Q^4)$ or N3LO. The magnitude of such errors depends on the momentum scale probed by the physical system of interest. For

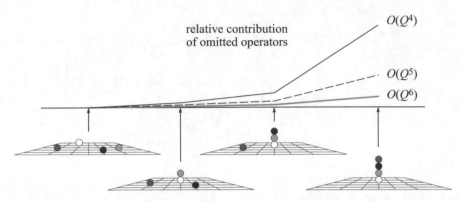

Fig. 4.4 Sketch of the relative contribution of omitted operators at $\mathscr{O}(Q^4)$, $\mathscr{O}(Q^5)$ and $\mathscr{O}(Q^6)$ for different types of nucleon configurations. The relative contribution is dominated by the case where four nucleons are close together

well-separated low-momentum nucleons, no significant deviations are expected to occur. For two nucleons in close proximity, the systematical error should also remain very small. The properties of the deuteron and soft nucleon-deuteron scattering are both accurately reproduced [48]. For three nucleons in close proximity, the error is somewhat increased, and for a tight cluster of four nucleons it is expected to be maximal. The expected trend for systematical errors is sketched qualitatively in Fig. 4.4. We also note that a localized collection of five or more nucleons (with vanishing relative orbital angular momentum) is forbidden by Fermi statistics.

As indicated in Fig. 4.4, the relative contribution is likely dominated by the case where four nucleons are close together. If this hypothesis is correct, then the contribution of higher-order operators to low-energy phenomena should be approximately universal. In other words, different higher-order operators would produce roughly the same effect on low-energy data. Such a situation is analogous to the difficulty one finds in resolving the value of c_D from low-energy three-nucleon data (see Chap. 5). One useful consequence of this universality is that most of the residual error can be cancelled by adjusting the coefficient of an "effective" four-nucleon contact term

$$\tilde{V}_{\mathrm{eff}}^{(4N)} \equiv \frac{1}{24} \tilde{D}_{\mathrm{eff}}^{(4N)} : \sum_{\mathbf{n}} \left[\rho^{a^\dagger,a}(\mathbf{n}) \right]^4 :, \tag{4.211}$$

which we emphasize should not be confused with the four-nucleon contact interaction that appears at $\mathscr{O}(Q^6)$ in the chiral EFT expansion. Thus, we are not suggesting a rearrangement of the chiral EFT power counting, but rather taking advantage of the expected universality of the omitted interactions at N3LO and higher orders.

If we denote a configuration of four nucleons on a single lattice site by

$$|4N\rangle \equiv a_{0,0}^\dagger(\mathbf{n}) a_{1,0}^\dagger(\mathbf{n}) a_{0,1}^\dagger(\mathbf{n}) a_{1,1}^\dagger(\mathbf{n}) |0\rangle, \tag{4.212}$$

then the potential energy of such a configuration is found to depend upon the three-nucleon contact operator and the local part of the three-nucleon OPE interaction, giving

$$\langle 4N | \, \tilde{V}_{\text{contact}}^{(3N)} \, | 4N \rangle = 4 \tilde{D}_{\text{contact}}^{(3N)}, \qquad (4.213)$$

and

$$\langle 4N | \, \tilde{V}_{\text{OPE}}^{(3N)} \, | 4N \rangle = \tilde{D}_{\text{OPE}}^{(3N)} \frac{12 g_A}{F_\pi} \frac{\alpha_t}{q_\pi} \sum_{S=1}^{3} G_{SS}^{(v)}(\mathbf{0}), \qquad (4.214)$$

such that if $\tilde{D}_{\text{contact}}^{(3N)}$ or $\tilde{D}_{\text{OPE}}^{(3N)}$ were sufficiently large and negative, a clustering instability can be produced in systems with four or more nucleons. This is a lattice artifact that appears on coarse lattices [10], and is similar to the clustering instability found with point-like two-nucleon contact interactions [49]. As we have shown in detail here, that problem was solved using an improved LO lattice action with momentum-dependent contact interactions. In principle, an analogous technique could be adopted for the three-nucleon interactions. Along with the non-perturbative treatment of the three-nucleon contact interactions and explicit inclusion of N3LO terms in lattice chiral EFT, this remains a topic of current investigation.

In practice, the coefficient of the 4N contact term (4.211) can be found by requiring that the binding energy of ^{4}He be accurately reproduced, once the 2NF and 3NF contributions have been fixed to available data for $A = 2$ and $A = 3$. Nuclei with $A > 4$ can then be considered as predictions of NLEFT, a situation which has so far prevailed for a majority of NLEFT calculations of light and medium-mass nuclei. We shall discuss this situation further in Chap. 7, in the context of the MC and Euclidean projection calculations of nuclear binding energies.

References

1. C. Körber, T. Luu, Applying twisted boundary conditions for few-body nuclear systems. Phys. Rev. C **93**, 054002 (2016)
2. K. Symanzik, Continuum limit and improved action in lattice theories. 1. Principles and ϕ^4 theory. Nucl. Phys. B **226**, 187 (1983)
3. K. Symanzik, Continuum limit and improved action in lattice theories. 2. O(N) nonlinear sigma model in perturbation theory. Nucl. Phys. B **226**, 205 (1983)
4. D. Lee, R. Thomson, Temperature-dependent errors in nuclear lattice simulations. Phys. Rev. C **75**, 064003 (2007)
5. C. Ordonez, L. Ray, U. van Kolck, The two nucleon potential from chiral Lagrangians. Phys. Rev. C **53**, 2086 (1996)
6. N. Kaiser, R. Brockmann, W. Weise, Peripheral nucleon-nucleon phase shifts and chiral symmetry. Nucl. Phys. A **625**, 758 (1997)
7. E. Epelbaum, W. Glöckle, U.-G. Meißner, Nuclear forces from chiral Lagrangians using the method of unitary transformation. 1. Formalism. Nucl. Phys. A **637**, 107 (1998)

8. E. Epelbaum, W. Glöckle, U.-G. Meißner, Nuclear forces from chiral Lagrangians using the method of unitary transformation. 2. The two nucleon system. Nucl. Phys. A **671**, 295 (2000)
9. H. Krebs, V. Bernard, U.-G. Meißner, Orthonormalization procedure for chiral effective nuclear field theory. Ann. Phys. **316**, 160 (2005)
10. D. Lee, Large-N droplets in two dimensions. Phys. Rev. A **73**, 063204 (2006)
11. C. Wu, J.P. Hu, S.C. Zhang, Exact SO(5) symmetry in spin 3/2 fermionic system. Phys. Rev. Lett. **91**, 186402 (2003)
12. C. Wu, Competing orders in one-dimensional spin 3/2 fermionic systems. Phys. Rev. Lett. **95**, 266404 (2005)
13. D. Lee, T. Schäfer, Cold dilute neutron matter on the lattice. II. Results in the unitary limit. Phys. Rev. C **73**, 015202 (2006)
14. L.H. Thomas, The interaction between a neutron and a proton and the structure of H^3. Phys. Rev. **47**, 903 (1935)
15. V.N. Efimov, Weakly-bound states Of 3 resonantly-interacting particles. Sov. J. Nucl. Phys. **12**, 589 (1971)
16. V. Efimov, Effective interaction of three resonantly interacting particles and the force range. Phys. Rev. C **47**, 1876 (1993)
17. P.F. Bedaque, H.-W. Hammer, U. van Kolck, Renormalization of the three-body system with short range interactions. Phys. Rev. Lett. **82**, 463 (1999)
18. P.F. Bedaque, H.-W. Hammer, U. van Kolck, The three boson system with short range interactions. Nucl. Phys. A **646**, 444 (1999)
19. P.F. Bedaque, H.-W. Hammer, U. van Kolck, Effective theory of the triton. Nucl. Phys. A **676**, 357 (2000)
20. E. Braaten, H.-W. Hammer, Universality in few-body systems with large scattering length. Phys. Rep. **428**, 259 (2006)
21. B. Borasoy, H. Krebs, D. Lee, U.-G. Meißner, The triton and three-nucleon force in nuclear lattice simulations. Nucl. Phys. A **768**, 179 (2006)
22. L. Platter, The three-nucleon system at next-to-next-to-leading order. Phys. Rev. C **74**, 037001 (2006)
23. L. Platter, H.-W. Hammer, U.-G. Meißner, The four boson system with short range interactions. Phys. Rev. A **70**, 052101 (2004)
24. J.A. Tjon, Bound states of 4He with local interactions. Phys. Lett. B **56**, 217 (1975)
25. A. Nogga, H. Kamada, W. Gloeckle, Modern nuclear force predictions for the alpha particle. Phys. Rev. Lett. **85**, 944 (2000)
26. L. Platter, H.-W. Hammer, U.-G. Meißner, On the correlation between the binding energies of the triton and the alpha-particle. Phys. Lett. B **607**, 254 (2005)
27. J.L. Friar, S.A. Coon, Non-adiabatic contributions to static two-pion-exchange nuclear potentials. Phys. Rev. C **49**, 1272 (1994)
28. S.R. Beane, P.F. Bedaque, M.J. Savage, U. van Kolck, Towards a perturbative theory of nuclear forces. Nucl. Phys. A **700**, 377 (2002)
29. A. Nogga, R.G.E. Timmermans, U. van Kolck, Renormalization of one-pion exchange and power counting. Phys. Rev. C **72**, 054006 (2005)
30. M.C. Birse, Power counting with one-pion exchange. Phys. Rev. C **74**, 014003 (2006)
31. M.C. Birse, Deconstructing triplet nucleon-nucleon scattering. Phys. Rev. C **76**, 034002 (2007)
32. E. Epelbaum, U.-G. Meißner, On the renormalization of the one-pion exchange potential and the consistency of Weinberg's power counting. Few Body Syst. **54**, 2175 (2013)
33. U. van Kolck, J.L. Friar, J.T. Goldman, Phenomenological aspects of isospin violation in the nuclear force. Phys. Lett. B **371**, 169 (1996)
34. U. van Kolck, M.C.M. Rentmeester, J.L. Friar, J.T. Goldman, J.J. de Swart, Electromagnetic corrections to the one pion exchange potential. Phys. Rev. Lett. **80**, 4386 (1998)
35. E. Epelbaum, U.-G. Meißner, Charge independence breaking and charge symmetry breaking in the nucleon-nucleon interaction from effective field theory. Phys. Lett. B **461**, 287 (1999)
36. J.L. Friar, U. van Kolck, Charge independence breaking in the two pion exchange nucleon-nucleon force. Phys. Rev. C **60**, 034006 (1999)

37. M. Walzl, U.-G. Meißner, E. Epelbaum, Charge dependent nucleon-nucleon potential from chiral effective field theory. Nucl. Phys. A **693**, 663 (2001)
38. J.L. Friar, U. van Kolck, G.L. Payne, S.A. Coon, Charge symmetry breaking and the two pion exchange two nucleon interaction. Phys. Rev. C **68**, 024003 (2003)
39. E. Epelbaum, U.-G. Meißner, J. E. Palomar, Isospin dependence of the three-nucleon force. Phys. Rev. C **71**, 024001 (2005)
40. E. Epelbaum, U.-G. Meißner, Isospin-violating nucleon-nucleon forces using the method of unitary transformation. Phys. Rev. C **72**, 044001 (2005)
41. J. Fujita, H. Miyazawa, Pion theory of three-body forces. Prog. Theor. Phys. **17**, 360 (1957)
42. U. van Kolck, Few nucleon forces from chiral Lagrangians. Phys. Rev. C **49**, 2932 (1994)
43. E. Epelbaum, A. Nogga, W. Glöckle, H. Kamada, U.-G. Meißner, H. Witala, Three nucleon forces from chiral effective field theory. Phys. Rev. C **66**, 064001 (2002)
44. V. Bernard, N. Kaiser, U.-G. Meißner, Chiral dynamics in nucleons and nuclei. Int. J. Mod. Phys. E **4**, 193 (1995)
45. P. Buettiker, U.-G. Meißner, Pion nucleon scattering inside the Mandelstam triangle. Nucl. Phys. A **668**, 97 (2000)
46. J.W. Chen, D. Lee, T. Schäfer, Inequalities for light nuclei in the Wigner symmetry limit. Phys. Rev. Lett. **93**, 242302 (2004)
47. D. Lee, Lattice simulations for few- and many-body systems. Prog. Part. Nucl. Phys. **63**, 117 (2009)
48. E. Epelbaum, H. Krebs, D. Lee, U.-G. Meißner, Lattice chiral effective field theory with three-body interactions at next-to-next-to-leading order. Eur. Phys. J. A **41**, 125 (2009)
49. B. Borasoy, E. Epelbaum, H. Krebs, D. Lee, U.-G. Meißner, Lattice simulations for light nuclei: chiral effective field theory at leading order. Eur. Phys. J. A **31**, 105 (2007)

Chapter 5
Two and Three Nucleons on the Lattice

Having familiarized ourselves with the two-nucleon and three-nucleon forces in NLEFT, we now turn to the numerical treatment of the nuclear two-body ($A = 2$) and three-body ($A = 3$) systems. The $A = 2$ system mostly serves to fix the unknown parameters of the two-nucleon force (2NF) in NLEFT, and the $A = 3$ system plays a similar role for the three-nucleon force (3NF), which first appears at NNLO in the chiral expansion. The NLEFT results for $A \geq 4$ can then be considered as predictions of NLEFT. For $A = 2$ and $A = 3$, the spectrum can advantageously be found by a direct solution of the Schrödinger equation and direct computation of the eigenvalues of the Hamiltonian (or transfer matrix) using sparse matrix diagonalization methods, though this does not preclude the use of Monte Carlo (MC) simulations either, which thus provides a useful consistency check for the MC results. However, the scaling of the computational effort and memory demands of non-MC methods typically scale exponentially or as a very high power of A, which in most cases precludes calculations beyond $A = 3$ (unless further simplifying assumptions are taken, such as truncation of the set of basis states). In general, for $A \geq 4$ MC calculations outperform sparse matrix diagonalization for physically reasonable system sizes.

The methods described in this chapter form the basis of a large number of studies in the literature, in particular for low-energy nuclear physics and for the physics of cold atomic gases. For a selection of recent lattice studies of low-energy nuclear physics in terms of effective interactions, see e.g. Refs. [1–14]. Similar lattice EFT techniques have been applied to fermionic gases of cold atoms, in particular in the limit of short-ranged interactions and large atom-atom scattering length (the so-called unitarity limit) [15–20]. The connection between the parameters of lattice interactions and physical observables (such as scattering phase shifts) is usually made using the method of Lüscher [21–23], which relates the energy levels of two-body states in a cubic box with periodic boundary conditions to the infinite-volume scattering matrix. On the other hand, in most NLEFT studies of nucleon-nucleon

© Springer Nature Switzerland AG 2019
T. A. Lähde, U.-G. Meißner, *Nuclear Lattice Effective Field Theory*,
Lecture Notes in Physics 957, https://doi.org/10.1007/978-3-030-14189-9_5

scattering, this connection has been made through the imposition of a hard wall in
the relative separation of the scattering particles (the "spherical wall" method).

We begin by recalling basic concepts of elastic scattering from a spherical
potential in Sect. 5.1, followed by a description of the partial wave expansion and
scattering phase shift in Sects. 5.2 and 5.3, respectively. The reader familiar with
this might want to start directly with Sect. 5.4, where we discuss lattice methods
for the calculation of phase shift and mixing angles, with emphasis on the Lüscher
and spherical wall methods. In Sect. 5.5, we present an overview of the treatment of
neutron-proton (np) scattering in NLEFT, and in Sect. 5.6, we apply our methods
to the Coulomb problem and proton-proton (pp) scattering. Before moving on to
Monte Carlo methods, we discuss in Sect. 5.7 the treatment of three-nucleon systems
in NLEFT using sparse matrix diagonalization.

5.1   Elastic Scattering from a Spherical Potential

We begin our treatment of the nucleon-nucleon scattering problem by assuming a
steady, continuous beam of particles which scatter elastically from a potential $V(\mathbf{r})$
that is spherically symmetric, and hence can be written as $V(r)$. Furthermore, we
shall assume that $V(r)$ has a finite range R_n, such that $V(r) = 0$ for $r \gg R_n$. This
excludes the special case of the Coulomb potential, which we shall treat separately
along similar lines in Sect. 5.6. While realistic particles should be described by wave
packets, we restrict ourselves to plane waves for simplicity. Also, wave packets
can be expanded in terms of plane waves, and any complications that could arise
are nevertheless already accounted for in the analysis of experimental results and
partial-wave analysis (PWA) studies.

For the moment, we shall ignore spin and isospin, which are key features of
realistic nucleon-nucleon interactions (such as the NLEFT potential). We shall
consider the time-independent Schrödinger equation for relative two-particle motion
with center-of-mass (or relative) energy E,

$$[\hat{T} + V - E]\psi(r, \theta, \phi) = 0, \tag{5.1}$$

in terms of the spherical coordinates $x = r \sin\theta \cos\phi$, $y = r \sin\theta \sin\phi$ and $z = r \cos\theta$. The kinetic energy operator in Eq. (5.1) is given by

$$\hat{T} = -\frac{\nabla^2}{2\mu} = \frac{1}{2\mu}\left[-\frac{1}{r^2}\frac{\partial}{\partial r}\left(r^2\frac{\partial}{\partial r}\right) + \frac{\hat{L}^2}{r^2}\right], \tag{5.2}$$

where $\mu \equiv m_1 m_2/(m_1 + m_2) = m_N/2$ is the reduced mass, and

$$\hat{L}^2 = -\left[\frac{1}{\sin^2\theta}\frac{\partial^2}{\partial\phi^2} + \frac{1}{\sin\theta}\frac{\partial}{\partial\theta}\left(\sin\theta\frac{\partial}{\partial\theta}\right)\right], \tag{5.3}$$

is the squared angular momentum operator. We also note the z component of angular momentum, $\hat{L}_z = -i\partial/\partial\phi$.

The problem of a spherical potential can be significantly simplified by taking the incoming particle beam to coincide with the z-axis, in which case the incoming plane wave $\sim \exp(ikz)$ has no dependence on ϕ, and is hence an eigensolution of $\hat{L}_z$ with eigenvalue $m = 0$. Also, since $V(r)$ commutes with $\hat{L}_z$, the eigenvalue $m = 0$ is conserved during the reaction, and hence the final state wave function cannot depend on ϕ either. These considerations reduce the problem to

$$[\hat{T} + V - E]\psi(r, \theta) = 0, \tag{5.4}$$

where we take the ansatz

$$\psi(r, \theta) \overset{kr \gg l}{\simeq} \psi_{\text{in}}(r, \theta) + \psi_{\text{sc}}(r, \theta) = C\left\{ \exp(ikz) + f(\theta, k)\frac{\exp(ikr)}{r} \right\}, \tag{5.5}$$

for the asymptotic solution of Eq. (5.4), which describes a superposition of an incoming plane wave and an outgoing spherical scattered wave proportional to the scattering amplitude $f(\theta, k)$. The essence of the problem is thus to find $f(\theta, k)$ for a given potential $V(r)$. Our criterion for the asymptotic region is $kr \gg l$, where l is the orbital angular momentum quantum number. This criterion will be discussed further in Sect. 5.6 when the Coulomb interaction is considered.

Before we discuss the determination of $f(\theta, k)$, it is useful to note how $f(\theta, k)$ is related to the differential cross section $d\sigma/d\Omega$, which is routinely measured in nuclear scattering experiments. For this purpose we equate the incoming flux of particles through an infinitesimal area element $d\sigma$ with the flux of particles scattered into an infinitesimal area element dA at a distance r from the scattering center. This gives

$$\mathbf{j} \cdot \hat{z}\, d\sigma = \mathbf{j} \cdot \hat{r}\, dA = \mathbf{j} \cdot \hat{r}\, r^2 d\Omega, \tag{5.6}$$

where

$$\mathbf{j} \equiv -\frac{i}{m}(\psi^* \nabla \psi + \psi \nabla \psi^*) = -\frac{i}{m}\text{Re}[\psi^* \nabla \psi] \tag{5.7}$$

$$\sim \text{Re}\left\{ \left(\exp(ikz) + f(\theta, k)\frac{\exp(ikr)}{r}\right)^* \nabla \left(\exp(ikz) + f(\theta, k)\frac{\exp(ikr)}{r}\right) \right\},$$

is the current associated with the ansatz (5.5). If fluctuating contributions (which average to zero) are neglected, we have

$$\mathbf{j} \sim \left(k\hat{z} + k\hat{r}\frac{|f(\theta, k)|^2}{r^2} \right) + \mathcal{O}(1/r^3), \tag{5.8}$$

and we note that only the asymptotic form is of interest here, since any experimental apparatus is likely to detect the scattered particles far away from the scattering center, such as an atomic nucleus. This gives

$$\mathbf{j} \cdot \hat{z} \sim k, \qquad \mathbf{j} \cdot \hat{r} \, r^2 \sim k \, |f(\theta, k)|^2 + \mathcal{O}(1/r), \tag{5.9}$$

and

$$\frac{d\sigma}{d\Omega} = |f(\theta, k)|^2, \tag{5.10}$$

for the differential cross section in terms of $f(\theta, k)$.

5.2 Partial Wave Expansion

As the nucleon-nucleon potential $V(r)$ is spherically symmetric (or nearly so to a good approximation), the results of nucleon-nucleon scattering experiments are conventionally given separately for each "channel" of orbital angular momentum l. We shall therefore expand the wave function $\psi(r, \theta)$ in terms of Legendre polynomials $P_l(\cos\theta)$, in what is known as a partial wave expansion. We note that the $P_l(\cos\theta)$ are eigenfunctions of $\hat{L}^2$ and $\hat{L}_z$, with eigenvalues $l(l+1)$ and $m = 0$, respectively. As V commutes with $\hat{L}$, solutions can be found for specific values of l.

We can express the wave function of a given partial wave l as the product of a Legendre polynomial and a radial part $u_l(r)/r$ that depends on r only. When this ansatz is operated on by ∇^2, we find

$$\nabla^2 P_l(\cos\theta) \frac{u_l(r)}{r} = \frac{1}{r} \left(\frac{d^2}{dr^2} - \frac{l(l+1)}{r^2} \right) u_l(r) P_l(\cos\theta), \tag{5.11}$$

and we also note that the $P_l(\cos\theta)$ provide a complete set of functions (i.e. a basis) which satisfies the orthonormality condition

$$\int_0^\pi P_l(\cos\theta) P_{l'}(\cos\theta) \sin\theta \, d\theta = \frac{2}{2l+1} \delta_{ll'}, \tag{5.12}$$

and thus we are free to express any function $F(r, \theta)$ as

$$F(r, \theta) = \sum_{l=0}^\infty b_l(r) P_l(\cos\theta), \tag{5.13}$$

where the dependence on r is contained in the coefficients $b_l(r)$. We will therefore expand the full wave function as

$$\psi(r, \theta) = \sum_{l=0}^{\infty} (2l + 1) i^l \frac{u_l(r)}{kr} P_l(\cos\theta), \tag{5.14}$$

where the factors of $(2l + 1)i^l$ have been made explicit in order to ensure a simple form for the functions $u_l(r)$ when $V(r) = 0$.

We are now in a position to substitute Eq. (5.14) into the Schrödinger equation, in order to find the equation satisfied by the "radial wave function" $u_l(r)$. After using Eq. (5.11) for the kinetic energy term, multiplying by $P_{l'}(\cos\theta)$ and integrating according to Eq. (5.12), we find

$$\left(\frac{d^2}{dr^2} - \frac{l(l+1)}{r^2} - 2\mu V(r) \right) u_l(r) = -k^2 u_l(r), \quad k = \sqrt{2\mu E}, \tag{5.15}$$

for each value of l separately, a consequence of the conservation of angular momentum by a spherically symmetric potential. It should be noted that Eq. (5.15) is a second order differential equation, and hence requires two boundary conditions in order to uniquely specify a solution. Since $\psi(r, \theta)$ is finite everywhere, the first boundary condition can be taken to be $u_l(0) = 0$. The second boundary condition is provided by the requirement that the asymptotic form is recovered as $r \to \infty$.

We shall next discuss the determination of $f(\theta, k)$ and the overall normalization of $u_l(r)$. Assuming that the potential is localized inside a given radius R_n, we may determine $u_l^{ex}(r)$ in the region $r \gg R_n$ external to the potential from

$$\left(\frac{d^2}{dr^2} - \frac{l(l+1)}{r^2} \right) u_l^{ex}(r) = -k^2 u_l^{ex}(r), \tag{5.16}$$

which excludes the important special case of the Coulomb potential. For consistency of notation, we shall nevertheless express the solution in terms of a linear combination of the Coulomb functions $F_l(\eta, kr)$ and $G_l(\eta, kr)$ for $\eta = 0$,

$$u_l^{ex}(r) = A F_l(0, kr) + B G_l(0, kr), \tag{5.17}$$

where the "Sommerfeld parameter" η is defined in Eq. (5.190). In Eq. (5.17), the constants A and B are determined by the boundary conditions. For $\eta = 0$, we may express

$$F_l(0, kr) = kr j_l(kr), \quad G_l(0, kr) = -kr y_l(kr), \tag{5.18}$$

in terms of the (regular) spherical Bessel functions $j_l(kr)$ and (irregular) spherical Neumann functions $y_l(kr)$. It is useful to note that the solution (5.17) may also be expressed in terms of the spherical Hankel functions

$$h_l^{(1)}(kr) \equiv j_l(kr) + i y_l(kr) \tag{5.19}$$

$$\overset{kr \gg l}{\simeq} (-i)^{l+1} \frac{\exp(ikr)}{kr} = -i \frac{\exp(ikr - il\pi/2)}{kr}, \tag{5.20}$$

and

$$h_l^{(2)}(kr) \equiv j_l(kr) - i y_l(kr) \tag{5.21}$$

$$\overset{kr \gg l}{\simeq} i^{l+1} \frac{\exp(-ikr)}{kr} = i \frac{\exp(-ikr + il\pi/2)}{kr}, \tag{5.22}$$

or as a linear combination of the Coulomb Hankel functions

$$H_l^+(0, kr) \equiv G_l(0, kr) + i F_l(0, kr) \tag{5.23}$$

$$\overset{kr \gg l}{\simeq} \exp(ikr - il\pi/2), \tag{5.24}$$

and

$$H_l^-(0, kr) \equiv G_l(0, kr) - i F_l(0, kr) \tag{5.25}$$

$$\overset{kr \gg l}{\simeq} \exp(-ikr + il\pi/2), \tag{5.26}$$

such that for $\eta = 0$, and we may again express

$$H_l^+(0, kr) = ikr \, h_l^{(1)}(kr), \quad H_l^-(0, kr) = -ikr \, h_l^{(2)}(kr), \tag{5.27}$$

in terms of the spherical Hankel functions. For reference, it is also useful to note the expressions for the spherical Bessel functions

$$j_l(kr) = \frac{h_l^{(1)}(kr) + h_l^{(2)}(kr)}{2} \overset{kr \gg l}{\simeq} \frac{\sin(kr - l\pi/2)}{kr}, \tag{5.28}$$

and the spherical Neumann functions

$$y_l(kr) = \frac{h_l^{(1)}(kr) - h_l^{(2)}(kr)}{2i} \overset{kr \gg l}{\simeq} -\frac{\cos(kr - l\pi/2)}{kr}, \tag{5.29}$$

in terms of the spherical Hankel functions, and the expressions for the regular Coulomb functions

$$F_l(0, kr) = \frac{H_l^+(kr) - H_l^-(kr)}{2i} \overset{kr \gg l}{\simeq} \sin(kr - l\pi/2), \tag{5.30}$$

and the irregular Coulomb functions

$$G_l(0, kr) = \frac{H_l^+(kr) + H_l^-(kr)}{2} \overset{kr \gg l}{\simeq} \cos(kr - l\pi/2), \tag{5.31}$$

in terms of the Coulomb Hankel functions, along with their asymptotic forms.

Given the radial momentum operator $\hat{p}_r = -i\partial/\partial r$ and the asymptotic expressions for the $H_l^{\pm}$, we find that these have radial momentum eigenvalues $\pm k$. Hence, the $H_l^{\pm}$ describe outgoing (5.24) and incoming (5.26) radial waves, respectively. In the asymptotic region $kr \gg l$, Eq. (5.16) can be further simplified to

$$\frac{d^2 u_l^{\mathrm{as}}(r)}{dr^2} = -k^2 u_l^{\mathrm{as}}(r), \tag{5.32}$$

as the centrifugal term is now also negligible. Clearly, the general solution is

$$u_l^{\mathrm{as}}(r) = C \exp(ikr) + D \exp(-ikr), \tag{5.33}$$

where C and D are again constants to be determined by the boundary conditions. As the scattered wave should consist of outgoing components only, we should take $D = 0$ for $\psi_{\mathrm{sc}}(r, \theta)$. On the other hand, the incoming plane wave $\psi_{\mathrm{in}}(r, \theta)$ can also be expressed in partial waves according to the Rayleigh expansion

$$\psi_{\mathrm{in}}(r, \theta) = \exp(ikz) = \exp(ikr \cos\theta) \tag{5.34}$$

$$= \sum_{l=0}^{\infty} (2l + 1) i^l j_l(kr) P_l(\cos\theta), \tag{5.35}$$

which is likewise of the form (5.17), provided that we set $B = 0$ to exclude the (irregular) spherical Neumann functions. In the plane wave limit, we have

$$u_l(r) = kr j_l(kr) = F_l(0, kr), \tag{5.36}$$

in Eq. (5.14), when the scattering potential $V(r) = 0$. Given the properties we have established for the spherical Coulomb functions $H_l^{\pm}$, the plane wave expansion can also be written as

$$\psi_{\mathrm{in}}(r, \theta) = \frac{i}{2} \sum_{l=0}^{\infty} (2l + 1) i^l \left(\frac{H_l^-(0, kr) - H_l^+(0, kr)}{kr} \right) P_l(\cos\theta) \tag{5.37}$$

$$\overset{kr \gg l}{\simeq} \frac{i}{2} \sum_{l=0}^{\infty} (2l + 1) i^l \left(\frac{\exp(-ikr + il\pi/2)}{kr} - \frac{\exp(ikr - il\pi/2)}{kr} \right)$$

$$\times P_l(\cos\theta), \tag{5.38}$$

$$= \sum_{l=0}^{\infty} (2l+1)i^l \left(\frac{\sin(kr - l\pi/2)}{kr} \right) P_l(\cos\theta), \tag{5.39}$$

which establishes that the incoming beam (plane wave) consists of a superposition of incoming and outgoing spherical wave components with equal and opposite amplitudes. We note that these considerations establish that the solution in the asymptotic region is indeed of the form (5.5), which describes the superposition of an incident beam and a scattered wave due to the potential $V(r)$.

5.3 Scattering Phase Shift

Next, we shall consider the case of $V(r) \neq 0$, and we note that while the general form of the radial wave function in the exterior region with $V(r) = 0$ is known, $f(\theta, k)$ remains unknown. At the origin where $V(r) \neq 0$, we know that $u_l(0) = 0$, but the derivative $u_l'(0)$ is undetermined. This suggests that we may integrate a "trial solution" $\tilde{u}_l(r)$ outwards, with starting conditions $\tilde{u}_l(0) = 0$ and $\tilde{u}_l'(0) \neq 0$ chosen arbitrarily. For instance, this can be achieved by integrating

$$\tilde{u}_l''(r) = \left(\frac{l(l+1)}{r^2} + 2\mu V(r) - k^2 \right) \tilde{u}_l(r), \tag{5.40}$$

using a numerical integration scheme for second-order differential equations, such as Runge-Kutta methods (or, in general, by means of any other suitable numerical techniques, such as sparse matrix diagonalization). Once the integration of the trial solution has progressed into the external region beyond the range of $V(r)$, it will become a superposition of the linearly independent solutions to Eq. (5.16). Thus

$$u_l(r) = B\tilde{u}_l(r), \tag{5.41}$$

and

$$u_l^{\text{ex}}(r) = A_l \left(H_l^-(0, kr) - S_l(k) H_l^+(0, kr) \right), \tag{5.42}$$

where A_l and $S_l(k)$ are complex-valued. We note that $S_l(k)$ is known as the partial wave S-matrix element, and $S_l(k) = 1$ for $V(r) = 0$, since the non-interacting solution should be proportional to $j_l(kr)$ only.

Given a trial solution $\tilde{u}_l(r)$, $S_l(k)$ can be determined by matching $\tilde{u}_l(r)$ and $u_l^{\text{ex}}(r)$ and their derivatives at $r = r_0$. This is equivalent to computing the inverse logarithmic derivative

$$r_0 R_l(k) = \frac{u_l(r_0)}{u_l'(r_0)} = \frac{\tilde{u}_l(r_0)}{\tilde{u}_l'(r_0)}, \tag{5.43}$$

or R-matrix element, at a suitably chosen radius $r_0 > R_n$, where $V(r)$ is negligible. Since $R_l(k)$ is independent of the (undetermined) normalization B, it is unambiguously determined by the trial solution. For Eq. (5.42), $R_l(k)$ is given by

$$r_0 R_l(k) = \frac{H_l^-(0, kr_0) - S_l(k) H_l^+(0, kr_0)}{H_l^{\prime -}(0, kr_0) - S_l(k) H_l^{\prime +}(0, kr_0)}, \qquad (5.44)$$

where $H_l^{\prime \pm}(0, z) \equiv \partial H_l^{\pm}(0, z)/\partial z$. Then

$$S_l(k) = \frac{H_l^-(0, kr_0) - r_0 R_l(k) H_l^{\prime -}(0, kr_0)}{H_l^+(0, kr_0) - r_0 R_l(k) H_l^{\prime +}(0, kr_0)}, \qquad (5.45)$$

which shows that $S_l(k)$ is uniquely determined by $V(r)$, once the matching with the inverse logarithmic derivative of $u_l^{ex}(r)$ has been achieved.

We shall now find the expression for the scattering amplitude $f(\theta, k)$ in terms of $S_l(k)$. By substituting Eq. (5.42) into the partial wave decomposition (5.14), we find

$$\psi_{ex}(r, \theta) = \sum_{l=0}^{\infty} (2l + 1) i^l A_l \left(\frac{H_l^-(0, kr) - S_l(k) H_l^+(0, kr)}{kr} \right) P_l(\cos\theta), \quad (5.46)$$

$$\overset{kr \gg l}{\simeq} \sum_{l=0}^{\infty} (2l + 1) i^l A_l \left(\frac{\exp(-ikr + il\pi/2)}{kr} - S_l(k) \frac{\exp(ikr - il\pi/2)}{kr} \right)$$

$$\times P_l(\cos\theta), \qquad (5.47)$$

using the asymptotic forms (5.24) and (5.26) for the spherical Hankel functions. Then, we can substitute the expression (5.38) for the incoming plane wave into the asymptotic expression (5.5), and equate the result with Eq. (5.47). This gives

$$\frac{i}{2} \sum_{l=0}^{\infty} (2l + 1) i^l \left(i^l \exp(-ikr) - i^{-l} \exp(ikr) \right) P_l(\cos\theta) + k f(\theta, k) \exp(ikr)$$

$$= \sum_{l=0}^{\infty} (2l + 1) i^l \frac{A_l}{C} \left(i^l \exp(-ikr) - S_l(k) i^{-l} \exp(ikr) \right) P_l(\cos\theta), \qquad (5.48)$$

which can be rearranged as

$$\exp(ikr) \left\{ k f(\theta, k) + \sum_{l=0}^{\infty} (2l + 1) i^l \left(\frac{A_l}{C} S_l(k) - \frac{i}{2} \right) i^{-l} P_l(\cos\theta) \right\}$$

$$- \exp(-ikr) \left\{ \sum_{l=0}^{\infty} (2l + 1) i^l \left(\frac{A_l}{C} - \frac{i}{2} \right) i^l P_l(\cos\theta) \right\} = 0, \qquad (5.49)$$

and since the factors $\exp(\pm ikr)$ are linearly independent, the expressions in curly brackets must equal zero. Furthermore, the orthogonality of the Legendre polynomials then implies that the factor in parentheses on the second line of Eq. (5.49) must vanish identically. Thus, we find

$$\frac{A_l}{C} = \frac{i}{2},$$ (5.50)

and

$$kf(\theta, k) + i \sum_{l=0}^{\infty} (2l+1) \left(\frac{S_l(k) - 1}{2} \right) P_l(\cos\theta) = 0,$$ (5.51)

giving

$$f(\theta, k) = \sum_{l=0}^{\infty} (2l+1) \left(\frac{S_l(k) - 1}{2ik} \right) P_l(\cos\theta),$$

$$= \sum_{l=0}^{\infty} (2l+1) \frac{\exp(i\delta_l(k)) \sin\delta_l(k)}{k} P_l(\cos\theta).$$ (5.52)

We define the partial wave scattering amplitude

$$f_l(k) \equiv \frac{S_l(k) - 1}{2ik},$$ (5.53)

and parameterize

$$S_l(k) \equiv \exp(2i\delta_l(k)),$$ (5.54)

with

$$\delta_l(k) = \frac{1}{2i} \log S_l(k) + n(k)\pi,$$ (5.55)

the scattering phase shift. Clearly, integer multiples of π can be added to $\delta_l(k)$ without affecting any physical observables (such as scattering cross sections) since $\exp(2i\pi n) \equiv 1$. Conventionally, we specify that $\delta_l(k)$ should be a continuous function of k, and that $\delta_l(k \to \infty) = 0$. Then $\delta_l(k = 0) = n_l\pi$, where n_l equals the number of bound states in partial wave l. This is referred to as Levinson's theorem, which can alternatively be given as $\delta_l(0) - \delta_l(\infty) = n_l\pi$. If we make use of the relations

$$\frac{S_l(k) - 1}{2i} = \exp(i\delta_l(k)) \sin \delta_l(k), \tag{5.56}$$

$$\frac{S_l(k) + 1}{2} = \exp(i\delta_l(k)) \cos \delta_l(k), \tag{5.57}$$

we can write Eq. (5.42) in the form

$$u_l^{ex}(r) \propto \frac{i}{2} \left(H_l^-(0, kr) - S_l(k) H_l^+(0, kr) \right) \tag{5.58}$$

$$= \exp(i\delta_l(k)) \left[\cos \delta_l(k) F_l(0, kr) + \sin \delta_l(k) G_l(0, kr) \right], \tag{5.59}$$

and with the asymptotic expressions for $F_l(0, kr)$ and $G_l(0, kr)$, we obtain

$$u_l^{as}(r) \propto \exp(i\delta_l(k)) \left[\cos \delta_l(k) \sin(kr - l\pi/2) + \sin \delta_l(k) \cos(kr - l\pi/2) \right],$$
$$= \exp(i\delta_l(k)) \sin(kr + \delta_l(k) - l\pi/2), \tag{5.60}$$

for $kr \gg l$, which illustrates the physical interpretation of the scattering phase shift: From Eq. (5.60), we find that the oscillations are pulled into the range of the potential for $\delta_l(k) > 0$, which is suggestive of an attractive interaction. Likewise, the oscillations are expelled from the range of the potential when $\delta_l(k) < 0$, which describes a repulsive interaction.

As the external form of the wave function can be expressed in many equivalent ways, it follows that several different conventions exist whereby the scattering information is encoded in the wave function external to $V(r)$. For instance, we can take

$$u_l^{ex}(r) \propto \left(F_l(0, kr) + T_l(k) H_l^+(0, kr) \right), \tag{5.61}$$

where

$$T_l(k) \equiv \exp(i\delta_l(k)) \sin \delta_l(k), \tag{5.62}$$

such that

$$S_l(k) = 1 + 2i T_l(k), \tag{5.63}$$

and hence Eq. (5.52) assumes a particularly simple form in terms of the so-called T-matrix element $T_l(k)$. Clearly, we have $T_l(k) = 0$ when $V(r) = 0$. Furthermore, from Eqs. (5.62) and (5.63),

$$f_l(k) = \frac{\exp(i\delta_l(k)) \sin \delta_l(k)}{k} = \frac{1}{k \cot \delta_l(k) - ik}, \tag{5.64}$$

and, as we shall find, the expression $k \cot \delta_l(k)$ plays a central role in descriptions of nucleon-nucleon scattering. Yet another form of $u_l^{ex}(r)$ is

$$u_l^{ex}(r) \propto \exp(i\delta_l(k)) \cos \delta_l(k) \left[F_l(0, kr) + K_l(k)G_l(0, kr) \right], \tag{5.65}$$

where the K-matrix element is

$$K_l(k) \equiv \tan \delta_l(k), \tag{5.66}$$

such that

$$S_l(k) = \frac{1 + iK_l(k)}{1 - iK_l(k)}, \tag{5.67}$$

and we note that $K_l(k)$ can be found from the R-matrix element $R_l(k)$ at a given matching radius r_0,

$$K_l(k) = -\frac{F_l(0, kr_0) - r_0 R_l(k) F_l'(0, kr_0)}{G_l(0, kr_0) - r_0 R_l(k) G_l'(0, kr_0)}, \tag{5.68}$$

and it should be emphasized that $\delta_l(k)$, as well $S_l(k)$, $T_l(k)$ and $K_l(k)$ are all independent of r_0, as long as $r_0 \gg R_n$.

Let us briefly consider the consequences of a real-valued scattering potential $V(r)$, in light of Eq. (5.68). Clearly, if $V(r)$ is real-valued, then our trial solution $\tilde{u}_l(r)$ along with the R-matrix of Eq. (5.43) will also be real-valued. Equations (5.68) and (5.66) then show that the phase shift $\delta_l(k)$ will likewise be real-valued. Then, a real-valued $V(r)$ leads to $|S_l| = 1$, and can be taken to describe scattering processes where no flux is lost. We note that $V(r)$ may be taken to include an imaginary (absorptive) component, which can be used to describe the loss of flux due to other reaction channels (such as recombination of particles into bound states). Since PWA analyses of nucleon-nucleon scattering typically provide experimental data for $\delta_l(k)$, where k is the center-of-mass momentum, our focus here shall be the calculation of $\delta_l(k)$ using the NLEFT potential for $V(r)$. In general, this will also couple scattering channels with different orbital angular momentum l (but equal total angular momentum J). We shall also discuss how the inclusion of an imaginary component into $V(r)$ can be used to simplify the numerical treatment of coupled channel processes.

It is also useful to note the relation between $\delta_l(k)$ and the differential cross section measured in scattering experiments. From Eq. (5.10), we have

$$\frac{d\sigma}{d\Omega} = |f(\theta, k)|^2 = \left| \sum_{l=0}^{\infty} (2l + 1) f_l(k) P_l(\cos \theta) \right|^2,$$

$$= \frac{1}{k^2} \left| \sum_{l=0}^{\infty} (2l + 1) \exp(i\delta_l(k)) \sin \delta_l(k) P_l(\cos \theta) \right|^2, \tag{5.69}$$

for the differential cross section, and

$$\sigma_{\text{tot}} = \sum_{l=0}^{\infty} \sum_{l'=0}^{\infty} (2l+1)(2l'+1) f_l^*(k) f_{l'}(k) \int d\Omega \, P_l(\cos\theta) P_{l'}(\cos\theta),$$

$$= 4\pi \sum_{l=0}^{\infty} (2l+1) |f_l(k)|^2 = \frac{4\pi}{k^2} \sum_{l=0}^{\infty} (2l+1) \sin^2 \delta_l(k), \qquad (5.70)$$

for the total cross section.

Next, we turn to the Effective Range Expansion (ERE), which is an important tool for low-energy nuclear physics. In the ERE $k^{2l+1} \cot \delta_l(k)$ is expanded as a power series in k^2 [24, 25]. For $l = 0$, the coefficients of the ERE are conventionally written as

$$k \cot \delta_0(k) = -\frac{1}{a_0} + \frac{1}{2} r_0 k^2 + \mathcal{O}(k^4), \qquad (5.71)$$

where a_0 is the S-wave scattering length, and r_0 the S-wave effective range. In general, for arbitrary partial waves, we have

$$k^{2l+1} \cot \delta_l(k) = -\frac{1}{a_l} + \frac{1}{2} r_l k^2 + \sum_{n=2}^{\infty} v_{l,n} k^{2n}, \qquad (5.72)$$

where the coefficients $v_{l,n}$ are referred to as shape parameters.

An instructive example is to consider hard-sphere scattering with radius R_n, which can be thought of as a highly simplified description of a repulsive "hard core" potential. As we have $V = 0$ for $r > R_n$ and the wave function must vanish for $r < R_n$, then the S-wave scattering solution must be proportional to $\sin(kr - kR_n)$, in other words $\delta_0(k) = -kR_n$ for an infinitely repulsive hard core potential (note that we cannot take $\delta_0(\infty) = 0$ in this case, unless the potential for $r < R_n$ is taken to be very large but finite). Returning to the ERE, we find

$$k \cot(-kR_n) = -\frac{1}{R_n} + \frac{1}{3} R_n k^2 + \cdots, \qquad (5.73)$$

such that $a_0 = R_n$ and $r_0 = 2R_n/3$ for the hard core potential. For more general cases, we have $r_0 \sim R_n$, however, a_s can vary over a large range. The case of $a_0 \simeq R_n$ is called "natural", while the "unnatural" case of $|a_0| \gg R_n$ is found to occur in many physically interesting systems composed of neutrons or cold atoms. If we specialize to the case of S-wave scattering at low energies, we find

$$f_0(k) = \frac{1}{-1/a_0 - ik}, \qquad (5.74)$$

and hence Eq. (5.70) gives

$$\sigma(k) = \frac{4\pi}{1/a_0^2 + k^2},\tag{5.75}$$

for the total S-wave cross section. For the natural case $|ka_s| \ll 1$, such that $\sigma = 4\pi a_0^2$. From Eq. (5.71) we have $\delta_0(k) \simeq -ka_0$, such that the slope of $\delta_0(k)$ for small k determines whether a shallow bound state ($a_0 > 0$) is encountered, or whether a nearly bound (virtual) state exists instead ($a_0 < 0$). The limit of a weakly bound state with vanishing binding energy $|a_0| \to \infty$ is referred to as the unitarity limit, in which case $\sigma = 4\pi/k^2$. For a review of the fascinating physics in the unitarity limit, see Ref. [26].

From the point of view of modern EFT, the ERE is a welcome result, at it shows that the low-energy theory of nuclear scattering is determined by a relatively small number of observables, which encode the (limited) high-energy features that affect the physics at low energies. Before we discuss the calculation of $\delta_l(k)$ in NLEFT, we note that the above considerations for $\delta_l(k)$ and the ERE are dependent on the assumption that the potential is short-ranged. For instance, while the summation over l in Eq. (5.70) is convergent for short-ranged interactions (such as the OPE component of the NLEFT potential), the presence of a long-range component (such as the Coulomb interaction between protons) necessitates a different approach. Similarly, the radius of convergence of the ERE becomes zero when the potential becomes unscreened (for instance, when the mass of a Yukawa potential is taken to zero). It is helpful to note that for the Coulomb interaction only, the Schrödinger equation can again be solved exactly in terms of Coulomb functions for $\eta \neq 0$, which then replace those for $\eta = 0$ in the asymptotic scattering wave function. It is also possible to derive an alternative ERE which is useful in the presence of a Coulomb potential. We shall return to these problems in Sect. 5.6.

5.4   Calculation of Phase Shifts and Mixing Angles

In our development of the formalism for the two-nucleon problem in NLEFT, we have thus far ignored issues related to the lattice regulator (lattice spacing a), and the fact that realistic nucleon-nucleon interactions are spin-dependent. As we shall find, such interactions couple channels with total spin $S = 1$ and orbital angular momentum $l = J \pm 1$. In order to introduce our computational methods, we shall first concentrate on a relatively simple trial potential, which includes the spin-dependence and partial-wave mixing encountered for the OPE potential, but which excludes the additional complications associated with ultraviolet divergences and singular interactions.

Our nucleon-nucleon model interaction for identical spin-1/2 nucleons of mass m_N is given in the continuum by the spin-dependent potential

$$V(\mathbf{r}) \equiv C\left[1 + \frac{r^2}{R_n^2} S_{12}(\hat{r})\right] \exp\left(-\frac{r^2}{2R_n^2}\right),\tag{5.76}$$

and since $V(\mathbf{r})$ is regular, we can take the lattice potential $V(\mathbf{n})$ to coincide with the continuum potential at $\mathbf{r} = a\mathbf{n}$. We have chosen Eq. (5.76) to produce a central force as well as a tensor force, with a Gaussian envelope that has a finite range given by R_n. The tensor force is produced by the operator

$$S_{12}(\hat{r}) \equiv 3(\hat{r} \cdot \boldsymbol{\sigma}_1)(\hat{r} \cdot \boldsymbol{\sigma}_2) - \boldsymbol{\sigma}_1 \cdot \boldsymbol{\sigma}_2,\tag{5.77}$$

which is of considerable general interest, as contributions proportional to S_{12} appear not only in the one-pion exchange (OPE) potential, but also in the electric and magnetic dipole interactions between atoms and molecules.

When the parameters C and R_n of Eq. (5.76) are appropriately chosen, a shallow deuteron-like bound state is produced in the $^3S(D)_1$ channel. For instance, with $C = -2$ MeV and $R_n = 2 \times 10^{-2}$ MeV$^{-1} \simeq 3.9$ fm, this bound state has an energy of $\simeq -0.155$ MeV. Note that the notation for channel mixing was established in Chap. 4. For total intrinsic spin $S = 0$, the tensor operator S_{12} vanishes. Hence, the "effective" potential is

$$V_{S=0}(\mathbf{r}) = \tilde{V}(\mathbf{r}) \equiv C \exp\left(-\frac{r^2}{2R_n^2}\right),\tag{5.78}$$

for $S = 0$ channels,

$$V_{S=1}(\mathbf{r}) = \left(1 + \frac{2r^2}{R_n^2}\right)\tilde{V}(\mathbf{r}),\tag{5.79}$$

for uncoupled $S = 1$ channels, and

$$V_J(r) = \left[1 + \frac{r^2}{R_n^2}\begin{pmatrix} -\frac{2(J-1)}{2J+1} & \frac{6\sqrt{J(J+1)}}{2J+1} \\ \frac{6\sqrt{J(J+1)}}{2J+1} & -\frac{2(J+2)}{2J+1} \end{pmatrix}\right]\tilde{V}(\mathbf{r}),\tag{5.80}$$

for coupled $S = 1$ channels, where the upper and lower components in the matrix notation refer to components with $l = J - 1$ and $l = J + 1$, respectively.

Since NLEFT calculations are usually performed using a transfer matrix with finite temporal lattice spacing δ, we shall next outline how our model potential can be accommodated within the transfer matrix formalism. For the two-nucleon system, the introduction of auxiliary fields is not necessary, since the problem can in general be solved without resorting to MC simulations. For free nucleons, we take the transfer matrix to be

$$M_{\text{free}} \equiv\, : \exp(-H_{\text{free}}\alpha_t)\, :, \tag{5.81}$$

and we recall that $\alpha_t \equiv \delta/a$ is the ratio of temporal to spatial lattice spacings, and colons indicate normal ordering. In this context, H_{free} can (for instance) be taken as the $\mathcal{O}(a^4)$-improved free lattice Hamiltonian of Chap. 4. As S_{12} gives no contribution for $S = 0$, the transfer matrix is

$$M_{S=0} \equiv\, : \exp\left[-H_{\text{free}}\alpha_t - \frac{\alpha_t}{2} \sum_{\mathbf{n}_1,\mathbf{n}_2} \tilde{V}(\mathbf{n}_1 - \mathbf{n}_2)\, \rho^{a^\dagger,a}(\mathbf{n}_1)\rho^{a^\dagger,a}(\mathbf{n}_2) \right] :, \tag{5.82}$$

where $\rho^{a^\dagger,a}(\mathbf{n})$ is the particle density operator

$$\rho^{a^\dagger,a}(\mathbf{n}) = \sum_{j=\uparrow,\downarrow} a_j^\dagger(\mathbf{n})a_j(\mathbf{n}), \tag{5.83}$$

and $\mathbf{n} \equiv (n_x, n_y, n_z)$ denotes coordinate vectors (with integer-valued components) on a three-dimensional spatial lattice. For $S = 1$, the tensor interaction in $V(\mathbf{r})$ should be included, and we find

$$M_{S=1} \equiv\, : \exp\left[-H_{\text{free}}\alpha_t - \frac{\alpha_t}{2} \sum_{\mathbf{n}_1,\mathbf{n}_2} \tilde{V}(\mathbf{n}_1 - \mathbf{n}_2)\, \rho^{a^\dagger,a}(\mathbf{n}_1)\rho^{a^\dagger,a}(\mathbf{n}_2) \right. \tag{5.84}$$

$$\left. -\frac{\alpha_t}{2R_n^2} \sum_{l,l'=1,2,3} \sum_{\mathbf{n}_1,\mathbf{n}_2} \tilde{V}(\mathbf{n}_1 - \mathbf{n}_2)T_{ll'}(\mathbf{n}_1 - \mathbf{n}_2)\rho_l^{a^\dagger,a}(\mathbf{n}_1)\rho_{l'}^{a^\dagger,a}(\mathbf{n}_2) \right] :,$$

where we have defined the spin densities

$$\rho_l^{a^\dagger,a}(\mathbf{n}) = \sum_{i,j=\uparrow,\downarrow} a_i^\dagger(\mathbf{n}) \left[\sigma_l\right]_{ij} a_j(\mathbf{n}), \tag{5.85}$$

and the lattice tensor operator

$$T_{ll'}(\mathbf{n}) = 3n_l n_{l'} - |\mathbf{n}|^2 \delta_{ll'}, \tag{5.86}$$

where $l, l' = 1, 2, 3$ refer to the spatial directions on the lattice.

When MC methods are not used, the eigenvalues of the transfer matrix M can be computed using sparse matrix diagonalization methods, which are widely available in standard implementations of numerical routines (sometimes referred to with nicknames such as "packages of canned eigenvector routines"). In the transfer matrix formalism, eigenvalues of the transfer matrix are interpreted as exponentials of the energy,

$$M|\psi\rangle = \lambda|\psi\rangle = \exp(-E_\lambda \alpha_t)|\psi\rangle, \tag{5.87}$$

which in general need to be extrapolated to the Hamiltonian limit $\delta \to 0$, as well as in the spatial lattice spacing $a \to 0$, if the intention is to study the continuum limit. In NLEFT, the potential becomes singular in the continuum limit, and the lattice spacing a is identified with the regulator of the EFT. At finite a, the breaking of rotational invariance due to the lattice regularization leads to a small splitting of the different SO(3, Z) representations comprising each orbital angular momentum multiplet. However, the elements of each SO(3, Z) representation remain exactly degenerate even at finite a.

5.4.1   Lüscher's Method

Given the energy eigenvalues of the two-nucleon system in a cubic volume, a practical way of extracting scattering phase shifts from this information is needed. If periodic boundary conditions are imposed on the cubic volume, Lüscher's method [21–23] can be used to extract scattering phase shifts from energy eigenvalues of two-nucleon states below inelastic thresholds. However, as noted in Chap. 3, the rotational symmetry is broken down to SO(3, Z), which means that there is mixing of the continuum phase shifts into the various irreps of SO(3, Z). In practice, this means that the determination of phase shifts beyond the lowest few partial waves becomes a rather complicated task [27]. We shall here briefly outline the procedure, following Ref. [28].

Before we consider the Lüscher formalism in three space dimensions, we discuss the one-dimensional case as an instructive example. Consider the scattering of two particles on a line segment of length L (the lattice size), with a finite interaction range $R \ll L$. For simplicity, we also assume an infinite time extent. As expected, outside the range of the interaction, the relative motion of the scattering particles should be described by a plane wave. Then, the interaction produces a finite phase shift $\delta(k)$, as explained in Sect. 5.3. As for the three-dimensional case, we impose periodic boundary conditions in the spatial dimension. Hence, for a plane wave with momentum k, we should have

$$\exp(ikL + 2i\delta(k)) = \exp(ik0) = 1, \tag{5.88}$$

which gives

$$k_n L + 2\delta(k_n) = 2\pi n, \quad n \in \mathbb{N}, \tag{5.89}$$

for the quantization of the momenta. This result is remarkable, as it relates energies computed on the lattice to the continuum phase shift. Given an appropriate (lattice or continuum) dispersion relation which relates the lattice energy levels $E_n(L)$ to the momentum modes $k_n(L)$, the continuum phase shift at discrete values of the scattering momentum can be reconstructed from the energy levels on the lattice. Also, the non-interacting limit $\delta(k_n) = 0$ is recovered for $k_n = 2\pi n/L$. For non-relativistic particles with mass μ, this method is applicable when $\mu L \gg 1$, so that further corrections are exponentially suppressed. Also, any inelasticity or coupling to other scattering channels introduces modifications.

Let us now return to the three-dimensional case. For simplicity, we shall restrict ourselves to the case of infinite lattice volume and uncoupled channels. Likewise, practical NLEFT calculations use a transfer matrix formalism with finite temporal lattice spacing δ, with the Hamiltonian limit recovered for $\delta \to 0$. The relationship between the energy eigenvalues of the Hamiltonian and the scattering phase shifts is found (at infinite lattice volume and for uncoupled channels) from the condition [28]

$$\det\left(\mathscr{C} - \mathscr{S} \cdot \mathscr{F}^{(FV)}\right) = 0, \tag{5.90}$$

with

$$\mathscr{C} \equiv \delta_{l_1,l_2}\delta_{m_1,m_2}\cos\delta_{l_1}(k),$$

$$\mathscr{S} \equiv \delta_{l_1,l_2}\delta_{m_1,m_2}\sin\delta_{l_1}(k),$$

$$\mathscr{F}^{(FV)} \equiv F_{l_1,m_1;l_2,m_2}^{(FV)}(k), \tag{5.91}$$

where the matrix elements of $\mathscr{F}^{(FV)}$ are given by

$$F_{l_1,m_1;l_2,m_2}^{(FV)}(k) \equiv \frac{(-1)^{m_2}}{\tilde{k}\pi^{3/2}}\sqrt{(2l_1+1)(2l_2+1)}\sum_{\bar{l}=|l_1-l_2|}^{|l_1+l_2|}\sum_{\bar{m}=-\bar{l}}^{\bar{l}}\frac{\sqrt{2\bar{l}+1}}{\tilde{k}^{\bar{l}}}$$

$$\times \begin{pmatrix} l_1 & \bar{l} & l_2 \\ 0 & 0 & 0 \end{pmatrix}\begin{pmatrix} l_1 & \bar{l} & l_2 \\ -m_1 & -\bar{m} & -m_2 \end{pmatrix}\mathscr{Z}_{\bar{l},\bar{m}}(1;\zeta), \tag{5.92}$$

which is a function of the dimensionless variable

$$\zeta \equiv \tilde{k}^2 \equiv \left(\frac{kL}{2\pi}\right)^2, \tag{5.93}$$

where L refers to the length dimension of the cubic lattice, while the various instances of l indicate orbital angular momentum quantum numbers. In Eq. (5.93), k is determined from the eigenvalues (energies) of the Hamiltonian using a suitable (discrete or continuum) dispersion relation. In Eq. (5.92), the arrays in parentheses are Wigner 3-j symbols, see e.g. Ref. [29], and the $\mathscr{Z}_{l,m}$ are the functions

$$\mathscr{Z}_{l,m}(s;\zeta) \equiv \sum_{\mathbf{n}} \frac{|\mathbf{n}|^l Y_{l,m}(\hat{n})}{(|\mathbf{n}|^2 - \zeta)^s},\tag{5.94}$$

introduced by Lüscher, where $\mathbf{n} \equiv (n_x, n_y, n_z)$ are triplets of integers.

In practice, the dimensionality of Eq. (5.90) needs to be truncated, such that the phase shifts are assumed to be non-vanishing for $l \leq l^{max}$, and vanishing for $l > l^{max}$. For instance, if we take $l^{max} = 6$, we are dealing with matrices no larger than 4×4. Nevertheless, the resulting expressions for the scattering phase shifts quickly become very complicated for high partial waves, and hence we shall here quote some of the more well-known ones for low values of l, along with the cubic irrep they apply to. Note that the cubic rotational group and its relationship to continuum angular momenta is discussed in Chap. 3. For instance, we have

$$\tilde{k} \cot \delta_0(k) = \frac{1}{\pi^{3/2}} \mathscr{Z}_{0,0}(1;\zeta),\tag{5.95}$$

which is the original formula given by Lüscher for $l = 0$ $[A_1^+]$, and

$$\tilde{k}^3 \cot \delta_1(k) = \frac{\zeta}{\pi^{3/2}} \mathscr{Z}_{0,0}(1;\zeta),\tag{5.96}$$

for $l = 1$ $[T_1^-]$,

$$\tilde{k}^5 \cot \delta_2(k) = \frac{1}{\pi^{3/2}} \left[\zeta^2 \mathscr{Z}_{0,0}(1;\zeta) + \frac{6}{7} \mathscr{Z}_{4,0}(1;\zeta) \right],\tag{5.97}$$

for $l = 2$ in the $[E^+]$ representation, and

$$\tilde{k}^5 \cot \delta_2(k) = \frac{1}{\pi^{3/2}} \left[\zeta^2 \mathscr{Z}_{0,0}(1;\zeta) - \frac{4}{7} \mathscr{Z}_{4,0}(1;\zeta) \right],\tag{5.98}$$

for $l = 2$ in the $[T_2^+]$ representation.

Efficient methods are also required for the numerical evaluation of the functions $\mathscr{Z}_{l,m}(1;\zeta)$. While the summations with $l \geq 1$ are convergent due to the spherical harmonic, the case of $l = 0$ requires UV regularization. Numerical evaluation by brute force of the $\mathscr{Z}_{0,0}(1;\zeta)$ can be achieved using the definition

$$\mathscr{Z}_{0,0}(1;\zeta) \equiv \frac{1}{2\pi^{1/2}} \lim_{\Lambda \to \infty} \left[\sum_{\mathbf{n}}^{\Lambda} \frac{\theta(\Lambda^2 - |\mathbf{n}|^2)}{|\mathbf{n}|^2 - \zeta} - 4\pi \Lambda \right],\tag{5.99}$$

which turns out to be rather inefficient due to slow convergence. By means of the Poisson resummation formula

$$\sum_{\mathbf{n}} \delta^{(3)}(\mathbf{x} - \mathbf{n}) = \sum_{\mathbf{m}} \exp(2\pi i \, \mathbf{m} \cdot \mathbf{x}), \qquad (5.100)$$

one obtains the alternative expression

$$\mathscr{Z}_{0,0}(1; \zeta) = \pi \exp(\zeta)(2\zeta - 1) + \frac{\exp(\zeta)}{2\pi^{1/2}} \sum_{\mathbf{n}} \frac{\exp(-|\mathbf{n}|^2)}{|\mathbf{n}|^2 - \zeta} \qquad (5.101)$$

$$- \frac{\pi}{2} \int_0^1 d\lambda \frac{\exp(\lambda\zeta)}{\lambda^{3/2}} \left(4\lambda^2\zeta^2 - \sum_{\mathbf{m}\neq 0} \exp(-\pi^2|\mathbf{m}|^2/\lambda) \right),$$

with exponentially accelerated convergence. For $l \neq 0$, a similar representation is given by [28]

$$\mathscr{Z}_{l,m}(1; \zeta) = \sum_{\mathbf{n}} \frac{|\mathbf{n}|^l Y_{l,m}(\hat{n}) \exp(-\Lambda(|\mathbf{n}|^2 - \zeta))}{|\mathbf{n}|^2 - \zeta} \qquad (5.102)$$

$$+ \sum_{\mathbf{p}} \int_0^\Lambda d\lambda \left(\frac{\pi}{\lambda} \right)^{l+3/2} \exp(\lambda\zeta)|\mathbf{p}|^l Y_{l,m}(\hat{p}) \exp(-\pi^2|\mathbf{p}|^2/\lambda),$$

and we also note that exact relations exist between the functions $\mathscr{Z}_{l,m}(1; \zeta)$ for fixed l and varying m [28]. While most NLEFT calculations make use of the spherical wall method which we describe next, the Lüscher method has also been applied at low momenta, in particular when the sensitivity of the NLEFT results to small changes in the light quark mass m_q and the fine-structure constant α_{EM} in Chap. 7. We therefore seek a straightforward way to compute the function $\mathscr{Z}_{0,0}(1; \zeta)$, which enables us to compute the S-wave phase shift. If we denote

$$k \cot \delta_0(k) = \frac{1}{\pi L} S(\zeta), \qquad (5.103)$$

then the function $S(\zeta) = 2\pi^{1/2} \mathscr{Z}_{0,0}(1; \zeta)$ (and its derivatives) may be computed numerically by means of a Laurent series expansion [30]

$$S(\zeta) \simeq -\frac{1}{\zeta} + S_0 + \sum_{n=1}^\infty S_n\zeta^n, \qquad (5.104)$$

with the first few coefficients S_n given by

$$S_0 = -8.913631, \quad S_1 = 16.532288, \quad S_2 = 8.401924, \quad S_3 = 6.945808,$$
$$S_4 = 6.426119, \quad S_5 = 6.202149, \quad S_6 = 6.098184, \quad S_7 = 6.048263,$$

(5.105)

where the accuracy of the expansion (5.104) can be increased, if necessary, by computing further coefficients S_n.

Lüscher's method has been extended in a number of different ways. Several studies have looked at higher partial waves and spins [27, 28, 31–34], moving frames [34–39], multi-channel scattering [40–45], including the use of unitarized chiral perturbation theory (and related methods) [46–51], asymmetric boxes [52, 53], as well as small volumes where the lattice length L is smaller than the scattering length [54]. For a recent review, see Ref. [55].

5.4.2 Spherical Wall Method

We shall next describe the "spherical wall method" for computing scattering phase shifts and mixing angles, which has proven to be of great usefulness in NLEFT [56–58]. As shown in Fig. 5.1, a hard spherical wall boundary is imposed on the relative separation of the two particles at a radius R_{wall}. In the center-of-mass frame, the Schrödinger equation is solved for spherical standing waves which vanish at $r = R_{\text{wall}}$. This presupposes that the cubic box is sufficiently large to contain a sphere of radius R_{wall}. While we shall defer the discussion of the relative merits of the Lüscher and spherical wall methods to the end of this section, we can immediately note that the spherical wall eliminates "copies" of the two-particle interactions due to the periodic boundary conditions of the box, which is particularly useful at high energies and angular momenta.

Fig. 5.1 Illustration of a hard spherical wall of radius R_{wall} imposed in the center-of-mass frame. The wall radius is somewhat ambiguous on the lattice, as the components of a given position vector $\mathbf{r}$ must equal integer multiples of the lattice spacing. The "physical" wall radius $\bar{R}_{\text{wall}} = R_{\text{wall}} + \epsilon$ in each channel is determined by requiring $\delta_l(k) = 0$ when $V = 0$

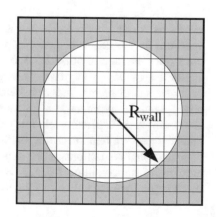

For simplicity, we shall introduce the spherical wall formalism for scattering channels with $S = 0$, and consider the case of $S = 1$ (where the tensor interaction induces partial-wave mixing) separately. As noted earlier, beyond the range of the interaction, the Schrödinger equation reduces to the Helmholtz equation. The radial solution can then be expressed in terms of spherical Bessel and Neumann functions, with the angular part given by the spherical harmonics $Y_{l,m}(\theta, \phi)$. This gives

$$\psi_l^{ex}(r, \theta, \phi) \propto \left[\cos \delta_l(k) j_l(kr) - \sin \delta_l(k) y_l(kr)\right] Y_{l,m}(\theta, \phi), \qquad (5.106)$$

for values of $r \gg R_n$, but interior to the asymptotic region $kr \gg l$. We shall now impose a spherical wall such that

$$V(\mathbf{n}_1 - \mathbf{n}_2) \rightarrow V(\mathbf{n}_1 - \mathbf{n}_2) + V_{wall}\theta(|\mathbf{n}_1 - \mathbf{n}_2| - \tilde{R}_{wall}), \qquad (5.107)$$

whereby the amplitude for two-nucleon separation greater than the wall radius $\tilde{R}_{wall}$ is strongly suppressed by a large potential energy V_{wall} multiplying the step function θ. Here, we have defined $\tilde{R}_{wall} \equiv R_{wall} + \epsilon$, which reflects the fact that there is some ambiguity (on the lattice) for the precise value of $\tilde{R}_{wall}$, since the components of $\mathbf{r}$ must be integer multiples of the lattice spacing. Also, since inevitably $V_{wall} < \infty$, the wave functions will always penetrate some small, but finite, distance into the spherical wall. One way to resolve this ambiguity is to determine ϵ in each channel (for a given, integer-valued R_{wall}) such that $\delta_l(k) = 0$ when the particles are non-interacting. With the spherical wall in place, the wave function (5.106) vanishes at the wall boundary, giving

$$\cos \delta_l(k) j_l(k\tilde{R}_{wall}) = \sin \delta_l(k) y_l(k\tilde{R}_{wall}), \qquad (5.108)$$

such that

$$\tan \delta_l(k) = \frac{j_l(k\tilde{R}_{wall})}{y_l(k\tilde{R}_{wall})}, \qquad (5.109)$$

allows us to determine $\delta_l(k)$. In Fig. 5.2, we show the free two-nucleon energy spectrum in the center-of-mass frame for a lattice spacing of $a = 1.97$ fm for our model interaction of Sect. 5.4, in the presence of a spherical wall at a relative separation of $R_{wall} = 10a$. This spectrum will give us access to scattering parameters at discrete values of k, determined from the energy E of each individual state, given an appropriate dispersion relation $E(k)$, which can be taken to be the continuum dispersion relation, or more accurately for higher momenta, the lattice dispersion relation corresponding to the free lattice Hamiltonian. Apart from computational restrictions on the maximum lattice size, the choice of R_{wall} should account for the condition $R_{wall} \gg R_n$, and be large enough to probe wavelengths much larger than the lattice spacing a.

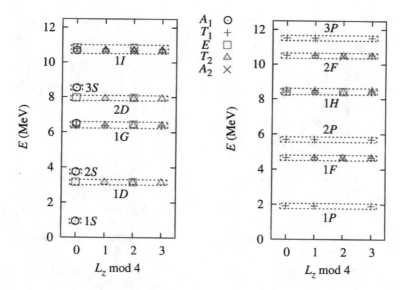

Fig. 5.2 Spectrum for $V = 0$ in the presence of a spherical wall, with $a = 1.97$ fm and $R_{wall} = 10a$ for our model interaction discussed in Sect. 5.4. The left panel shows states with even parity, and the right panel states with odd parity. The SO(3, Z) multiplets are enclosed by dotted lines, and the individual states are classified according to SO(3, Z) irreps and $L_z \equiv m$. For $V = 0$, identification of the continuum analogs is straightforward. For strong, singular interactions with mixing effects, the identification of states can be highly problematic

Next, we shall illustrate the spherical wall method with a practical example for the $^1S_0$ channel, where the tensor interaction does not contribute. The interacting energy spectrum for $S = 0$ channels with a spherical wall applied is shown in Fig. 5.3. The interaction is attractive in each of the $S = 0$ channels, as expected since $V_{S=0}$ is negative definite. Let us now consider the calculation of $\delta_l(k)$ using the $1\,^1S_0$ state (the ground state in the $^1S_0$ channel). For the free particle system, the $1\,^1S_0$ state has energy $E_{free} \simeq 0.9280$ MeV for $R_{wall} = 10a$, which corresponds to

$$k_{free} = \sqrt{2\mu E_{free}} = \sqrt{m_N E_{free}} \simeq 29.52 \text{ MeV}, \tag{5.110}$$

using the continuum dispersion relation $E(k) = k^2/2\mu = k^2/m_N$. For $\delta_{1\,S_0}(k) = 0$, Eq. (5.108) reduces to

$$j_0(k_{free} \tilde{R}_{wall}) - 0, \tag{5.111}$$

and since the ground state in each channel has exactly one extremum, we obtain

$$\tilde{R}_{wall} = \frac{\pi}{k_{free}} \simeq 0.1064 \text{ MeV}^{-1} = 0.1064 \times 100a = 10.64a, \tag{5.112}$$

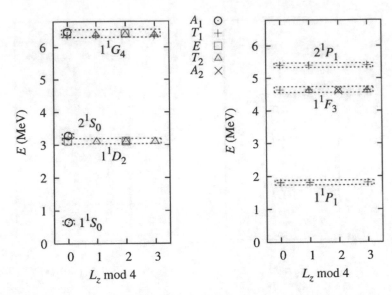

Fig. 5.3 Spectrum for $S = 0$ channels and $V \neq 0$, with $a = 1.97$ fm and $R_{\text{wall}} = 10a$. The left panel shows states with even parity, and the right panel states with odd parity. Individual states are denoted as in Fig. 5.2, and the corresponding phase shifts for the $^1S_0$ and $^1P_1$ channels are shown in Fig. 5.4

which shows that the "physical" hard wall radius $\tilde{R}_{\text{wall}} \simeq 10.64a$ is found at a slightly larger separation than $10a$ lattice units. Having determined $\tilde{R}_{\text{wall}}$ for the $^1S_0$ channel, we may now consider the case where the interaction is turned on. For the interacting case, Ref. [57] found $E_{1S_0} \simeq 0.6445$ MeV, which gives $k_{1S_0} \simeq 24.60$ MeV. Using Eq. (5.109), the $^1S_0$ phase shift at $k = 24.60$ MeV is

$$\delta_{1S_0}(k = 24.60\,\text{MeV}) = \tan^{-1}\left[\frac{j_0(k\tilde{R}_{\text{wall}})}{y_0(k\tilde{R}_{\text{wall}})}\right] \simeq 30.0^\circ, \tag{5.113}$$

and we may then proceed in this manner for the excited states in the $^1S_0$ channel, followed by the other channels with $S = 0$, in which case we again determine $\tilde{R}_{\text{wall}}$ using the lowest-energy state in each channel. In cases where the spin-multiplet is slightly non-degenerate due to rotational symmetry breaking effects on the lattice, we average over the energies of a given multiplet.

The phase shifts in the $^1S_0$ and $^1P_1$ channels are shown in Fig. 5.4 as a function of center-of-mass momentum [57]. The comparison of lattice and continuum results in Fig. 5.4 is performed by numerical solution of the Lippmann-Schwinger equation in momentum space with the same potential. We note that the continuum limit poses no particular difficulties for our trial potential, which (unlike the realistic NLEFT potential) is regular and free of UV singularities. The lattice results are found to be within a few percent of the exact results for momenta below $k \simeq 80$ MeV. The

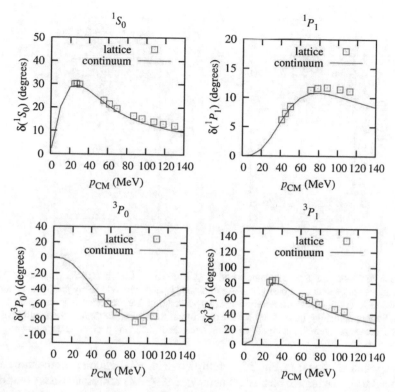

Fig. 5.4 Lattice phase shifts for uncoupled channels, using the model interaction of Sect. 5.4 for $a = 1.97$ fm. Upper panels: $^1S_0$ and $^1P_1$ channels with $S = 0$. Lower panels: $^3P_0$ and $^3P_1$ channels with $S = 1$. Note that the (diagonal) contribution from the tensor interaction in the $^3P_0$ channel is sufficiently strong to overcome the attraction of the central component, and make the net contribution repulsive. In order of increasing momentum, the lattice data points correspond to the first radial excitation for spherical wall radii $R_{wall} = 10a$, $9a$ and $8a$ units, followed by the same sequence for the second radial excitation, and so on. Very large wall radii are required to access low momenta, and many different radii are needed to cover the full range of lattice momenta. The lines labeled "continuum" are obtained by numerical solution of the Lippmann-Schwinger equation in momentum space with the same potential

discrepancies increase to 10–15% for momenta near $k \simeq 120$ MeV. This is as good as can be expected without further fine-tuning of the lattice action (remember that $a = 1.97$ fm corresponds to a momentum cutoff $\pi/a = 314$ MeV).

5.4.3 Partial-Wave Mixing

Next, we shall consider the uncoupled $S = 1$ channels, in preparation for the cases where the tensor operator S_{12} induces mixing between partial waves with $S = 1$ and $l = J \pm 1$. The interacting energy spectrum for $S = 1$ channels with a spherical

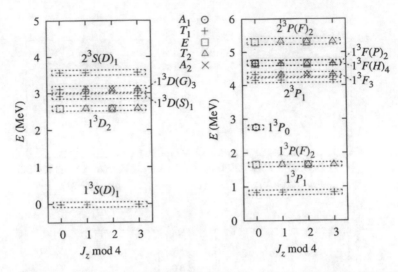

Fig. 5.5 Spectrum for $S = 1$ channels and $V \neq 0$, for $a = 1.97$ fm and $R_{\text{wall}} = 10a$. The left panel shows states with even parity, and the right panel states with odd parity. Individual states are denoted as in Fig. 5.2, with the notation $1^3S(D)_1$ indicating that the state becomes a pure S-wave in the limit of vanishing mixing. With the exception of the lowest multiplets, the identification of states becomes problematic at higher energies, because of the many accidental degeneracies

wall applied is shown in Fig. 5.5. Multiplets with total angular momentum J are deduced from the approximate degeneracy of $SO(3, Z)$ representations comprising the multiplet decompositions in Table 3.2. In some cases, the accidental degeneracy of different multiplets makes this process difficult. For instance, as shown in Fig. 5.5 for $R_{\text{wall}} = 10a$, the $1^3F(P)_2$ and $1^3F(H)_4$ multiplets are nearly degenerate. In such cases, further information can be obtained by calculating the inner product of the solution with the spherical harmonics $Y_{l,m}(\theta, \phi)$. Once the $^{2S+1}l_J$ multiplets are identified, the procedure for the uncoupled $S = 1$ channels equals that for the case of $S = 0$. In Fig. 5.4, we compare lattice results and exact continuum results for the $^3P_0$ and $^3P_1$ phase shifts, according to Ref. [57]. The diagonal component of the tensor interaction produces significant repulsion in the $^3P_0$ channel, sufficient to overcome the attraction from the central force interaction, as indicated by the negative phase shift. The overall agreement with continuum calculations is similar to the case of $S = 0$. The $1^3S(D)_1$ bound state has an energy of $\simeq -0.170$ MeV in the infinite-volume limit, which is close to the continuum result $\simeq -0.155$ MeV mentioned earlier.

We shall now discuss how the spherical wall formalism can account for channels coupled by the tensor interaction [59, 60]. Instead of the spherical harmonics used in Eq. (5.106), we introduce the vector spherical harmonics

$$Y_l^{J,J_z}(\theta, \phi) \equiv \sum_{m=-l}^{l} \sum_{S_z=-1}^{1} Y_{l,m}(\theta, \phi)|1, S_z\rangle\langle l, m; 1, S_z|J, J_z\rangle, \qquad (5.114)$$

for $S = 1$, in order to account for the mixing of partial waves with $l = J \pm 1$. Here, the $\langle l, m; S, S_z | J, J_z \rangle$ denote Clebsch-Gordan coefficients for the addition of orbital angular momentum l and spin S to total angular momentum J, and the spin functions $|S, S_z\rangle$ satisfy $\hat{S}^2 |S, S_z\rangle = S(S + 1)|S, S_z\rangle$ and $\hat{S}_z |S, S_z\rangle = S_z |S, S_z\rangle$. Then, we take the full wave function (in continuous space) to be

$$\psi(r, \theta, \phi) = \psi_{J-1}(r) \, Y_{J-1}^{J,J_z}(\theta, \phi) + \psi_{J+1}(r) \, Y_{J+1}^{J,J_z}(\theta, \phi), \tag{5.115}$$

for $S = 1$, and given J and J_z. For a potential proportional to S_{12}, the ansatz (5.115) leads to a radial equation with two components, coupled by the effective potential of Eq. (5.80). If we organize the wave function into the two-component vector

$$\psi(r) = \begin{bmatrix} \psi_{J-1}(r) \\ \psi_{J+1}(r) \end{bmatrix}, \tag{5.116}$$

the S-matrix becomes a 2×2 matrix, which can be parameterized in different ways. In the "Stapp parameterization" [59], the S-matrix is given by

$$S \equiv \begin{bmatrix} \exp(i\delta_{J-1}) & \\ & \exp(i\delta_{J+1}) \end{bmatrix} \begin{bmatrix} \cos(2\epsilon_J) & i \sin(2\epsilon_J) \\ i \sin(2\epsilon_J) & \cos(2\epsilon_J) \end{bmatrix} \begin{bmatrix} \exp(i\delta_{J-1}) & \\ & \exp(i\delta_{J+1}) \end{bmatrix}, \tag{5.117}$$

in terms of the phase shifts $\delta_{J\pm1}$ and the mixing angle ϵ_J. For simplicity, we shall suppress the momentum arguments of phase shifts and mixing angles in the intermediate steps, and reintroduce them as needed. As for the uncoupled case in Eq. (5.47), we take the incoming component (at asymptotically large r) to be

$$\psi_{\text{in}}(r) = -\frac{\exp(-ikr + iJ\pi/2)}{2ikr} \begin{bmatrix} \exp(-i\pi/2) & \\ & \exp(i\pi/2) \end{bmatrix} \begin{bmatrix} C \\ D \end{bmatrix}, \tag{5.118}$$

for real-valued constants C and D. Similarly, we may obtain the scattered component by operating on the outgoing component with S, as in Eq. (5.47). In order to work with a convenient (and real-valued) representation, we instead let the asymptotic incoming wave be

$$\tilde{\psi}_{\text{in}}(r) = -\frac{\exp(-ikr + iJ\pi/2)}{2ikr} \begin{bmatrix} \exp(-i\pi/2) & \\ & \exp(i\pi/2) \end{bmatrix} W \begin{bmatrix} C \\ D \end{bmatrix}, \tag{5.119}$$

where

$$W = \begin{bmatrix} \exp(-i\delta_{J-1}) & \\ & \exp(-i\delta_{J+1}) \end{bmatrix} \begin{bmatrix} \cos \varepsilon_J & -i \sin \varepsilon_J \\ -i \sin \varepsilon_J & \cos \varepsilon_J \end{bmatrix}, \tag{5.120}$$

which satisfies $WW^\dagger = W^\dagger W = 1$ and $W^* W^\dagger = S$. This gives

$$\tilde{\psi}_{\text{out}}(r) = \frac{\exp(ikr - iJ\pi/2)}{2ikr} \begin{bmatrix} \exp(i\pi/2) \\ & \exp(-i\pi/2) \end{bmatrix} W^* \begin{bmatrix} C \\ D \end{bmatrix}, \qquad (5.121)$$

for the outgoing component (since $W^* = SW$). Then, we find

$$\tilde{\psi}(r) = \tilde{\psi}_{\text{in}}(r) + \tilde{\psi}_{\text{out}}(r) = 2\,\text{Re}\,\tilde{\psi}_{\text{out}}(r) \qquad (5.122)$$

$$= \frac{1}{kr} \left\{ \begin{bmatrix} C\sin(kr - (J-1)\pi/2 + \delta_{J-1}) \\ D\sin(kr - (J+1)\pi/2 + \delta_{J+1}) \end{bmatrix} \cos\varepsilon_J \right.$$

$$\left. + \begin{bmatrix} D\cos(kr - (J-1)\pi/2 + \delta_{J-1}) \\ C\cos(kr - (J+1)\pi/2 + \delta_{J+1}) \end{bmatrix} \sin\varepsilon_J \right\}, \qquad (5.123)$$

for the (real-valued) solution in the asymptotic regime where $kr \gg l$.

We now introduce a hard spherical wall boundary at $r = \tilde{R}_{\text{wall}}$, where the "physical" wall radius has again been determined by requiring a vanishing phase shift when $V = 0$. Both partial waves vanish at the wall boundary, and we define the angle

$$-\Delta \equiv k\tilde{R}_{\text{wall}} - (J-1)\pi/2, \qquad (5.124)$$

in terms of which

$$\frac{C}{D}\tan(-\Delta + \delta_{J-1}) = -\tan\varepsilon_J, \qquad (5.125)$$

and

$$\frac{D}{C}\tan(-\Delta - \pi + \delta_{J+1}) = \frac{D}{C}\tan(-\Delta + \delta_{J+1}) = -\tan\varepsilon_J, \qquad (5.126)$$

using Eq. (5.123). In general, there are two solutions for Δ per angular interval π. In the asymptotic case where $k\tilde{R}_{\text{wall}} \gg 1$, the spacing between energy levels becomes infinitesimal. We can therefore find two independent solutions Δ^{I} and Δ^{II} (with nearly the same k), and we neglect the difference in k in the following steps. These two solutions satisfy

$$\tan(-\Delta^{\text{I,II}} + \delta_{J-1})\tan(-\Delta^{\text{I,II}} + \delta_{J+1}) = \tan^2\varepsilon_J, \qquad (5.127)$$

such that Eq. (5.123) can be recast as

$$\tilde{\psi}(r) \equiv \frac{C}{kr} \begin{bmatrix} \tilde{\psi}_{J-1}(r)\cos\varepsilon_J \\ \tilde{\psi}_{J+1}(r)\sin\varepsilon_J \end{bmatrix}, \qquad (5.128)$$

where

$$\tilde{\psi}_{J-1}(r) = \sin(kr - (J-1)\pi/2 + \Delta)/\cos(-\Delta + \delta_{J-1}), \quad (5.129)$$

$$\tilde{\psi}_{J+1}(r) = -\sin(kr - (J+1)\pi/2 + \Delta)/\sin(-\Delta + \delta_{J+1}), \quad (5.130)$$

and if we parameterize the two linearly independent, real-valued solutions as

$$\psi^{I,II}(r) \propto \frac{1}{kr} \begin{bmatrix} A_{J-1}^{I,II} \sin(kr - (J-1)\pi/2 + \Delta^{I,II}) \\ A_{J+1}^{I,II} \sin(kr - (J+1)\pi/2 + \Delta^{I,II}) \end{bmatrix}, \quad (5.131)$$

we find

$$A_{J-1}^{I,II} \tan \varepsilon_J = -A_{J+1}^{I,II} \frac{\sin(-\Delta^{I,II} + \delta_{J+1})}{\cos(-\Delta^{I,II} + \delta_{J-1})}, \quad (5.132)$$

as additional constraints. Of the four constraints (5.127) and (5.132), only three are needed to determine δ_{J-1}, δ_{J+1}, and ε_J. For instance

$$A_{J-1}^{II} \tan \varepsilon_J = -A_{J+1}^{II} \frac{\sin(-\Delta^{II} + \delta_{J+1})}{\cos(-\Delta^{II} + \delta_{J-1})}, \quad (5.133)$$

must be redundant, which makes sense in light of the symmetry associated with the interchange of the solutions Δ^I and Δ^{II}. Furthermore, the orthogonality of ψ^I and ψ^{II} implies $A_{J-1}^I A_{J-1}^{II} + A_{J+1}^I A_{J+1}^{II} = 0$, and hence

$$\frac{A_{J+1}^I}{A_{J-1}^I} = -\frac{A_{J-1}^{II}}{A_{J+1}^{II}}, \quad (5.134)$$

which, together with Eq. (5.127), provides a relation between the constraints (5.127) for the solutions Δ^I and Δ^{II}.

It should be noted that the constraints (5.127) and (5.132) hold if the spherical wall is located at a sufficiently large radius, $k\tilde{R}_{wall} \gg l$. In particular, the nodes of the spherical Bessel functions $j_{J-1}(kr)$ and $j_{J+1}(kr)$ then coincide, and it suffices to find one angle Δ for both partial wave components, according to Eq. (5.124). Due to computational constraints on the lattice dimensions, it is often more feasible to work with smaller values of $k\tilde{R}_{wall} \sim l$, and the coincidence of nodes for $l = J - 1$ and $l = J + 1$ at $\tilde{R}_{wall}$ then no longer implies coincidence of nodes for $kr \gg l$. Instead of Eq. (5.131), we determine solutions $(\Delta_{J-1}^I, \Delta_{J+1}^I, k^I)$ and $(\Delta_{J-1}^{II}, \Delta_{J+1}^{II}, k^{II})$ with separate values of $\Delta_{J\pm1}$ for the partial wave components with $l = J \pm 1$, according to

$$\psi^I(r) \propto \frac{1}{k^I r} \begin{bmatrix} A_{J-1}^I \sin(k^I r - (J-1)\pi/2 + \Delta_{J-1}^I) \\ A_{J+1}^I \sin(k^I r - (J+1)\pi/2 + \Delta_{J+1}^I) \end{bmatrix}, \quad (5.135)$$

for $k^{\mathrm{I}} = \sqrt{m_N E^{\mathrm{I}}}$, and

$$\psi^{\mathrm{II}}(r) \propto \frac{1}{k^{\mathrm{II}} r} \left[\begin{matrix} A_{J-1}^{\mathrm{II}} \sin(k^{\mathrm{II}} r - (J-1)\pi/2 + \Delta_{J-1}^{\mathrm{II}}) \\ A_{J+1}^{\mathrm{II}} \sin(k^{\mathrm{II}} r - (J+1)\pi/2 + \Delta_{J+1}^{\mathrm{II}}) \end{matrix} \right], \tag{5.136}$$

for $k^{\mathrm{II}} = \sqrt{m_N E^{\mathrm{II}}}$. The set of equations (5.127) through (5.132) can now be generalized according to

$$\tan^2 \varepsilon_J(k^{\mathrm{I}}) = \tan(-\Delta_{J-1}^{\mathrm{I}} + \delta_{J-1}(k^{\mathrm{I}})) \tan(-\Delta_{J+1}^{\mathrm{I}} + \delta_{J+1}(k^{\mathrm{I}})), \tag{5.137}$$

$$\tan^2 \varepsilon_J(k^{\mathrm{II}}) = \tan(-\Delta_{J-1}^{\mathrm{II}} + \delta_{J-1}(k^{\mathrm{II}})) \tan(-\Delta_{J+1}^{\mathrm{II}} + \delta_{J+1}(k^{\mathrm{II}})), \tag{5.138}$$

and

$$A_{J-1}^{\mathrm{I}} \tan \varepsilon_J(k^{\mathrm{I}}) = - A_{J+1}^{\mathrm{I}} \frac{\sin(-\Delta_{J+1}^{\mathrm{I}} + \delta_{J+1}(k^{\mathrm{I}}))}{\cos(-\Delta_{J-1}^{\mathrm{I}} + \delta_{J-1}(k^{\mathrm{I}}))}, \tag{5.139}$$

$$A_{J-1}^{\mathrm{II}} \tan \varepsilon_J(k^{\mathrm{II}}) = - A_{J+1}^{\mathrm{II}} \frac{\sin(-\Delta_{J+1}^{\mathrm{II}} + \delta_{J+1}(k^{\mathrm{II}}))}{\cos(-\Delta_{J-1}^{\mathrm{II}} + \delta_{J-1}(k^{\mathrm{II}}))}, \tag{5.140}$$

where Eqs. (5.137) and (5.139) provide two constraints for $k = k^{\mathrm{I}}$, while Eqs. (5.138) and (5.140) provide two constraints for $k = k^{\mathrm{II}}$. In general, this situation requires interpolation between the momenta, unless closely spaced pairs of solutions with $k^{\mathrm{I}} \approx k^{\mathrm{II}}$ can be found. In such cases, we can take

$$\tan^2 \varepsilon_J(k^{\mathrm{I}}) = \tan(-\Delta_{J-1}^{\mathrm{I}} + \delta_{J-1}(k^{\mathrm{I}})) \tan(-\Delta_{J+1}^{\mathrm{I}} + \delta_{J+1}(k^{\mathrm{I}})), \tag{5.141}$$

$$\tan^2 \varepsilon_J(k^{\mathrm{I}}) \approx \tan(-\Delta_{J-1}^{\mathrm{II}} + \delta_{J-1}(k^{\mathrm{I}})) \tan(-\Delta_{J+1}^{\mathrm{II}} + \delta_{J+1}(k^{\mathrm{I}})), \tag{5.142}$$

$$A_{J-1}^{\mathrm{I}} \tan \varepsilon_J(k^{\mathrm{I}}) = - A_{J+1}^{\mathrm{I}} \frac{\sin(-\Delta_{J+1}^{\mathrm{I}} + \delta_{J+1}(k^{\mathrm{I}}))}{\cos(-\Delta_{J-1}^{\mathrm{I}} + \delta_{J-1}(k^{\mathrm{I}}))}, \tag{5.143}$$

which provides three constraints for the phase shifts and the mixing angle at $k = k^{\mathrm{I}}$, and

$$\tan^2 \varepsilon_J(k^{\mathrm{II}}) \approx \tan(-\Delta_{J-1}^{\mathrm{I}} + \delta_{J-1}(k^{\mathrm{II}})) \tan(-\Delta_{J+1}^{\mathrm{I}} + \delta_{J+1}(k^{\mathrm{II}})),$$
$$\tag{5.144}$$

$$\tan^2 \varepsilon_J(k^{\mathrm{II}}) = \tan(-\Delta_{J-1}^{\mathrm{II}} + \delta_{J-1}(k^{\mathrm{II}})) \tan(-\Delta_{J+1}^{\mathrm{II}} + \delta_{J+1}(k^{\mathrm{II}})),$$
$$\tag{5.145}$$

$$A_{J-1}^{\mathrm{II}} \tan \varepsilon_J(k^{\mathrm{II}}) = - A_{J+1}^{\mathrm{II}} \frac{\sin(-\Delta_{J+1}^{\mathrm{II}} + \delta_{J+1}(k^{\mathrm{II}}))}{\cos(-\Delta_{J-1}^{\mathrm{II}} + \delta_{J-1}(k^{\mathrm{II}}))}, \tag{5.146}$$

for $k = k^{\mathrm{II}}$.

A moderately successful prescription (which largely eliminates the need for interpolation in the momenta) is to choose the "pairs of closely spaced solutions" as the $(n + 1)$st radial excitation for $l = J - 1$, and the nth radial excitation for $l = J + 1$. In Fig. 5.6, we show the lattice and continuum results for the $^3S_1$ and $^3D_1$ phase shifts, and the $J = 1$ mixing angle ε_1 of Ref. [57]. The lattice results for coupled channels appear somewhat less accurate than those for uncoupled channels, a major source of systematical error being the remaining discrepancy $k^{\mathrm{I}} \approx k^{\mathrm{II}}$. Another significant source of error may be mixing with unphysical channels due to the broken rotational invariance on the lattice. For instance, the results shown in Fig. 5.6 may have a small but non-negligible contamination from a $^3D_3$ component, as noted in Chap. 4 in the discussion of rotational-symmetry breaking operators.

As the tensor force of our model potential is (purposefully) very strong, the mixing effects are also large. For realistic nucleon-nucleon interactions, all of the mixing angles in the Stapp parameterization are numerically small when $kM_\pi \ll 1$. If $\Delta^{\mathrm{I}}_{J-1} \to \delta_{J-1}(k^{\mathrm{I}})$ and $\Delta^{\mathrm{II}}_{J+1} \to \delta_{J+1}(k^{\mathrm{II}})$ for $\varepsilon_J = 0$, then the expressions in

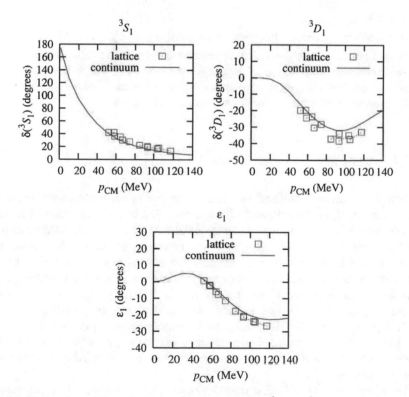

Fig. 5.6 Lattice phase shifts and mixing angle for the coupled $^3S_1$ and $^3D_1$ channels with $J = 1$ of our model interaction for $a = 1.97\,\mathrm{fm}$. Notation and conventions are as for Fig. 5.4. Note the pairs of solutions with $k = k^{\mathrm{I}}$ and $k = k^{\mathrm{II}}$, which form the leading source of systematical error in the coupled-channel analysis, as k^{I} and k^{II} are only approximately equal

Eqs. (5.141) through (5.146) can be expanded for small ε_J, giving

$$\delta_{J-1}(k^{\mathrm{I}}) = \Delta^{\mathrm{I}}_{J-1} + \frac{\varepsilon_J^2(k^{\mathrm{I}})}{\tan(-\Delta^{\mathrm{I}}_{J+1} + \delta_{J+1}(k^{\mathrm{I}}))} + \mathscr{O}(\varepsilon_J^4), \tag{5.147}$$

$$\varepsilon_J(k^{\mathrm{I}}) = -\frac{A^{\mathrm{I}}_{J+1}}{A^{\mathrm{I}}_{J-1}} \sin(\Delta^{\mathrm{II}}_{J+1} - \Delta^{\mathrm{I}}_{J+1}) + \mathscr{O}(\varepsilon_J^3), \tag{5.148}$$

for $k = k^{\mathrm{I}}$, and

$$\delta_{J+1}(k^{\mathrm{II}}) = \Delta^{\mathrm{II}}_{J+1} + \frac{\varepsilon_J^2(k^{\mathrm{II}})}{\tan(-\Delta^{\mathrm{II}}_{J-1} + \delta_{J-1}(k^{\mathrm{II}}))} + \mathscr{O}(\varepsilon_J^4), \tag{5.149}$$

$$\varepsilon_J(k^{\mathrm{II}}) = \frac{A^{\mathrm{II}}_{J-1}}{A^{\mathrm{II}}_{J+1}} \sin(\Delta^{\mathrm{II}}_{J-1} - \Delta^{\mathrm{I}}_{J-1}) + \mathscr{O}(\varepsilon_J^3), \tag{5.150}$$

for $k = k^{\mathrm{II}}$.

As a final computational note, we remark that $A^{\mathrm{I}}_{J+1}/A^{\mathrm{I}}_{J-1}$ and $A^{\mathrm{II}}_{J-1}/A^{\mathrm{II}}_{J+1}$ can be determined by computing the inner product of $\psi(r)$ in the vicinity of $\tilde{R}_{\mathrm{wall}}$ with the spherical harmonics $Y_{l,m}(\hat{r})$. Testing for orthogonality with different spherical harmonics may also help with state identification, when many closely spaced levels with different angular momentum quantum numbers appear.

5.4.4 Auxiliary Potential Method

In Sect. 5.4.3, we have outlined the basics of the spherical wall method, including the treatment of coupled-channel effects due to the tensor interaction. The most significant challenge was found to be the need for two independent solutions with similar energies, in order to unambiguously determine scattering phase shifts and mixing angles in coupled $S = 1$ channels. On lattices of a practical size, such a situation can only be approximately achieved. In view of these shortcomings, the "auxiliary potential method" was put forward in Ref. [58], which is a refinement of the original spherical wall method. As we shall see, the auxiliary potential method greatly simplifies the calculation of mixing angles, and also allows for a continuous adjustment of the energy eigenvalues. Hence, phase shifts and mixing angles can be obtained with good accuracy for any desired value of the center-of-mass momentum.

For simplicity, we shall first consider uncoupled channels, where the nuclear potential $V(r)$ vanishes for $r \gg R_n$. As already noted, a hard spherical wall at radius R_{wall} provides access to discrete energy eigenvalues only, and a very large box is needed for low energies (and hence center-of-mass momenta). In the auxiliary

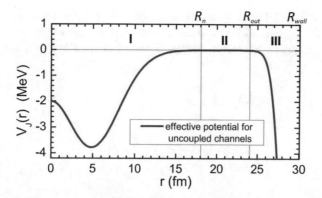

Fig. 5.7 Illustration of the radial potential $V_J(r)$ for $S = 1$, as applied in the auxiliary potential method. The auxiliary potential of Eq. (5.151) is shown for $C_{aux} = -25$ MeV, and only contributes for $R_{out} < r < R_{wall}$. Likewise, the nuclear potential is negligible for $r < R_n$, such that only the centrifugal barrier contributes in the range $R_n < r < R_{out}$. In the absence of a Coulomb potential, the radial wave function for $R_n < r < R_{out}$ is given by Eq. (5.152). Figure courtesy of Bingnan Lu

potential method, we introduce a Gaussian potential

$$V_{aux}(r) \equiv C_{aux} \exp\left[-\frac{(r - R_{wall})^2}{a_W^2}\right], \tag{5.151}$$

with $C_{aux} < 0$, which only contributes beyond a chosen "outer" radius R_{out}, intermediate between R_n and R_{wall}, as shown in Fig. 5.7. In Eq. (5.151), the Gaussian width a_W is chosen such that V_{aux} is negligible for $r < R_{out}$. Clearly, the full potential $V(r) + V_{aux}(r)$ vanishes for $R_n < r < R_{out}$, and hence shifting C_{aux} allows for a continuous adjustment of the energy eigenvalues, without appreciable distortion of the solutions to the Schrödinger equation.

Since $V(r)$ vanishes (except for the centrifugal barrier) when $R_n < r < R_{out}$, we may extract the scattering phase shifts in that region. We recall from the treatment in Sect. 5.3 that the wave function for $r > R_n$ can be expressed as

$$u_L^{ex}(r) = A H_l^-(0, kr) - B H_l^+(0, kr), \tag{5.152}$$

in terms of the Coulomb functions with $\eta = 0$ and $k = \sqrt{2\mu E}$. Once the solution is found using sparse matrix diagonalization, a least-squares fit of Eq. (5.152) to the solution can be performed. This yields the (complex-valued) coefficients A and B,

$$B = A S_l(k) = A \exp(2i\delta_l(k)), \tag{5.153}$$

from which the phase shift $\delta_l(k)$ can be obtained. This formalism can be straightforwardly generalized to the case of coupled channels (including the tensor component of the nuclear force). For coupled channels with $S = 1$, $\psi_l^{\mathrm{ex}}(r)$ has two components, $u_{J-1}^{\mathrm{ex}}(r)$ and $u_{J+1}^{\mathrm{ex}}(r)$, with $l = J \pm 1$. Both of these are of the form (5.152) for $R_n < r < R_{\mathrm{out}}$, and we recall that the S-matrix in the Stapp parameterization is given by

$$
S_J(k) \equiv \begin{bmatrix} \exp(i\delta_{J-1}) & \\ & \exp(i\delta_{J+1}) \end{bmatrix}
$$

$$
\times \begin{bmatrix} \cos(2\epsilon_J) & i\sin(2\epsilon_J) \\ i\sin(2\epsilon_J) & \cos(2\epsilon_J) \end{bmatrix} \begin{bmatrix} \exp(i\delta_{J-1}) & \\ & \exp(i\delta_{J+1}) \end{bmatrix},
$$

(5.154)

and by introducing the notation

$$
A_J \equiv \begin{bmatrix} A_{J-1} \\ A_{J+1} \end{bmatrix}, \quad B_J \equiv \begin{bmatrix} B_{J-1} \\ B_{J+1} \end{bmatrix}, \tag{5.155}
$$

we may again solve for $S_J(k)$ from Eq. (5.153) along with

$$
A^* = S_J(k)B^*, \tag{5.156}
$$

which can be inferred using the expression (5.67) for $S_l(k)$ in terms of the (real-valued) K-matrix. We find

$$
S_J(k) = \begin{pmatrix} B_{J+1} & A_{J+1}^* \\ B_{J-1} & A_{J-1}^* \end{pmatrix} \begin{pmatrix} A_{J+1} & B_{J+1}^* \\ A_{J-1} & B_{J-1}^* \end{pmatrix}^{-1}, \tag{5.157}
$$

from (5.153) and (5.156), where the inverse can clearly only be computed if

$$
A_{J+1}B_{J-1}^* - B_{J+1}^*A_{J-1} \neq 0, \tag{5.158}
$$

such that we are again faced with the issue that two linearly independent vectors A and B are needed for an unambiguous determination of $S_J(k)$. For a simple hard wall boundary, only one independent solution per lattice energy eigenvalue is obtained.

The auxiliary potential method provides a more elegant and more accurate solution to this ambiguity, which was previously circumvented by taking two eigenfunctions with approximately the same energy and neglecting their energy difference. As the model potential (5.80) is real and Hermitian, an exact symmetry under time reversal results. We add to the coupled-channel potential $V_J(r)$ for $S = 1$ an imaginary component

$$V_J(r) \rightarrow V_J(r) + \begin{pmatrix} & iU_{\text{aux}}(r) \\ -iU_{\text{aux}}(r) & \end{pmatrix}, \tag{5.159}$$

where $U_{\text{aux}}(r)$ is an arbitrary, real-valued function with support for $r > R_{\text{out}}$ only. This leaves $V_J(r)$ Hermitian and the energy eigenvalues real, while the time-reversal symmetry is broken. Also, the solutions $\psi(r)$ and $\psi^*(r)$ are now linearly independent, and satisfy the radial Schrödinger equation for $r < R_{\text{out}}$ with identical energy eigenvalues. It should be noted that the inverse in Eq. (5.157) cannot be computed for $U_{\text{aux}}(r) = 0$, since then $A = -B^*$. In order to circumvent this problem, it suffices to take

$$U_{\text{aux}}(r) \equiv U_0 \, \delta_{r,r_0}, \tag{5.160}$$

where r_0 is an arbitrarily chosen radial mesh point for $R_{\text{out}} < r_0 < R_{\text{wall}}$, and U_0 is an arbitrary (real-valued) constant. The distortion of the energy eigenvalues and wave functions introduced by this choice is found to be minimal. The resulting phase shifts and mixing angles for the Gaussian model interaction are shown in Fig. 5.8 for different choices of C_{aux}. In particular, the mixing angles at higher center-of-mass momenta are very accurately reproduced.

Before we move on to the application of the NLEFT potential to the problem of neutron-proton scattering, let us conclude this section by a brief discussion of the relative merits of the Lüscher and spherical wall methods. As we have seen, for the scattering of particles on a cubic lattice, Lüscher's finite-volume method uses periodic boundary conditions to find relationships between elastic scattering phase shifts and energy eigenvalues. An important advantage of Lüscher's method is that periodic boundary conditions are commonly used in a wide range of lattice calculations of nuclear, hadronic, atomic and condensed matter systems. Hence, the method is straightforward to apply to a wide class of problems.

In spite of these attractive features, Lüscher's method does have a number of significant drawbacks. As we have noted, the relationship between scattering phase shifts and energy eigenvalues quickly becomes complicated at higher energies and higher orbital angular momenta. Similarly, the accurate treatment of spin-orbit coupling and partial-wave mixing have proven difficult within the Lüscher formalism, because of artifacts generated by the periodic cubic boundary. Any method overcoming these theoretical problems would likely not be numerically practical when applied to Lattice QCD, as each individual hadron must be constructed as a bound state of quark and gluon fields. Hence, the MC signal for hadron-hadron scattering states well above threshold would be very weak. However, for an effective lattice field theory where nucleons are treated as "fundamental" point particles, alternative techniques such as the spherical wall method have

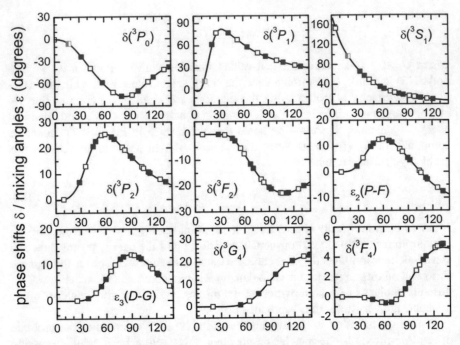

Fig. 5.8 Phase shifts and mixing angles for the Gaussian model potential, computed using the auxiliary potential method for $S = 1$ (coupled and uncoupled channels). Results are shown for $C_{aux} = 0$ (red/filled squares) and $C_{aux} = -25$ MeV (white/hollow squares). The model interaction used here has a much stronger tensor component than the NLEFT interaction. Note how $C_{aux} < 0$ provides access to much lower center-of-mass momenta than is possible with the standard spherical wall method. The solid lines are solutions to the Lippmann-Schwinger equation in the continuum

proven viable, since the full two-particle spectrum is relatively easy to compute.

The spherical wall method, and in particular its recent development into the "auxiliary potential method" described above, provides a general technique for measuring phase shifts and mixing angles for two-particle scattering on the lattice. In particular, the auxiliary field method provides an efficient and reliable way to compute phase shifts and mixing angles for coupled channels. For the Gaussian model interaction considered here, we were able to identify multiplets of total angular momentum J from the approximate degeneracy of the $SO(3, Z)$ representations, however for the realistic (singular) NLEFT potential, these multiplets (beyond the first few excited states) are typically split apart much further. As we shall see in the next Section, the auxiliary potential method can be combined with a "radial Hamiltonian" for a specific set of angular momentum quantum numbers. This eliminates the problem of correct identification of states, and also makes the removal of unphysical mixings due to broken rotational symmetry straightforward.

It should also be noted that Lüscher's method requires that finite-volume energy levels can be accurately determined, with errors small compared to the separation between adjacent energy levels. This is not practical in cases such as nucleus-nucleus scattering, where the separation between finite-volume energy levels is many orders of magnitude smaller than the total energy of the system. Fortunately, this problem has been solved using an alternative approach called the "adiabatic projection method", whereby initial cluster states composed of nucleons are evolved using Euclidean time projection, and used to calculate an "effective" two-cluster Hamiltonian (or transfer matrix). In the limit of large projection time, the spectral properties of such an effective two-cluster Hamiltonian coincide with those of the original underlying theory. We shall present the adiabatic projection method in Chap. 6.

5.5 Neutron-Proton Scattering in NLEFT

We shall next describe the calculation of neutron-proton (np) phase shifts and mixing angles using the actual NLEFT nucleon-nucleon potential. This also serves the purpose of fixing the unknown constants in the short-distance contributions in the NLEFT two-nucleon force (2NF). Along with pp and nn data which fixes the short-distance terms in the EM interaction, this enables predictive calculations for $A \geq 4$, once the case of $A = 3$ has been used to fix any unknown coefficients in the NLEFT three-nucleon force (3NF). We shall discuss these issues in Sects. 5.6 and 5.7, respectively. In our treatment of np scattering in NLEFT, we shall first focus on the issue of operator smearing, followed by the introduction of a Hamiltonian of reduced dimensionality, expressed in radial coordinates only. The np potential is given by the NLEFT expressions for the 2NF in Chap. 4, for the case without auxiliary fields and pion fields. These are needed for the MC simulations, which is the method of choice for $A \geq 4$. Hence, the expressions for the potential will only be minimally repeated here.

For np scattering, we shall use the NLEFT transfer matrix described in Chap. 4. Note that the Hamiltonian can also be used (which equals the transfer matrix in the limit $\delta \to 0$) if the purpose is not to establish contact with MC simulations at nonzero δ. However, an important objective of the NLEFT treatment of np scattering is the determination of unknown EFT coefficients at LO and NLO, thereby fixing the 2NF for predictive calculations of heavier nuclei. Thus far, we have introduced the spherical wall method by computing phase shifts and mixing angles for an attractive Gaussian potential which shares many features with the realistic NLEFT potential. For instance, both feature a tensor component proportional to S_{12}, and for a Gaussian envelope of characteristic size $R_n = 2 \times 10^{-2}$ MeV^{-1}, the strength of the potential can be tuned to produce a shallow, deuteron-like $^3S(D)_1$ bound state. At LO, the long-range part of the NLEFT transfer matrix is given by the OPE component, which in the continuum reduces to the spin- and isospin-dependent

Yukawa potential

$$V_\pi(\mathbf{r}) = \frac{g_A^2}{4F_\pi^2}\left\{\frac{M_\pi^2 \exp(-M_\pi r)}{12\pi r}\left[\left(1 + \frac{3}{M_\pi r} + \frac{3}{M_\pi^2 r^2}\right) S_{12}(\hat{r}) + \sigma_1 \cdot \sigma_2\right]\right.$$

$$\left. - \frac{1}{3}\sigma_1 \cdot \sigma_2\, \delta^{(3)}(\mathbf{r})\right\}\tau_1 \cdot \tau_2, \qquad\qquad (5.161)$$

which is structurally similar to the Gaussian model interaction considered earlier. An important conceptual difference is that the OPE potential is divergent at short distances, and moreover the NLEFT potential contains two contact terms at LO, the coefficients of which need to be fitted to low-energy S-wave scattering parameters (preferentially the scattering lengths in the $^1S_0$ and $^3S_1$ channels). As the lattice spacing a also functions as the UV regulator of the EFT, the singularity at $r \to 0$ is never reached. Hence one should not take the limit $a \to 0$, but rather study the behavior of the theory under variation of the lattice spacing down to $a \simeq 1$ fm. Also, the singular short-distance interactions are likely to cause larger breaking of rotational invariance than that found for the UV-bounded model interaction. In such cases, it may be useful to improve the description by means of higher-order derivative interactions, specifically designed to cancel lattice artifacts.

Before we review the application of the auxiliary potential method to np scattering in NLEFT, we briefly discuss the qualitative effects of the smearing of the LO contact interactions, which are always treated non-perturbatively in the sparse matrix diagonalization of the transfer matrix. The resulting energy spectra for the LO_1 (unsmeared contact interactions) and LO_2 (Gaussian smeared contact interactions) transfer matrices are shown in Fig. 5.9 according to Ref. [61]. Given a lattice dispersion relation corresponding to the improved free lattice Hamiltonian H_{free}, one obtains the "target energies" for each state, indicated by black lines in Fig. 5.9. Once the computed energies coincide with the target energies, the phase shifts will also coincide with the Nijmegen PWA. Clearly, the LO_1 interaction is too attractive in the S-wave channels, which requires compensation by a sizable (repulsive) NLO contribution. Such large, repulsive contributions are potentially detrimental to MC calculations, see e.g. the discussion in Chap. 8. A further problem is that the LO_2 interaction produces unphysical attractive forces in the P-wave channels. As shown in Ref. [62], the smeared LO_3 transfer matrix circumvents that problem, at the price of introducing unphysical forces in the (presumably less important) D-wave channels. The LO_3 transfer matrix forms the basis of our following treatment of np scattering.

5.5.1 Radial Hamiltonian Method

So far, we have found that the auxiliary potential method greatly simplifies the use of a spherical wall to determine scattering phase shifts and mixing angles.

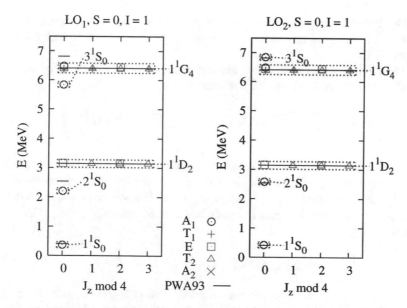

Fig. 5.9 Spectrum for np with $S = 0, I = 1$, using the (unsmeared) lattice action LO_1, and the LO_2 lattice action (with Gaussian smearing), for $a = 1.97$ fm and $R_{wall} = 10a$. The solid lines indicate "target energies" which reproduce exactly the data from the Nijmegen PWA. The excess attraction at LO in the $^1S_0$ channel disappears when smearing is applied to the LO contact terms. The solutions are classified according to the irreps of the cubic group, and the dotted lines enclose continuum angular momentum multiplets

Still, the computational complexity of the sparse matrix diagonalization using the full three-dimensional basis remains problematic. We shall therefore simplify the problem further by means of the "radial Hamiltonian method", where a lattice Hamiltonian (or transfer matrix) of reduced dimensionality is constructed in radial coordinates [63]. Hence, we seek a transformation

$$M(\mathbf{r}', \mathbf{r}) \rightarrow \tilde{M}(r', r), \tag{5.162}$$

where $r \equiv |\mathbf{r}| = a|\mathbf{n}|$. This entails grouping the lattice points (n_x, n_y, n_z) into "shells" according to the radial coordinate r, and by weighting their relative contribution with the spherical harmonics $Y_{l,m}(\hat{r})$. Thus, instead of working with the full three-dimensional basis $|\mathbf{r}\rangle$, we have the reduced basis

$$|r; l, m\rangle = \sum_{\mathbf{r}'} Y_{l,m}(\hat{r}')\delta_{r,r'}|\mathbf{r}'\rangle, \tag{5.163}$$

which can be reduced further into segments $|r - a|\mathbf{n}|| < a_r$, where the bin width a_r is a small, positive parameter [64]. Such binning of the radial coordinates is useful in particular for arbitrary, non-square lattice geometries. As the components of m which fall into different irreps of the cubic group are (unphysically) split apart when $a \neq 0$, it is useful to note that weighted averages over the multiplets show much less dependence on a, as we shall discuss in Appendix C.

When spin degrees of freedom are included, we project onto states with total angular momentum (J, J_z) in the continuum limit,

$$|r; l, S, J, J_z\rangle \equiv \sum_{\mathbf{r}'} \sum_{m, S_z} \langle l, m; S, S_z | J, J_z \rangle Y_{l,m}(\hat{r}') \delta_{r,r'} |\mathbf{r}'\rangle \otimes |S, S_z\rangle, \qquad (5.164)$$

where (as discussed before) we use the continuum notation. Due to rotational symmetry breaking effects, the angular momentum quantum numbers are not exactly good quantum numbers when $a \neq 0$. The states of Eq. (5.164) form a complete basis on the lattice, which however is not orthonormal. Hence, we calculate

$$N_{l,S,J,J_z}(r) \equiv \langle r; l, S, J, J_z | r; l, S, J, J_z \rangle, \qquad (5.165)$$

with respect to the number of lattice points per radial shell, and the angular momentum quantum numbers. Clearly, for $|\mathbf{r}| = 0$ we have only one lattice point, while for $|\mathbf{r}| = \sqrt{2}a$ we have six points, for $|\mathbf{r}| = \sqrt{3}a$ there are 12 points, and so on. These multiplicities are given for the case of no binning, or when the bin width $a_r \ll a$, see Ref. [64] for more details. In the continuum, the radial norm would be proportional to $\sim r^2$. Assuming a local potential $V(\mathbf{r})$, we obtain

$$V_{l,S,J,J_z}(r) = N_{l,S,J,J_z}^{-1/2}(r) \langle l, S, J, J_z; r | V(\mathbf{r}) | r; l, S, J, J_z \rangle N_{l,S,J,J_z}^{-1/2}(r) \qquad (5.166)$$

using Eqs. (5.164) and (5.165). In the radial Hamiltonian method, the potential in a given scattering channel is given by Eq. (5.166). For each partial wave, the eigenvectors of the radial Hamiltonian are identified with a superposition of regular and irregular solutions to the Helmholtz equation, in the region beyond the range of the interaction. The reduced memory allocation requirements for sparse matrix solvers are particularly useful in time-consuming calculations, where the unknown coefficients are fitted by chi-square minimization to empirical scattering data [65]. We also note that Eq. (5.166) eliminates the state identification problem, which was due to the many accidental degeneracies of states with different angular momenta in the spectrum of the full three-dimensional Hamiltonian.

As an illustrative example, consider np scattering using the radial transfer matrix formalism. For instance, we may compute the contributions from the set of NLO contact operators introduced in Chap. 4, to each partial wave channel. From a total of seven operators, only three distinct ones remain when the dependence on spin and isospin has been accounted for. In the spin-singlet $^1S_0$ channel, we have

$$\langle {}^1S_0|V^{(2)}|{}^1S_0\rangle = \widetilde{C}_1 \langle {}^1S_0|\mathscr{O}_{q^2}|{}^1S_0\rangle, \tag{5.167}$$

such that the dependence on the NLO contact terms is reduced to a single amplitude with coupling constant $\widetilde{C}_1$. Here, $\mathscr{O}_{q^2}$ is defined by

$$\widetilde{V}_{q^2} \equiv C_{q^2}\mathscr{O}_{q^2}, \tag{5.168}$$

with similar definitions for the remaining NLO contact operators. For the $^1P_1$ channel, we find a similar expression

$$\langle {}^1P_1|V^{(2)}|{}^1P_1\rangle = \widetilde{C}_4 \langle {}^1P_1|\mathscr{O}_{q^2}|{}^1P_1\rangle, \tag{5.169}$$

while for the uncoupled spin-triplet P-wave channels, we have

$$\langle {}^3P_0|V^{(2)}|{}^3P_0\rangle = \widetilde{C}_5 \langle {}^3P_0|\mathscr{O}_{q^2}|{}^3P_0\rangle + \widetilde{C}_6 \langle {}^3P_0|\mathscr{O}_{(q\cdot S)^2}|{}^3P_0\rangle$$
$$+\widetilde{C}_7 \langle {}^3P_0|\mathscr{O}^{I=1}_{(iq\times S)\cdot k}|{}^3P_0\rangle, \tag{5.170}$$

and

$$\langle {}^3P_1|V^{(2)}|{}^3P_1\rangle = \widetilde{C}_5 \langle {}^3P_1|\mathscr{O}_{q^2}|{}^3P_1\rangle + \widetilde{C}_6 \langle {}^3P_1|\mathscr{O}_{(q\cdot S)^2}|{}^3P_1\rangle$$
$$+\widetilde{C}_7 \langle {}^3P_1|\mathscr{O}^{I=1}_{(iq\times S)\cdot k}|{}^3P_1\rangle, \tag{5.171}$$

which also involve $\mathscr{O}_{(q\cdot S)^2}$ and the $I=1$ spin-orbit operator $\mathscr{O}^{I=1}_{(iq\times S)\cdot k}$. For channels coupled by the tensor operator S_{12}, we find

$$\langle {}^3S(D)_1|V^{(2)}|{}^3S(D)_1\rangle = \widetilde{C}_2 \langle {}^3S(D)_1|\mathscr{O}_{q^2}|{}^3S(D)_1\rangle + \widetilde{C}_3 \langle {}^3S(D)_1|\mathscr{O}_{(q\cdot S)^2}|{}^3S(D)_1\rangle, \tag{5.172}$$

for the $^3S(D)_1$ channel, and

$$\langle {}^3P(F)_2|V^{(2)}|{}^3P(F)_2\rangle = \widetilde{C}_5 \langle {}^3P(F)_2|\mathscr{O}_{q^2}|{}^3P(F)_2\rangle$$
$$+\widetilde{C}_6 \langle {}^3P(F)_2|\mathscr{O}_{(q\cdot S)^2}|{}^3P(F)_2\rangle$$
$$+ \widetilde{C}_7 \langle {}^3P(F)_2|\mathscr{O}^{I=1}_{(iq\times S)\cdot k}|{}^3P(F)_2\rangle, \tag{5.173}$$

for the $^3P(F)_2$ channel. Clearly, only specific combinations of the NLO contact operators contribute to each partial wave, which simplifies their determination from fits to PWA data for np scattering. Once the $\widetilde{C}_i$ have been determined, the NLO coupling constants of Chap. 4 can be found from the relation

$$
\begin{bmatrix} \tilde{C}_1 \\ \tilde{C}_2 \\ \tilde{C}_3 \\ \tilde{C}_4 \\ \tilde{C}_5 \\ \tilde{C}_6 \\ \tilde{C}_7 \end{bmatrix} = \begin{pmatrix} 1 & 1 & -3 & -3 & -1 & -1 & 0 \\ 1 & -3 & 1 & -3 & 0 & 0 & 0 \\ 0 & 0 & 0 & 0 & 1 & -3 & 0 \\ 1 & -3 & -3 & 9 & -1 & 3 & 0 \\ 1 & 1 & 1 & 1 & 0 & 0 & 0 \\ 0 & 0 & 0 & 0 & 1 & 1 & 0 \\ 0 & 0 & 0 & 0 & 0 & 0 & 1 \end{pmatrix} \begin{bmatrix} C_{q^2} \\ C_{I^2,q^2} \\ C_{S^2,q^2} \\ C_{S^2,I^2,q^2} \\ C_{(q\cdot S)^2} \\ C_{I^2,(q\cdot S)^2} \\ C^{I=1}_{(iq\times S)\cdot k} \end{bmatrix}, \tag{5.174}
$$

which summarizes the dependence of the NLO contact terms on spin and isospin. The NLEFT potentials are defined in momentum space and Fourier transformed to coordinate space. In particular, the Gaussian smearing of the LO_3 operators is most easily treated in momentum space.

Let us now consider two different ways of treating the contributions beyond LO in NLEFT. Firstly, all orders up to NNLO can be treated non-perturbatively, which is similar to the solution of the Lippmann-Schwinger equation in the continuum. In this case, the full transfer matrix is given by

$$
M \equiv\ :\exp\left[-\alpha_t \left(H_{\text{free}} + V_{\text{LO}} + V_{\text{NLO}} + V_{\text{NNLO}} \right) \right] :, \tag{5.175}
$$

where, according to Chap. 4, the potential terms are

$$
V_{\text{LO}} = C_{S=0,I=1} \tilde{V}_{S=0,I=1} + C_{S=1,I=0} \tilde{V}_{S=1,I=0} + \tilde{V}_\pi, \tag{5.176}
$$

at LO,

$$
V_{\text{NLO}} = C_{q^2} \mathscr{O}_{q^2} + C_{I^2,q^2} \mathscr{O}_{I^2,q^2} + C_{S^2,q^2} \mathscr{O}_{S^2,q^2} + C_{S^2,I^2,q^2} \mathscr{O}_{S^2,I^2,q^2}
$$
$$
+ C_{(q\cdot S)^2} \mathscr{O}_{(q\cdot S)^2} + C_{I^2,(q\cdot S)^2} \mathscr{O}_{I^2,(q\cdot S)^2} + C^{I=1}_{(iq\times S)\cdot k} \mathscr{O}^{I=1}_{(iq\times S)\cdot k} + V^{(2)}_{\text{TPE}}, \tag{5.177}
$$

at NLO, and

$$
V_{\text{NNLO}} = V^{(3)}_{\text{TPE}}, \tag{5.178}
$$

at NNLO, where all the TPE contributions given in Chap. 4 have been retained, as is appropriate when $a \simeq 1$ fm. Secondly, we consider np scattering where NLO and higher orders are treated perturbatively. This coincides with the approach taken in MC simulations for light and medium-mass nuclei, and avoids potentially severe sign oscillations due to strong, repulsive NLO contributions. Instead of Eq. (5.175), in the perturbative analysis the transfer matrix is

$$M^{\text{pert}} = M_{\text{LO}} - \alpha_t : (V_{\text{NLO}} + \Delta V_{\text{LO}} + V_{\text{NNLO}}) M_{\text{LO}} :, \tag{5.179}$$

where

$$M_{\text{LO}} =: \exp\left[-\alpha_t \left(H_{\text{free}} + V_{\text{LO}}\right)\right] :, \tag{5.180}$$

is treated non-perturbatively. In Eq. (5.179), the term ΔV_{LO} describes shifts of the LO coefficients and contributions from perturbative improvement of the OPE derivative couplings (see Chap. 4 for details). As the LO constants are kept fixed when the perturbative NLO and NNLO contributions are added, it follows that inclusion of ΔV_{LO} equals a refit of the LO constants (as in the non-perturbative case). Also, the contact terms in ΔV_{LO} absorb a sizable part of the short-distance contributions from TPE at NLO and NNLO. As shown in Chap. 4, rotational symmetry-breaking terms also may be included in ΔV_{LO}. The treatment of electromagnetic effects will be discussed in Sect. 5.6.

5.5.2 Variation of the Lattice Spacing

We shall conclude the section on np scattering by discussing the systematical study of lattice spacing effects, along with a determination of the LO and NLO coupling constants using chi-square minimization to the Nijmegen PWA [66]. In early NLEFT work, the most commonly used lattice spacing was $a = 1.97$ fm, which corresponds to a momentum cutoff $\Lambda \simeq \pi/a = 314$ MeV. As Λ is relatively low, significant lattice artifacts may be induced, in particular at high momenta. With this in mind, Ref. [65] studied np scattering for lattice spacings as small as $a = (200\,\text{MeV})^{-1} = 0.98$ fm, with a temporal lattice spacing of $\delta = (600\,\text{MeV})^{-1}$. The physical volume was taken to be $V = (La)^3 \simeq (63\,\text{fm})^3$. We denote by p_{CM} the relative momentum of the scattering neutron and proton in the center-of-mass frame. For such relatively large V, finite volume effects are expected to be insignificant as long as $p_{\text{CM}} \lesssim 200$ MeV. Because of the decreased lattice spacing a, the full structure of the TPE potential up to NNLO was included explicitly for the first time.

Since the long-range OPE and TPE contributions do not contain parameters to be determined from the empirical np scattering phase shifts, we can take the chi-square function

$$\chi^2 \equiv \chi^2(\{b_i\}, C_{S=0,I=1}, C_{S=1,I=0}, \tilde{C}_1, \ldots, \tilde{C}_7), \tag{5.181}$$

as the basis for judging the level of agreement between PWA and NLEFT scattering phase shifts and mixing angles. Hence, in Eq. (5.181), the adjustable parameters are the coefficients of the short-range LO and NLO contributions, which will be refitted for each value of a and δ. In Eq. (5.181), $\{b_i\}$ denotes collectively any parameters that affect the smearing of the LO contact interactions. When fitting the Nijmegen PWA, we note certain simplifying features. For instance, at LO only the S-wave phase shifts are fitted, which determines $C_{S=0,I=1}$ and $C_{S=1,I=0}$, and fixes the smearing parameters of the LO contact terms (which are subsequently not changed at higher orders). The deuteron binding E_d energy is not fitted at LO, but is included as an additional constraint at higher orders. At NLO, $C_{S=0,I=1}$ and $\tilde{C}_1$ are determined from the $^1S_0$ phase shift, while $C_{S=1,I=0}$, $\tilde{C}_2$ and $\tilde{C}_3$ are constrained by the $^3S_1$ phase shift and ϵ_1. The remaining NLO parameters are determined by the P-wave channels. As no additional fit parameters appear at NNLO, those fits only differ by the inclusion of the NNLO TPE potential. At LO, the range of the fit was taken to be $p_{CM}^{max} = 100$ MeV, and was extended at higher orders to $p_{CM}^{max} = 150$ MeV.

If we denote the fit parameters collectively by $\{C_i\}$, then

$$\chi^2 = \chi_{min}^2 + \frac{1}{2}\sum_{i,j} h_{ij}(C_i - C_i^{min})(C_j - C_j^{min}) + \dots, \qquad (5.182)$$

where $\{C_i^{min}\}$ denotes the set of fit parameters that minimizes Eq. (5.181). Given that χ^2 reaches its minimum value for $C_i = C_i^{min}$, the terms with one derivative vanish, and we obtain the covariance matrix

$$\mathscr{E}_{ij} \equiv \frac{1}{2}\big[h^{-1}\big]_{ij}, \quad h_{ij} \equiv \frac{\partial^2 \chi^2}{\partial C_i \partial C_j}, \qquad (5.183)$$

in terms of the Hessian matrix h_{ij}. Given Eq. (5.183), we can express the standard deviations of the C_i^{min} as $\sigma_i = \sqrt{\mathscr{E}_{ii}}$, using the diagonal elements of (5.183). In the absence of systematical errors, we expect to find a normalized chi-square of $\tilde{\chi}^2 \equiv \chi^2/N_{dof} \approx 1$, where N_{dof} is the number of degrees of freedom (number of fitted data $-$ number of free parameters) of the minimization. However, in our analysis $\tilde{\chi}^2 > 1$ in most cases, particularly at LO and for larger values of a. Such a systematical error suggests that the uncertainties computed from Eq. (5.183) are underestimated.

In order to make use of the covariance matrix, we require $\tilde{\chi}^2 \simeq 1$. Therefore, we rescale the input errors by [67]

$$\Delta_i \to \Delta_i\sqrt{\tilde{\chi}_{min}^2}, \quad \chi^2 \to \frac{\chi^2}{\tilde{\chi}_{min}^2} = N_{dof}\frac{\chi^2}{\chi_{min}^2}, \qquad (5.184)$$

such that $\chi^2/N_{dof} = 1$ for $C_i = C_i^{min}$. This has the effect of masking the systematical error, which we identify as mainly arising from two sources. Firstly, our procedure of fitting the Nijmegen PWA rather than the actual np scattering data appears to give an overly stringent measure for goodness of the fit. Moreover, certain channels (such as $^3S_1$) have much smaller input uncertainties from the

PWA, and would therefore have a very large relative weight in the minimization procedure. Secondly, while rescaling of the PWA errors is needed at NNLO due to discrepancies between the NLEFT description and the PWA, such discrepancies are expected to be much reduced at N3LO and higher orders, thereby eliminating the need for such rescaling. Once the covariance matrix has been obtained from the chi-square minimization, we may use this information to form error estimates for more complicated observables, such as nuclear binding energies. For a given observable $\mathscr{O}$, we assign an uncertainty according to

$$\Delta\mathscr{O} \equiv \sqrt{(J_{\mathscr{O}}^T)_i \mathscr{E}_{ij} (J_{\mathscr{O}})_j}, \quad (J_{\mathscr{O}})_i \equiv \frac{\partial \mathscr{O}}{\partial C_i}, \tag{5.185}$$

in terms of the Jacobian vector of $\mathscr{O}$ with respect to the C_i. This prescription accounts for errors due to variances as well as covariances between the set of LO and NLO coupling constants. Hence, the nuclear binding energies receive contributions from statistical MC errors, as well as from the "theoretical error" associated with the short-distance contributions to the NLEFT potential.

The np scattering results of Ref. [65] for the smallest lattice spacing ($a = 0.98$ fm) are shown in Figs 5.10 and for the coarse a in Fig. 5.11 (non-perturbative and perturbative treatment of the NLO and NNLO contributions, respectively). The description of the S-waves is quite good even at LO, particularly for the spin-triplet channel. Significant improvements are found at NLO, in particular for the spin-singlet and spin-triplet P-waves, as well as for the mixing angle ϵ_1. The TPE contribution at NNLO appears to have an overall minor effect, being to a large extent absorbed into the NLO contact terms. For the D-wave channels, we recall that the current way of smearing the LO contact interactions produces unwanted residual attractive forces. The D-wave channels should be dominated by the OPE contribution, and the detrimental effect of the LO smearing is noticeable. This situation is expected to improve upon addition of the N3LO contributions to the 2NF, which have only recently been included in NLEFT, see Chap. 8. While the NLEFT results for $a = 1.97$ fm provides a good overall description up to $p_{CM} \simeq 100$ MeV, a decrease in the lattice spacing to $a = 0.98$ fm allows for center-of-mass momenta up to $p_{CM} \simeq 200$ MeV. While the systematical errors appear much reduced as a is decreased, the convergence of NLEFT could be accelerated for large p_{CM} by improving the lattice momenta in the OPE potential and in the NLO contact interactions. Most important, however, is the finding that for $p_{CM} \lesssim 100$ MeV, the physics of the two-nucleon system is largely cutoff-independent when 1 fm $\leq a \leq 2$ fm.

5.6 Proton-Proton Scattering in NLEFT

We have so far developed our formalism for np scattering, without consideration of electromagnetic or isospin-breaking effects. As we have seen in Chap. 4, the inclusion of electromagnetic effects entails contact interactions for pp and nn, as

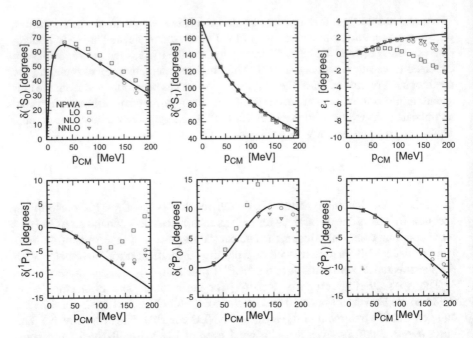

Fig. 5.10 Phase shifts and mixing angles for np scattering, using a radial Hamiltonian with the auxiliary potential method. The upper row shows the S-wave channels and the mixing angle ε_1 due to the tensor interaction, and the lower row shows the phase shifts in some of the P-wave channels. Results are given for the smallest lattice spacing of $a = (200\,\text{MeV})^{-1} = 0.98$ fm, with a temporal lattice spacing of $\delta = (600\,\text{MeV})^{-1}$. The (blue) squares, (green) circles and (red) triangles denote LO, NLO and NNLO results, respectively, all of which were treated non-perturbatively in the numerical diagonalization. The 2NF at each order is fixed by a chi-square fit to the Nijmegen PWA (solid black line)

well as a long-range Coulomb tail. In a perturbative treatment similar to that used for np scattering, this gives

$$M^{\text{pert}} = M_{\text{LO}} - \alpha_t : (V_{\text{NLO}} + \Delta V_{\text{LO}} + V_{\text{NNLO}} + V_{\text{EMIB}}) M_{\text{LO}} :, \qquad (5.186)$$

and we note that a non-perturbative description can be constructed along similar lines. In Eq. (5.186), the EMIB contribution is

$$V_{\text{EMIB}} \equiv \tilde{V}_\pi^{\text{IB}} + \tilde{V}_{\text{EM}} + \tilde{V}_{\text{nn}} + \tilde{V}_{\text{pp}}, \qquad (5.187)$$

where we use the term "EMIB" to refer specifically to the strong and electro-magnetic isospin-breaking terms. The contact terms $\tilde{V}_{\text{nn}}$ and $\tilde{V}_{\text{pp}}$ account for the observed inequality of the nn and pp scattering lengths, to which their respective coefficients are fitted. The long-range Coulomb tail $\tilde{V}_{\text{EM}}$ only contributes to pp scattering, for which it introduces complications due to the long-range (unscreened) Coulomb potential.

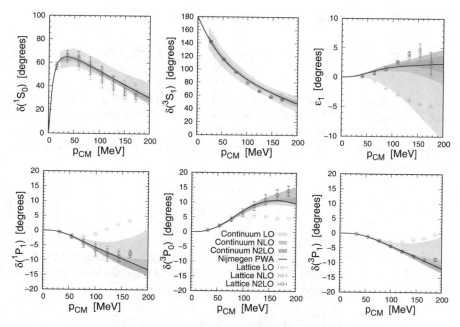

Fig. 5.11 Phase shifts and mixing angles for np scattering, using a radial Hamiltonian with the auxiliary potential method. Notation and conventions are as for Fig. 5.10, except that the NLO and NNLO contributions are treated as perturbations. All NLEFT results are for a coarse lattice spacing of $a = 1.97$ fm. The shaded bands denote continuum results obtained by solution of a Lippmann-Schwinger equation [68], and the Nijmegen PWA is given by the black line

We shall now revisit the treatment of the scattering problem and derive an expression for the wave function beyond the range R_n of the nuclear interaction, which can be used in the presence of a Coulomb term in the potential. So far, we have only concerned ourselves with the case where the scattering potential $V(r)$ is short-ranged, in other words the radial wave function has been taken to satisfy Eq. (5.16), which can be written in the form

$$\left(\frac{d^2}{d\rho^2} - \frac{l(l+1)}{\rho^2} + 1\right) u_l^{\text{ex}}(\rho/k) = 0, \qquad (5.188)$$

for $\rho \equiv kr$. We shall first develop a "non-interacting" description for a pure $\sim 1/r$ potential, and then consider the realistic case where the full potential consists of a sum of Coulomb and (finite-range) nuclear components. The Coulomb potential between two point-like particles with charges $Z_1 e$ and $Z_2 e$ is

$$V_c(r) \equiv \frac{Z_1 Z_2 e^2}{r}, \qquad (5.189)$$

and we define the dimensionless "Sommerfeld parameter"

$$\eta \equiv \frac{Z_1 Z_2 e^2}{v} = \frac{Z_1 Z_2 e^2 \mu}{k} = Z_1 Z_2 e^2 \sqrt{\frac{\mu}{2E}} = \frac{Z_1 Z_2 \alpha_{\text{EM}} m_N}{2k}, \tag{5.190}$$

for scattering with relative velocity v, where $E = k^2/2\mu$. For pp scattering, we take $Z_1 = Z_2 = 1$. We note that η appears naturally in the classical description of hyperbolic scattering orbits in a Coulomb potential. For instance, the distance of closest approach in a head-on collision equals $2\eta/k$. More generally, the scattering angle θ is given by

$$\tan\left(\frac{\theta}{2}\right) = \frac{\eta}{bk}, \tag{5.191}$$

where b is the impact parameter.

The scattering formalism developed for $\eta = 0$ can be generalized to the Coulomb case without much difficulty. We note that Eq. (5.188) is a special case for $\eta = 0$ of the Coulomb wave equation

$$\left(\frac{d^2}{d\rho^2} - \frac{l(l+1)}{\rho^2} - \frac{2\eta}{\rho} + 1\right) u_l^{\text{ex}}(\eta, \rho) = 0, \tag{5.192}$$

which is solved by the linear combination of Coulomb functions

$$u_l^{\text{ex}}(\eta, \rho) = A F_l(\eta, \rho) + B G_l(\eta, \rho), \tag{5.193}$$

and we recall that

$$F_l(0, \rho) = \rho j_l(\rho), \quad G_l(0, \rho) = -\rho y_l(\rho), \tag{5.194}$$

for $\eta = 0$. We also recall the Coulomb Hankel functions

$$H_l^{\pm}(\eta, \rho) \equiv G_l(\eta, \rho) \pm i F_l(\eta, \rho), \tag{5.195}$$

for which

$$H_l^+(0, \rho) = i\rho h_l^{(1)}(\rho), \quad H_l^-(0, \rho) = -i\rho h_l^{(2)}(\rho), \tag{5.196}$$

in terms of the spherical Hankel functions given in Eqs. (5.19) and (5.21), respectively.

We shall briefly discuss the relationship of the Coulomb functions to standard transcendental functions (see e.g. Ref. [69] for an overview). In terms of the Kummer confluent hypergeometric function of the first kind $_1F_1(a; b; z)$ [69],

$F_l(\eta, \rho)$ can be expressed as

$$F_l(\eta, \rho) = \rho^{l+1} \exp(\mp i\rho)\, C_l(\eta) \,_1F_1(l + 1 \mp i\eta;\, 2l + 2;\, \pm 2i\rho), \qquad (5.197)$$

where

$$C_l(\eta) \equiv \frac{2^l \exp(-\pi\eta/2)\, |\Gamma(l + 1 + i\eta)|}{\Gamma(2l + 2)}, \qquad (5.198)$$

is referred to as the "Coulomb constant", and

$$\Gamma(z) \equiv \int_0^\infty dx\, x^{z-1} \exp(-x), \qquad (5.199)$$

is the Euler gamma function [69]. Either choice of sign in Eq. (5.197) gives the same result, and is thus a matter of convention. The functions $H_l^\pm(\eta, \rho)$ are given by

$$H_l^\pm(\eta, \rho) = \exp(\pm i\, Q)(\mp 2i\rho)^{1+l\pm i\eta} U(1 + l \pm i\eta;\, 2l + 2;\, \mp 2i\rho), \qquad (5.200)$$

where $U(a; b; z)$ denotes the Kummer confluent hypergeometric function of the second kind [69]. We have

$$Q \equiv \rho - l\pi/2 + \sigma_l(\eta) - \eta \ln(2\rho), \qquad (5.201)$$

in Eq. (5.200), where

$$\sigma_l(\eta) \equiv \arg \Gamma(1 + l + i\eta), \qquad (5.202)$$

is referred to as the Coulomb "phase shift". In other words, the shift $\sigma_l(\eta)$ (relative to the asymptotic form for $\eta = 0$) remains, even when the short-range nuclear potential vanishes.

It is also useful to consider the limiting behavior of the Coulomb functions for small and large values of ρ. For small ρ, we have

$$F_l(\eta, \rho) \sim C_l(\eta)\rho^{l+1}, \qquad G_l(\eta, \rho) \sim \frac{1}{(2l + 1)C_l(\eta)\rho^l}, \qquad (5.203)$$

in terms of the Coulomb constant (5.198). This may be computed recursively for different l using

$$C_l(\eta) = \frac{\sqrt{l^2 + \eta^2}}{l(2l + 1)} C_{l-1}(\eta), \qquad C_0^2(\eta) = \frac{2\pi\eta}{\exp(2\pi\eta) - 1}, \qquad (5.204)$$

and we may also generalize the behavior of the Coulomb functions for $\eta = 0$ to

$$F_l(\eta, \rho) \sim \sin Q, \qquad G_l(\eta, \rho) \sim \cos Q, \tag{5.205}$$

and

$$H_l^{\pm}(\eta, \rho) \sim \exp(\pm i Q), \tag{5.206}$$

for asymptotically large values of ρ. More specifically, the asymptotic behavior sets in for $\rho \gg \rho_t$, where the "turning point" ρ_t is defined by

$$2\eta/\rho_t + l(l + 1)/\rho_t^2 = 1, \tag{5.207}$$

which gives

$$\rho_t = \eta \pm \sqrt{\eta^2 + l(l + 1)}, \tag{5.208}$$

and we note that only one (physical value of the) turning point exists when $\eta > 0$, for purely Coulomb and centrifugal potentials. Classically, ρ_t corresponds to the distance of closest approach between projectile and target. Also, for $\eta = 0$, we can take $kr \gg l$ as the criterion for the asymptotic region.

Given Eq. (5.188) for $\eta \neq 0$, we shall now determine the contribution of the Coulomb potential to the scattering amplitude $f(\theta, k)$. It is necessary to account for the distortion of the incoming plane wave $\exp(i\mathbf{k} \cdot \mathbf{r})$ by the Coulomb interaction. The equivalent solution $\psi_{c,\text{in}}(r, \theta)$ to the Schrödinger equation for $\eta \neq 0$ is given in terms of the hypergeometric function by

$$\psi_c, \text{in}(r, \theta) = \exp(i\mathbf{k} \cdot \mathbf{r}) \exp(-\pi \eta/2) \, \Gamma(1 + i\eta) \, {}_1F_1(-i\eta; 1; i(kr - \mathbf{k} \cdot \mathbf{r})), \tag{5.209}$$

which for $i\mathbf{k} \cdot \mathbf{r} = ikz = ikr \cos \theta$ becomes

$$\psi_c, \text{in}(r, \theta) = \exp(ikr \cos \theta) \exp(-\pi \eta/2) \, \Gamma(1 + i\eta) \, {}_1F_1(-i\eta; 1; ikr(1 - \cos \theta)), \tag{5.210}$$

and we note the partial wave expansion

$$\psi_{c,\text{in}}(r, \theta) = \sum_{l=0}^{\infty} (2l + 1)i^l \frac{F_l(\eta, kr)}{kr} P_l(\cos \theta), \tag{5.211}$$

which for $\eta \to 0$ reduces to Eq. (5.35). Asymptotically,

$$F_l(\eta, kr) \simeq \sin(kr - l\pi/2 + \sigma_l(\eta) - \eta \ln(2kr)), \tag{5.212}$$

which differs from the case of $\eta = 0$ by the logarithmic term, which is generated by the Coulomb potential. For $kr \gg \eta \ln(2kr)$, the solution is shifted relative to the case of $\eta = 0$ by $\sigma_l(\eta)$.

We shall denote the Coulomb contribution to the scattering amplitude by $f_c(\theta, k)$. This can be determined analogously to the case of $\eta = 0$, provided that Eq. (5.5) for the scattering amplitude is modified to

$$\psi_c(r, \theta) \overset{\rho \gg \rho_t}{\simeq} \psi_{c,in}(r, \theta) + \psi_{c,sc}(r, \theta) \tag{5.213}$$

$$= C\left\{ \exp(ikz + i\eta \ln(k(r-z))) + f_c(\theta, k)\frac{\exp(ikr - i\eta \ln(2kr))}{r} \right\},$$

with C a constant. This leads to

$$f_c(\theta, k) = \sum_{l=0}^{\infty} (2l+1)\left(\frac{\exp(2i\sigma_l(\eta)) - 1}{2ik}\right) P_l(\cos\theta), \tag{5.214}$$

whereby $f_c(\theta, k)$ is expressed as a sum over partial-wave components in terms of the Coulomb phase shift $\sigma_l(\eta)$. However, as alluded to earlier in our discussion for $\eta = 0$, Eq. (5.214) is divergent, since the phase shifts $\sigma_l(\eta)$ do not fall off sufficiently fast for large l. One possible approach is to consider a screened Coulomb potential, in which case Eq. (5.214) is convergent and well-defined, after which the limit of large screening radius can be taken. However, it is also possible to match Eq. (5.209) directly to Eq. (5.213) without the intermediate step of a partial-wave expansion, in which case one finds

$$f_c(\theta, k) = -\frac{\eta}{2k\sin^2(\theta/2)}\exp(-i\eta\log(\sin^2(\theta/2)) + 2i\sigma_0(\eta)), \tag{5.215}$$

which is called the "point Coulomb" scattering amplitude. This corresponds to the differential cross section

$$\frac{d\sigma_R}{d\Omega} = |f_c(\theta, k)|^2 = \frac{\eta^2}{4k^2\sin^4(\theta/2)}, \tag{5.216}$$

which is also referred to as the Rutherford cross section, since it coincides with the differential cross section of classical point-Coulomb scattering.

In realistic descriptions of nucleon-nucleon scattering, the phase shift receives contributions from both the Coulomb potential and the (short-ranged) nuclear potential. We shall now use our knowledge of the pure Coulomb problem to disentangle the two contributions. The full potential can be written as $V(r) = V_c(r) + V_n(r)$, where $V_c(r)$ is given by Eq. (5.189). Again, we shall assume that the nuclear potential $V_n(r)$ is spherically symmetric. Since the full phase shift

δ_l (relative to the free solution) due to $V(r)$ will no longer equal $\sigma_l(\eta)$, we define

$$\delta_l \equiv \sigma_l(\eta) + \delta_l^n, \qquad (5.217)$$

where δ_l^n is the "Coulomb-modified" nuclear phase shift due to $V_n(r)$. By means of Eq. (5.52), the full scattering amplitude $f_{nc}(\theta, k)$ due to $V(r)$ can be decomposed using

$$\exp(2i\delta_l) - 1 = \left[\exp(2i\sigma_l(\eta)) - 1\right] + \exp(2i\sigma_l(\eta))\left[\exp(2i\delta_l^n) - 1\right], \qquad (5.218)$$

into Coulomb and nuclear scattering amplitudes according to

$$f_{nc}(\theta, k) \equiv f_c(\theta, k) + f_n(\theta, k), \qquad (5.219)$$

where $f_n(\theta, k)$ is the "Coulomb-modified" nuclear scattering amplitude, and $f_c(\theta, k)$ is given by Eqs. (5.214) and (5.215).

In order to find the Coulomb-modified nuclear phase shift δ_l^n (or the equivalent nuclear S-matrix element $S_l^n \equiv \exp(2i\delta_l^n)$), we shall generalize the procedure for $\eta = 0$ slightly. Instead of Eq. (5.42), we take

$$u_l^{ex}(r) = A_l \left(H_l^-(\eta, kr) - S_l^n(k) H_l^+(\eta, kr) \right), \qquad (5.220)$$

where we use Eq. (5.190) instead of taking $\eta = 0$. Also, instead of Eq. (5.59), we now have

$$u_l^{ex}(\eta, \rho) \propto \exp(i\delta_l^n(k)) \left[\cos \delta_l^n(k) F_l(\eta, \rho) + \sin \delta_l^n(k) G_l(\eta, \rho)\right], \qquad (5.221)$$

for $r \gg R_n$, in terms of δ_l^n and the Coulomb functions with $\eta \neq 0$. Given the generalization of Eq. (5.42) to Eq. (5.220), we are now in a position to apply the frameworks of the spherical wall method, the auxiliary potential method, and the radial Hamiltonian method to Coulomb problems such as pp scattering. The S-wave pp scattering phase shift, calculated up to NNLO with non-perturbative treatment of Coulomb effects [70], is shown in Fig. 5.12. The description of the pp and np phase shifts at NNLO are of similar quality, and moreover a significant systematical improvement is found at high energies, when a is decreased.

To compute the nuclear scattering amplitude, we note that the second term of Eq. (5.218) gives

$$f_n(\theta, k) = \sum_{l=0}^{\infty} (2l+1) \exp(2i\sigma_l(\eta)) \left(\frac{S_l^n(k) - 1}{2ik}\right) P_l(\cos \theta), \qquad (5.222)$$

such that this Coulomb-modified nuclear scattering amplitude $f_n(\theta, k)$ is not only due to short-range forces, but also receives a contribution from $\sigma_l(\eta)$. We also have

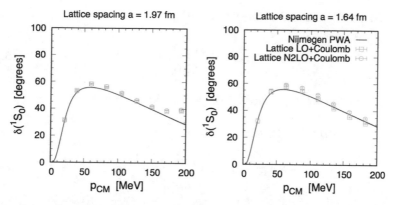

Fig. 5.12 Phase shifts in the $S = 0$ channel for pp scattering, using the auxiliary potential method with a radial Hamiltonian, for $a = 1.97$ fm and $a = 1.64$ fm [70]. As $I = 1$ for the pp system, one only has $S = 0$ channels because of the Pauli principle. The long-range Coulomb interaction between protons is treated non-perturbatively using Eq. (5.220), while the pp contact term is treated perturbatively and fitted to the pp phase shift. The coefficients of the NLO potential are fitted to np scattering data and the pp scattering length

the total (Coulomb+nuclear) cross section

$$\frac{d\sigma_{nc}}{d\Omega} \equiv |f_{nc}(\theta, k)|^2 = |f_c(\theta, k) + f_n(\theta, k)|^2, \tag{5.223}$$

from which we find that either $f_c(\theta, k)$ of Eq. (5.215) or $f_n(\theta, k)$ of Eq. (5.222) can be multiplied by a phase factor such as $\exp(-2i\sigma_0(\eta))$, as the total cross section is evidently only affected by the relative phase of $f_c(\theta, k)$ and $f_n(\theta, k)$. A standard way to present elastic scattering cross sections is

$$\frac{d\sigma_{nc}/d\Omega}{d\sigma_R/d\Omega} = \frac{|f_c(\theta, k) + f_n(\theta, k)|^2}{|f_c(\theta, k)|^2}, \tag{5.224}$$

such that the divergence of the Rutherford cross section (5.216) at small scattering angles θ is factored out.

To conclude our treatment of Coulomb modifications to the two-nucleon scattering problem, let us consider how the effective-range expansion (ERE) can be extended to account for the long-range Coulomb interaction. We recall that the radius of convergence of the ERE is controlled by the screening mass of a Yukawa-type potential, and hence it cannot be used without modification for power-law potentials. A modified ERE, which circumvents the problem of a vanishing radius of convergence, can be written as [63, 71]

$$C_l^2(\eta)k^{2l+1}\cot\delta_l^n(k) + 2k\eta h_l(k) = -\frac{1}{a_l^n} + \frac{1}{2}r_l^n k^2 + \sum_{i=2}^{\infty} v_{l,i}^n k^{2i}, \tag{5.225}$$

such that the Coulomb effects are described by

$$h_l(k) \equiv k^{2l} \frac{C_l^2(\eta)}{C_0^2(\eta)} h(\eta), \tag{5.226}$$

and the Coulomb constants $C_l(\eta)$, given in Eq. (5.204). Moreover

$$h(\eta) \equiv -\log|\eta| + \mathrm{Re}\,\psi(i\eta) \tag{5.227}$$

$$= -\log|\eta| - \gamma_E + \eta^2 \sum_{n=1}^{\infty} \frac{1}{n(n^2 + \eta^2)}, \tag{5.228}$$

with $\psi(z) \equiv \Gamma'(z)/\Gamma(z)$ the digamma function [69], and $\gamma_E \simeq 0.5772$ is the Euler-Mascheroni constant. This Coulomb-modified ERE will be especially useful for our treatment of α-α scattering in the adiabatic projection formalism, which we shall return to in Chap. 7. Besides the Coulomb interaction, similar modified ERE expressions can be obtained for arbitrary power-law potentials [71].

5.7   Three Nucleons in NLEFT

So far, we have focused on developing the formalism for the two-nucleon problem in NLEFT, with the objective of fixing the unknown coefficients due to the NLO contact interactions in the 2NF. This task requires not only a calculation of the spectrum of the two-nucleon system, but also a workable method which allows us to relate the two-nucleon energy levels to scattering phase shifts and mixing angles. At NNLO, we need to also deal with the lowest-order contribution to the 3NF, which we have established to be of the form

$$V_{\mathrm{NNLO}}^{(3N)} \equiv \tilde{V}_{\mathrm{contact}}^{(3N)} + \tilde{V}_{\mathrm{OPE}}^{(3N)} + \tilde{V}_{\mathrm{TPE1}}^{(3N)} + \tilde{V}_{\mathrm{TPE2}}^{(3N)} + \tilde{V}_{\mathrm{TPE3}}^{(3N)}, \tag{5.229}$$

such that the perturbative transfer matrix can again be constructed along the lines of Eq. (5.186). Of these, the contact and OPE terms depend on the unknown coefficients c_E and c_D respectively, while the TPE terms depend on the c_i which are fixed by analyses of pion-nucleon scattering data. Hence, c_E and c_D should be determined from suitable observables (preferentially for $A = 3$), such that NLEFT for $A \geq 4$ is predictive at NNLO. For $A = 3$, it turns out that sparse matrix diagonalization is the most practical and straightforward approach. Nevertheless, a significant drawback is that the radial Hamiltonian method is specific to the two-nucleon problem. Instead of working in radial coordinates, one calculates the spectrum in three dimensions with inclusion of the 3NF potential (5.229). For sparse matrix calculations, the potential (5.229) should be expressed in terms of the two-pion correlator rather than the pion field (as shown in Chap. 4). On the other hand, when MC methods are used for $A \geq 4$, the equivalent formulation with pion fields should be used. We shall now turn to a discussion of the calculation of the triton (^{3}H) and the ^{3}He binding energies as well as the neutron-deuteron (nd) scattering phase shift, and show to what extent c_E and c_D can be determined from such data.

5.7.1   Binding Energies of Three-Nucleon Bound States

The binding energy of the triton at infinite volume, $E_3(\infty) \simeq -8.48$ MeV is the most obvious observable to calculate in the three-nucleon sector. At NNLO, this can be used for instance to fix the coefficient c_E as a function of c_D. In Fig. 5.13, we show the triton energy calculated with sparse matrix diagonalization for $L = 8a$ to $L = 14a$ up to NNLO in NLEFT. Such a calculation turns out to be numerically challenging for Euclidean projection Monte Carlo, due to sign oscillations and relatively slow convergence with L. In order to extract triton binding energy in the infinite volume limit, an extrapolation $L \to \infty$ using the lattice data is required. One possibility is to use Lüscher's asymptotic parameterization [21],

$$E_3(L) = E_3(\infty) + \frac{\mathscr{A}_0}{L} \exp(-L/L_0), \qquad (5.230)$$

Fig. 5.13 Triton binding energy as a function of L, computed by sparse matrix diagonalization, for a lattice spacing of $a = 1.32$ fm [70]. Results are shown at LO, using the 2NF only at NNLO ("N2LO 2N"), using the 2NF at NNLO with EMIB effects added ("N2LO 2N + EM"), and finally with the 3NF included ("N2LO 2N + EM + 3N"). Note that when the 3NF contribution is included, c_E is fitted to the empirical triton binding energy (black horizontal line) while keeping $c_D = -0.79$ fixed, as in earlier NLEFT work. Results for $L > 10a$ are included in an extrapolation to infinite volume using Eq. (5.231)

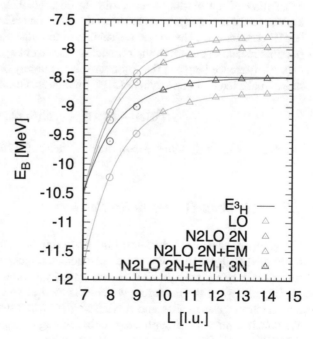

where L_0 is a length scale associated with the physical size of the triton wave function, and $\mathscr{A}_0$ is a constant. Another possibility is to compute finite-volume corrections for the three-particle system in the unitary limit, as well as for the case when one particle is loosely bound to a tightly bound dimer [72–74]. While the physical triton is intermediate between these two cases, when L is large enough the difference becomes negligible, and we take the formula

$$E_3(L) = E_3(\infty) + \frac{\mathscr{A}_\kappa}{\kappa L^{3/2}} \exp(2\kappa L/\sqrt{3}), \quad \kappa \equiv -\sqrt{m_N E_3(\infty)}, \qquad (5.231)$$

(where $\mathscr{A}_\kappa$ is a constant) which is valid in the unitary limit, in order to extrapolate the LO triton energies at finite L to $L \to \infty$. The perturbative higher-order operators are taken to receive a finite-volume correction of the same form as Eq. (5.231). An extrapolation of the triton binding energy to infinite volume using Eq. (5.231) is shown in Fig. 5.13. In this context, we note that the "pseudo-triton" energy, which is an estimate of the triton binding energy when nn interactions are replaced by np interactions [75], is also of some interest. The difference between the triton and pseudo-triton energies is indicative of the systematic error in the (isospin-symmetric) calculations with the 2NF matched to np scattering data.

The calculation of the triton binding energy can only provide us with c_D as a function of c_E, which means that some other quantity, preferably a three-nucleon observable, is needed to completely fix the NNLO contribution to the 3NF. A calculation of the binding energy of ^{3}He does not help in fixing c_D once c_E is known, as the energy splitting between the triton and ^{3}He is due to the perturbative EMIB contribution. Therefore, we shall turn to a discussion of nd scattering as a possible means to constrain the remaining coefficient c_D. However, as the binding energy difference between the triton and ^{3}He is purely of EMIB origin, this allows one to make the first non-trivial NLEFT prediction. One finds

$$E(^3\mathrm{He}) - E(^3\mathrm{H}) = 0.78(5) \ \mathrm{MeV}, \qquad (5.232)$$

which compares well with the experimental value of 0.76 MeV [76].

5.7.2 Neutron-Deuteron Scattering

The problem of neutron-deuteron (nd) scattering has been considered in Ref. [77] up to NNLO in NLEFT. For the three-nucleon sector, we have so far focused on the triton binding energy. This is conceptually similar to the calculation of the binding energy of the deuteron, which was used as an additional constraint when np scattering phase shifts and mixing angles were fitted using the NLEFT 2NF. We note that nd scattering proceeds either through the spin-quartet ($^4S_{3/2}$) or spin-doublet ($^2S_{1/2}$) channels. As we are dealing with S-waves only, and as the available experimental data for nd and pd scattering [78] is relatively limited, the Lüscher

Fig. 5.14 nd scattering in the $^2S_{1/2}$ channel, to NNLO in NLEFT. $p \cot \delta(p)$ is shown as a function of p^2, in the center-of-mass frame. The experimental results are nd and pd scattering data from the PWA of Ref. [78]. The dashed line represents an empirical model with a pole singularity just below $p^2 = 0$. Note the (deeply bound) triton and ^{3}He bound states for $p^2 < 0$

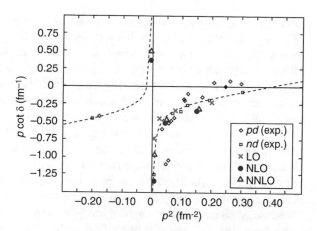

method clearly provides the simplest way to relate the energies in a finite periodic cube to scattering phase shifts at infinite volume. For instance, we may solve the three-nucleon problem on a cubic lattice for $L = 4a$ to $L = 8a$ using sparse matrix diagonalization, and compute the three-nucleon energy levels relative to the threshold energy for a non-interacting nd system in the same volume.

In Fig. 5.14, we show the effective range function $p \cot \delta(p)$ for the $^2S_{1/2}$ channel, as a function of p^2 in the center-of-mass frame, along with available experimental data from Ref. [78]. Here, p denotes the relative momentum of the scattering neutron and deuteron in the center-of-mass frame. At NNLO, results are given for $c_D = 1$, with c_E fitted to the physical triton binding energy. Below the threshold for deuteron breakup at $p^2 \simeq 0.07$ fm^{-2}, the NLEFT prediction agrees closely with experiment. Above the breakup energy, the Lüscher analysis does not account for mixing between nd and three-nucleon (nnp) states, and hence the level of agreement is expected to deteriorate. Such effects appear to be more prevalent for the $^4S_{3/2}$ channel, see Ref. [77] for a detailed description. As can be seen from Fig. 5.14, both the NLEFT calculation and the experimental data show evidence of non-trivial behavior in the immediate vicinity of the nd threshold, consistent with a pole singularity just below $p^2 \simeq 0$. Such behavior may be connected to the Efimov effect [79–81].

NLEFT appears able to describe nd scattering with similar accuracy as the triton binding energy, at least below the deuteron break-up threshold. The question remains to what extent c_D can be constrained using available nd scattering data. For instance, c_D could be determined from the physical scattering length [82]

$$^2a_{nd} = (-0.645 \pm 0.003_{\text{exp}} \pm 0.007_{\text{th}}) \text{ [fm]} , \tag{5.233}$$

or from the "pseudo" scattering length

$$^2\tilde{a}_{nd} = -0.45(4) \text{ [fm]} , \tag{5.234}$$

which is obtained by adjusting the strength of the nn interactions to match those of the np interactions [75, 83]. If we consider the ERE, we find that the matching to the empirical nd scattering length should be done in the vicinity of $p^2 = 0$. However, any value between $c_D \simeq -6$ and $c_D \simeq 6$ is found to produce a result consistent with the pseudo-scattering length (5.234). As such a weak constraint is not sufficient, other methods for determining c_D are called for. A natural candidate observable, which might make it possible to pin down c_D, appears to be the binding energy of ^{4}He. However, as shown e.g. in Ref. [70], this observable is again not very sensitive to c_D, and is hence of limited usefulness. Several other ways of determining c_D have been suggested so far, such as the beta decay rate of the triton [84], or the SU(2) axial coupling to two-nucleon states, using pion production data in pp scattering [85]. Due to the relatively high energy of the pion production threshold, such calculations are likely to prove challenging for NLEFT. Ultimately, one expects that Lattice QCD one will be able to pin down c_D more precisely. For the state-of-the-art determination in the continuum, see Ref. [86]. As shown there, the strongest constraint on c_D results from the minimum of the cross section in proton-deuteron scattering at the energy of $E = 70$ MeV. Clearly, the determination of the three-nucleon force to high precision is a continuing challenge for low-energy nuclear physics.

References

1. H.M. Muller, S.E. Koonin, R. Seki, U. van Kolck, Nuclear matter on a lattice. Phys. Rev. C **61**, 044320 (2000)
2. T. Abe, R. Seki, A.N. Kocharian, A mean field calculation of thermal properties of simple nucleon matter on a lattice. Phys. Rev. C **70**, 014315 (2004) [Erratum: Phys. Rev. C **71**, 059902 (2005)]
3. S. Chandrasekharan, M. Pepe, F.D. Steffen, U.J. Wiese, Lattice theories with nonlinearly realized chiral symmetry. Nucl. Phys. Proc. Suppl. **129**, 507 (2004)
4. S. Chandrasekharan, M. Pepe, F.D. Steffen, U.J. Wiese, Nonlinear realization of chiral symmetry on the lattice. J. High Energy Phys. **0312**, 035 (2003)
5. D. Lee, B. Borasoy, T. Schäfer, Nuclear lattice simulations with chiral effective field theory. Phys. Rev. C **70**, 014007 (2004)
6. D. Lee, T. Schäfer, Neutron matter on the lattice with pionless effective field theory. Phys. Rev. C **72**, 024006 (2005)
7. M. Hamilton, I. Lynch, D. Lee, Lattice gas models derived from effective field theory. Phys. Rev. C **71**, 044005 (2005)
8. R. Seki, U. van Kolck, Effective field theory of nucleon-nucleon scattering on large discrete lattices. Phys. Rev. C **73**, 044006 (2006)
9. D. Lee, T. Schäfer, Cold dilute neutron matter on the lattice. I. Lattice virial coefficients and large scattering lengths. Phys. Rev. C **73**, 015201 (2006)
10. D. Lee, T. Schäfer, Cold dilute neutron matter on the lattice. II. Results in the unitary limit. Phys. Rev. C **73**, 015202 (2006)
11. B. Borasoy, H. Krebs, D. Lee, U.-G. Meißner, The Triton and three-nucleon force in nuclear lattice simulations. Nucl. Phys. A **768**, 179 (2006)
12. F. de Soto, J. Carbonell, Low energy scattering parameters from the solutions of the non-relativistic Yukawa model on a 3-dimensional lattice (2006). hep-lat/0610040

13. B. Borasoy, E. Epelbaum, H. Krebs, D. Lee, U.-G. Meißner, Lattice simulations for light nuclei: chiral effective field theory at leading order. Eur. Phys. J. A **31**, 105 (2007)
14. D. Lee, R. Thomson, Temperature-dependent errors in nuclear lattice simulations. Phys. Rev. C **75**, 064003 (2007)
15. J.W. Chen, D.B. Kaplan, A lattice theory for low-energy fermions at finite chemical potential. Phys. Rev. Lett. **92**, 257002 (2004)
16. M. Wingate, Critical temperature for fermion pairing using lattice field theory (2005). cond-mat/0502372 [cond-mat.stat-mech]
17. A. Bulgac, J.E. Drut, P. Magierski, Spin 1/2 fermions in the unitary regime: a Superfluid of a new type. Phys. Rev. Lett. **96**, 090404 (2006)
18. D. Lee, Ground state energy of spin-1/2 fermions in the unitary limit. Phys. Rev. B **73**, 115112 (2006)
19. E. Burovski, N. Prokof'ev, B. Svistunov, M. Troyer, Critical temperature and thermodynamics of attractive fermions at unitarity. Phys. Rev. Lett. **96**, 160402 (2006)
20. E. Burovski, N. Prokof'ev, B. Svistunov, M. Troyer, The Fermi-Hubbard model at unitarity. New J. Phys. **8**, 153 (2006)
21. M. Lüscher, Volume dependence of the energy spectrum in massive quantum field theories. 1. Stable particle states. Commun. Math. Phys. **104**, 177 (1986)
22. M. Lüscher, Volume dependence of the energy spectrum in massive quantum field theories. 2. Scattering states. Commun. Math. Phys. **105**, 153 (1986)
23. M. Lüscher, Two particle states on a torus and their relation to the scattering matrix. Nucl. Phys. B **354**, 531 (1991)
24. H.A. Bethe, Theory of the effective range in nuclear scattering. Phys. Rev. **76**, 38 (1949)
25. J.D. Jackson, J.M. Blatt, The interpretation of low energy proton-proton scattering. Rev. Mod. Phys. **22**, 77 (1950)
26. E. Braaten, H.-W. Hammer, Universality in few-body systems with large scattering length. Phys. Rep. **428**, 259 (2006)
27. V. Bernard, M. Lage, U.-G. Meißner, A. Rusetsky, Resonance properties from the finite-volume energy spectrum. J. High Energy Phys. **0808**, 024 (2008)
28. T. Luu, M.J. Savage, Extracting scattering phase-shifts in higher partial-waves from lattice QCD calculations. Phys. Rev. D **83**, 114508 (2011)
29. M.E. Rose, *Elementary Theory of Angular Momentum* (Wiley, New York, 1957)
30. D. Lee, The Symmetric heavy-light ansatz. Eur. Phys. J. A **35**, 171 (2008)
31. S. Koenig, D. Lee, H.-W. Hammer, Volume dependence of bound states with angular momentum. Phys. Rev. Lett. **107**, 112001 (2011)
32. S. Koenig, D. Lee, H.-W. Hammer, Non-relativistic bound states in a finite volume. Ann. Phys. **327**, 1450 (2012)
33. R.A. Briceño, Z. Davoudi, Moving multichannel systems in a finite volume with application to proton-proton fusion. Phys. Rev. D **88**, 094507 (2013)
34. M. Göckeler, R. Horsley, M. Lage, U.-G. Meißner, P.E.L. Rakow, A. Rusetsky, G. Schierholz, J.M. Zanotti, Scattering phases for meson and baryon resonances on general moving-frame lattices. Phys. Rev. D **86**, 094513 (2012)
35. K. Rummukainen, S.A. Gottlieb, Resonance scattering phase shifts on a nonrest frame lattice. Nucl. Phys. B **450**, 397 (1995)
36. S. Bour, S. Koenig, D. Lee, H.-W. Hammer, U.-G. Meißner, Topological phases for bound states moving in a finite volume. Phys. Rev. D **84**, 091503 (2011)
37. Z. Davoudi, M.J. Savage, Improving the volume dependence of two-body binding energies calculated with lattice QCD. Phys. Rev. D **84**, 114502 (2011)
38. Z. Fu, Rummukainen-Gottlieb's formula on two-particle system with different mass. Phys. Rev. D **85**, 014506 (2012)
39. L. Leskovec, S. Prelovsek, Scattering phase shifts for two particles of different mass and non-zero total momentum in lattice QCD. Phys. Rev. D **85**, 114507 (2012)
40. C. Liu, X. Feng, S. He, Two particle states in a box and the S-matrix in multi-channel scattering. Int. J. Mod. Phys. A **21**, 847 (2006)

41. M. Lage, U.-G. Meißner, A. Rusetsky, A method to measure the antikaon-nucleon scattering length in lattice QCD. Phys. Lett. B **681**, 439 (2009)
42. V. Bernard, M. Lage, U.-G. Meißner, A. Rusetsky, Scalar mesons in a finite volume. J. High Energy Phys. **1101**, 019 (2011)
43. M. Döring, J. Haidenbauer, U.-G. Meißner, A. Rusetsky, Dynamical coupled-channel approaches on a momentum lattice. Eur. Phys. J. A **47**, 163 (2011)
44. N. Li, C. Liu, Generalized Lüscher formula in multichannel baryon-meson scattering. Phys. Rev. D **87**, 014502 (2013)
45. P. Guo, J. Dudek, R. Edwards, A.P. Szczepaniak, Coupled-channel scattering on a torus. Phys. Rev. D **88**, 014501 (2013)
46. M. Döring, U.-G. Meißner, E. Oset, A. Rusetsky, Unitarized chiral perturbation theory in a finite volume: scalar meson sector. Eur. Phys. J. A **47**, 139 (2011)
47. A. Martinez Torres, L.R. Dai, C. Koren, D. Jido, E. Oset, The KD, ηD_s interaction in finite volume and the nature of the $D_{s^*0}(2317)$ resonance. Phys. Rev. D **85**, 014027 (2012)
48. M. Döring, U.-G. Meißner, Finite volume effects in pion-kaon scattering and reconstruction of the $\kappa(800)$ resonance. J. High Energy Phys. **1201**, 009 (2012)
49. M. Albaladejo, J.A. Oller, E. Oset, G. Rios, L. Roca, Finite volume treatment of pi pi scattering and limits to phase shifts extraction from lattice QCD. J. High Energy Phys. **1208**, 071 (2012)
50. J.J. Wu, T.-S.H. Lee, A.W. Thomas, R.D. Young, Finite-volume Hamiltonian method for coupled-channels interactions in lattice QCD. Phys. Rev. C **90**, 055206 (2014)
51. B. Hu, R. Molina, M. Döring, A. Alexandru, Two-flavor simulations of the $\rho(770)$ and the role of the $K\bar{K}$ channel. Phys. Rev. Lett. **117**, 122001 (2016)
52. X. Li, C. Liu, Two particle states in an asymmetric box. Phys. Lett. B **587**, 100 (2004)
53. X. Feng, X. Li, C. Liu, Two particle states in an asymmetric box and the elastic scattering phases. Phys. Rev. D **70**, 014505 (2004)
54. S.R. Beane, P.F. Bedaque, A. Parreno, M. J. Savage, Two nucleons on a lattice. Phys. Lett. B **585**, 106 (2004)
55. R.A. Briceno, J.J. Dudek, R.D. Young, Scattering processes and resonances from lattice QCD. Rev. Mod. Phys. **90**, 025001 (2018)
56. J. Carlson, V.R. Pandharipande, R.B. Wiringa, Variational calculations of resonant states in ^{4}He. Nucl. Phys. A **424**, 47 (1984)
57. B. Borasoy, E. Epelbaum, H. Krebs, D. Lee, U.-G. Meißner, Two-particle scattering on the lattice: phase shifts, spin-orbit coupling, and mixing angles. Eur. Phys. J. A **34**, 185 (2007)
58. B.N. Lu, T.A. Lähde, D. Lee, U.-G. Meißner, Precise determination of lattice phase shifts and mixing angles. Phys. Lett. B **760**, 309 (2016)
59. H.P. Stapp, T.J. Ypsilantis, N. Metropolis, Phase shift analysis of 310-MeV proton proton scattering experiments. Phys. Rev. **105**, 302 (1957)
60. J.M. Blatt, L.C. Biedenharn, Neutron-proton scattering with spin-orbit coupling. 1. General expressions. Phys. Rev. **86**, 399 (1952)
61. B. Borasoy, E. Epelbaum, H. Krebs, D. Lee, U.-G. Meißner, Chiral effective field theory on the lattice at next-to-leading order. Eur. Phys. J. A **35**, 343 (2008)
62. E. Epelbaum, H. Krebs, D. Lee, U.-G. Meißner, Lattice calculations for A=3,4,6,12 nuclei using chiral effective field theory. Eur. Phys. J. A **45**, 335 (2010)
63. S. Elhatisari, D. Lee, G. Rupak, E. Epelbaum, H. Krebs, T.A. Lähde, T. Luu, U.-G. Meißner, Ab initio alpha-alpha scattering. Nature **528**, 111 (2015)
64. S. Elhatisari, D. Lee, U.-G. Meißner, G. Rupak, Nucleon-deuteron scattering using the adiabatic projection method. Eur. Phys. J. A **52**(6), 174 (2016)
65. J.M. Alarcón et al., Neutron-proton scattering at next-to-next-to-leading order in nuclear lattice effective field theory. Eur. Phys. J. A **53**(5), 83 (2017)
66. V.G.J. Stoks, R.A.M. Klomp, M.C.M. Rentmeester, J.J. de Swart, Partial wave analaysis of all nucleon-nucleon scattering data below 350-MeV. Phys. Rev. C **48**, 792 (1993)
67. R. Navarro Perez, J.E. Amaro, E. Ruiz Arriola, Statistical error analysis for phenomenological nucleon-nucleon potentials. Phys. Rev. C **89**(6), 064006 (2014)

68. E. Epelbaum, H. Krebs, U.-G. Meißner, Improved chiral nucleon-nucleon potential up to next-to-next-to-next-to-leading order. Eur. Phys. J. A **51**(5), 53 (2015)
69. M. Abramowitz, I. Stegun, *Handbook of Mathematical Functions with Formulas, Graphs and Mathematical Tables* (U.S. National Bureau of Standards, Washington, DC, 1964)
70. N. Klein, S. Elhatisari, T.A. Lähde, D. Lee, U.-G. Meißner, The Tjon band in nuclear lattice effective field theory. Eur. Phys. J. A **54**, 121 (2018)
71. H. van Haeringen, L.P. Kok, Modified effective range function. Phys. Rev. A **26**, 1218 (1982)
72. U.-G. Meißner, G. Ríos, A. Rusetsky, Spectrum of three-body bound states in a finite volume. Phys. Rev. Lett. **114**, 091602 (2015) [Erratum: Phys. Rev. Lett. **117**, 069902 (2016)]
73. H.W. Hammer, J.Y. Pang, A. Rusetsky, Three-particle quantization condition in a finite volume: 1. The role of the three-particle force. J. High Energy Phys. **1709**, 109 (2017)
74. S. König, D. Lee, Volume dependence of N-body bound states. Phys. Lett. B **779**, 9 (2018)
75. E. Epelbaum, A. Nogga, W. Gloeckle, H. Kamada, U.-G. Meißner, H. Witala, Three nucleon forces from chiral effective field theory. Phys. Rev. C **66**, 064001 (2002)
76. E. Epelbaum, H. Krebs, D. Lee, U.-G. Meißner, Lattice effective field theory calculations for A = 3,4,6,12 nuclei. Phys. Rev. Lett. **104**, 142501 (2010)
77. E. Epelbaum, H. Krebs, D. Lee, U.-G. Meißner, Lattice chiral effective field theory with three-body interactions at next-to-next-to-leading order. Eur. Phys. J. A **41**, 125 (2009)
78. W.T.H. van Oers, J.D. Seagrave, The neutron-deuteron scattering lengths. Phys. Lett. **24B**, 562 (1967)
79. V.N. Efimov, Weakly-bound states Of 3 resonantly-interacting particles. Sov. J. Nucl. Phys. **12**, 589 (1971)
80. V. Efimov, Effective interaction of three resonantly interacting particles and the force range. Phys. Rev. C **47**, 1876 (1993)
81. S. Kreuzer, H.-W. Hammer, Efimov physics in a finite volume. Phys. Lett. B **673**, 260 (2009)
82. K. Schoen et al., Precision neutron interferometric measurements and updated evaluations of the n p and n d coherent neutron scattering lengths (2003). nucl-ex/0306012
83. H. Witala, A. Nogga, H. Kamada, W. Glöckle, J. Golak, R. Skibinski, Modern nuclear force predictions for the neutron deuteron scattering lengths. Phys. Rev. C **68**, 034002 (2003)
84. D. Gazit, S. Quaglioni, P. Navratil, Three-nucleon low-energy constants from the consistency of interactions and currents in chiral effective field theory. Phys. Rev. Lett. **103**, 102502 (2009)
85. C. Hanhart, U. van Kolck, G.A. Miller, Chiral three nucleon forces from p wave pion production. Phys. Rev. Lett. **85**, 2905 (2000)
86. E. Epelbaum et al. [LENPIC Collaboration], Few- and many-nucleon systems with semilocal coordinate-space regularized chiral two- and three-body forces. Phys. Rev. C **99**, 024313 (2019)

Chapter 6
Lattice Monte Carlo

Thus far, we have developed the path integral and transfer matrix in Chap. 3 and shown how to apply this formalism to the two-nucleon and three-nucleon forces in NLEFT in Chap. 4. In Chap. 5, we have discussed the calculation of observables in the two- and three-nucleon sectors, and shown how these can be used to fix the free parameters of the NLEFT interactions, which allows the theory to be predictive for $A \geq 4$. However, we are still faced with the difficulty that most analytical and numerical methods become impractical for the multi-nucleon problem (for systems of a realistic size) due to adverse exponential scaling of the computational effort with nucleon number A. In this chapter, we shall establish viable computational methods for $A \geq 4$ which, given sufficiently realistic nuclear interactions, can be applied to a large range of heavier nuclei as well. In the Euclidean time projection technique, the energy eigenvalues and wave functions of the low-lying spectrum of the Hamiltonian are computed by repeatedly operating on a suitably chosen "trial state" by the operator $\exp(-\delta H)$, which acts as a "low-energy filter". An advantageous feature is the great freedom in the choice of the trial state, as long as it has a finite overlap with the true ground state of H. Moreover, Euclidean time projection can be straightforwardly combined with Monte Carlo (MC) methods, where importance sampling according to the Boltzmann weight is used as a stochastic estimator for the NLEFT path integrals over products of transfer matrices. Within the context of NLEFT, Euclidean time projection has been further developed into the "adiabatic projection" method, which allows for an a priori treatment of scattering processes or reactions, where the constituents are themselves composite objects, such as α particles or heavier nuclei. The adiabatic projection method provides a "cluster Hamiltonian" which can be found by Euclidean time projection Monte Carlo (for instance for α-α or ^{12}C-α scattering). The spherical wall method can then be applied to the cluster Hamiltonian, in order extract scattering observables such as phase shifts.

© Springer Nature Switzerland AG 2019
T. A. Lähde, U.-G. Meißner, *Nuclear Lattice Effective Field Theory*,
Lecture Notes in Physics 957, https://doi.org/10.1007/978-3-030-14189-9_6

The application of MC methods to theories with dynamical fermions is particularly challenging, because of the highly non-local structure of the fermion action. Hence, MC algorithms where the auxiliary or gauge field variables are updated locally suffer from adverse cubic scaling of the computational effort with system size. While the standard Metropolis algorithm satisfies detailed balance, a high acceptance rate can only be maintained for local updates. A lot of effort has been put toward the development of algorithms which allow for efficient global updates of the auxiliary fields on the lattice. However, such algorithms (notably the Langevin and Molecular Dynamics (MD) algorithms) typically require extrapolation to recover detailed balance, which makes them unattractive for systematical, large-scale computational work. An elegant solution to this problem, which has been pioneered in the field of Lattice QCD, is the so-called Hybrid Monte Carlo (HMC) algorithm, which uses the Hybrid update (a combination of Langevin and MD updates) as the proposed update in the Metropolis algorithm. If the Hamiltonian equations of motion in the MD update are integrated sufficiently accurately, a high acceptance rate can be maintained for global updates as well, as changes in the Euclidean action need not be small during the MD update. Moreover, if the MD integration is performed using a reversible and area-conserving (symplectic) integrator, then the HMC update also satisfies detailed balance, and a high acceptance rate can be maintained for arbitrarily large lattices.

When an Euclidean time projection amplitude is evaluated with MC methods, we speak of Projection Monte Carlo (PMC) methods. In NLEFT, the PMC calculations are performed using HMC updates. This provides an algorithm, for which the computational effort scales roughly as $\sim A^2$ with nucleon number A. In principle, this establishes the method as a viable means for approaching the limit of large number of nucleons, $A \to \infty$, without the need to introduce additional approximations. However, as is often the case for MC methods, the PMC approach to NLEFT is also not entirely free of oscillations of the average sign (or complex phase) of the MC probability weight (the so-called "sign problem", familiar from many other physical systems, such as Lattice QCD at finite chemical potential). However, due to the approximate Wigner SU(4) symmetry of the NLEFT potential, the sign problem is, in most cases, relatively mild and can be controlled by clever choices of operator smearing and perturbative treatment of higher-order interactions. In this chapter, we shall first introduce the method of Euclidean time projection, followed by the development of the HMC algorithm. Finally, we shall discuss the so-called "adiabatic projection" method, whereby PMC is extended to describe nuclear scattering and reaction processes.

6.1 Euclidean Time Projection

Thus far, we have elaborated on the structure of the NLEFT Hamiltonian and transfer matrix, and discussed practical methods for computing nuclear spectra and structure, that involved the use of sparse matrix algorithms such as Lanczos. We shall now describe the method of Euclidean time projection, which can in principle be applied to systems of arbitrarily many nucleons [1]. We denote the total number of nucleons as $A = Z + N$, where Z is the number of protons and N is the number of neutrons. As in Chap. 4, we denote by H_{LO} the (continuum) NLEFT Hamiltonian with the full LO structure (for instance the H_{LO_3} Hamiltonian), and by $H_{SU(4)}$ a simplified form of the LO Hamiltonian, where $g_A = 0$ and with all the spin- and isospin-dependent terms set to zero. Note that $H_{SU(4)}$ corresponds to the limit where the spin and isospin degrees of freedom are interchangeable, such that the $SU(2) \times SU(2)$ spin-isospin symmetry becomes the Wigner $SU(4)$ symmetry.

Let us define a "trial wave function"

$$|\Psi(t')\rangle \equiv \exp(-H_{SU(4)}t')|\Psi_{Z,N}^{\text{free}}\rangle, \tag{6.1}$$

where $|\Psi_{Z,N}^{\text{free}}\rangle$ is a Slater determinant of free-nucleon standing waves in a periodic cube, for Z protons and N neutrons, and t' is the Euclidean time associated with evolution by $H_{SU(4)}$. With the trial wave function (6.1), we define the Euclidean time projection amplitude

$$Z(t) \equiv \langle \Psi(t')| \exp(-H_{LO}t)|\Psi(t')\rangle, \tag{6.2}$$

and the "transient energy"

$$E(t) = -\frac{\partial}{\partial t} \log Z(t), \tag{6.3}$$

such that in the limit of large t,

$$\lim_{t \to \infty} E(t) = E_0, \tag{6.4}$$

where E_0 is the energy of the lowest eigenstate $|\Psi_0\rangle$ of H_{LO} which has a non-vanishing overlap with the trial state $|\Psi(t')\rangle$. In order to compute expectation values of (normal ordered) operators $\mathcal{O}$, we take

$$Z_{\mathcal{O}}(t) \equiv \langle \Psi(t')| \exp(-H_{LO}t/2)\, \mathcal{O} \, \exp(-H_{LO}t/2)|\Psi(t')\rangle, \tag{6.5}$$

such that the expectation value of $\mathcal{O}$ for $|\Psi_0\rangle$ is given by

$$\langle \Psi_0| \mathcal{O} |\Psi_0\rangle = \lim_{t \to \infty} \frac{Z_{\mathcal{O}}(t)}{Z(t)}, \tag{6.6}$$

in the limit of large t. It should be noted that the $SU(4)$ Hamiltonian in (6.1) is computationally inexpensive, and can be thought of as an approximate "low-energy filter". Moreover, this low-energy filter is free of MC sign oscillations, in other words the probability measure is strictly positive for even A [2, 3]. In general,

Euclidean time projection with H_{LO} introduces a sign problem, the severity of which depends on A and the extent of the projection time t.

In NLEFT, we often encounter situations where the trial wave function consists of more than one component, for example in the study of excited states and transitions between them and the ground state, or in the description of nuclear scattering processes using the adiabatic projection method. A more general form of the trial state is thus

$$|\Psi\rangle \equiv \sum_i^{N_{ch}} c_i |\Psi_i\rangle, \tag{6.7}$$

in terms of the "channel" number N_{ch}. Here, the coefficients c_i denote the relative weights of the contributions to the trial state (such as Clebsch-Gordan coefficients). The Euclidean projection amplitude then receives contributions from N_{ch}^2 channels, according to

$$Z(t) = \langle \Psi | \exp(-Ht) | \Psi \rangle = \sum_{i,j}^{N_{ch}} c_i c_j A_{ij}, \tag{6.8}$$

where

$$A_{ij} \equiv \langle \Psi_i | \exp(-Ht) | \Psi_j \rangle, \tag{6.9}$$

such that the computational size of the problem quickly becomes large, as the number of channels grows. We note that MC methods can also be used to randomly sample the various channels that contribute to Eq. (6.8). We illustrate the random sampling of the amplitudes (6.9) in more detail in Appendix D.

6.1.1 Transfer Matrix Formalism

We shall now apply the Euclidean time projection method to the transfer matrix formalism. We note that the auxiliary field path integral and the LO action of NLEFT have already been treated in Chaps. 3 and 4, respectively. The path integrals involve an integration over the pion and auxiliary field variables σ on each spatial and temporal lattice site. Here, σ is again used to collectively denote the set of auxiliary fields s, s_S, s_I, and $s_{S,I}$ (or a subset of them, in the case where not all of them appear in a given LO action). For brevity of notation, we shall in most cases suppress the dependence of the transfer matrix M on σ, and only indicate the "slice" of Euclidean time to which a particular instance of M is referring.

In the transfer matrix formalism, the trial wave function (6.1) is given by

$$|\Psi(t')\rangle \equiv \int \mathcal{D}s \, \exp\left(-S_{ss}(s)\right) M_{\mathrm{SU}(4)}(N_{t'}-1)\cdots M_{\mathrm{SU}(4)}(0)|\Psi_{Z,N}^{\mathrm{free}}\rangle, \quad (6.10)$$

where $t' \equiv \delta N_{t}'$, and the transfer matrices M have been labeled according to "time slices" of Euclidean projection time. Since

$$M_{\mathrm{SU}(4)}(s,t') = \, : \exp\left(-\tilde{H}_{\mathrm{free}}\alpha_t + \sqrt{-\tilde{C}_{\mathrm{SU}(4)}}\alpha_t \sum_{\mathbf{n}} s(\mathbf{n},t)\,\rho^{a^\dagger,a}(\mathbf{n})\right) :,$$

$$(6.11)$$

the dependence on π_I', s_S, s_I, and $s_{S,I}$ drops out of Eq. (6.10). The Euclidean time projection amplitude then becomes

$$Z(t) = \int \mathcal{D}\pi_I' \mathcal{D}\sigma \, \exp\left(-S_{\pi\pi}(\pi_I') - S_{ss}(\sigma)\right) \langle\Psi(t')|M(N_t-1)\cdots M(0)|\Psi(t')\rangle,$$

$$(6.12)$$

in terms of the full LO transfer matrix M, where $t \equiv \delta N_t$. Similarly, expectation values of normal-ordered operators $\mathcal{O}$ are computed from

$$Z_{\mathcal{O}}(t) = \int \mathcal{D}\pi_I' \mathcal{D}\sigma \, \exp\left(-S_{\pi\pi}(\pi_I') - S_{ss}(\sigma)\right)$$
$$\times \langle\Psi(t')|M(N_t-1)\cdots M(N_t/2)\,\mathcal{O}\,M(N_t/2-1)\cdots M(0)|\Psi(t')\rangle,$$

$$(6.13)$$

where $\mathcal{O}$ has been inserted at the midpoint of the product of time slices.

In order to make further progress, we shall make use of the fact that in the auxiliary-field formalism, no direct interactions between nucleons appear, as these have been replaced by a description in terms of single-nucleon operators, which couple to the auxiliary- and pion background fields. For the moment, we shall assume that each nucleon is distinguishable (for instance as a "billiard ball" labeled by a "nucleon number" from $1 \to A$). Hence, we can form the matrix

$$\mathcal{M}_{kl}(\pi_I', \sigma, t', t)$$
$$= \langle\psi_k| M_{\mathrm{SU}(4)}(2N_{t'} + N_t - 1)\cdots M_{\mathrm{SU}(4)}(N_{t'} + N_t)$$
$$\times M(N_{t'} + N_t - 1)\cdots M(N_{t'}) M_{\mathrm{SU}(4)}(N_{t'} - 1)\cdots M_{\mathrm{SU}(4)}(0)|\psi_l\rangle,$$

$$(6.14)$$

of dimension $A \times A$, where $|\psi_k\rangle$ denotes the single-particle orbitals comprising the Slater determinant $|\Psi_{Z,N}^{\mathrm{free}}\rangle$. The calculation of the Euclidean projection amplitude then reduces to

$$Z(t) = \int \mathscr{D}\pi'_I \mathscr{D}\sigma \, \exp\left(-S_{\pi\pi}(\pi'_I) - S_{ss}(\sigma)\right) \det \mathscr{M}(\pi'_I, \sigma, t), \qquad (6.15)$$

where the determinant implements the proper antisymmetrization, in spite of our treatment of the nucleons as distinguishable particles in the Euclidean time evolution. For instance, if any of the nucleons in the trial state are placed in identical orbitals $|\psi_k\rangle$, then the time evolution of these nucleons will also be identical. This would give a vanishing contribution to the amplitude (6.15). We shall return to this issue when we discuss suitable trial wave functions for systems with $A > 4$, as more than four nucleons cannot be placed on the same spatial lattice site due to the Pauli principle (with spin and isospin accounted for). Explicit examples of how the Euclidean action (and functional derivatives thereof) can be implemented, are given in Appendix E.

Given the expression for $Z(t)$, the most convenient and efficient way to compute $Z_\mathscr{O}(t)$ depends on the exact structure of the operator $\mathscr{O}$ under consideration. As an illustrative example, let us consider the spatial nucleon density correlation

$$\mathscr{O} \equiv \, : \rho^{a^\dagger, a}(\mathbf{n}')\rho^{a^\dagger, a}(\mathbf{n}) \, :, \qquad (6.16)$$

which can also be expressed as

$$\mathscr{O} = \lim_{\epsilon_1, \epsilon_2 \to 0} \frac{\partial^2}{\partial\epsilon_1 \partial\epsilon_2} M_\mathscr{O}(\epsilon_1, \epsilon_2), \qquad (6.17)$$

where

$$M_\mathscr{O}(\epsilon_1, \epsilon_2) \equiv \, : \exp\left(\epsilon_1 \sum_{i,j} a^\dagger_{i,j}(\mathbf{n}')a_{i,j}(\mathbf{n}') + \epsilon_2 \sum_{i,j} a^\dagger_{i,j}(\mathbf{n})a_{i,j}(\mathbf{n})\right) :, \qquad (6.18)$$

which turns out to be computationally convenient, as it resembles a transfer matrix composed of single-nucleon operators only. Similarly to Eq. (6.14), we define

$$\begin{aligned}
\mathscr{M}_{kl}&(\pi'_I, \sigma, t', t, \epsilon_1, \epsilon_2) \\
&= \langle\psi_k| M_{\mathrm{SU}(4)}(2N_{t'} + N_t - 1) \cdots M_{\mathrm{SU}(4)}(N_{t'} + N_t) \\
&\quad \times M(N_{t'} + N_t - 1) \cdots M(N_{t'} + N_t/2) M_\mathscr{O}(\epsilon_1, \epsilon_2) \\
&\quad \times M(N_{t'} + N_t/2 - 1) \cdots M(N_{t'})M_{\mathrm{SU}(4)}(N_{t'} - 1) \cdots M_{\mathrm{SU}(4)}(0)|\psi_l\rangle,
\end{aligned}$$
$$(6.19)$$

in terms of which

$$Z_{\mathscr{O}}(t) = \lim_{\epsilon_1, \epsilon_2 \to 0} \frac{\partial^2}{\partial \epsilon_1 \partial \epsilon_2} \int \mathscr{D}\pi'_I \mathscr{D}\sigma \, \exp\left(-S_{\pi\pi}(\pi'_I) - S_{ss}(\sigma)\right)$$

$$\times \det \mathscr{M}(\pi'_I, \sigma, t', t, \epsilon_1, \epsilon_2), \tag{6.20}$$

where in practice, the derivatives with respect to the source fields ϵ_i can be evaluated as numerical finite differences. Some explicit examples of how different types of observables can be implemented are given in Appendix F.

6.1.2 Trial States

We shall now turn to the construction of trial wave functions suitable for Euclidean time projection in NLEFT. A significant advantage of this method is that we only need to specify the single-particle components $|\psi_i\rangle$ of the full Slater state $|\Psi_{Z,N}^{\text{free}}\rangle$, and in principle we only need to make sure that the trial state has non-vanishing overlap with the true ground state. However, the choice of the $|\psi_i\rangle$ nevertheless requires some physical intuition, since the Euclidean time projection can usually only be performed up to $t < t_{\max}$, due to computational constraints and possible sign oscillations in the MC calculation of $Z(t)$. Moreover, the trial wave functions should preferentially have zero momentum as well as specific spin, angular momentum and parity quantum numbers. These points are most easily understood by considering a few examples.

Arguably, the simplest possible choice for nuclei up to $A = 4$ is to simply take set the spatial parts of the $|\psi_i\rangle$ to a (suitably normalized) constant,

$$\langle 0|a_{i,j}(\mathbf{n})|\psi_1\rangle = \frac{1}{\sqrt{L^3}} \delta_{i,0}\delta_{j,1} \equiv |\uparrow, n\rangle,$$

$$\langle 0|a_{i,j}(\mathbf{n})|\psi_2\rangle = \frac{1}{\sqrt{L^3}} \delta_{i,0}\delta_{j,0} \equiv |\uparrow, p\rangle,$$

$$\langle 0|a_{i,j}(\mathbf{n})|\psi_3\rangle = \frac{1}{\sqrt{L^3}} \delta_{i,1}\delta_{j,1} \equiv |\downarrow, n\rangle,$$

$$\langle 0|a_{i,j}(\mathbf{n})|\psi_4\rangle = \frac{1}{\sqrt{L^3}} \delta_{i,1}\delta_{j,0} \equiv |\downarrow, p\rangle, \tag{6.21}$$

which describes a ^{4}He system with $J = 0$, $J_z = 0$, and zero total momentum. Since further nucleons cannot be placed into the same orbitals due to the Pauli principle, one possibility to add more nucleons is to place these into the standing wave orbitals in a cubic box. For instance, we may describe ^{8}Be by taking

$$\langle 0|a_{i,j}(\mathbf{n})|\psi_5\rangle = \sqrt{\frac{2}{L^3}} \cos\left(\frac{2\pi n_3}{L}\right) \delta_{i,0}\delta_{j,1} \equiv (0,0,2\pi/L) \otimes |\uparrow, n\rangle,$$

$$\langle 0|a_{i,j}(\mathbf{n})|\psi_6\rangle = \sqrt{\frac{2}{L^3}} \cos\left(\frac{2\pi n_3}{L}\right) \delta_{i,0}\delta_{j,0} \equiv (0,0,2\pi/L) \otimes |\uparrow, p\rangle,$$

$$\langle 0|a_{i,j}(\mathbf{n})|\psi_7\rangle = \sqrt{\frac{2}{L^3}} \cos\left(\frac{2\pi n_3}{L}\right) \delta_{i,1}\delta_{j,1} \equiv (0,0,2\pi/L) \otimes |\downarrow, n\rangle,$$

$$\langle 0|a_{i,j}(\mathbf{n})|\psi_8\rangle = \sqrt{\frac{2}{L^3}} \cos\left(\frac{2\pi n_3}{L}\right) \delta_{i,1}\delta_{j,0} \equiv (0,0,2\pi/L) \otimes |\downarrow, p\rangle, \qquad (6.22)$$

where we recall that $\mathbf{n} = (n_1, n_2, n_3)$. Again, four nucleons may be placed into a specific cubic box orbital. Note that for $A > 4$, the momentum of the nucleus is no longer exactly zero, but the state is rather described by a wave packet with a small spread, centered around zero momentum. For heavier nuclei, it can be more convenient to instead "inject" the first four nucleons and evolve these for one step in Euclidean projection time, before injecting the next four nucleons at the second time step, and so on. This avoids the restrictions due to the Pauli principle on the number of nucleons that can be simultaneously placed on a single lattice site. Hence, trial wave functions for heavier nuclei such as ^{12}C can be constructed with zero momentum.

A more sophisticated choice of trial wave function is to use nuclear shell model states as a starting point for Euclidean time projection. We recall that in the nuclear shell model, protons and neutrons are taken to move independently in a (three-dimensional) harmonic oscillator potential $V_{\mathrm{HO}}(r) = \mu\omega^2 r^2/2$, where μ is a parameter with dimension of mass, and ω is the oscillator frequency. The energies of the nuclear levels are given by

$$E_{klm} = \omega(2k + l + 3/2) \equiv \omega(n + 3/2), \qquad (6.23)$$

where the quantum number n, k, l are positive integers (including zero) and m takes on values between $-l$ and l in integer steps. When supplemented by a (perturbatively treated) spin-orbit component, the harmonic oscillator model reproduces the observed "magic numbers" for protons and neutrons, at which nuclei are exceptionally stable. The spatial part of the shell-model wave function can be decomposed into angular and radial parts. The angular part is given by the spherical harmonics $Y_{l,m}(\theta, \phi)$, and the radial part $R_{kl}(r)$ is expressed in terms of Laguerre polynomials. For instance, we have

$$R_{00}(r) = \frac{2\nu^{3/4}}{\pi^{1/4}} \exp\left(-\frac{\nu r^2}{2}\right), \qquad R_{01}(r) = \frac{2\sqrt{6}\nu^{5/4}}{3\pi^{1/4}} r \exp\left(-\frac{\nu r^2}{2}\right), \qquad (6.24)$$

for the lowest shells with ($k = 0, l = 0$) and ($k = 0, l = 1$). Here, $\nu \equiv \mu\omega$ can be adjusted to optimize the agreement between empirical and shell-model nuclear energy levels. For the purposes of Euclidean time projection, ν can instead be optimized to accelerate the convergence to the ground state.

Let us now consider how to construct trial wave functions for nuclei in terms of single-nucleon shell-model states. Without the spin-orbit interaction, the lowest shell is denoted $1s$, with a degeneracy of 2, followed by the $1p$ shell with degeneracy of 8, followed in turn by the closely spaced $1d$ and $2s$ shells with a total degeneracy of 12. In the presence of a spin-orbit coupling, the $1s$ shell becomes the $1s_{1/2}$ shell, the $1p$ shell is split into the $1p_{1/2}$ and $1p_{3/2}$ shells with degeneracies of 2 and 4, respectively, and so on. For the single-nucleon states, we use the notation $|k, l, j, j_z\rangle_p$ for protons and $|k, l, j, j_z\rangle_n$ for neutrons. We note that the states $|l, j, j_z\rangle$ can be expressed as linear combinations of the uncoupled states $|l, m, s_z\rangle$.

When the trial wave functions for larger nuclei are constructed, it is helpful to make use of the concept of completely filled shells, as these do not contribute to the overall angular momentum of the nucleus, which is due to the nucleons outside of the filled shells. For instance, in the ^{4}He ground state, the $1s_{1/2}$ shell ($k = 0, l = 0, j = 1/2$) is completely occupied, which means that we can take

$$|^4\mathrm{He}\rangle = \left|0, 0, \frac{1}{2}, \frac{1}{2}\right\rangle_p \left|0, 0, \frac{1}{2}, \frac{1}{2}\right\rangle_n \left|0, 0, \frac{1}{2}, -\frac{1}{2}\right\rangle_p \left|0, 0, \frac{1}{2}, -\frac{1}{2}\right\rangle_n, \quad (6.25)$$

which has total angular momentum $J = 0$ and positive parity. We may now proceed to describe larger nuclei, such as ^{6}Li (note that there is no stable nucleus with $A = 5$) which consists of 3 protons and 3 neutrons. As for the ground state of ^{4}He, the first four nucleons form a closed s shell, such that the spin and parity of the ^{6}Li ground state is given by the remaining proton and neutron in the p shell. The shell model with spin-orbit coupling suggests that the $1p_{3/2}$ shell is lower in energy. We note that the parity of two nucleons in a p shell is even, and we wish to couple the p shell nucleons to total angular momentum $J = 1$. Using the Clebsch-Gordan coefficients for $J_z = 1$, we find

$$|^6\mathrm{Li}, J = 1, J_z = 1\rangle = \sqrt{\frac{3}{10}}|^4\mathrm{He}\rangle \times \left|0, 1, \frac{3}{2}, \frac{3}{2}\right\rangle_p \left|0, 1, \frac{3}{2}, -\frac{1}{2}\right\rangle_n$$

$$- \sqrt{\frac{2}{5}}|^4\mathrm{He}\rangle \times \left|0, 1, \frac{3}{2}, \frac{1}{2}\right\rangle_p \left|0, 1, \frac{3}{2}, \frac{1}{2}\right\rangle_n$$

$$+ \sqrt{\frac{3}{10}}|^4\mathrm{He}\rangle \times \left|0, 1, \frac{3}{2}, -\frac{1}{2}\right\rangle_p \left|0, 1, \frac{3}{2}, \frac{3}{2}\right\rangle_n, \quad (6.26)$$

and similarly for $J_z = 0$ and $J_z = -1$. Hence, it is possible to construct a trial wave function for any given nucleus, by successively filling up the states in the shell model with the required number of nucleons. For instance,

$$|^{16}O\rangle = |^4He\rangle \times \left|0, 1, \frac{3}{2}, \frac{3}{2}\right\rangle_p \left|0, 1, \frac{3}{2}, \frac{3}{2}\right\rangle_n \left|0, 1, \frac{3}{2}, \frac{1}{2}\right\rangle_p \left|0, 1, \frac{3}{2}, \frac{1}{2}\right\rangle_n$$

$$\times \left|0, 1, \frac{3}{2}, -\frac{1}{2}\right\rangle_p \left|0, 1, \frac{3}{2}, -\frac{1}{2}\right\rangle_n \left|0, 1, \frac{3}{2}, -\frac{3}{2}\right\rangle_p \left|0, 1, \frac{3}{2}, -\frac{3}{2}\right\rangle_n$$

$$\times \left|0, 1, \frac{1}{2}, \frac{1}{2}\right\rangle_p \left|0, 1, \frac{1}{2}, \frac{1}{2}\right\rangle_n \left|0, 1, \frac{1}{2}, -\frac{1}{2}\right\rangle_p \left|0, 1, \frac{1}{2}, -\frac{1}{2}\right\rangle_n, \tag{6.27}$$

provides the shell-model trial wave function for ^{16}O.

Another important class of trial wave function is the so-called "alpha cluster" wave function, where the $|\psi_i\rangle$ are taken to be localized wave packets, centered on a given lattice site. Again, up to four nucleons can be placed on a single lattice site, with spin and isospin treated as for the plane wave trial wave functions. As for the plane wave trial states, further α clusters can be injected on the same lattice sites at different Euclidean time steps. The alpha cluster wave functions are optimal for nuclei where α clustering is likely to be significant (such as ^{12}C, ^{16}O, and other "alpha nuclei"), as well as for the description of α-α scattering in NLEFT. For instance, the alpha cluster states can be taken to have a Gaussian radial distribution

$$R_\alpha(r) \sim \exp\left(-\frac{r_d^2}{2\Gamma_\alpha}\right), \tag{6.28}$$

where Γ_α is a (suitably chosen) width parameter for the wave packet, and the squared distance is defined as

$$r_d^2 \equiv \min(x^2, (L-x)^2) + \min(y^2, (L-y)^2) + \min(z^2, (L-z)^2), \tag{6.29}$$

for a square cubic lattice of linear size L. A natural choice of Γ_α is to relate it to the empirical root-mean-square radius of the α particle, $r_\alpha \simeq 1.68$ fm [4]. Alternatively, it may be chosen (as for the parameter ν in the shell-model description) to optimize convergence in Euclidean time or overlap with the ground state of the Hamiltonian.

For alpha cluster and shell model trial wave functions, the Euclidean projection amplitude receives contributions from all possible locations of the center-of-mass. The different center-of-mass positions of the trial states are therefore also sampled in the MC algorithms, as explained in more detail in Appendix D. Regardless of the choice of trial wave function, the coupling $\tilde{C}_{SU(4)}$ of Eq. (6.11) can be adjusted to further accelerate convergence in Euclidean projection time. This optimal value is expected to depend on the specific observable being computed. Furthermore, one can make use of the fact that the computed expectation values of observables should become independent of the choice of trial wave function, when t is sufficiently large. This can be used to check for convergence, or to perform constrained extrapolations to the limit $t \to \infty$.

6.1.3 Grand Canonical Ensemble

We have thus far developed the PMC formalism within the canonical ensemble, since this gives us the advantage of working with a fixed nucleon number A. Also, the great flexibility in the choice of trial wave function is an attractive feature. Furthermore, the matrix $\mathcal{M}$ encountered in the canonical calculations is of dimension $A \times A$, which contributes to the efficiency of the method. In particular, most of the computational effort is spent in calculating the matrix elements $\mathcal{M}_{kl}$ rather than $\det(\mathcal{M})$, as the numerical evaluation of small determinants by LU decomposition is straightforward. Nevertheless, the auxiliary field formalism can easily be adapted to grand canonical MC simulations of NLEFT, where the particle number is allowed to fluctuate. We shall now briefly discuss the grand canonical case, along with some useful algorithmic techniques similar to Lattice QCD.

Instead of the Euclidean time projection amplitude in the canonical formalism, we evaluate the grand canonical partition function

$$\mathcal{Z}(\mu, \beta) \equiv \int \mathcal{D}\pi'_I \mathcal{D}\sigma \, \exp\left(-S_{\pi\pi}(\pi'_I) - S_{ss}(\sigma)\right)$$

$$\times \, \text{Tr}\left\{ M(\pi'_I, \sigma, N_t - 1, \mu) \cdots M(\pi'_I, \sigma, 0, \mu) \right\}, \qquad (6.30)$$

with

$$M(\pi'_I, \sigma, t, \mu) \equiv M(\pi'_I, \sigma, t) \, \exp\left\{ \tilde{\mu}\alpha_t \sum_{\mathbf{n}} \rho^{a^\dagger, a}(\mathbf{n}) \right\}, \qquad (6.31)$$

where $\tilde{\mu}$ is the chemical potential in units of the lattice spacing a, and $\beta \equiv \delta N_t$ is the inverse temperature. Noting the antiperiodic boundary conditions in Euclidean time, we may use the operator identity

$$\text{Tr}\left\{ \exp(-\hat{H}_1) \exp(-\hat{H}_2) \cdots \exp(-\hat{H}_n) \right\}$$

$$= \det \begin{pmatrix} \mathbb{1} & -\exp(-H_1) & 0 & \cdots \\ 0 & \mathbb{1} & -\exp(-H_2) & \cdots \\ \vdots & & & \ddots \\ \exp(-H_n) & 0 & \cdots & \mathbb{1} \end{pmatrix}, \qquad (6.32)$$

$$= \det\left(\mathbb{1} + \exp(-H_1)\exp(-H_2) \cdots \exp(-H_n)\right), \qquad (6.33)$$

for even n, where

$$\hat{H}_k = [H_k]_{pq} \, a^\dagger_p a_q, \qquad (6.34)$$

to express the grand canonical partition function $\mathscr{Z}$ similarly as the Euclidean projection amplitude encountered earlier, in terms of a functional fermion determinant. Again, this makes use of the fact that the explicit interactions between nucleons have been decoupled by means of a suitable Hubbard-Stratonovich transformation. On the one hand, Eq. (6.32) is a determinant of an $(N_t \times N) \times (N_t \times N)$ matrix, and is referred to as the Hirsch-Fye (HF) formulation of the fermion determinant. On the other hand, Eq. (6.33) is a determinant of an $N \times N$ matrix, and is referred to as the Blankenbecler-Sugar-Scalapino (BSS) formulation. Clearly, various intermediate forms can also be constructed, where the non-diagonal elements of Eq. (6.32) consist of products of more than one time slice. If we introduce the single-nucleon amplitudes

$$\left[\mathscr{M}(\pi'_I, \sigma, \beta)\right]_{\mathbf{n}', \mathbf{n}} \equiv \langle \mathbf{n}' | M(\pi'_I, \sigma, N_t - 1) \cdots M(\pi'_I, \sigma, 0) | \mathbf{n} \rangle, \qquad (6.35)$$

then we can use Eq. (6.33) to express

$$\text{Tr}\left\{ M(\pi'_I, \sigma, N_t - 1, \mu) \cdots M(\pi'_I, \sigma, 0, \mu) \right\} \equiv \det\left(\mathbb{1} + \exp(\tilde{\mu}\alpha_t) \mathscr{M}(\pi'_I, \sigma, \beta) \right)^2, \qquad (6.36)$$

where we have assumed that the interactions for spin-up and spin-down nucleons are identical. The square of the determinant then guarantees that the sign problem is absent (due to the action being the square of a real-valued determinant). However, this assumption does not hold for the more advanced cases of operator smearing in the NLEFT action at LO.

The grand canonical version of NLEFT [5] is closely related to similar formulations of Lattice QCD, of Hubbard Models in condensed matter physics [6–11], and of ultracold atomic systems such as the Unitary Fermi gas [12–16]. As noted earlier, in distinction to the $A \times A$ matrix of the canonical formalism, we are now dealing with the determinant of an $N \times N$ matrix which in itself is a product of time slices, according to Eq. (6.33). Such a product can accumulate excessive numerical round-off error, unless special techniques such as singular value decomposition (SVD) or Gram-Schmidt orthogonalization are employed. This product of time slices can be avoided by switching to the even larger matrix (6.32), which nevertheless can be quite ill-conditioned (especially at large β), and therefore difficult to invert efficiently using iterative solvers. For these reasons, one may use the representation

$$\det\left(\mathscr{S}(\pi'_I, \sigma) \right)^2 \equiv \int \mathscr{D}\phi^\dagger \mathscr{D}\phi \, \exp(-S_p(\phi, \pi'_I, \sigma)), \qquad (6.37)$$

where the matrix $\mathscr{S}(\pi'_I, \sigma)$ can be found using Eq. (6.32). The "pseudofermion" field ϕ is a complex scalar field which (similarly to Grassmann fields) satisfies antiperiodic boundary conditions. The pseudofermion action is given by

$$S_p(\phi, \pi'_I, \sigma) \equiv \sum_{\mathbf{n}',\mathbf{n},t',t} \phi^\dagger(\mathbf{n}, t) \left[\mathscr{S}^{-1\dagger}(\pi'_I, \sigma)\mathscr{S}^{-1}(\pi'_I, \sigma) \right]_{\mathbf{n}',t';\mathbf{n},t} \phi(\mathbf{n}', t'),$$

(6.38)

where the matrix $\mathscr{S}^\dagger\mathscr{S}$ should be positive definite, hence the square of the determinant in Eq. (6.37).

The pseudofermion trick (6.37) can be thought of as a stochastic evaluation of the fermion determinant, whereby ϕ is sampled from a random, Gaussian distribution. Instead of a full inversion of $\mathscr{S}$, it suffices to apply $\mathscr{S}$ to a given vector ϕ in order to evaluate the pseudofermion action (6.38), which leads to a dramatic reduction in the computational effort. To summarize, the grand canonical partition function expressed in terms of pseudofermions becomes

$$\mathscr{Z}(\mu, \beta) = \int \mathscr{D}\phi^\dagger \mathscr{D}\phi \mathscr{D}\pi'_I \mathscr{D}\sigma \, \exp\left(-S_{\pi\pi}(\pi'_I) - S_{ss}(\sigma) - S_p(\phi, \pi'_I, \sigma)\right),$$

(6.39)

where S_p is given by Eq. (6.38). However, as the determinant of the $A \times A$ matrix of single-nucleon amplitudes in the canonical formalism is much easier to evaluate numerically, pseudofermions are rarely needed in PMC calculations.

6.1.4 Application to NLEFT

We recall that the expressions for the NLEFT transfer matrices at LO, NLO and NNLO have been derived in Chap. 4. Having established the Euclidean time projection formalism, we may turn to the computation of the ground state energies of nuclei at each order in NLEFT. At LO, we take

$$Z(t) = \langle \Psi(t')|M_{\mathrm{LO}}(N_t - 1) \cdots M_{\mathrm{LO}}(0)|\Psi(t')\rangle, \qquad (6.40)$$

for the Euclidean time projection amplitude. For instance, the transient energy $E_{\mathrm{LO}}(t + \alpha_t/2)$ can be computed from the ratio

$$\exp(-\alpha_t E_{\mathrm{LO}}(t + \alpha_t/2)) = \frac{Z(t + \alpha_t)}{Z(t)}, \qquad (6.41)$$

of amplitudes for Euclidean projection time t and $t+\alpha_t$. The LO ground state energy $E_{0,\mathrm{LO}}$ is obtained (asymptotically) from

$$E_{0,\mathrm{LO}} = \lim_{t\to\infty} E_{\mathrm{LO}}(t + \alpha_t/2), \qquad (6.42)$$

where the transient energy refers to the midpoint between time steps N_t and $N_t + 1$. We recall that while $E_{0,\text{LO}}$ is computed non-perturbatively, the NLO and NNLO shifts are computed as first-order perturbations, as we have established in Chaps. 4 and 5. Computing them non-perturbatively would require the introduction of significantly more auxiliary-field components. This would increase the statistical error and, since the NLO contribution is in general repulsive, the sign problem which limits the accessible range in t.

In order to obtain the ground-state energies at NLO and NNLO, we first compute the expectation value of M_{LO} from

$$Z_{M_{\text{LO}}}(t) \equiv \langle \Psi(t') | M_{\text{LO}}(N_t - 1) \cdots M_{\text{LO}}(N_t/2)$$
$$\times M_{\text{LO}}\, M_{\text{LO}}(N_t/2 - 1) \cdots M_{\text{LO}}(0) | \Psi(t') \rangle, \qquad (6.43)$$

where M_{LO} has been inserted as an additional time step in the middle of a string of LO transfer matrices. Similarly, for the perturbative higher-order terms we compute

$$Z_{M_{\text{NLO}}}(t) \equiv \langle \Psi(t') | M_{\text{LO}}(N_t - 1) \cdots M_{\text{LO}}(N_t/2)$$
$$\times M_{\text{NLO}}\, M_{\text{LO}}(N_t/2 - 1) \cdots M_{\text{LO}}(0) | \Psi(t') \rangle, \qquad (6.44)$$

for M_{NLO}, and

$$Z_{M_{\text{NNLO}}}(t) \equiv \langle \Psi(t') | M_{\text{LO}}(N_t - 1) \cdots M_{\text{LO}}(N_t/2)$$
$$\times M_{\text{NNLO}}\, M_{\text{LO}}(N_t/2 - 1) \cdots M_{\text{LO}}(0) | \Psi(t') \rangle, \qquad (6.45)$$

for M_{NNLO}. Details of the practical implementation of these expressions can be found in Appendix F. The ratio of amplitudes

$$\frac{Z_{M_{\text{NLO}}}(t)}{Z_{M_{\text{LO}}}(t)} = 1 - \alpha_t \Delta E_{\text{NLO}}(t) + \cdots, \qquad (6.46)$$

provides the (transient) NLO energy shift $\Delta E_{\text{NLO}}(t)$, where the above expression has been truncated to exclude contributions beyond first order in the NLO coupling constants. The shifted ground state energy is then found asymptotically from

$$E_{0,\text{NLO}} = E_{0,\text{LO}} + \lim_{t \to \infty} \Delta E_{\text{NLO}}(t), \qquad (6.47)$$

at NLO. Similarly, at NNLO we compute

$$\frac{Z_{M_{\text{NNLO}}}(t)}{Z_{M_{\text{LO}}}(t)} = 1 - \alpha_t \Delta E_{\text{NNLO}}(t) + \cdots, \qquad (6.48)$$

and

$$E_{0,\text{NNLO}} = E_{0,\text{LO}} + \lim_{t \to \infty} \Delta E_{\text{NNLO}}(t), \tag{6.49}$$

which gives the perturbative NNLO energy shift.

The above considerations establish that Euclidean projection calculations refer to the limit $t \to \infty$, which in practice means that calculations are performed for $t \leq t_{\text{max}}$, where t_{max} represents the largest Euclidean projection time which can be accessed with given computational resources. In many cases, the extent of t_{max} is limited by the sign oscillations of $Z(t)$, which may grow uncontrollably large with increasing t. This means that extrapolation is often required to reach the limit $t \to \infty$. Before we turn to the description of MC algorithms, we shall discuss suitable methods for such extrapolations.

6.1.5 Extrapolation in Euclidean Time

In a practical Euclidean projection calculation, we obtain a set of data points for the LO transient energy $E(t)$ at discrete projection times (along with any other observables calculated simultaneously, such as expectation values of higher-order operators), up to a maximum feasible projection time t_{max}. Given the limited extent of the data (and statistical errors in the case of MC simulations), there are uncertainties in the extrapolation $t_{\text{max}} \to \infty$. At large Euclidean times, the behavior of $Z(t)$ and $Z_{\mathscr{O}}(t)$ is controlled by the low-energy spectrum of the Hamiltonian. Let us label by $|E\rangle$ the eigenstates of the Hamiltonian with energy E, and let $\rho_A(E)$ denote the density of states for a system of A nucleons (for simplicity, we omit possible complications due to degenerate states). We can then construct the spectral representation

$$Z_A(t) = \int dE \, \rho_A(E) \left| \langle E | \Psi_A(t') \rangle \right|^2 \exp(-Et), \tag{6.50}$$

of $Z(t)$, where ψ_A is a trial state for A nucleons which is assumed to have a non-vanishing overlap with the ground state. Similarly, we have

$$Z_A^{\mathscr{O}}(t) = \int dEdE' \, \rho_A(E)\rho_A(E') \, \langle \Psi_A(t')|E\rangle \langle E|\mathscr{O}|E'\rangle \langle E'|\Psi_A(t')\rangle$$

$$\times \exp(\,(E + E')t/2), \tag{6.51}$$

for $Z_{\mathscr{O}}(t)$. We can obtain arbitrary accuracy over any finite range in t by taking

$$\rho_A(E) \simeq \sum_{k=0}^{k_{\text{max}}} c_{A,k} \delta(E - E_{A,k}), \tag{6.52}$$

as a sum of delta functions (typically three to four). By means of Eqs. (6.50) and (6.52), we obtain the extrapolation formula

$$E_A^j(N_t) = E_{A,0} + \sum_{k=1}^{k_{\max}} |c_{A,j,k}| \exp\left(-\frac{\Delta_{A,k} N_t}{\Lambda_t}\right),$$

(6.53)

for the LO energy $E_{A,0}$, where $t = N_t/\Lambda_t$. For instance, $\Lambda_t = 150$ MeV corresponds to $\delta = 1.32$ fm. The energy gaps in Eq. (6.53) are defined as $\Delta_{A,k} \equiv E_{A,k} - E_{A,0}$, and the index j denotes a specific choice of t' and $\tilde{C}_{SU(4)}$ in the trial wave function $|\Psi_A(t')\rangle$. We take $\Delta_A^{k+1} > \Delta_A^k$, and $k_{\max} = 3$ for ^{4}He ($A = 4$) and $k_{\max} = 2$ for $A \geq 8$ [17, 18]. For the operator matrix elements that make up the perturbative NLO and NNLO corrections, we use Eqs. (6.51) and (6.52) to find

$$X_A^{\mathcal{O},j}(N_t) = X_{A,0}^{\mathcal{O}} + \sum_{k=1}^{k_{\max}} x_{A,j,k} \exp\left(-\frac{\Delta_{A,k} N_t}{2\Lambda_t}\right),$$

(6.54)

where the dominant contributions are taken to be due to transition amplitudes involving the ground state and excited states.

In order for the extrapolation formulas (6.53) and (6.54) to be reliable, it is necessary that the overlap between our trial state and the ground state not be small compared to the overlap with the low-lying excited states. It should be noted that the coefficients $x_{A,j,k}$ can be positive as well as negative, which gives us the possibility of "triangulating" the asymptotic values $X_{A,0}^{\mathcal{O}}$ from above and below. For this purpose, the parameters $t' = N_t'/\Lambda_t$ and $\tilde{C}_{SU(4)}$ should be optimally chosen for each value of A. For Eq. (6.53), the dependence on t is monotonically decreasing and thus no triangulation of the asymptotic value from above and below is possible. However, we are helped by the fact that the rate of convergence is twice that of Eq. (6.54). In order to determine $E_{A,0}$ and $X_{A,0}^{\mathcal{O}}$, a correlated χ^2 fit to the LO energy and all NLO and NNLO matrix elements in NLEFT is performed for each value of A. This procedure also determines the coefficients $c_{A,j,k}$, $x_{A,j,k}$ and $\Delta_{A,k}$. We find that using two to six distinct trial states for each A allows for a significantly more accurate and stable determination of $E_{A,0}$ and $X_{A,0}^{\mathcal{O}}$ than would be possible with a single trial state. Note that the energy gaps $\Delta_{A,k}$ in the extrapolation functions are taken to be independent of the trial wave function j, which gives an additional consistency criterion. We find that a simultaneous description using Eqs. (6.53) and (6.54) accounts for all of the PMC data have been obtained for different $|\Psi_A(t')\rangle$ so far.

6.2   Monte Carlo Methods

As we have established for Euclidean time projection calculations in NLEFT (and in general, for a large class of other theories such as Lattice QCD and the Hubbard model), the expectation values of observables $\mathcal{O}$ are expressed as path integrals of the form

$$\langle \mathscr{O} \rangle = \frac{1}{\mathscr{Z}} \int \mathscr{D}s \, \mathscr{O}[s] \exp(-S_E[s, \beta]), \tag{6.55}$$

where

$$\mathscr{Z} = \int \mathscr{D}s \, \exp(-S_E[s, \beta]), \tag{6.56}$$

is the partition function, and s collectively denotes the auxiliary (or gauge) fields that appear in a specific theory.

The Boltzmann factor $\exp(-S_E[s, \beta])$ is given by the Euclidean action functional $S_E[s, \beta]$, which is assumed to be real-valued and bounded from below. Here, β denotes the set of parameters (such as temperature or coupling constants) relevant to the problem at hand, which we shall henceforth suppress for simplicity. As we have also established, the path integral (6.55) is formulated on a space-time lattice, where (assuming that a non-compact Hubbard-Stratonovich transformation has been applied) the field variables are integrated on each lattice site from $-\infty$ to ∞. As we are dealing with a highly multi-dimensional integration problem, numerical quadrature methods are out of the question. Instead, we approximate the path integral by the "ensemble average"

$$\langle \mathscr{O} \rangle \approx \frac{\sum_{\{s\}} \mathscr{O}[s] \exp(-S_E[s])}{\sum_{\{s\}} \exp(-S_E[s])}, \tag{6.57}$$

where a limited sample (or "ensemble") of N_s configurations $\{s\}$ are selected with MC methods. For instance, the value of s on each lattice site can be drawn from a standard random number generator which produces uniformly distributed random numbers in the range $[0, 1]$, and rescaling them using a suitable change of variables such that the entire integration interval is sampled for each auxiliary field component on each site. For the case of NLEFT, we have

$$S_E[s, \beta] = S_{\pi\pi}(\pi_I') + S_{ss}(\sigma) - \log(\det \mathscr{M}(\pi_I', \sigma, t)), \tag{6.58}$$

where we need to consider that the fermion determinant is in general not positive definite. Hence, we are in practice computing

$$\langle \mathscr{O} \rangle \approx \frac{\sum_{\{s\}} \mathscr{O}[s] \exp(i\theta[s]) \exp(-\tilde{S}_E[s])}{\sum_{\{s\}} \exp(i\theta[s]) \exp(-\tilde{S}_E[s])}, \tag{6.59}$$

where we sample according to

$$\tilde{S}_E[s] = S_{\pi\pi}(\pi_I') + S_{ss}(\sigma) - \log(|\det \mathscr{M}(\pi_I', \sigma, t)|), \tag{6.60}$$

and define the complex phase

$$\exp(i\theta[s]) \equiv \frac{\exp(-S_E[s])}{\exp(-\tilde{S}_E[s])} = \frac{\det \mathcal{M}(\pi'_I, \sigma, t)}{|\det \mathcal{M}(\pi'_I, \sigma, t)|}, \tag{6.61}$$

which effectively becomes a part of the observable $\mathcal{O}$. Moreover, all observables are divided by the expectation value of the complex phase. Hence, the expectation value of $\exp(i\theta[s])$ is a measure of the severity of the sign oscillations. If sign oscillations were absent, this factor would equal unity. Thus, in cases where the phase factor is strongly oscillating, the MC sampling becomes extremely inefficient. For simplicity, we shall henceforth suppress the complex phase factor in our development of the MC formalism. For a discussion of the causes and possible solutions to the sign problem in theories with fermions, the reader is referred to Refs. [19, 20]. Also, a more detailed discussion of the sign problem in NLEFT is given in Chap. 8.

Let us briefly consider the computational complexity of the numerical evaluation of Eq. (6.55) using the method of Eq. (6.57). In a typical NLEFT calculation, we may have $N = 6$ spatial and $N_t = 32$ temporal lattices sites, for a total of $6^3 \times 32 = 6912$ sites, with roughly ~ 10 auxiliary fields on each site (depending on the exact version of the LO action used) which gives roughly 7×10^4 degrees of freedom which should be integrated over. Even if a rather modest 10 mesh points are used for each integration, the full multi-dimensional integral would be approximated by $\sim 10^{70000}$ terms. Using random sampling to cover such a huge configuration space efficiently is a daunting task, which is not made easier by the observation that most randomly generated configurations would have a very large action and hence contribute only negligibly to $\langle \mathcal{O} \rangle$. As an illustrative example, an Ising model at low temperatures exhibits a ferromagnetic phase, where most of the spins are aligned in the same direction. For any lattice of a substantial size, the likelihood of a random sampling method generating such a configuration is clearly very small. In general, the integrand is likely to be non-negligible only for a relatively small subset of configurations for which S_E is close to its minimum value.

6.2.1   Importance Sampling

These observations suggest that random sampling of Eq. (6.55) is likely to be highly inefficient and dominated by statistical noise. Hence, we shall turn to a technique known as "importance sampling", which is based on the generation of configurations with a non-uniform distribution. For instance, in one dimension, we can introduce a "weighted" random number generator, which is characterized by a density function

$$W(x) \geq 0, \qquad \int_0^1 dx \, W(x) = 1, \tag{6.62}$$

where $W(x)$ is normalized in some interval of integration, in this case $x \in [0, 1]$. Note that $dx\, W(x)$ equals the fraction of random numbers generated in the interval $[x, x + dx]$. Then, one-dimensional integrals can be approximated by

$$I = \int_0^1 dx\, f(x) \approx \frac{1}{N_s} \sum_{i=1}^{N_s} \frac{f(x_n)}{W(x_n)}, \tag{6.63}$$

where the samples x_n have been randomly generated according to the distribution $W(x)$. It should be noted that while we may have $W(x_i) = 0$ for some discrete values x_i (typically at the interval boundaries), such values would never be generated by a practical random number generator. The usefulness of importance sampling is somewhat restricted by the feasibility of generators which provide random numbers distributed according to the chosen $W(x)$. As a simple example, consider

$$I = \int_0^\infty dx\, f(x) = \int_0^\infty dx\, \exp(-x^2) g(x), \tag{6.64}$$

for which we choose a Gaussian weight function

$$W(x) = \sqrt{\frac{2}{\pi}} \exp(-x^2), \tag{6.65}$$

giving

$$I \approx \frac{1}{N_s} \sum_{i=1}^{N_s} \frac{f(x_n)}{W(x_n)} = \frac{1}{N_s} \sqrt{\frac{\pi}{2}} \sum_{i=1}^{N_s} g(x_n), \tag{6.66}$$

where the integration range is infinite, and the function $g(x)$ is assumed to be slowly varying. In this case, the required x_n can easily be generated using any of the widely available numerical routines that produce random numbers with a Gaussian distribution. If the samples used to evaluate the integral are uncorrelated, then we expect the error to decrease as the square root of the number of sampled points $\sigma_I \sim 1/\sqrt{N_s}$, where σ_I^2 is the variance of the mean.

The path integral formulation of NLEFT can easily be reformulated using the technique of importance sampling. Instead of uniform sampling of the integrand, we sample according to

$$\langle \mathcal{O} \rangle \approx \frac{\sum_{\{s\}} \mathcal{O}[s] \exp(-S_E[s]) W^{-1}[s]}{\sum_{\{s\}} \exp(-S_E[s]) W^{-1}[s]}, \tag{6.67}$$

where each configuration in the ensemble $\{s\}$ occurs with probability $W[s]$. The most natural choice of $W[s]$ accounts for the Boltzmann factor in the path integral, giving

$$W[s] \equiv \frac{\exp(-S_E[s])}{\mathscr{Z}}, \tag{6.68}$$

such that most of the generated configurations are now clustered around the minimum of $S_E[s]$, while the appearance of configurations with a large action is now highly unlikely. Using Eq. (6.68), the expectation value (6.67) reduces to

$$\langle \mathscr{O} \rangle \approx \frac{1}{N_s} \sum_{\{s\}} \mathscr{O}[s], \tag{6.69}$$

where the number terms in the expression for $\mathscr{Z}$ equals N. Assuming that we have the ability to generate ensembles of auxiliary field configurations $\{s\}$ according to Eq. (6.68), then we could use that as a basis for a MC calculation which efficiently samples the path integral, with emphasis on configurations that give a large contribution to $\langle \mathscr{O} \rangle$. We are thus faced with the problem of constructing an algorithm which allows us to generate configurations according to Eq. (6.68), and for different choices of $S_E[s]$. As we shall see, practical algorithms can be obtained by considering so-called Markov chains of field configurations.

6.2.2 Markov Chains

Let us consider a process which starts from a configuration $\{s_1\}$ at time $\tau = 0$, and generates a new configuration $\{s_2\}$ at time $\tau + \epsilon$, and so on,

$$\{s_1\} \rightarrow \{s_2\} \rightarrow \{s_3\} \rightarrow \cdots, \tag{6.70}$$

in such a way that each transition only depends on the current state of the system, and not on the prior history (known as the "Markov property"). For NLEFT, the $\{s_i\}$ with $i = 1, 2, \ldots$ correspond to different configurations of the auxiliary and pion fields. For the moment, we shall neglect the fact that the auxiliary fields in NLEFT are continuous variables, and make the appropriate generalizations as needed. When referring to a specific degree of freedom (such as an auxiliary field component or lattice site), we shall use the index k, such that s_k refers to component k of the configuration $\{s\}$, and $s_{i,k}$ refers to component k of the configuration $\{s_i\}$. As the evolution of the Markov chain in Eq. (6.70) is governed by transition probabilities $\Omega(\{s_i\} \rightarrow \{s_j\}) \geq 0$ which satisfy the normalization condition

$$\sum_j \Omega(\{s_i\} \rightarrow \{s_j\}) = 1, \tag{6.71}$$

the Markov chain is referred to as a stochastic process. The Markov process can be thought of as describing the evolution of the system in "Monte Carlo (MC) time" τ (that should not be confused with the Euclidean time discussed before).

We shall now describe the evolution of the system during the Markov process. At any given MC time τ, the system is described by the probabilities W_i of being in the state $\{s_i\}$. Then the probability of finding the system in the state $\{s_j\}$ is given by

$$W_j(\tau + \epsilon) = \sum_i W_i(\tau)\Omega(\{s_i\} \to \{s_j\}), \tag{6.72}$$

at a later MC time $\tau + \epsilon$. In matrix notation, Eq. (6.72) becomes

$$W(\tau + \epsilon) = \Omega W(\tau), \tag{6.73}$$

where Ω has elements

$$\Omega_{ij} \equiv \Omega(\{s_i\} \to \{s_j\}), \tag{6.74}$$

and W is a vector with dimension equal to the total number of configurations available to the system. In the limit $\epsilon \to 0$, Eq. (6.73) becomes a differential equation known as the "Master equation". Here, W is referred to as a stochastic vector if

$$W_i \geq 0 \quad \forall i, \qquad \sum_i W_i = 1, \tag{6.75}$$

and Ω is similarly referred to as a stochastic matrix, if it is square with non-negative elements, such that the sum of each column of Ω equals unity, which follows from Eq. (6.71). In general,

$$\Omega(\{s_i\} \to \{s_j\}) \neq \Omega(\{s_j\} \to \{s_i\}), \tag{6.76}$$

such that a stochastic matrix does not need to be symmetric.

Given a matrix Ω that defines the transition probabilities, it is of particular interest for our MC calculation to find eigenvectors of Ω that correspond to eigenvalues of unity, as such eigenvectors describe a stationary (or equilibrium) state of the system. From Eq. (6.73), we find

$$W_{eq}(\tau + \epsilon) = \Omega W_{eq}(\tau) = W_{eq}(\tau), \tag{6.77}$$

in other words the probability of finding the system in any given configuration does not evolve in MC time. The convergence of the Markov process to an equilibrium distribution W_{eq} is dependent on a number of properties of the Markov chain. Firstly, the Markov chain should be irreducible, which means that any final configuration should be reachable from any initial configuration in a finite number of steps. In other words,

$$\Omega_{ij}^{(N_s)} \equiv \sum_{i_1} \sum_{i_2} \cdots \sum_{i_{N_s-1}} \Omega_{ii_1} \Omega_{i_1 i_2} \cdots \Omega_{i_{N_s-1} j} \neq 0, \tag{6.78}$$

for a finite number of Markov steps N_s. In Eq. (6.78), all possible intermediate states are summed over at each step in the chain. Secondly, the Markov chain should be aperiodic, which means that

$$\Omega_{ii}^{(N)} \neq 0, \tag{6.79}$$

for any value of N. Thirdly, the states comprising the Markov chain should be positive, which means that their mean recurrence time should be finite. For a more detailed treatment of the theory of Markov chains and the properties that we have stated above without further derivation, the reader is referred to Refs. [21, 22].

Finally, in order for a MC calculation to converge to the correct answer, the entire manifold of available configurations of the system should be explored by any computational algorithm we might choose to implement. This property of ergodicity means that each configuration should be reachable from any other state of the system. In other words, we require that the equilibrium distribution W_{eq} is unique, and independent of the starting configuration of the Markov chain. While it is allowable that some of the transition rates to and from a given configuration vanish, there should exist at least one path, such as

$$\{s_1\} \rightarrow \{s_{32}\} \rightarrow \{s_5\} \rightarrow \{s_{12}\} \rightarrow \cdots \rightarrow \{s_2\}, \tag{6.80}$$

along which all the transition probabilities are non-zero. Hence, the arbitrarily chosen states $\{s_1\}$ and $\{s_2\}$ are connected, even though a direct transition between them might be forbidden. In practice, it turns out that the issue of ergodicity is not merely a feature of the Euclidean action, but also of the specific MC algorithm used to effect the transitions between different states of the system.

6.2.3 Detailed Balance

From our discussion of importance sampling in Sect. 6.2.1, it is clear that the equilibrium (or stationary) probability distribution W_{eq} should equal the Boltzmann distribution. In order to turn this prescription into a practical computational algorithm, we need to consider the "inverse problem" to our prior discussion of Markov chains. In other words, how do the transition probabilities Ω need to be chosen in order for W_{eq} to equal Eq. (6.68)?

Again, we require that the stationary stochastic vector is an eigenvector of Ω with an eigenvalue of unity, in other words $\Omega W_{eq} = W_{eq}$. Hence, we can write

$$W_{eq}[s] = \sum_{\{s'\}} W_{eq}[s']\Omega(\{s'\} \to \{s\}), \tag{6.81}$$

and by means of Eq. (6.71), we find

$$\sum_{\{s'\}} W_{eq}[s]\Omega(\{s\} \to \{s'\}) = \sum_{\{s'\}} W_{eq}[s']\Omega(\{s'\} \to \{s\}), \tag{6.82}$$

which is referred to as the "global balance" condition. The physical interpretation of Eq. (6.82) is that the total probability flux from a given configuration $\{s\}$ to any other configuration of the system equals the flux from any other configuration to $\{s\}$. A more restrictive condition is

$$W_{eq}[s]\Omega(\{s\} \to \{s'\}) = W_{eq}[s']\Omega(\{s'\} \to \{s\}), \tag{6.83}$$

which is known as "detailed balance". If Eq. (6.83) holds, then Eq. (6.82) must also be satisfied, although the reverse is not true.

We are now in a position to formulate the detailed balance condition for the generation of configurations according to the normalized Boltzmann distribution

$$W_{eq}[s] \equiv W[s] = \frac{\exp(-S_E[s])}{\sum_{\{s'\}} \exp(-S_E[s'])}, \tag{6.84}$$

by means of a MC calculation. If Eq. (6.84) is substituted into Eq. (6.81), we find

$$\sum_{\{s'\}} W_{eq}[s']\Omega(\{s'\} \to \{s\}) = \sum_{\{s'\}} \frac{\exp(-S_E[s'])}{\sum_{\{s''\}} \exp(-S_E[s''])}\Omega(\{s'\} \to \{s\}), \tag{6.85}$$

which should equal $W_{eq}[s]$. Similarly, if Eq. (6.84) is substituted into Eq. (6.83), then we find

$$\exp(-S_E[s])\Omega(\{s\} \to \{s'\}) = \exp(-S_E[s'])\Omega(\{s'\} \to \{s\}), \tag{6.86}$$

for the detailed balance condition. Substitution of Eq. (6.86) into Eq. (6.85) gives

$$\sum_{\{s'\}} W_{eq}[s']\Omega(\{s'\} \to \{s\}) = \sum_{\{s'\}} \frac{\exp(-S_E[s])}{\sum_{\{s''\}} \exp(-S_E[s''])}\Omega(\{s\} \to \{s'\}), \tag{6.87}$$

and by means of the normalization condition (6.71), one has

$$\sum_{\{s'\}} W_{\text{eq}}[s']\Omega(\{s'\} \to \{s\}) = \frac{\exp(-S_E[s])}{\sum_{\{s''\}} \exp(-S_E[s''])} = W_{\text{eq}}[s], \tag{6.88}$$

and hence it is sufficient to require (assuming ergodicity holds) that the transition probability density satisfies the detailed balance condition of Eq. (6.86), in order to establish Eq. (6.84) as the unique equilibrium of the Markov process. Let us assume that we have succeeded in generating a set of N_s such configurations, or elements of a Markov chain which has "thermalized", i.e. reached its unique equilibrium probability distribution. We can then compute

$$\langle \mathcal{O} \rangle_{\text{Markov}} \equiv \frac{1}{N_s} \sum_{i=1}^{N_s} \mathcal{O}[s(\tau_i)] \tag{6.89}$$

for an observable $\mathcal{O}[s(\tau_i)]$, which can be evaluated for all of the configurations in the Markov chain. Such an average over τ for N_s configurations can be thought of as an average over temporal "snapshots" of the relevant field variables. When we perform a MC "simulation", we are referring to the use of Eq. (6.89) to compute expectation values of observables according to Eq. (6.69), again with a statistical uncertainty of $\mathcal{O}(1/\sqrt{N_s})$, assuming uncorrelated samples. We shall return to the issue of autocorrelations in MC time momentarily.

While we have shown that the requirement of detailed balance is key to generating a MC simulation where the sampled configurations are eventually distributed according to the desired Boltzmann distribution, it is not sufficient by itself to uniquely determine the transition probability. This freedom of choice can be exploited to construct different classes of algorithms, which are intended to be optimal for different physical systems. As we are mainly interested in applications to NLEFT, we shall focus on the development of the HMC algorithm, which is well suited to global updates in theories with dynamical fermions, to which the important case of Lattice QCD also belongs.

6.2.4 Metropolis Algorithm

One of the simplest and most straightforward ways to generate ensembles of configurations using importance sampling is the Metropolis algorithm [23]. We shall use the term "update" to indicate trial changes that take us from step n to step $n + 1$ in a Markov chain. In the Metropolis algorithm, such trial changes are accepted with a conditional probability, designed to achieve detailed balance. We shall first state the algorithm, after which we show that it generates ensembles of configurations distributed according to the Boltzmann distribution. If the initial field configuration is denoted $\{s\}$, updates are proposed with a probability $\Omega_0(\{s\} \to \{s'\})$, which satisfies

$$\Omega_0(\{s\} \to \{s'\}) = \Omega_0(\{s'\} \to \{s\}), \tag{6.90}$$

which is known as "micro-reversibility". We may randomly select a degree of freedom s_k (such as a specific lattice site, spin and isospin), or alternatively "sweep" the lattice in a systematical way, by updating each and every site in turn. We then have

$$s'_k = s_k + \Delta s_k, \tag{6.91}$$

which constitutes a "local" update, as only a single degree of freedom is modified. As we shall see, global lattice updates are also possible, but not straightforward.

In the Metropolis algorithm, proposed updates are conditionally accepted, in a way which depends on the action $S_E[s]$ of the initial configuration, and the action $S_E[s']$ of the proposed configuration. Specifically, if the proposed update decreases the action, then

$$\exp(-S_E[s']) > \exp(-S_E[s]), \tag{6.92}$$

in which case the proposed configuration is always accepted, and becomes the initial configuration for the next step in the Markov process. If the action has increased instead, the proposed configuration may still be accepted with a certain probability, which depends on how much the action increased. We generate a random number $\xi \in [0, 1]$, and accept the proposed configuration only if

$$\xi \leq \frac{\exp(-S_E[s'])}{\exp(-S_E[s])}, \tag{6.93}$$

which is referred to as the "Metropolis test". If the test is not passed, the proposed configuration is rejected, after which a new update is proposed with the initial configuration retained as the starting point. As the classical trajectory of the system corresponds to the equilibrium of minimum action, such conditional acceptance of configurations with increased action builds in the expected quantum fluctuations. In practice, the acceptance rate (or likelihood that a proposed update is accepted) is determined by the magnitude of Δs_k in Eq. (6.91). The optimal value of the acceptance rate is algorithm-dependent and given by computational considerations, and is usually taken to be $\sim70\%$. We shall return to this question in our treatment of algorithms for global lattice updates.

Let us now show that the Metropolis algorithm satisfies the detailed balance condition (6.86). If we denote the amplitude for the transition $\{s\} \to \{s'\}$ by $\Omega(\{s\} \to \{s'\})$, we note that this is given by the product of the probability $\Omega_0(\{s\} \to \{s'\})$ of suggesting $\{s'\}$ as the new configuration, and the probability of accepting it. Thus, the Metropolis algorithm is equivalent to

$$\Omega(\{s\} \to \{s'\}) = \Omega_0(\{s\} \to \{s'\}), \tag{6.94}$$

$$\Omega(\{s'\} \to \{s\}) = \Omega_0(\{s'\} \to \{s\}) \frac{\exp(-S_E[s])}{\exp(-S_E[s'])}, \tag{6.95}$$

for $\exp(-S_E[s']) > \exp(-S_E[s])$, and

$$\Omega(\{s\} \to \{s'\}) = \Omega_0(\{s\} \to \{s'\}) \frac{\exp(-S_E[s'])}{\exp(-S_E[s])}, \tag{6.96}$$

$$\Omega(\{s'\} \to \{s\}) = \Omega_0(\{s'\} \to \{s\}), \tag{6.97}$$

for $\exp(-S_E[s']) < \exp(-S_E[s])$. As long as we propose updates according to the micro-reversibility condition (6.90), the detailed balance condition (6.86) is also satisfied. After a sufficient number of thermalization steps (assuming ergodicity), the dependence of the Markov chain on the starting configuration is lost, and the ensemble generated by the Metropolis algorithm would be distributed according to the Boltzmann distribution.

The Metropolis algorithm is attractive because of its simplicity and applicability to a large number of problems. However, the highly local and random method of proposing updates according to Eq. (6.91) is also potentially a major stumbling block, especially if the action is highly non-local, as is the case for theories with dynamical fermions. In this context, "non-local" means that any lattice update requires a new computation of the fermion determinant. This is often problematic, as the decorrelation time of a given observable is measured in numbers of full "sweeps" of the lattice. If we denote by $\sigma_0^2 \equiv \langle \mathcal{O}^2 \rangle - \langle \mathcal{O} \rangle^2$ the sample variance assuming statistically uncorrelated configurations, then we can compute the autocorrelation coefficients $C(\tilde{\tau}) \equiv \langle x_i x_j \rangle$, with $x_i \equiv \mathcal{O}[s_i] - \langle \mathcal{O} \rangle / \sigma_0$. Here, the time lag $\tilde{\tau} \equiv \epsilon(i - j)$ refers to configurations at MC times τ_i and τ_j, with equally spaced intervals ϵ between successive configurations. If successive configurations are uncorrelated, we have $C(\tilde{\tau}) = 0$ when $\tilde{\tau} \neq 0$. When autocorrelations exist, they typically fall off exponentially with increasing $\tilde{\tau}$, with the autocorrelation time defined by the rate of exponential decay. In order to achieve good computational scaling with lattice size, it would be preferable to perform global (rather than local) updates, where all V sites are updated at once. Here, V equals the space-time volume, or total number of lattice sites of the system.

In general, global (random) updates of the field variables are incompatible with the Metropolis accept/reject step, for any substantially sized system. This is because random changes in a substantial number of usually leads to a large and unpredictable change in the action. Hence, the likelihood of the proposed configuration being accepted would be very small. As already noted, the restriction to local updates introduces an additional factor of $\sim V$ to the computational scaling for theories of dynamical fermions (such as NLEFT or Lattice QCD). In this sense, Euclidean projection calculations for fixed nucleon number A have a significant advantage over grand canonical calculations, as the dimensionality of the fermion determinant in the former is given by A, rather than the space-time lattice volume V as in the latter. Still, in the thermodynamic limit, the efficient updating of the fermion determinant may again become an issue.

6.2.5   Langevin Algorithm

We shall now turn to the construction of algorithms, which allow for global updates of the lattice field variables. First, we consider the stochastic Langevin and deterministic Molecular Dynamics (MD) algorithms. Both of these provide a solution to the problem of global updates, at the cost of introducing systematical errors into the MC calculation. Second, we consider the HMC algorithm, which combines elements of the Langevin, MD, and Metropolis algorithms. We shall find that HMC provides efficient global updates for non-local lattice actions, is free of systematical errors, and maintains a high acceptance rate.

In global algorithms, the auxiliary field variables s_k are continuously evolved using a suitably chosen set of differential equations. Such global updates can be viewed as "trajectories" in a time variable τ, which describes the evolution of the field variables in MC time. For practical purposes, the evolution in τ is often discretized, such that each consecutive step τ_n can be thought of as labeling the elements of a Markov chain. In the Langevin algorithm [24, 25], the configuration $\{s(\tau_{n+1})\}$ at MC time $\tau_{n+1} = (n+1)\epsilon_L$ is obtained from the configuration $\{s(\tau_n)\}$ at $\tau_n = n\epsilon_L$ by means of the difference equation

$$s_k(\tau_{n+1}) = s_k(\tau_n) + \epsilon_L \left(-\frac{\partial S_E[s]}{\partial s_k(\tau_n)} + \eta_k(\tau_n) \right), \qquad (6.98)$$

where ϵ_L is the Langevin time step. As in NLEFT, $S_E[s]$ is here assumed to be a functional of a set of continuous (rather than discrete) auxiliary fields. The components of the field variables evolve stochastically in MC time, as the η_k are independent random variables drawn from the Gaussian distribution

$$W_g[\eta(\tau_n)] \equiv \left(\prod_k \sqrt{\frac{\epsilon_L}{4\pi}} \right) \exp\left(-\sum_k \frac{\epsilon_L}{4} \eta_k(\tau_n)^2 \right), \qquad (6.99)$$

and we note that in the limit $\epsilon_L \to 0$, Eq. (6.98) reduces to

$$\frac{ds_k}{d\tau} = -\frac{\partial S_E[s]}{\partial s_k} + \eta_k(\tau), \qquad (6.100)$$

which is known as the Langevin equation. The Langevin algorithm is relatively simple to implement, as for instance Eq. (6.98) can be solved with standard methods for integrating differential equations, such as Runge–Kutta iteration with time step ϵ_L. As the Metropolis test is omitted, global updates pose no particular problem, except possibly in the calculation of the functional derivative $\partial S_E[s]/\partial s_k$. For instance, assuming that the action depends on a single auxiliary field component, so that

$$S_E[s] = \frac{1}{2}\sum_k s_k^2 - \log(\det \mathcal{M}(s,t)), \qquad (6.101)$$

then

$$
\begin{aligned}
\frac{\partial S_E[s]}{\partial s_k} &= s_k - \frac{1}{\det \mathcal{M}(s,t)} \sum_{i,j} \frac{\partial \det(\mathcal{M}(s,t))}{\partial \mathcal{M}_{ij}} \frac{\partial \mathcal{M}_{ij}(s,t)}{\partial s_k}, \\
&= s_k - \sum_{i,j} \left[M^{-1}(s,t) \right]_{j,i} \frac{\partial \mathcal{M}_{ij}(s,t)}{\partial s_k}, \qquad (6.102)
\end{aligned}
$$

where the second term (the "fermion force term") is of particular interest for the development of efficient algorithms, since its numerical evaluation in many cases determines the computational scaling for theories with dynamical fermions. More details on the fermion force term in NLEFT are given in Appendix E.

The lack of a Metropolis accept/reject step in the Langevin algorithm suggests that we need to carefully consider under which conditions the method satisfies detailed balance. In order to investigate this point, let us perform the change of variables

$$\eta_k(\tau) \equiv \sqrt{2\epsilon_L}\,\tilde{\eta}_k(\tau_n), \qquad (6.103)$$

which gives

$$s_k(\tau_{n+1}) = s_k(\tau_n) - \epsilon_L \frac{\partial S_E[s]}{\partial s_k(\tau_n)} + \sqrt{2\epsilon_L}\,\tilde{\eta}_k(\tau_n), \qquad (6.104)$$

such that the Gaussian probability distribution is now

$$W_g(\{\tilde{\eta}(\tau_n)\}) = \left(\prod_k \frac{1}{\sqrt{2\pi}} \right) \exp\left(-\sum_k \frac{\tilde{\eta}_k(\tau_n)^2}{2} \right), \qquad (6.105)$$

where the orthonormality condition

$$\langle \tilde{\eta}_k(\tau_n)\tilde{\eta}_{k'}(\tau_{n'}) \rangle = \delta_{kk'}\delta_{nn'}, \qquad (6.106)$$

is satisfied. In order to determine the transition probability from configuration $\{s\}$ at MC time τ_n to $\{s'\}$ at MC time τ_{n+1} by means of a Langevin update, we note from Eq. (6.104) that this will happen when a random vector is generated such that

$$\tilde{\eta}_k \equiv \tilde{\eta}_k[s', s] = \frac{s'_k - s_k}{\sqrt{2\epsilon_L}} + \frac{1}{2}\sqrt{2\epsilon_L}\frac{\partial S_E[s]}{\partial s_k}, \qquad (6.107)$$

and hence the transition probability is

$$\Omega(\{s\} \to \{s'\}) = N_0 \exp\left(-\sum_k \frac{\tilde{\eta}_k[s', s]^2}{2}\right), \qquad (6.108)$$

with N_0 a normalization constant. In order to obtain an expression which resembles the detailed balance condition of Eq. (6.86), we need to consider the limit $\epsilon_L \to 0$, where the Langevin step size is small. Since $s'_k - s_k \sim \mathcal{O}(\sqrt{\epsilon_L})$ for the Langevin equation, then

$$\lim_{\epsilon_L \to 0} \frac{\Omega(\{s\} \to \{s'\})}{\Omega(\{s'\} \to \{s\})} \simeq \exp\left[-\sum_i (s'_i - s_i)\frac{\partial S_E[s]}{\partial s_i}\right] \to \frac{\exp(-S_E[s'])}{\exp(-S_E[s])}, \qquad (6.109)$$

which establishes that the Langevin transition amplitude satisfies detailed balance only in the limit $\epsilon_L \to 0$. As practical calculations are performed at finite ϵ_L, systematical errors are introduced that need to be controlled.

The Langevin algorithm solves the problem of performing global updates without the need to maintain a large Metropolis acceptance rate, and is therefore an attractive choice for systems with non-local actions, such as NLEFT or Lattice QCD with dynamical fermions. However, the systematical error at finite step size ϵ_L is troublesome, as it necessitates calculations for multiple values of ϵ_L, followed by an extrapolation to $\epsilon_L \to 0$ (in addition to any other necessary extrapolations, such as the continuum limit or the limit infinite Euclidean projection time). At the same time, it would be desirable to maximize ϵ_L in order to maximize the decorrelation between successive field configurations, and hence make efficient use of available CPU time. These limitations have led to a search for alternative methods which evolve the field variables more effectively, and with less systematical error. Still, it should be noted that the complex Langevin method [26, 27] has been put forward as a possible solution to the sign problem in QCD with non-zero baryon chemical potential $\mu > 0$, or with non-zero vacuum angle θ. For a recent review of this active field of study, see Ref. [28].

6.2.6 Molecular Dynamics Algorithm

The Molecular Dynamics (MD) algorithm is based on the idea that $\mathscr{Z}$ of Eq. (6.56) can be expressed as the partition function of a classical Hamiltonian system, by introducing a set of canonically conjugate momenta [29–32]. The resulting MD Hamiltonian determines the evolution of the system in MC time τ. We recall that τ describes the evolution of the system in the sense of a Markov chain, and is not to be confused with the Euclidean projection time t. In contrast to the stochastic Làngevin method, the MD updates are deterministic. While the MD algorithm was originally put forward as an alternative formulation of lattice gauge theory in terms of a "microcanonical" Hamiltonian formalism [29, 31], we focus instead on the development of the MD update as a proposed global update, to be used in the Metropolis algorithm instead of a local, random update.

Let us introduce a set of momentum degrees of freedom p_k, which are canonically conjugate to the auxiliary field components s_k. If we assume that the observables $\mathscr{O}$ are independent of the p_k, we have in the grand canonical formalism

$$\langle \mathscr{O} \rangle = \frac{1}{\mathscr{Z}} \int \mathscr{D}s \mathscr{D}p \; \mathscr{O}[s] \exp(-H[s, p, \beta]), \tag{6.110}$$

and

$$\mathscr{Z} = \int \mathscr{D}s \mathscr{D}p \; \exp(-H[s, p, \beta]), \tag{6.111}$$

where the MD Hamiltonian

$$H[s, p, \beta] \equiv \sum_k \frac{p_k^2}{2} + S_E[s, \beta], \tag{6.112}$$

determines the evolution in τ. Since the original problem has been reformulated in terms of classical statistical mechanics, we may now use known results from there, in order to construct an evolution equation in MC time τ. In particular, we may rewrite Eq. (6.110) as a canonical ensemble average

$$\langle \mathscr{O} \rangle_{\text{can}} = \frac{1}{\mathscr{Z}_{\text{can}}} \int \mathscr{D}s \mathscr{D}p \; \mathscr{O}[s] \int dE \, \delta(H[s, p, \beta] - E) \exp(-E), \tag{6.113}$$

with

$$\mathscr{Z}_{\text{can}} = \int \mathscr{D}s \mathscr{D}p \int dE \, \delta(H[s, p, \beta] - E) \exp(-E), \tag{6.114}$$

where $\langle \mathcal{O} \rangle_{\text{can}}$ is a function of β. Similarly, we may define the microcanonical ensemble average

$$\langle \mathcal{O} \rangle_{\text{mic}} = \frac{1}{\mathscr{Z}_{\text{mic}}} \int \mathscr{D}s \mathscr{D}p \, \mathcal{O}[s] \int dE \, \delta(H[s, p, \beta] - E), \tag{6.115}$$

with

$$\mathscr{Z}_{\text{mic}} = \int \mathscr{D}s \mathscr{D}p \int dE \, \delta(H[s, p, \beta] - E), \tag{6.116}$$

which is evaluated for fixed β on the "energy surface" $H = E$.

The canonical and microcanonical ensembles can be related to each other, if we introduce the entropy

$$h(E, \beta) \equiv \log \mathscr{Z}_{\text{mic}}(E, \beta), \tag{6.117}$$

in terms of which

$$\langle \mathcal{O} \rangle_{\text{can}}(\beta) = \frac{\int dE \, \langle \mathcal{O} \rangle_{\text{mic}}(E, \beta) \exp(-E + h(E, \beta))}{\int dE \, \exp(-E + h(E, \beta))}, \tag{6.118}$$

where it is useful to note that in the thermodynamic limit, the exponential factors in Eq. (6.118) become strongly peaked around an energy $\hat{E}$ (which depends on β), such that

$$\langle \mathcal{O} \rangle_{\text{can}}(\beta) \simeq \langle \mathcal{O} \rangle_{\text{mic}}(E = \hat{E}(\beta), \beta), \tag{6.119}$$

when the number of lattice degrees of freedom becomes very large. In a practical MD simulation, the relationship between $\hat{E}$ and β can be determined, for instance, using the equipartition theorem [29]. The approximate correspondence (6.119), which becomes exact in the thermodynamic limit, provides the justification for using the Hamiltonian equations of motion

$$\dot{s}_k = \frac{\partial H[s, p]}{\partial p_k}, \qquad \dot{p}_k = -\frac{\partial H[s, p]}{\partial s_k}, \tag{6.120}$$

to generate ensembles of "phase space" configurations (of s and p) with constant energy $\hat{E}$. Here, the dots indicate derivatives with respect to τ, and the dependence of H and S_E on β will henceforth be suppressed. The MD equations of motion (6.120) define a "trajectory" in phase space, and we note that since S_E in Eq. (6.112) is independent of p, the Gaussian integral over p can easily be performed, giving a result proportional to the original quantum partition function. Hence, provided that the evolution in MC time generated by the Hamiltonian (6.112) is ergodic, we may compute

$$\langle \mathcal{O} \rangle_{\mathrm{mic}} (E = \hat{E}(\beta), \beta) \simeq \lim_{\bar{\tau} \to \infty} \frac{1}{\bar{\tau}} \int_{\tau_0}^{\tau_0 + \bar{\tau}} d\tau \, \mathcal{O}[s(\tau)], \tag{6.121}$$

by averaging over MD "trajectories" of length $\bar{\tau}$, where τ_0 is the thermalization time of the system, which the MC simulation should be allowed to reach before sampling is carried out. When the Hamiltonian equations of motion are integrated along a trajectory $[s(\tau_0), p(\tau_0)] \to [s(\tau), p(\tau)]$, the probability of encountering a configuration $\{s\}$ is given by $\exp(-S_E[s])$. Since

$$\dot{s}_k(\tau) = p_k(\tau), \tag{6.122}$$

the trajectories in Eq. (6.121) are solutions to

$$\dot{p}_k(\tau) = \ddot{s}_k(\tau) = -\frac{\partial S_E[s]}{\partial s_k}, \tag{6.123}$$

with the constraint $E = \hat{E}$. It should be noted that while the energy remains fixed, S_E is allowed to fluctuate during the MD trajectory, and hence quantum fluctuations are accounted for in the MC simulation. In practice, for sufficiently large lattices, the microcanonical ensemble is found to be a good approximation to the canonical one, provided that β is tuned appropriately [29].

It should be noted that the MD equation of motion (6.123) is a deterministic second order differential equation, while the Langevin equation (6.100) is stochastic and has a first order time derivative. If we apply the symmetric finite difference

$$\ddot{s}_k(\tau_n) \simeq \frac{s_k(\tau_{n+1}) + s_k(\tau_{n-1}) - 2s_k(\tau_n)}{\epsilon_{\mathrm{MD}}^2}, \tag{6.124}$$

to Eq. (6.123), we find

$$s_k(\tau_{n+1}) \simeq s_k(\tau_n) - \frac{\epsilon_{\mathrm{MD}}^2}{2} \frac{\partial S_E[s(\tau_n)]}{\partial s_k} + \epsilon_{\mathrm{MD}} \, p_k(\tau_n), \tag{6.125}$$

and

$$p_k(\tau_n) \simeq \frac{s_k(\tau_{n+1}) - s_k(\tau_{n-1})}{2\epsilon_{\mathrm{MD}}}, \tag{6.126}$$

where $\epsilon_{\mathrm{MD}} \equiv \tau_{n+1} - \tau_n$ is the MD (or microcanonical) time step. In fact, we can obtain the discretized Langevin equation from Eq. (6.125) by drawing the momenta p_k from a random Gaussian distribution, and equating the Langevin and MD time steps according to $\epsilon_L = \epsilon_{\mathrm{MD}}^2 / 2$.

A significant advantage of MD over the Langevin algorithm is its faster decorrelation in MC time. Specifically, the evolved MC time scales as $\sim \mathcal{O}(n\epsilon_{\text{MD}})$ for MD updates, to be compared with $\sim \mathcal{O}(\sqrt{n}\epsilon_{\text{L}})$ for Langevin, due to the stochastic nature of the latter. Still, the MD algorithm has a number of potentially serious drawbacks. Firstly, the replacement of the canonical ensemble with the microcanonical one is only strictly valid for an infinitely large lattice (which may become a problem for practical NLEFT calculations on moderately sized lattices). Secondly, it is not immediately obvious how to build in thermalization and ergodicity into an MD algorithm. Hence, we shall next consider "hybrid" algorithms that seek to combine the attractive features of the Langevin, MD and Metropolis algorithms.

6.2.7 Hybrid Monte Carlo Algorithm

As we have seen in our treatments of the stochastic Langevin algorithm and the deterministic MD algorithm, both contain some desirable features, which we shall now seek to combine in a single formalism. Such methods are referred to as "hybrid" algorithms [33–35]. We have already noted that the MD equation of motion reduces to the Langevin equation, if we identify the respective MC time steps as $\epsilon_L = \epsilon_{\text{MD}}^2/2$, provided that the momentum components p_k are drawn from the random Gaussian ensemble

$$W_g[p] = \left(\prod_k \frac{1}{\sqrt{2\pi}} \right) \exp\left(-\sum_k \frac{p_k^2}{2} \right), \tag{6.127}$$

which also equals the distribution of the momenta in the MD Hamiltonian. Assuming that the evolution driven by the MD Hamiltonian is ergodic, the momenta would in fact eventually be distributed according to Eq. (6.127), while the auxiliary field components s_k would be distributed according to the Boltzmann distribution. In the hybrid algorithm, the integration of the Hamiltonian equations of motion along a trajectory with energy $\hat{E}$ is periodically interrupted, and a new set of momenta p_k is drawn from the distribution (6.127). The MD integration then resumes, along a trajectory with a different value of $\hat{E}$. This periodic "refreshing" of the momenta equals performing a Langevin step. Such a "Hybrid Molecular Dynamics" (HMD) update can be thought of as a method, which interpolates between pure Langevin and MD updates. On the one hand, if the canonical momenta are refreshed after each step in the integration of the MD equations of motion, the HMD methods reduces to Langevin. On the other hand, if the momenta are refreshed very infrequently, HMD becomes equivalent to a pure MD update.

Hybrid algorithms still suffer from systematical errors due to the finite MD step size ϵ_{MD}, since the energy $\hat{E}$ is only conserved exactly in the limit $\epsilon_{\text{MD}} \to 0$. However, it turns out to be relatively straightforward to combine the hybrid algorithm with a Metropolis accept/reject decision, based on the size of the energy violation δH. This HMC algorithm [36] uses HMD trajectories as the proposed (global) updates in the Metropolis algorithm, which are then accepted or rejected based on the amount of energy violation incurred due to the integration of the Hamiltonian equations of motion with finite accuracy in ϵ_{MD}. The great advantage of HMC is that detailed balance is satisfied at finite ϵ_{MD}, which eliminates the need for a costly and time-consuming extrapolation. Naturally, one may ask whether the reintroduction of the Metropolis accept/reject decision risks putting us back in the original situation, where the low acceptance rate prohibited large (random) updates. For instance, even though detailed balance is satisfied without extrapolation in ϵ_{MD}, there is a priori no guarantee that a high acceptance rate can be maintained. For the HMC algorithm to be a viable option for practical calculations, the dependence on the acceptance probability on ϵ_{MD} should be understood. We shall return to this question in Sect. 6.2.9.

Let us now consider how the HMC algorithm satisfies detailed balance. Firstly, we take $\{s, p\}$ as the initial configuration. Here $\{s\}$ is arbitrarily chosen, and $\{p\}$ is drawn from Eq. (6.127) with zero mean and unit variance. Secondly, the starting configuration $\{s, p\}$ is evolved through a number of MD integration steps to give the updated configuration $\{s', p'\}$. In principle, since the MD equation of motion involves a second time derivative, we should be able to reverse the motion by simply changing the sign of all the momentum components in the initial and final configurations. Hence

$$\Omega_{MD}(\{s, p\} \to \{s', p'\}) = \Omega_{MD}(\{s', -p'\} \to \{s, -p\}), \tag{6.128}$$

which states that the probability of the update $\{s, p\} \to \{s', p'\}$ equals that of $\{s', -p'\} \to \{s, -p\}$. We shall momentarily return to this important point, as not every numerical finite difference formula with MD time step ϵ_{MD} satisfies this property of reversibility. Lastly, the probability of accepting the proposed update is

$$\Omega_A(\{s, p\} \to \{s', p'\}) = \min\big(1, \exp(-H[s', p'])/\exp(-H[s, p])\big), \tag{6.129}$$

according to the Metropolis algorithm, applied to the HMC Hamiltonian. While the MD evolution formally conserves energy, the numerical finite difference method introduces a truncation error, which translates into a violation

$$\delta H \equiv H[s', p'] - H[s, p], \tag{6.130}$$

of the energy conservation during an MD trajectory.

We may now express the HMC amplitude for the transition $\{s\} \to \{s'\}$ as

$$\Omega(\{s\} \to \{s'\}) = \int \mathscr{D}p\,\mathscr{D}p'\, W_g[p]$$

$$\times \Omega_{MD}(\{s, p\} \to \{s', p'\})\, \Omega_A(\{s, p\} \to \{s', p'\}), \tag{6.131}$$

in terms of the product of the amplitudes for generating the initial momentum vector p, performing the MD evolution $\{s, p\} \to \{s', p'\}$, and accepting the updated configuration. If we multiply Eq. (6.131) with $\exp(-S_E[s])$, we find

$$\exp(-S_E[s])\,\Omega(\{s\} \to \{s'\}) = \int \mathscr{D}p\mathscr{D}p'\,\exp(-H[s, p])$$
$$\times\, \Omega_{\mathrm{MD}}(\{s, p\} \to \{s', p'\})\,\Omega_A(\{s, p\} \to \{s', p'\}),$$
$$(6.132)$$

and

$$\exp(-H[s, p])\,\Omega_A(\{s, p\} \to \{s', p'\})$$
$$= \exp(-H[s', p'])\,\Omega_A(\{s', p'\} \to \{s, p\})$$
$$= \exp(-H[s', -p'])\,\Omega_A(\{s', -p'\} \to \{s, -p\}), \qquad (6.133)$$

where we have used the fact that the Metropolis accept-reject step satisfies detailed balance, and the property $H[s, p] = H[s, -p]$ of the HMC Hamiltonian. If we substitute Eq. (6.133) into Eq. (6.132), and assume that the reversibility condition (6.128) holds, we have

$$\exp(-S_E[s])\,\Omega(\{s\} \to \{s'\}) = \exp(-S_E[s'])$$
$$\times \int \mathscr{D}p\mathscr{D}p'\, W_g[p']\,\Omega_{\mathrm{MD}}(\{s', -p'\}$$
$$\to \{s, -p\})\,\Omega_A(\{s', -p'\} \to \{s, -p\})$$
$$= \exp(-S_E[s'])\,\Omega(\{s'\} \to \{s\}), \qquad (6.134)$$

where the last equality follows from the invariance of the integration measure when the signs of all the momenta are changed. This shows that the HMC algorithm indeed satisfies the detailed balance condition and produces Markov chains of field configurations $\{s\}$ distributed according to the Boltzmann factor $\exp(-S_E[s])$, provided that the reversibility of the MD update is not violated.

6.2.8 Integration of the HMC Equations of Motion

We shall now turn to the problem of finding a suitable integration scheme for the MD equations of motion. The "leap-frog" integration rule is a finite-difference approximation to the MD equations of motion, which satisfies the reversibility condition of Eq. (6.128). We shall now discretize the MD equations of motion in

MC time τ using a time step $\epsilon = \epsilon_{MD}$, by means of the Taylor series expansions

$$s_k(\tau + \epsilon) \simeq s_k(\tau) + \epsilon \, \dot{s}_k(\tau) + \frac{\epsilon^2}{2} \ddot{s}_k(\tau) + \mathcal{O}(\epsilon^3), \tag{6.135}$$

$$p_k(\tau + \epsilon) \simeq p_k(\tau) + \epsilon \, \dot{p}_k(\tau) + \frac{\epsilon^2}{2} \ddot{p}_k(\tau) + \mathcal{O}(\epsilon^3), \tag{6.136}$$

for $s_k(\tau + \epsilon)$ and $p_k(\tau + \epsilon)$. From the Hamiltonian equations of motion, we find

$$\dot{s}_k(\tau) = p_k(\tau), \qquad \ddot{s}_k(\tau) = \dot{p}_k(\tau) = -\frac{\partial S_E[s(\tau)]}{\partial s_k}, \tag{6.137}$$

and one may further construct

$$\ddot{p}_k(\tau) \simeq \frac{1}{\epsilon} \left(\frac{\partial S_E[s(\tau)]}{\partial s_k} - \frac{\partial S_E[s(\tau + \epsilon)]}{\partial s_k} \right) + \mathcal{O}(\epsilon), \tag{6.138}$$

such that

$$s_k(\tau + \epsilon) \simeq s_k(\tau) + \epsilon \left(p_k(\tau) - \frac{\epsilon}{2} \frac{\partial S_E[s(\tau)]}{\partial s_k} \right) + \mathcal{O}(\epsilon^3), \tag{6.139}$$

for the Taylor expansion of s_k, and

$$\left(p_k(\tau + \epsilon) - \frac{\epsilon}{2} \frac{\partial S_E[s(\tau + \epsilon)]}{\partial s_k} \right) \simeq \left(p_k(\tau) - \frac{\epsilon}{2} \frac{\partial S_E[\tau]}{\partial s_k} \right)$$
$$- \epsilon \frac{\partial S_E[s(\tau + \epsilon)]}{\partial s_k} + \mathcal{O}(\epsilon^3), \tag{6.140}$$

for the Taylor expansion of p_k. We may summarize these results as

$$s_k(\tau + \epsilon) \simeq s_k(\tau) + \epsilon p_k(\tau + \epsilon/2), \tag{6.141}$$

$$p_k(\tau + 3\epsilon/2) \simeq p_k(\tau + \epsilon/2) - \epsilon \frac{\partial S_E[s(\tau + \epsilon)]}{\partial s_k}, \tag{6.142}$$

and we note that iteration of these field and momentum updates amounts to integrating the Hamilton equations of motion in MC time τ.

The term "leap-frog" refers to the fact that the coordinates s_k are evaluated at time steps $\tau_n \equiv n\epsilon$, while the momenta $p_k = \dot{s}_k$ are evaluated at the midpoints between consecutive MD steps. These "half-steps" are given by

$$p_k(\tau + \epsilon/2) \simeq p_k(\tau) - \frac{\epsilon}{2} \frac{\partial S_E[s(\tau)]}{\partial s_k} + \mathcal{O}(\epsilon^2), \tag{6.143}$$

where the Taylor expansion of p_k has been truncated at $\mathcal{O}(\epsilon^2)$. During the iterative application of the leap-frog rule, the error increases as $\mathcal{O}(\epsilon^3)$ during each step. Assuming that the number of MD steps per trajectory N_{MD} scales as $\mathcal{O}(\epsilon^{-1})$, the change in the total energy during a given MD trajectory is of $\mathcal{O}(\epsilon^2)$. In practical terms, the length of an MD trajectory is taken to be $\bar{\tau} = N_{\mathrm{MD}}\epsilon \lesssim 1$. As we shall find in Sect. 6.2.9, the leap-frog step ϵ should be adjusted such that the average Metropolis acceptance rate is $\simeq 60\%$. Also, the special case of $N_{\mathrm{MD}} = 1$, where the p_k are refreshed after every step, equals a pure Langevin update.

We may now summarize the implementation of HMC with leap-frog integration:

1. Select an initial field configuration $\{s\}$, for instance one generated by a previous step in a MC calculation, or a randomly chosen starting configuration for the Markov chain.
2. Randomly select the momentum configuration $\{p\}$ from the Gaussian ensemble (6.127).
3. Generate an updated trial configuration $\{s', p'\}$ from the initial configuration $\{s, p\}$ by iterating

$$s_k(n+1) = s_k(n) + \epsilon \tilde{p}_k(n), \tag{6.144}$$

and

$$\tilde{p}_k(n+1) = \tilde{p}_k(n) - \epsilon \frac{\partial S_E[s(n+1)]}{\partial s_k}, \tag{6.145}$$

for N_{MD} steps. Here, the $s_k(n)$ are field variables at MD time steps τ_n, and the $\tilde{p}_k(n)$ are momentum variables at intermediate time steps $\tau_{n+1/2} \equiv (n+1/2)\epsilon$. These are obtained from the $p_k(n)$ using

$$\tilde{p}_k(n) = p_k(n) - \frac{\epsilon}{2} \frac{\partial S_E[s(n)]}{\partial s_k}. \tag{6.146}$$

4. Accept or reject the trial configuration $\{s', p'\}$ according to the Metropolis criterion of Eq. (6.129).
5. If the Metropolis test was passed, $\{s'\}$ becomes the new initial field configuration $\{s\}$ in the next HMC update (using a different set of random Gaussian momenta). In the Metropolis test was not passed, the configuration $\{s'\}$ is rejected and the original $\{s\}$ is retained as the initial configuration in the next HMC update, again using a new set of Gaussian momenta.

6.2.9 Acceptance Rate and Scaling of HMC

Our treatment of the HMC algorithm has included the reversible leap-frog integration rule, which is a standard choice for the MD equations of motion. As long as the MD integration is reversible to a high degree, the Metropolis step eliminates the need for extrapolation in the HMD step ϵ. Nevertheless, if ϵ is too large, the acceptance rate may still be adversely affected. It is therefore of some interest to ask whether more accurate integration rules could be derived in a systematical manner, and to study the dependence of the acceptance rate on ϵ, as well as on the type of integrator used.

The leap-frog integration rule is a special case of so-called "symplectic" (or area-preserving) integrators. Symplectic integrators conserve the differential volume element $\mathscr{D}s\,\mathscr{D}p$ during the MD trajectory. A convenient way to describe symplectic integrators is by means of the Lie algebra formalism of [37–39], where Hamilton's equations can be expressed as

$$\dot{f} = \{f, H\}, \tag{6.147}$$

where $f = s$ or p, H is the MD Hamiltonian, and

$$\{f, H\} \equiv \sum_k \left(\frac{\partial f}{\partial s_k} \frac{\partial H}{\partial p_k} - \frac{\partial f}{\partial p_k} \frac{\partial H}{\partial s_k} \right), \tag{6.148}$$

is the Poisson bracket. If we define the linear operator

$$L(H)f \equiv \{f, H\}, \tag{6.149}$$

then the formal solution of Hamilton's equations is given by

$$f(\tau + \epsilon) = \exp\left[\epsilon L(H)\right] f(\tau), \tag{6.150}$$

where the exponential is typically replaced by an approximation, which is accurate to a given order in ϵ. Then, we find

$$L(H) = L(\sum_k p_k^2/2) + L(S_E[s]) \equiv K + V, \tag{6.151}$$

given the MD Hamiltonian (6.112). We can then factorize the MD evolution operator using the Trotter decomposition of exponential operators [40], which gives

$$\exp\left[\epsilon(K + V)\right] = \exp(V\epsilon/2)\exp(K\epsilon)\exp(V\epsilon/2) + \mathcal{O}(\epsilon^3), \tag{6.152}$$

where the leap-frog integrator is

$$I_{2LF}(\epsilon) \equiv T_V(\epsilon/2)T_K(\epsilon)T_V(\epsilon/2) = \exp(V\epsilon/2)\exp(K\epsilon)\exp(V\epsilon/2), \qquad (6.153)$$

which we refer to as a second-order method, as the error term (of each consecutive MD step) is of $\mathcal{O}(\epsilon^3)$. Here, $T_K(\epsilon)$ effects the update of Eq. (6.144) and $T_V(\epsilon)$ the update of Eq. (6.145). In other words, T_K only updates the field variables and leaves the momentum variables unchanged, while T_V updates the momentum variables only. When this process is repeated, successive momentum half-steps combine into a full step, with the exception of the first and last steps. Clearly, an equivalent leap-frog scheme is obtained by interchanging T_K and T_V in Eq. (6.153). The 2LF integrator can also be expressed in terms of the mapping

$$\begin{pmatrix} s(\tau+\epsilon) \\ p(\tau+\epsilon) \end{pmatrix} = \begin{pmatrix} 1 & \epsilon/2 \\ 0 & 1 \end{pmatrix} \begin{pmatrix} 1 & 0 \\ -\epsilon\frac{\partial S_E[s]}{s\,\partial s} & 1 \end{pmatrix} \begin{pmatrix} 1 & \epsilon/2 \\ 0 & 1 \end{pmatrix} \begin{pmatrix} s(\tau) \\ p(\tau) \end{pmatrix}, \qquad (6.154)$$

$$\equiv I_{2LF}(\epsilon) \begin{pmatrix} s(\tau) \\ p(\tau) \end{pmatrix}, \qquad (6.155)$$

where each of the matrices representing the elementary substeps are triangular with unit determinant, and hence the mapping is symplectic [41]. Moreover,

$$I_{2LF}(\epsilon)I_{2LF}(-\epsilon) = \mathbb{1}, \qquad (6.156)$$

and hence the 2LF integrator is also exactly time-reversible.

Another class of symplectic integrators is referred to as "minimum norm" integrators, which seek to minimize the magnitude of the error term (which consists of commutators of K and V) at a given order. A prominent example is the Omelyan (or 2MN) integrator [42, 43]

$$I_{2MN}(\epsilon) \equiv \exp(\lambda\epsilon K)\exp(V\epsilon/2)\exp((1-2\lambda)\epsilon K)\exp(V\epsilon/2)\exp(\lambda\epsilon K), \qquad (6.157)$$

where $\lambda \simeq 0.193$ is the canonical value of the adjustable parameter λ. The Omelyan integrator is also of second order, and we note that the limits $\lambda = 0$ and $\lambda = 1/2$ correspond to the two different formulations of the leap-frog integrator. As each application of (6.157) requires two evaluations of the fermion force term (since V occurs twice), the Omelyan integrator would seem to have twice the computational cost. However, the norm of the leading error coefficient is expected to be smaller than that of the leap-frog method by roughly an order of magnitude [42], which allows for a significant increase in ϵ. The overall outcome is that Omelyan outperforms the leap-frog integrator, and has thus become the method of choice for modern Lattice QCD simulations. For an overview of higher-order leap-frog and minimum norm integrators, see Ref. [41].

Let us now consider the effects of using a symplectic integrator with HMC. The computational efficiency of the algorithm requires that a high acceptance rate can be maintained, without a strong dependence on the system size V (since physically interesting results are usually obtained at large V). In order to evaluate the HMC acceptance rate as a function of ϵ and V for symplectic integrators, we consider

$$\mathcal{L} = \int \mathcal{D}s' \mathcal{D}p' \, \exp(-H[s', p']) = \int \mathcal{D}s \mathcal{D}p \, \exp(-H[s, p] - \delta H), \quad (6.158)$$

where the energy violation δH is defined in Eq. (6.130). Hence, symplecticity implies that

$$\langle \exp(-\delta H) \rangle = 1, \quad (6.159)$$

which is known as Creutz's equality [44], which gives

$$\langle \delta H \rangle = \frac{1}{2} \langle (\delta H)^2 \rangle + \mathcal{O}(\delta H)^3, \quad (6.160)$$

when expanded to second order in δH. Moreover, we have

$$\langle \exp(-\delta H) \rangle \geq \exp(-\langle \delta H \rangle), \quad (6.161)$$

from Jensen's inequality for convex functions, which implies $\langle \delta H \rangle \geq 0$, with an equal sign for the case when energy is exactly conserved. The requirement that the average energy violation be positive provides a useful check on the correctness of HMC implementations.

From our earlier treatment of the leap-frog integrator, we concluded that the energy violation of a given MD trajectory behaves as $\sim \epsilon^2$. As we are integrating V independent variables, we expect the error from each degree of freedom to add in quadrature. Together with Eq. (6.160), this gives [44]

$$\langle \delta H \rangle \sim \langle (\delta H)^2 \rangle \propto V \epsilon^{2n}, \quad (6.162)$$

where n denotes the order of the symplectic integrator ($n = 2$ for the 2LF and 2MN integrators). Then, the root-mean-square energy violation is given by

$$\langle (\delta H)^2 \rangle^{1/2} \simeq A_n V^{1/2} \epsilon^n, \quad (6.163)$$

where the constant A_n depends on the specific Hamiltonian and integrator. Provided that the acceptance rate for $V \to \infty$ is controlled by the variance of δH (assuming that higher cumulants vanish), then the acceptance probability is given by [45]

$$\langle \Omega_A \rangle = \langle \min[1, \exp(-x)] \rangle$$

$$= \frac{1}{\sqrt{4\pi \langle \delta H \rangle}} \int_{-\infty}^{\infty} dx \, \min[1, \exp(-x)] \exp \left(-\frac{(x - \langle \delta H \rangle)^2}{4 \langle \delta H \rangle} \right)$$

$$= \text{erfc} \left(\langle (\delta H)^2 / 8 \rangle^{1/2} \right), \tag{6.164}$$

in terms of the complementary error function $\text{erfc}(x) \equiv 1 - \text{erf}(x)$, which can be simplified to

$$\langle \Omega_A \rangle \simeq \exp \left(-\frac{1}{\sqrt{2\pi}} \langle (\delta H)^2 \rangle^{1/2} \right), \tag{6.165}$$

when the acceptance rate is greater than $\simeq 20\%$ [46]. A measure of the efficiency of the HMC algorithm is the (inverse of the) computational effort expended per accepted trajectory. The best performance is obtained when the product $\langle \Omega_A \rangle \times \epsilon$ is maximized. Using Eqs. (6.163) and (6.165), one obtains [46]

$$\langle \Omega_A \rangle^{\text{opt}} = \exp(-1/n), \tag{6.166}$$

for the optimal acceptance rate, which depends only on the order of the integrator, and not on any details of H.

In order to maintain a constant acceptance rate for $V \rightarrow \infty$, Eq. (6.163) suggests that the leap-frog and Omelyan ($n = 2$) integration steps ϵ should scale as $\epsilon \sim V^{-1/4}$. This (relatively mild) scaling shows that the original problem of performing large global updates without destroying the acceptance rate has been solved. It should also be noted that there is no a priori requirement that the "evolution Hamiltonian" (which controls the HMD evolution) equal the "acceptance Hamiltonian" of the Metropolis accept/reject step. In practice, they are often taken to be identical, since a large difference between the evolution and acceptance Hamiltonians would again lead to a low acceptance probability. Still, one may attempt to optimize the HMC acceptance rate by slight adjustments of the parameters in the evolution Hamiltonian, while keeping those of the acceptance Hamiltonian at their physical values. Alternatively, one may introduce additional parameters into the former.

For HMC, the dominant contribution to the computational scaling is normally due to the inversion of the fermion operator when the fermion force term is calculated repeatedly during each MD evolution. As noted earlier, for PMC calculations this contribution is subleading, since the fermion matrix is merely of dimension $A \times A$. However, for grand canonical calculations, the computational cost of inverting the fermion matrix scales as $\sim V^3$ using direct solver techniques such as LU decomposition, where V equals the number of spatial lattice sites. When pseudofermions are applied to HMC (the "phi-algorithm"), the scaling is improved to roughly $\sim V^2$. With iterative solvers such as Conjugate Gradient (CG), the scaling is further reduced by a factor of V, given the (optimistic) assumption that CG will reliably converge within an acceptable number of iterations. Combined with the expected dependence of the acceptance rate on lattice size, the scaling of HMC (with pseudofermions) is often given as $\sim V^{5/4}$ [44].

The computational scaling of the PMC algorithm is somewhat different, and has been investigated empirically in Ref. [47] as a function of A, the spatial lattice volume $L^3 = (Na)^3$, and the (total) number of Euclidean time steps N_t. The multiplication of transfer matrix time slices for each of the single-particle states of the trial wave function scales as $\sim A \times N^3 \times N_t$, and the construction of the fermion matrix from inner products of single-particle states scales as $\sim A^2 \times N^3$. The calculation of the fermion determinant using LU decomposition scales as $\sim A^3$. As already noted for HMC, the largest amount of CPU time is spent computing various functional derivatives with respect to the auxiliary fields and their conjugate momenta, which are needed for the MD update. In Ref. [47], this was found to involve operations that scaled as $\sim N^3 \times N_t$, $\sim N^6 \times N_t$, and $\sim A \times N^3 \times N_t$. However, many of these operations can be performed much more efficiently by means of the Fast Fourier Transform (FFT) algorithm, which greatly improves on the computational scaling with N^3. The average sign also impacts the computational scaling of HMC as applied to NLEFT. For instance, the sign problem was found to be particularly mild when A is a multiple of four, an effect attributed to clustering of nucleons into α particles [47]. As will be discussed in Chap. 8, recent developments in the smearing of the LO terms in NLEFT has greatly lessened the sign problem for all nuclei.

6.3   Adiabatic Projection Method

So far, we have developed the ability to perform Euclidean projection calculations with arbitrary, multi-nucleon trial wave functions, and introduced MC methods for evaluating time projection amplitudes, which are applicable to large numbers of nucleons. Here, we shall discuss how these method can be extended into an *ab initio* framework for nuclear scattering and reactions [48, 49], called the "adiabatic projection method". The general strategy in this formalism is to separate the calculation into two main parts. In the first part, Euclidean time projection is used to determine a so-called "adiabatic Hamiltonian" for the participating nuclei. If we define a set of "cluster states" $|\mathbf{R}\rangle$ which are labeled by their separation vector $\mathbf{R}$, we may form "dressed" cluster states

$$|\mathbf{R}\rangle_t \equiv \exp(-Ht)|\mathbf{R}\rangle, \qquad (6.167)$$

by evolving them in Euclidean time t with the full microscopic Hamiltonian H (such as the NLEFT Hamiltonian at LO). The evolution in Euclidean time with

H can be thought of as dynamically "cooling" the initial trial cluster states to the physical state, such that deformations and polarizations of the interacting clusters are incorporated automatically. An "adiabatic Hamiltonian" matrix defined by the microscopic Hamiltonian is then computed, which is restricted to the subspace of dressed cluster states, obtained in the limit of large Euclidean time. Though restricted to a description of cluster configurations, such a construction remains *ab initio*. For inelastic processes, one may construct dressed cluster states for each of the possible scattering channels, and calculate matrix elements for all operators relevant to the reaction process. For instance, in the case of radiative capture reactions, one would calculate the adiabatic Hamiltonian and matrix elements of one-photon vertex operators. In the second part of the adiabatic projection method, the adiabatic Hamiltonian and operator matrix elements for dressed cluster states are used to calculate scattering amplitudes. For instance, elastic phase shifts can be obtained using Lüscher's method or the spherical wall method as discussed in Chap. 5. For inelastic reactions, additional calculational steps are required. Since the problem has been reduced to a few-body system of nuclear clusters, one possibility is to resort to Green's functions defined in Minkowski time. Such methods are demonstrated in Ref. [48] for the radiative neutron capture process $n + p \rightarrow d + \gamma$ using pionless EFT.

We shall introduce the adiabatic projection method by first considering the construction of dressed cluster states and the adiabatic Hamiltonian in Sect. 6.3.1. For this purpose, we consider the problem of elastic fermion-dimer scattering with two-component fermions, in the limit of attractive zero-range interactions. This equals the problem of nd scattering in the spin-quartet channel at LO in pionless EFT. In Sect. 6.3.3, we analyze the spectrum of the adiabatic Hamiltonian, and show that it matches the spectrum of the original microscopic Hamiltonian to arbitrary accuracy (as required for an *ab initio* method) below the inelastic threshold. We also show that the S-wave phase shift is reproduced in good agreement with the prediction of the Skorniakov-Ter-Martirosian (STM) integral equation in the continuum and infinite-volume limits [50]. Finally, in Sect. 6.3.4 we discuss how the adiabatic projection method can be extended to inelastic reactions and radiative capture processes.

6.3.1 Cluster States and Adiabatic Hamiltonian

As an introduction to the adiabatic projection method, we consider the problem of dimer-fermion scattering for two-component (spin up and spin-down) fermions [49], where the bound dimer is composed of a spin-up and a spin-down fermion. This allows us to introduce the central concepts of the adiabatic projection method without complications such as auxiliary fields and MC calculations. The interaction is taken to be attractive, and we consider the limit where the effective range of the interaction is negligible. It should be noted that at LO in pionless EFT, nd scattering in the spin-quartet channel is completely equivalent to our fermion-dimer system. For the *nd* problem, the two fermion components correspond to isospin, while the nucleon spins are fully symmetrized into a spin-quartet state. We take the lattice Hamiltonian to be

$$H = \frac{1}{2\tilde{m}_N} \sum_{\mathbf{n}} \sum_{i=\uparrow,\downarrow} \sum_{l=1}^{3} \left[2a_i^\dagger(\mathbf{n})a_i(\mathbf{n}) - a_i^\dagger(\mathbf{n})a_i(\mathbf{n}+\hat{e}_l) + a_i^\dagger(\mathbf{n})a_i(\mathbf{n}-\hat{e}_l) \right]$$
$$+ \tilde{C}_0 \sum_{\mathbf{n}} a_\downarrow^\dagger(\mathbf{n})a_\downarrow(\mathbf{n})a_\uparrow^\dagger(\mathbf{n})a_\uparrow(\mathbf{n}), \tag{6.168}$$

where $\mathbf{n}$ labels the lattice sites, $\tilde{C}_0$ is the (lattice-regularized) coupling in units of the (spatial) lattice spacing, and the $\hat{e}_l$ are unit lattice vectors. We note that (6.168) corresponds to the continuum Hamiltonian

$$H = -\frac{1}{2m_N} \sum_{i=\uparrow,\downarrow} \int d^3r\, a_i^\dagger(\mathbf{r})\nabla^2 a_i(\mathbf{r})$$
$$+ \int d^3r\, d^3r'\, a_\downarrow^\dagger(\mathbf{r})a_\downarrow(\mathbf{r})V(\mathbf{r}-\mathbf{r}')\, a_\uparrow^\dagger(\mathbf{r}')a_\uparrow(\mathbf{r}'), \tag{6.169}$$

where

$$V(\mathbf{r}-\mathbf{r}') \to C_0 \delta^{(3)}(\mathbf{r}-\mathbf{r}'), \tag{6.170}$$

in the zero-range limit. Motivated by the *nd* system, we take the fermions mass to equal the nucleon mass $m_N \simeq 939\,\text{MeV}$, and we tune the coupling constant $\tilde{C}_0$ to match the binding energy of the deuteron $E_B \simeq 2.2246\,\text{MeV}$. For such a shallow dimer, the fermion-dimer scattering problem is known to be strongly coupled even at low momenta [51, 52].

We shall first describe the initial cluster states to be used in the adiabatic projection. Here, we shall illustrate the method for a fermion-dimer system, which consists of two spin-up fermions and one spin-down fermion. As we are working in the center-of-mass frame, particle locations are measured relative to the position of the spin-down fermion. Hence, we shall assume that the spin-down fermion is "anchored" at $\mathbf{R}_0$, which can be taken as lattice position $\mathbf{R}_0 \equiv \mathbf{0} = (0, 0, 0)$. Except for the constraints imposed by the Pauli principle, the locations of the two spin-up fermions are arbitrary. We may take the initial cluster states to be

$$|\mathbf{R}\rangle \equiv a_\uparrow^\dagger(\mathbf{R})a_\uparrow^\dagger(\mathbf{0})a_\downarrow^\dagger(\mathbf{0})|0\rangle, \tag{6.171}$$

for any $\mathbf{R} \neq \mathbf{0}$, which is illustrated in Fig. 6.1. As for the other types of trial wave functions discussed so far, the trial cluster state for adiabatic projection is also constructed as a Slater determinant of single-nucleon states. Also, on a cubic lattice with periodic boundary conditions, $\mathbf{R}$ takes on $N^3 - 1$ possible values. The initial states $|\mathbf{R}\rangle$ are then evolved in Euclidean time t using the microscopic Hamiltonian

Fig. 6.1 Illustration of a
fermion-dimer cluster initial
state $|\mathbf{R}\rangle$, with fermion and
dimer separated by a
displacement vector $\mathbf{R}$.
Figure courtesy of Dean Lee

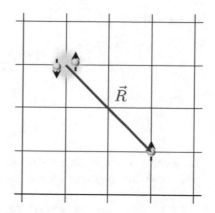

H, in order to produce the dressed cluster states $|\mathbf{R}\rangle_t$ of Eq. (6.167). Here, we have
chosen a particularly simple form for the initial cluster states in Eq. (6.171), in
order to develop the adiabatic projection method without complications of technical
nature. It should be note that the convergence in Euclidean time can be accelerated
by choosing an initial cluster state that better reproduces the physical dimer wave
function. Also, the Euclidean time evolution is done by exact matrix multiplication
using the Trotter decomposition

$$\exp(-Ht) \approx \left(1 - \frac{t}{N_t}H\right)^{N_t}, \tag{6.172}$$

for a large number of time slices N_t, where $t \equiv \delta N_t$.

The initial state $|\mathbf{R}\rangle$ can be thought of as analogous to an interpolating field,
which roughly approximates the desired continuum state. Then, the approximation
is systematically improved by Euclidean time projection, which accounts for
all possible deformations and polarizations due to the interactions between the
composite bodies. In the limit of large projection time t, the set of dressed cluster
states $|\mathbf{R}\rangle_t$ spans the low-energy spectrum of the microscopic Hamiltonian H. As
we shall find in Chap. 7, the technique of generating cluster scattering states using
Euclidean time projection is suggested by PMC simulations of the ^{12}C nucleus,
where two characteristic time scales appear [53, 54]. The "fast" time scale is
associated with the formation of alpha clusters. Starting from an arbitrary trial state
for ^{12}C, individual alpha clusters emerge quickly as a function of t. The "slow"
time scale is associated with the overall structure of the alpha clusters relative to
each other, which develops only much later in t. The adiabatic projection formalism
uses this separation of time scales to represent the low-energy continuum states
efficiently as superpositions of dressed cluster states. In general, the "dressed"
cluster states $|\mathbf{R}\rangle_t$ are not orthogonal. Therefore, it is convenient to define the "dual"

vector $_t(\mathbf{R}|$ as

$$_t(\mathbf{R}|v) \equiv \sum_{\mathbf{R}'} \left[N_t^{-1}\right]_{\mathbf{R},\mathbf{R}'} \, _t\langle\mathbf{R}'|v\rangle, \tag{6.173}$$

where

$$\left[N_t\right]_{\mathbf{R},\mathbf{R}'} \equiv \, _t\langle\mathbf{R}|\mathbf{R}'\rangle_t, \tag{6.174}$$

is the norm matrix, constructed from inner products of cluster states. The dual vector $_t(\mathbf{R}|$ annihilates any vector v orthogonal to all dressed cluster states,

$$_t\langle\mathbf{R}|v\rangle = 0, \forall\mathbf{R} \quad \Rightarrow \quad _t(\mathbf{R}|v) = 0, \forall\mathbf{R}, \tag{6.175}$$

and it also serves as a "dual basis",

$$_t(\mathbf{R}|\mathbf{R}'\rangle_t = \delta_{\mathbf{R},\mathbf{R}'}, \tag{6.176}$$

within the linear subspace of dressed cluster states. Let H_t^a be the matrix representation of the Hamiltonian operator H projected onto the set of dressed cluster states,

$$\left[H_t^a\right]_{\mathbf{R},\mathbf{R}'} \equiv \, _t(\mathbf{R}|H|\mathbf{R}'\rangle_t, \tag{6.177}$$

where H_t^a is referred to as the "adiabatic Hamiltonian", an effective two-body matrix Hamiltonian describing the fermion-dimer system. Using Eq. (6.173), we can write H_t^a as

$$\left[H_t^a\right]_{\mathbf{R},\mathbf{R}'} = \sum_{\mathbf{R}''} \left[N_t^{-1}\right]_{\mathbf{R},\mathbf{R}''} \, _t\langle\mathbf{R}''|H|\mathbf{R}'\rangle_t, \tag{6.178}$$

and if we apply the similarity transform

$$\left[\tilde{H}_t^a\right]_{\mathbf{R},\mathbf{R}'} \equiv \sum_{\mathbf{R}'',\mathbf{R}'''} \left[N_t^{-1/2}\right]_{\mathbf{R},\mathbf{R}''} \, _t\langle\mathbf{R}''|H|\mathbf{R}'''\rangle_t \left[N_t^{-1/2}\right]_{\mathbf{R}''',\mathbf{R}'}, \tag{6.179}$$

then the resulting adiabatic Hamiltonian $\tilde{H}_t^a$ is Hermitian.

Before continuing with an explicit application, it is important to understand more precisely what we mean by widely-separated clusters at large distances [55]. Note that the Hamiltonian contains the lattice version of the kinetic energy operator $-\nabla^2/(2m)$, with m the pertinent particle mass. Euclidean time-evolution thus corresponds to a diffusion operator with the diffusion constant inversely proportional to the particle mass. Because of this, we need to be more precise in the definition of well-separated clusters. For doing so, let us introduce an error tolerance ε, which measures the contamination due to excited cluster states, and let t_ε be the time, for which this contamination is less than ε. Consider now Euclidean time projection for the time duration t_ε, during which excited states are removed. During this time, each cluster $i = 1, 2$ of mass M_i undergoes spatial diffusion with an average distance (diffusion length) of $d_{\varepsilon,i} \equiv \sqrt{t_\varepsilon/M_i}$. In order for the clusters to be considered well separated, the box size L should satisfy $L \gg d_{\varepsilon,i}$. For $|\mathbf{R}| \gg d_{\varepsilon,i}$, the dressed cluster states $|\mathbf{R}\rangle_{t_\varepsilon}$ then consist of non-overlapping clusters. This allows one to define the asymptotic radius R_ε. For $|\mathbf{R}| > R_\varepsilon$, the overlap between the clusters is less than ε. In this region, the dressed clusters are well separated and interact only through the long-range Coulomb force. From this, we can immediately conclude that the effective cluster Hamiltonian H_{eff} coincides with the free Hamiltonian for two point particles, supplemented by the long-range interactions inherited from the microscopic Hamiltonian. As the adiabatic Hamiltonian coincides with the effective cluster Hamiltonian in the asymptotic region, one finds the result

$$\left[H_t^a\right]_{\mathbf{R},\mathbf{R}'} = [H_{\mathrm{eff}}]_{\mathbf{R},\mathbf{R}'}\,, \tag{6.180}$$

the significance of which should not be underestimated. Despite the diffusion generated by the Euclidean time projection, in the asymptotic region one is left with an effective cluster Hamiltonian in a coordinate space basis. Therefore, the scattering states of the adiabatic Hamiltonian can be described by either Bessel or Coulomb functions, depending on the nature of the underlying forces.

Having introduced the concepts of cluster states and the adiabatic Hamiltonian, we next illustrate the adiabatic projection method by calculating the low-energy spectrum and S-wave phase shift for the fermion-dimer scattering problem. The application of adiabatic projection to the general problem of A nucleons requires combination with MC methods. This will be discussed in Chap. 7, when we apply adiabatic projection to α-α scattering.

6.3.2 Low-Energy Spectrum

As a first example, we consider how the low-energy spectrum $E_0 \leq E_1 \leq E_2 \cdots$ of the microscopic Hamiltonian H is described by the adiabatic Hamiltonian H_t^a, defined in the subspace spanned by N_R cluster states $|\mathbf{R}\rangle$. For practical computational reasons, the number of cluster states N_R on the lattice is limited, as is the accessible range in Euclidean projection time t. At finite volume and above the threshold for three-body states, the eigenstates of H will in general be a mixture of two-body and three-body states. However, at large volumes we can still classify which energy eigenstates are predominantly of two-body or three-body character. Typically, the two-body trial cluster states $|\mathbf{R}\rangle$ will be nearly orthogonal to the three-body continuum states. In order to reproduce the three-body continuum spectrum of H with adiabatic projection, it would be necessary to include initial states with better overlap with the three-body states. Let us first consider the (idealized) case where the initial trial states $|\mathbf{R}\rangle$ are completely orthogonal to all three-body and higher-body states. Let us denote by

$$E_0^{(2)} \leq E_1^{(2)} \leq \cdots \leq E_{N_R-1}^{(2)}, \tag{6.181}$$

the energy-ordered spectrum of H when only two-body states are included. In the asymptotic limit $t \to \infty$, it is straightforward to show that the two-body spectrum of H_t^a will be reproduced with error

$$\Delta E_j^{(2)} \sim \mathcal{O}\left[\exp\left(-2(E_{N_R}^{(2)} - E_j^{(2)})t\right)\right], \tag{6.182}$$

such that the convergence in Euclidean projection time is controlled by the lowest energy level not accounted for by the chosen number of initial cluster states. We may also consider the more realistic situation where some small (but non-zero) overlap exists between the initial cluster states $|\mathbf{R}\rangle$ and higher-body states of H. Such states introduce a small additional error

$$\Delta E_j^{(2)} \to \Delta E_j^{(2)} + \sum_E \rho_j^{(\geq 3)}(E) \exp\left(-2(E - E_j^{(2)})t\right), \tag{6.183}$$

to the determination of the two-body energy levels, where $\rho_j^{(\geq 3)}(E)$ is an energy-dependent spectral function which encodes the residual overlap between the higher-body states and our two-body cluster states $|\mathbf{R}\rangle$.

The optimal choice of N_R should be carefully considered. Retaining only a small number of lattice separation vectors $\mathbf{R}$ in the set of initial cluster states $|\mathbf{R}\rangle$ reduces the numerical task of computing $|\mathbf{R}\rangle_t$ and avoids numerical stability problems due to ill-conditioned norm matrices. On the other hand, a reduction in N_R will in turn decrease $E_{N_R}^{(2)}$, and the effect of this on the convergence with t is evident from the error estimate of Eq. (6.182). Roughly speaking, the most efficient approach is to work with a value for N_R which is somewhat larger than the number of scattering states one wishes to compute. Then, the strategy then is to choose t sufficiently large so that the desired accuracy is achieved. In Fig. 6.2, we show the low-lying energy levels of H_t^a as a function of Euclidean projection time t. These calculations are for $L = 13.8$ fm, using initial cluster states separated by at least two lattice sites in each direction. As the system is evolved in Euclidean time, the lowest energy levels of H_t^a converge exponentially to the corresponding levels of H. For this particular value of L, the spectrum of (predominantly) three-body continuum states starts at $E \simeq 8.0$ MeV, shown by the uppermost horizontal line in Fig. 6.2. It should be noted that different types of continuum states can be identified by studying the spatial correlations among the three fermions. In particular, fermion-dimer states are readily distinguished by their high probability for a spin-up and spin-down fermion occupying the same lattice site. Unlike the fermion-dimer states, the three-body continuum states are not accurately reproduced for the values of t shown in Fig. 6.2. As L is increased, the number of two-body continuum states below the three-body

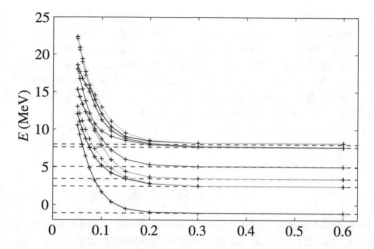

Fig. 6.2 Energy levels as a function of Euclidean projection time t, for the adiabatic Hamiltonian in an $L = 13.8$ fm periodic cubic box, compared to the energy levels of the microscopic Hamiltonian H (horizontal lines). The lowest 10 energy levels of the adiabatic Hamiltonian are found to converge exponentially onto the equivalent levels of H. With degeneracies accounted for, these 10 eigenvalues yield 5 distinct energy levels. For the $N = 7$ periodic box used here, the spectrum of predominantly three-body states starts at $E \simeq 8.0$ MeV, shown by the uppermost horizontal line

threshold also increases. Hence, more two-body continuum states can be reproduced using the two-body adiabatic Hamiltonian at larger lattice volumes.

6.3.3 Elastic Phase Shift

As a second example, we show how adiabatic projection can be applied to the calculation of the fermion-dimer elastic S-wave phase shift $\delta_0(p)$. The physics of the scattering process is encoded in the (discrete) values of the relative momentum p between the scattering bodies, which can be found from the energy levels of the adiabatic Hamiltonian in the periodic cube. The energy of an S-wave fermion-dimer scattering state in a cubic box of size $L = Na$ can be expressed as [56–58]

$$E_{fd}(p, L) = E_{fd}(p, \infty) + \tau_d(\zeta)\Delta E_{0,d}(L), \tag{6.184}$$

where ζ is defined in Eq. (5.93). The first term in Eq. (6.184) is the fermion-dimer energy at infinite volume,

$$E_{fd}(p, \infty) = \frac{p^2}{2m_d} + \frac{p^2}{2m_N} - B(\infty), \tag{6.185}$$

where m_d is the effective mass of the dimer, and $B(\infty) \simeq 2.2246$ MeV equals the infinite-volume dimer (deuteron) binding energy. In the non-relativistic continuum limit, the effective dimer mass reduces to $m_d = 2m_N$. In order to reduce the systematical error in the lattice calculation, the renormalization of the effective dimer mass at non-zero a should be accounted for. For this purpose, the dimer dispersion relation is computed numerically on a very large lattice (for instance $N = 60$) in order to minimize finite-volume effects.

The second term in Eq. (6.184) describes the finite-volume corrections to the fermion-dimer energy due to the dimer wave function wrapping around the periodic boundary [59–61]. Here

$$\Delta E_{0,d}(L) = B(\infty) - B(L), \tag{6.186}$$

is the finite-volume energy shift for the bound dimer in the two-body center-of-mass frame. The "topological" factor $\tau_d(\zeta)$, which accounts for the (single) winding of the S-wave dimer wave function around the cubic periodic boundary, is given by [56, 57]

$$\tau_d(\zeta) \equiv C_\tau^{-1} \sum_k \frac{\tau(\mathbf{k}, 1/2)}{(\mathbf{k}^2 - \zeta^2)^2}, \quad C_\tau^{-1} \equiv \sum_k \frac{1}{(\mathbf{k}^2 - \zeta^2)^2}, \tag{6.187}$$

where

$$\tau(\mathbf{k}, \xi) \equiv \frac{1}{3} \sum_{i=1}^{3} \cos(\xi k_i L), \tag{6.188}$$

such that

$$\Delta E_{\mathbf{k},d}(L) = \Delta E_{0,d}(L)\tau(\mathbf{k}, 1/2), \tag{6.189}$$

accounts for the finite-volume energy shift of a dimer moving with center-of-mass momentum $\mathbf{k}$ in the fermion-dimer system. The topological factor $\tau_d(\zeta)$, as given by Eq. (6.187), neglects the higher-order finite-volume effects due to multiple windings of the dimer wave function around the cubic boundary, as well as the finite-volume corrections to the fermion-dimer interactions.

The low-energy spectrum of the adiabatic Hamiltonian, as calculated on the lattice, provides $E_{fd}(L)$ and $\Delta E_{0,d}(L)$. The corresponding lattice momentum is then obtained as the value of p (and of ζ) which satisfies Eq. (6.184) with these energies. The solution for p is found iteratively by taking $\tau_d(\zeta) = 1$ as a starting guess. One can then solve recursively for $\tau_d(\zeta)$ and p until convergence is found. It should be noted that $\tau_d(\zeta) = 1$ implies $\Delta E_{\mathbf{k},d}(L) = \Delta E_{0,d}(L)$, such that

$$E_{fd}(p, L) \approx \frac{p^2}{2m_d} + \frac{p^2}{2m_N} - B(L), \tag{6.190}$$

which is an adequate initial guess for p, given $E_{fd}(L)$ and $B(L)$. In Ref. [57], convergence was found in ~ 10 iterations. Once p has been found, the S-wave phase shift $\delta_0(p)$ is calculated from the Lüscher formula

$$p \cot \delta_0(p) = \frac{1}{\pi L} S(\zeta), \tag{6.191}$$

where the function $S(\zeta)$ is defined (and its numerical evaluation discussed) in Chap. 5. Having found the lattice phase shift through adiabatic projection, we are in a position to compare it with the exact result in the continuum and infinite-volume limits, as obtained from the Skornyakov–Ter-Martirosian (STM) integral equation. For instance, the half-off-shell fermion-dimer T-matrix for S-wave scattering is given by [62–64]

$$T(k, p) = -\frac{4\pi \gamma}{m_N kp} \log \left(\frac{k^2 + p^2 + kp - m_N E}{k^2 + p^2 - kp - m_N E} \right) - \frac{1}{\pi} \int_0^\infty dq \left(\frac{q}{p} \right)$$
$$\times \frac{T(k, q)}{-\gamma + \sqrt{3q^2/4 - m_N E} - i\,0+} \log \left(\frac{q^2 + p^2 + qp - m_N E}{q^2 + p^2 - qp - m_N E} \right), \tag{6.192}$$

where $\gamma \equiv \sqrt{m_N B}$ is the dimer binding momentum, and $E \equiv 3p^2/(4m_N) - B$. Then, the on-shell T-matrix

$$T(p, p) = \frac{3\pi}{m_N} \frac{1}{p \cot \delta_0(p) - ip}, \tag{6.193}$$

determines the S-wave fermion-dimer phase shift $\delta_0(p)$.

A comparison between the adiabatic projection and STM equation results for $\delta_0(p)$ is shown in Fig. 6.3 for three different lattice spacings. Each of the adiabatic projection results use $N_R \simeq 30$ initial cluster states. At low momenta, the agreement with STM is excellent, with small deviations showing up around the dimer breakup momentum of $\simeq 52.7$ MeV. While the tree-body breakup amplitude can also be calculated using adiabatic projection, this requires inclusion of low-lying three-body states, which was not attempted in Ref. [49]. It should be noted that the fermion-dimer breakup amplitude happens to be numerically small at low momenta, which is why the agreement between adiabatic projection and STM results remains relatively good also above the breakup momentum.

6.3.4 Inelastic Processes

It is also of interest to generalize the adiabatic Hamiltonian formalism to account for two-body scattering processes with charge or mass transfer. For this purpose,

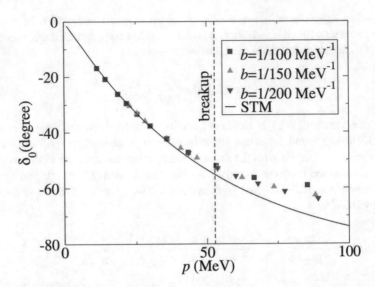

Fig. 6.3 S-wave elastic fermion-dimer phase shift from adiabatic projection (colored symbols), compared with the STM result (solid curve). The adiabatic projection results are shown for three different spatial lattice spacings. The dimer breakup momentum $\simeq 52.7$ MeV is shown by the dashed vertical line

the adiabatic formalism should be generalized to include an additional scattering channel. This can be accomplished by considering two sets of initial cluster states $|\mathbf{R}_1\rangle$ and $|\mathbf{R}_2\rangle$. In order to account for the dressed cluster states in each channel as well as their mixing, we take

$$[N_t]_{\mathbf{R}_i, \mathbf{R}_j'} \equiv {}_t\langle \mathbf{R}_i | \mathbf{R}_j' \rangle_t, \tag{6.194}$$

for the multi-channel norm matrix, and

$$[H_t^a]_{\mathbf{R}_i, \mathbf{R}_j'} = \sum_{\mathbf{R}_k''} [N_t^{-1}]_{\mathbf{R}_i, \mathbf{R}_k''} \, {}_t\langle \mathbf{R}_k'' | H | \mathbf{R}_j' \rangle_t, \tag{6.195}$$

for the multi-channel adiabatic Hamiltonian.

A possible application of the multi-channel adiabatic projection method is radiative capture processes, which are of great relevance to our understanding of hydrogen and helium burning in stars. The determination of astrophysical S-factors and asymptotic normalization coefficients currently require additional theoretical input. For instance, analyses of processes such as $^7\mathrm{Li}(n, \gamma)^8\mathrm{Li}$ and $^7\mathrm{Be}(p, \gamma)^8\mathrm{B}$ at low energies indicate that the (strong) nuclear interaction component of such processes are sensitive to the elastic n-$^7\mathrm{Li}$ and p-$^7\mathrm{Be}$ scattering parameters at LO, which are not well constrained experimentally [65, 66]. In order to incorporate radiative capture reactions in the adiabatic projection formalism, we need to add a one-body cluster state $|X\rangle$ to the set of initial cluster states. This state describes the outgoing nucleus after capture. As a first step, the multi-channel norm matrix and adiabatic Hamiltonian are computed. However, we also need to compute one-photon transition matrix elements

$$_t\langle X|\mathscr{O}_\gamma|\mathbf{R}\rangle_t,\tag{6.196}$$

between the dressed cluster states, where $\mathscr{O}_\gamma$ is a pertinent one-photon multipole operator. After computing all quantities involving the dressed cluster states, the problem has been reduced to a description of radiative capture which only involves two incoming bodies. Hence, the capture amplitude can be calculated in the same manner as the radiative capture process $n + p \rightarrow d + \gamma$ (see Ref. [48] for a detailed lattice calculation using infrared-regulated Green's function methods).

References

1. D. Lee, Ground state energy of spin-1/2 fermions in the unitary limit. Phys. Rev. B **73**, 115112 (2006)
2. D. Lee, Pressure inequalities for nuclear and neutron matter. Phys. Rev. C **71**, 044001 (2005)
3. J.W. Chen, D. Lee, T. Schäfer, Inequalities for light nuclei in the Wigner symmetry limit. Phys. Rev. Lett. **93**, 242302 (2004)
4. I. Sick, Precise root-mean-square radius of He-4. Phys. Rev. C **77**, 041302 (2008)
5. D. Lee, B. Borasoy, T. Schäfer, Nuclear lattice simulations with chiral effective field theory. Phys. Rev. C **70**, 014007 (2004)
6. R.T. Scalettar, D.J. Scalapino, R.L. Sugar, D. Toussaint, Hybrid molecular-dynamics algorithm for the numerical simulation of many-electron systems. Phys. Rev. B **36**, 8632 (1987)
7. R. Brower, C. Rebbi, D. Schaich, Hybrid Monte Carlo simulation on the graphene hexagonal lattice, in *Proceedings of the XXIX International Symposium on Lattice Field Theory (Lattice 2011)*, 056 (2011)
8. P.V. Buividovich, M.I. Polikarpov, Monte-Carlo study of the electron transport properties of monolayer graphene within the tight-binding model. Phys. Rev. B **86**, 245117 (2012)
9. D. Smith, L. von Smekal, Monte-Carlo simulation of the tight-binding model of graph-ene with partially screened Coulomb interactions. Phys. Rev. B **89**(19), 195429 (2014)
10. T. Luu, T.A. Lähde, Quantum Monte Carlo calculations for carbon nanotubes. Phys. Rev. B **93**(15), 155106 (2016)
11. S. Beyl, F. Goth, F.F. Assaad, Revisiting the hybrid quantum Monte Carlo method for Hubbard and electron-phonon models. Phys. Rev. B **97**(8), 085144 (2018)
12. A. Bulgac, J.E. Drut, P. Magierski, Spin 1/2 Fermions in the unitary regime: a superfluid of a new type. Phys. Rev. Lett. **96**, 090404 (2006)
13. A. Bulgac, J.E. Drut, P. Magierski, Quantum Monte Carlo simulations of the BCS-BEC crossover at finite temperature. Phys. Rev. A **78**, 023625 (2008)
14. J.E. Drut, T.A. Lähde, T. Ten, Momentum distribution and contact of the unitary fermi gas. Phys. Rev. Lett. **106**, 205302 (2011)
15. J.E. Drut, T.A. Lähde, G. Wlazłowski, P. Magierski, The equation of state of the unitary fermi gas: an update on lattice calculations. Phys. Rev. A **85**, 051601 (2012)
16. G. Wlazłowski, P. Magierski, J.E. Drut, Shear viscosity of a unitary fermi gas. Phys. Rev. Lett. **109**, 020406 (2012)

17. T.A. Lähde, E. Epelbaum, H. Krebs, D. Lee, U.G. Meißner, G. Rupak, Lattice effective field theory for medium-mass nuclei. Phys. Lett. B **732**, 110 (2014)

18. T.A. Lähde, E. Epelbaum, H. Krebs, D. Lee, U.G. Meißner, G. Rupak, Uncertainties of Euclidean time extrapolation in lattice effective field theory. J. Phys. G **42**(3), 034012 (2015)

19. E.Y. Loh, J.E. Gubernatis, R.T. Scalettar, S.R. White, D.J. Scalapino, R.L. Sugar, Sign problem in the numerical simulation of many-electron systems. Phys. Rev. B **41**, 9301 (1990)

20. S. Chandrasekharan, U.J. Wiese, Meron cluster solution of a fermion sign problem. Phys. Rev. Lett. **83**, 3116 (1999)

21. J.M. Hammersley, D.C. Handscomb, *Monte Carlo Methods*. Methuen's Monographs (Methuen, London, 1975)

22. A.B. Clarke, R.L. Disney, *Probability and Random Processes: A First Course with Applications* (Wiley, New York, 1985)

23. N. Metropolis, A. Rosenbluth, M. Rosenbluth, A. Teller, E. Teller, Equation of state calculations by fast computing machines. J. Chem. Phys. **21**, 1087 (1953)

24. A. Ukawa, M. Fukugita, Langevin simulation including dynamical quark loops. Phys. Rev. Lett. **55**, 1854 (1985)

25. G.G. Batrouni, G.R. Katz, A.S. Kronfeld, G.P. Lepage, B. Svetitsky, K.G. Wilson, Langevin simulations of lattice field theories. Phys. Rev. D **32**, 2736 (1985)

26. G. Parisi, On complex probabilities. Phys. Lett. **131B**, 393 (1983)

27. J.R. Klauder, Stochastic quantization. Acta Phys. Austriaca Suppl. **25**, 251 (1983)

28. E. Seiler, Status of complex langevin. EPJ Web Conf. **175**, 01019 (2018)

29. D.J.E. Callaway, A. Rahman, The microcanonical ensemble: a new formulation of lattice gauge theory. Phys. Rev. Lett. **49**, 613 (1982)

30. D.J.E. Callaway, A. Rahman, Lattice gauge theory in microcanonical ensemble. Phys. Rev. D **28**,1506 (1983)

31. M. Creutz, Microcanonical Monte Carlo simulation. Phys. Rev. Lett. **50**, 1411 (1983)

32. J. Polonyi, H.W. Wyld, Microcanonical simulation of fermionic systems. Phys. Rev. Lett. **51**, 2257 (1983) [Erratum: Phys. Rev. Lett. **52** (1984) 401]

33. S. Duane, Stochastic quantization versus the microcanonical ensemble: getting the best of both worlds. Nucl. Phys. B **257**, 652 (1985)

34. S. Duane, J.B. Kogut, The theory of hybrid stochastic algorithms. Nucl. Phys. B **275**, 398 (1986)

35. S.A. Gottlieb, W. Liu, D. Toussaint, R.L. Renken, R.L. Sugar, Hybrid molecular dynamics algorithms for the numerical simulation of quantum chromodynamics. Phys. Rev. D **35**, 2531 (1987)

36. S. Duane, A.D. Kennedy, B.J. Pendleton, D. Roweth, Hybrid Monte Carlo. Phys. Lett. B **195**, 216 (1987)

37. E. Forest, R.D. Ruth, Fourth order symplectic integration. Physica D **43**, 105 (1990)

38. H. Yoshida, Construction of higher order symplectic integrators. Phys. Lett. A **150**, 262 (1990)

39. J.C. Sexton, D.H. Weingarten, Hamiltonian evolution for the hybrid Monte Carlo algorithm. Nucl. Phys. B **380**, 665 (1992)

40. M. Suzuki, Generalized Trotter's formula and systematic approximants of exponential operators and inner derivations with applications to many body problems. Commun. Math. Phys. **51**, 183 (1976)

41. T. Takaishi, P. de Forcrand, Testing and tuning new symplectic integrators for hybrid Monte Carlo algorithm in lattice QCD. Phys. Rev. E **73**, 036706 (2006)

42. I.P. Omelyan, I.M. Mryglod, R. Folk, Optimized Verlet-like algorithms for molecular dynamics simulations. Phys. Rev. E **65**, 056706 (2002)

43. I.P. Omelyan, I.M. Mryglod, R. Folk, Symplectic analytically integrable decomposition algorithms: classification, derivation, and application to molecular dynamics, quantum and celestial mechanics simulations. Comput. Phys. Commun. **151**, 272 (2003)

44. M. Creutz, Global Monte Carlo algorithms for many-fermion systems. Phys. Rev. D **38**, 1228 (1988)

45. S. Gupta, A. Irback, F. Karsch, B. Petersson, The acceptance probability in the hybrid Monte Carlo method. Phys. Lett. B **242**, 437 (1990)
46. T. Takaishi, Choice of integrator in the hybrid Monte Carlo algorithm. Comput. Phys. Commun. **133**, 6 (2000)
47. B. Borasoy, E. Epelbaum, H. Krebs, D. Lee, U.-G. Meißner, Lattice simulations for light nuclei: chiral effective field theory at leading order. Eur. Phys. J. A **31**, 105 (2007)
48. G. Rupak, D. Lee, Radiative capture reactions in lattice effective field theory. Phys. Rev. Lett. **111**(3), 032502 (2013)
49. M. Pine, D. Lee, G. Rupak, Adiabatic projection method for scattering and reactions on the lattice. Eur. Phys. J. A **49**, 151 (2013)
50. G.V. Skorniakov, K.A. Ter-Martirosian, Three body problem for short range forces. 1. Scattering of low energy neutrons by deuterons. Sov. Phys. JETP **4**, 648 (1957) [J. Exp. Theor. Phys. (U.S.S.R.) **31**, 775 (1956)]
51. P.F. Bedaque, U. van Kolck, Nucleon deuteron scattering from an effective field theory. Phys. Lett. B **428**, 221 (1998)
52. P.F. Bedaque, H.W. Hammer, U. van Kolck, Effective theory for neutron deuteron scattering: energy dependence. Phys. Rev. C **58**, R641 (1998)
53. E. Epelbaum, H. Krebs, D. Lee, U.-G. Meißner, Ab initio calculation of the Hoyle state. Phys. Rev. Lett. **106**, 192501 (2011)
54. E. Epelbaum, H. Krebs, T.A. Lähde, D. Lee, U.-G. Meißner, Structure and rotations of the Hoyle state. Phys. Rev. Lett. **109**, 252501 (2012)
55. A. Rokash, M. Pine, S. Elhatisari, D. Lee, E. Epelbaum, H. Krebs, Scattering cluster wave functions on the lattice using the adiabatic projection method. Phys. Rev. C **92**, 054612 (2015)
56. S. Bour, S. Koenig, D. Lee, H.-W. Hammer, U.-G. Meißner, Topological phases for bound states moving in a finite volume. Phys. Rev. D **84**, 091503 (2011)
57. S. Bour, H.-W. Hammer, D. Lee, U.-G. Meißner, Benchmark calculations for elastic fermion-dimer scattering. Phys. Rev. C **86**, 034003 (2012)
58. A. Rokash, E. Epelbaum, H. Krebs, D. Lee, U.-G. Meißner, Finite volume effects in low-energy neutron-deuteron scattering. J. Phys. G **41**, 015105 (2014)
59. M. Lüscher, Volume dependence of the energy spectrum in massive quantum field theories. 1. Stable particle states. Commun. Math. Phys. **104**, 177 (1986)
60. S. Koenig, D. Lee, H.-W. Hammer, Volume dependence of bound states with angular momentum. Phys. Rev. Lett. **107**, 112001 (2011)
61. S. Koenig, D. Lee, H.-W. Hammer, Non-relativistic bound states in a finite volume. Ann. Phys. **327**, 1450 (2012)
62. P.F. Bedaque, H.W. Griesshammer, Quartet S wave neutron deuteron scattering in effective field theory. Nucl. Phys. A **671**, 357 (2000)
63. F. Gabbiani, P.F. Bedaque, H.W. Griesshammer, Higher partial waves in an effective field theory approach to nd scattering. Nucl. Phys. A **675**, 601 (2000)
64. G. Rupak, X.W. Kong, Quartet S-wave p-d scattering in EFT. Nucl. Phys. A **717**, 73 (2003)
65. G. Rupak, R. Higa, Model-independent calculation of radiative neutron capture on lithium-7. Phys. Rev. Lett. **106**, 222501 (2011)
66. L. Fernando, R. Higa, G. Rupak, Resonance contribution to radiative neutron capture on lithium-7. Eur. Phys. J. A **48**, 24 (2012)

Chapter 7
Light and Medium-Mass Nuclei on the Lattice

The success of the NLEFT description of light and medium-mass nuclei (with $A \geq 4$) is dependent on the theoretical formalism and Monte Carlo (MC) techniques developed in the previous chapters. First and foremost, these include the EFT description of the nuclear force in terms of nucleons and pions (rather than quarks and gluons), and the expression of the transfer matrix formalism in terms of a path integral over auxiliary fields. The application of this theoretical framework to nuclei requires accurate computations of scattering phase shifts using the spherical wall method, and the use of an efficient Hybrid Monte Carlo (HMC) algorithm with favorable computational scaling in A. Observables are then computed by means of Euclidean time projection and effective cluster Hamiltonians. As we have shown in Chaps. 5 and 6, these methods overcome significant obstacles to a successful a priori treatment of multi-nucleon systems on the lattice. In this chapter, we will show how these components of NLEFT, when taken together, allow some of the key problems in nuclear theory to be addressed. It should be clear from our earlier presentation of the EFT framework and computational methods that NLEFT is a vigorously developing field of study where theory as well as algorithms are subject to continuous refinement. In this chapter, we shall therefore avoid putting focus on any given version of the NLEFT action and results, but rather focus on the general methods with which NLEFT can be applied to systems with $A \geq 4$, and on the issues and problems encountered along the way.

We structure this chapter as follows. In Sect. 7.1, we elaborate on the NLEFT calculation of spectra and nuclear structure, using the methods of auxiliary field lattice MC and Euclidean time projection. Our main focus is on the spectrum of the $A = 12$ system (^{12}C), but we also discuss the cases of $A = 16$ (^{16}O) and heavier nuclei. We also outline how the nuclear charge density distribution is obtained in NLEFT, and how observables such as charge radii, quadrupole moments, and electromagnetic transition amplitudes are computed by considering

© Springer Nature Switzerland AG 2019
T. A. Lähde, U.-G. Meißner, *Nuclear Lattice Effective Field Theory*,
Lecture Notes in Physics 957, https://doi.org/10.1007/978-3-030-14189-9_7

appropriate moments of the charge distribution. We compare NLEFT predictions with experiment for the ground state and the excited "Hoyle state" of ^{12}C, along with the analogous states in ^{16}O, and discuss evidence for α clustering as obtained from NLEFT. In Sect. 7.2, we show how NLEFT can be used to determine the dependence of nuclear energy levels on the fundamental parameters of nature, such as the light quark mass m_q and the electromagnetic fine-structure constant α_{EM}. We also elaborate on the astrophysical and anthropic implications of such NLEFT calculations, by considering the triple alpha (3α) reaction rate and its sensitivity to shifts in the spectrum of ^{12}C. Finally, in Sect. 7.3 we return to the adiabatic projection formalism by applying it to the problem of α-α scattering. We show how MC simulations using "dressed" cluster states can be used to reduce the $A = 8$ problem into an "effective" two-body problem, which allows for scattering phase shifts to be computed using the spherical wall method.

7.1 Medium-Mass Nuclei in NLEFT

The extension of NLEFT to medium-mass nuclei has proven to be far from trivial. The general strategy has been to use the $A = 2$ and $A = 3$ systems to fix the free parameters of the action up to NNLO, after which systems with $A \geq 4$ can be regarded as genuine predictions of NLEFT. However, already on a purely two-body level the LO action was found to give large overbinding of ^{4}He and heavier nuclei. This is the clustering instability problem which we encountered in Chap. 4. The solution was found to be smearing of the LO contact operators, a prescription which has been successively refined. Such smearing regularizes the potential at short distances, at the price of introducing an additional parameter, which can, however, be fixed by the S-wave effective ranges in the neutron-proton scattering channels. Such regularization of the potential can be regarded as similar to the cutoff regularization employed in continuum Chiral EFT descriptions of NN scattering, where it can be shown that the dependence on the cutoff decreases as the EFT expansion is extended to successively higher orders. If operator smearing were systematically applied to the entire NLEFT potential, the lattice spacing would merely act as a regulator for the path integral, and could be taken to zero. Without operator smearing, the large overbinding at LO would have to be absorbed by an unacceptably large (in particular for the purposes of MC simulation) repulsive NLO two-nucleon contribution.

An accurate description of ^{3}H and ^{3}He requires the three-nucleon force (3NF) which first appears at NNLO, in addition to the electromagnetic and isospin-breaking contributions which are counted as NLO in the NLEFT power counting. As we have seen in Chap. 5, the coupling constants c_D and c_E turn out to be strongly correlated and difficult to separate with available three-nucleon data. However, a more serious problem was the lack of good agreement for the ^{4}He binding energy once the $A = 2$ and $A = 3$ systems were optimally described (violation of the so-called Tjon line). The introduction of a four-nucleon contact term designed to mimic the effects of neglected contributions at N3LO and higher orders led to the first successful description of medium-mass nuclei such as ^{12}C and ^{16}O. Such lessening of the strong four-nucleon correlations can be viewed as conceptually similar to the smearing of the LO contact operators, all of which accelerate the convergence of the NLEFT expansion at large lattice spacings of $a \simeq 2$ fm. Such advances in the NLEFT description enable a wide range of studies of the spectra, structure and transitions of the medium-mass nuclei, especially the so-called "alpha nuclei" (nuclei with $Z = N$ and A a multiple of four), which exhibit strong effects due to the clustering of α particles.

The extension of NLEFT to a complete, well-rounded description of nuclei up to ^{40}Ca (and beyond) is an active, ongoing field of study. The most recent NLEFT lattice actions use a combination of local and non-local smearing of the LO operators (see Chap. 8). This makes it much easier to simultaneously obtain an optimal description of NN scattering, of α nuclei and neutron-rich halo nuclei, and of α-α scattering phase shifts. Such non-locally smeared actions are also found to introduce much milder sign oscillations in the MC probability measure. This has recently enabled studies with lattice spacings as low as $a \simeq 1$ fm. In turn, this leads to dramatically reduced lattice artifacts due to rotational symmetry breaking, and allows (together with a careful treatment of the $\delta \to 0$ limit) for a good description of the Tjon line without the need for additional four-nucleon operators. With the extension of the NLEFT expansion to N3LO, along with increased EFT cutoff momenta, the range of applicability of NLEFT has also been dramatically improved (see Chap. 8).

7.1.1 Trial States and Euclidean Time Projection

We shall begin our treatment of light and medium-mass nuclei in NLEFT with the choice of trial wave functions for Euclidean time projection. Trial wave functions based on plane wave states have already been introduced in Chap. 6. As a suitable check on the equivalence of the sparse eigenvector and auxiliary field MC methods, we also perform MC simulations for the $A = 3$ system (^{3}H and ^{3}He). For $A = 3$, a lattice volume of $\simeq 16$ fm^3 was found to render finite-volume corrections small. Similarly to the α cluster trial states presented in Chap. 6, for the ground state calculations at $A = 3$ we take the trial states to be

$$|\Psi_{1,2}\rangle \equiv \sum_{\mathbf{n},\mathbf{n}',\mathbf{n}'',\mathbf{n}'''} \exp(-\alpha_\gamma |\mathbf{n} - \mathbf{n}'|) \exp(-\alpha_\gamma |\mathbf{n} - \mathbf{n}''|) \exp(-\alpha_\gamma |\mathbf{n} - \mathbf{n}'''|)$$

$$\times \, a^\dagger_{\uparrow,p}(\mathbf{n}') a^\dagger_{\uparrow,n}(\mathbf{n}'') a^\dagger_{\downarrow,n}(\mathbf{n}''')|0\rangle, \tag{7.1}$$

for ^{3}H, and

$$|\Psi_{2,1}\rangle \equiv \sum_{\mathbf{n,n',n'',n'''}} \exp(-\alpha_\gamma |\mathbf{n}-\mathbf{n'}|) \exp(-\alpha_\gamma |\mathbf{n}-\mathbf{n''}|) \exp(-\alpha_\gamma |\mathbf{n}-\mathbf{n'''}|)$$

$$\times\, a^\dagger_{\uparrow,n}(\mathbf{n'}) a^\dagger_{\uparrow,p}(\mathbf{n''}) a^\dagger_{\downarrow,p}(\mathbf{n'''})|0\rangle, \tag{7.2}$$

for ^{3}He. The value $\alpha_\gamma = 2$ was (arbitrarily) chosen for the width of the wave packets centered around lattice site $\mathbf{n}$. This type of trial state provides smaller overlap with excited states, and hence more rapid convergence in Euclidean projection time, as compared to the plane-wave trial states.

For the ground-state calculations of the $A \geq 8$ systems, a somewhat smaller lattice volume of $\simeq 12\,\text{fm}^3$ has been found adequate. For the construction of the trial wave functions for such nuclei, we introduce the operators $\tilde{a}^\dagger(0)$ to denote zero-momentum creation operators that correspond to the creation operators in Eqs. (7.1) and (7.2). We also introduce the shorthand notation

$$\tilde{a}^\dagger \equiv \tilde{a}^\dagger_{\uparrow,p}(0)\tilde{a}^\dagger_{\downarrow,p}(0)\tilde{a}^\dagger_{\uparrow,n}(0)\tilde{a}^\dagger_{\downarrow,n}(0), \tag{7.3}$$

for the case where a maximum of four nucleons have been placed in the same zero-momentum state. The initial states are then given by

$$|\Psi_{4,4}\rangle \equiv \tilde{a}^\dagger\, M_{\text{SU}(4)}(0)\, \tilde{a}^\dagger|0\rangle, \tag{7.4}$$

for the $A = 8$ system with $Z = 4$ and $N = 4$ (^{8}Be),

$$|\Psi_{6,6}\rangle \equiv \tilde{a}^\dagger\, M_{\text{SU}(4)}(1)\, \tilde{a}^\dagger\, M_{\text{SU}(4)}(0)\, \tilde{a}^\dagger|0\rangle, \tag{7.5}$$

for the $A = 12$ system with $Z = 6$ and $N = 6$ (^{12}C),

$$|\Psi_{8,8}\rangle \equiv \tilde{a}^\dagger\, M_{\text{SU}(4)}(2)\, \tilde{a}^\dagger\, M_{\text{SU}(4)}(1)\, \tilde{a}^\dagger\, M_{\text{SU}(4)}(0)\, \tilde{a}^\dagger|0\rangle, \tag{7.6}$$

for the $A = 16$ system with $Z = 8$ and $N = 8$ (^{16}O),

$$|\Psi_{10,10}\rangle \equiv \tilde{a}^\dagger\, M_{\text{SU}(4)}(3)\, \tilde{a}^\dagger\, M_{\text{SU}(4)}(2)\, \tilde{a}^\dagger\, M_{\text{SU}(4)}(1)\, \tilde{a}^\dagger\, M_{\text{SU}(4)}(0)\, \tilde{a}^\dagger|0\rangle, \tag{7.7}$$

for the $A = 20$ system with $Z = 10$ and $N = 10$ (^{20}Ne), and so on. Here, the argument of the SU(4) symmetric transfer matrix labels slices of Euclidean projection time. The number of "outer" time slices $N_{t'}$ should be sufficiently large to allow for all the nucleons to be "injected" in this manner, before the application of the full LO transfer matrix begins. The results of a ground-state energy calculation with multiple trial states that differ in the coupling constant of $M_{\text{SU}(4)}$ are shown in Figs. 7.1 and 7.2. For the case of ^{16}O, the complete results are shown, with the non-perturbative LO energy in Fig. 7.1 and the perturbative higher-order contributions

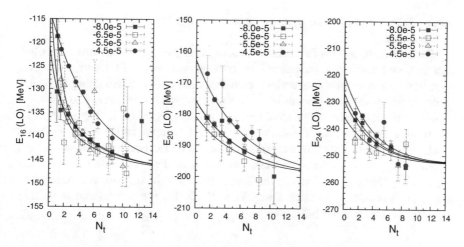

Fig. 7.1 Monte Carlo results for the LO transient energy $E_A(t)$ for $A = 16, 20$, and 24, for different values of the SU(4) coupling in each of the trial states. The curves show a fit using a spectral density $\rho_A(E)$ given by a sum of three energy delta functions. The fit for $A = 16$ is correlated with the higher-order corrections shown in Fig. 7.2

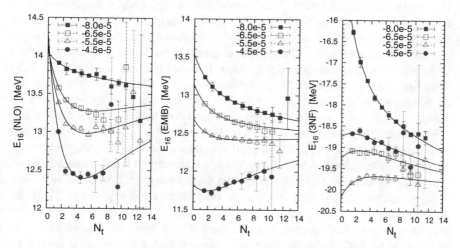

Fig. 7.2 Monte Carlo results for the higher-order corrections to the $A = 16$ LO binding energy, for different values of the SU(4) coupling in each of the trial states. From left to right are shown the total isospin-symmetric correction (NLO), the electromagnetic and isospin-breaking corrections (EMIB), and the three-nucleon force correction (3NF). The curves show a fit with $\rho_A(E)$ given by the sum of three energy delta functions, correlated with Fig. 7.1

up to NNLO in Fig. 7.2. The results are extrapolated to infinite Euclidean projection time using the method of multi-exponential fits described in Chap. 6. In this case, all trial states were simultaneously fitted by a density of states $\rho_A(E)$ given by a sum of three energy delta functions. The ^{16}O fit shown in Fig. 7.1 is correlated with those of Fig. 7.2, with same spectral density $\rho_A(E)$. The fits for each A are treated as independent.

The latest MC results for nuclear binding energies (as of the time of writing) can be found in Chap. 8. In general, simulations at $a = 1.97$ fm with exclusively locally smeared contact interactions exhibit an increasing overbinding at NNLO, which amounts to a few percent for ^{12}C, and increases systematically to $\simeq 30\%$ for ^{28}Si. On the other hand, as we shall find in Chap. 8, purely non-locally smeared LO interactions generate a weakly interacting gas of α particles, and in order to generate nuclei with realistic binding energies, a mixture of the two types of smearing is required. Before we move on to the treatment of nuclear structure and transitions, we note that similar problems in the description of light and medium-mass nuclei are often encountered by other ab initio methods that use soft potentials with the same set of interactions [1–3].

7.1.2   Low-Energy Spectrum of Carbon-12

While an accurate description of the ground-state energies across the nuclear chart is an important first test for NLEFT, a good description of excited states and electromagnetic transitions is also required. For this purpose, we shall focus on the low-lying excited states of ^{12}C and generalize the Euclidean time projection method to a multi-channel calculation, similar to that already encountered for shell-model wave functions and the adiabatic projection method in Chap. 6. We will make use of the spectroscopic notation J_n^π, where J is the total angular momentum, π is the parity, and n labels the excitation starting from 1 for the lowest level. In terms of this notation, the ground state is 0_1^+, the "Hoyle state" is 0_2^+, and the lowest state with $J = 2$ is denoted 2_1^+. As we shall find later on in this chapter, these states play an important role in nuclear astrophysics, and many of their properties can be understood in terms of strong clustering of α particles in the ^{12}C nucleus. We follow Ref. [4], and take the trial state for the ^{12}C calculation which consists of 24 single-nucleon components in a periodic cube. From these, trial states consisting of six protons and six neutrons each can be constructed, such that four orthogonal energy levels with the desired quantum numbers are obtained. All four states have even parity and total momentum equal to zero, while three states have $J_z = 0$ and one has $J_z = 2$. Naturally, the identification of the states is complicated by the reduction of the rotational group to the cubic subgroup on the lattice. Still, the degeneracy (or lack thereof) of energy levels for $J_z = 0$ and $J_z = 2$ makes it possible to distinguish states with $J = 0$ from states with $J = 2$. As will become clear later, the 2_2^+ state is found to be a rotational excitation of the Hoyle state.

In Fig. 7.3, we illustrate the extraction of excited state energies for ^{12}C at LO. For each excited state k, we compute

$$A_k(t) \sim \log \frac{Z_k(t)}{Z_{0_1^+}(t)}, \tag{7.8}$$

Fig. 7.3 NLEFT calculation at LO of the two lowest positive-parity excited states of ^{12}C. Shown are logarithmic projection amplitudes $A_k(t)$ according to Eq. (7.8). The slope of $A_k(t)$ at large projection time t determines the excitation energy

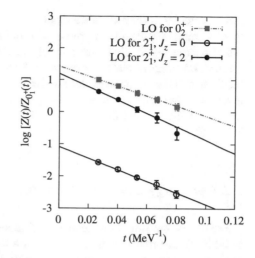

Fig. 7.4 Comparison of the experimental low-energy spectrum of ^{12}C with the NLEFT calculation. Energies are shown at NNLO for the even-parity 0^+ and 2^+ states. Uncertainties are not shown, but are $\simeq 300$ keV, as detailed in Ref. [5]

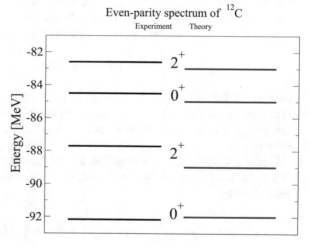

where 0_1^+ denotes the ^{12}C ground state. The slope $A_k(t)$ in the limit $t \rightarrow \infty$ provides the energy of state k relative to the ground state. As shown schematically in Fig. 7.4, the NNLO results from NLEFT are in general agreement with the experimental values for the Hoyle state and the first two $J = 2$ states of ^{12}C. It is remarkable to find this level of agreement on the absolute energies, because in various many-body methods, excitation energies are easily obtained, but convergence for the ground state is often problematic. Furthermore, while the ground state and the $J = 2$ states have been calculated with other methods [6–8], the NLEFT results are the first ab initio calculations where the Hoyle state was found with an energy close to the phenomenologically important ^{8}Be-α threshold at -84.8 MeV. In this respect, a noteworthy level crossing occurs, as the Hoyle state is lower in energy than the 2_1^+ state at LO. The empirical level ordering is restored at NLO, as the NLO interactions provide increased short-distance repulsion. The 0_1^+ and 0_2^+ states are more sensitive to short-range repulsion than those with higher total angular momentum, as their radial charge distributions are strongly peaked at short distances. However, this finding also suggests that some degree of fine-tuning is necessary in order to obtain a Hoyle state which is situated close to the ^{8}Be-α threshold. On the most fundamental level, there are only a few parameters which control such fine-tuning, such as the masses of the up and down quarks [9–12]. We shall return to this question in Sect. 7.2.

7.1.3 Charge Density Distribution

We shall next consider the calculation of nuclear structure and transition amplitudes in NLEFT. Given that such observables are typically expressed in terms of various moments of the nucleon (or charge density) distribution, we first describe how NN correlation functions are computed from MC simulations. For instance, for a liquid of nucleons (such as a neutron star), we can express the density distribution as

$$\rho(\mathbf{r}) = \rho g_2(\mathbf{r}), \qquad (7.9)$$

in terms of the nucleon pair correlation function g_2, which is conventionally normalized such that $g_2(\mathbf{r}) \to 1$ at asymptotically large distances. For such a macroscopic liquid, ρ equals the number of nucleons in the system divided by the lattice volume.

A related quantity is the normalized proton density correlation function, which we use to describe finite nuclei on the lattice. For a system consisting of Z protons, this is given by

$$f_{pp}(\mathbf{n}) \equiv \frac{1}{Z-1} \sum_{\mathbf{n'}} \langle \Psi_A | : \rho_p^{a^\dagger,a}(\mathbf{n}+\mathbf{n'}) \rho_p^{a^\dagger,a}(\mathbf{n'}) : | \Psi_A \rangle, \qquad (7.10)$$

$$= \frac{L^3}{Z-1} \langle \Psi_A | : \rho_p^{a^\dagger,a}(\mathbf{n}) \rho_p^{a^\dagger,a}(\mathbf{0}) : | \Psi_A \rangle, \qquad (7.11)$$

where we have made use of translational invariance and fixed one proton at the origin. We recall that the numerical evaluation of such density-density operators in NLEFT has been discussed in Chap. 6. Also, we denote $A = Z + N$, where N is the number of neutrons. The proton and neutron density operators in Eq. (7.11) are defined as

$$\rho_p^{a^\dagger,a}(\mathbf{n}) \equiv \frac{1}{2}(\rho^{a^\dagger,a}(\mathbf{n}) + \rho_{I=3}^{a^\dagger,a}(\mathbf{n})), \quad \rho_n^{a^\dagger,a}(\mathbf{n}) \equiv \frac{1}{2}(\rho^{a^\dagger,a}(\mathbf{n}) - \rho_{I=3}^{a^\dagger,a}(\mathbf{n})), \qquad (7.12)$$

similarly to the EM contact terms introduced in Chap. 4. Furthermore, we have

$$\sum_{\mathbf{n}} \langle \Psi_A | \rho_p^{a^\dagger, a}(\mathbf{n}) | \Psi_A \rangle = Z, \tag{7.13}$$

and

$$\sum_{\mathbf{n}} \langle \Psi_A | : \rho_p^{a^\dagger, a}(\mathbf{n}) \rho_p^{a^\dagger, a}(\mathbf{0}) : | \Psi_A \rangle = \sum_{\mathbf{n}} \langle \Psi_A | \rho_p^{a^\dagger, a}(\mathbf{n}) \rho_p^{a^\dagger, a}(\mathbf{0}) | \Psi_A \rangle$$

$$- \langle \Psi_A | \rho_p^{a^\dagger, a}(\mathbf{0}) | \Psi_A \rangle = \frac{Z(Z-1)}{L^3}, \tag{7.14}$$

thus the integral of $f_{pp}(\mathbf{n})$ over the lattice volume equals Z. Using any given proton as a reference point, the function $f_{pp}(\mathbf{n})$ is proportional to the probability of finding a second proton at a distance $\mathbf{n}$, as measured from the origin. The equivalent correlation function for all nucleons can be obtained by setting $Z \rightarrow A$, and by dropping the proton subscript p of the density operators throughout. In Fig. 7.5, we show the radial projections $f_{pp}(r)$ of $f_{pp}(\mathbf{n})$ for the low-lying states of ^{12}C. The density distributions of the $J = 0$ states are compact and strongly peaked for $r \simeq 0$, while the $J = 2$ states show a maximum around $r \simeq 1.5$ fm. This finding is consistent with the relative sensitivity of the Hoyle state to the repulsive NLO contribution. A secondary maximum at $r \simeq 4$ fm is prominent in the ground state, and likely arises from configurations where three α clusters are

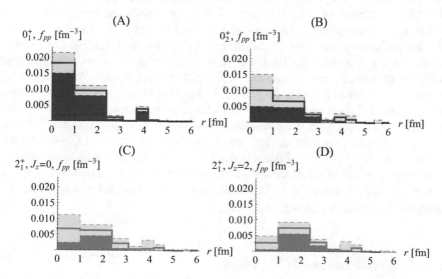

Fig. 7.5 Radial distribution functions $f_{pp}(r)$ for the ground state (**a**), Hoyle state (**b**), and the $J_z = 0$ (**c**) and $J_z = 2$ (**d**) components of the $J = 2$ state. Yellow bands show the MC error

arranged approximately linearly. As we shall find in Chap. 8, methods now exist to calculate such distributions, whose resolution is not limited by the lattice spacing (here $a = 1.97$ fm) as in Fig. 7.5, but rather by a/A, which equals $a/12 \simeq 0.16$ fm for the ^{12}C nucleus.

We shall now consider some observables, which are central to the description of nuclear structure in NLEFT. This can be accomplished using the normalized density-density correlation function $f_{pp}(\mathbf{n})$, which is obtained from MC simulations. For instance, the nuclear root-mean-square (rms) charge radius r_{EM} is found from

$$r_{EM}^2 = \sum_{\mathbf{n}} \mathbf{r}^2 f_{pp}(\mathbf{n}), \qquad (7.15)$$

where we note that

$$\int d^3 r \, \rho(\mathbf{r}) \rightarrow \sum_{\mathbf{n}} f_{pp}(\mathbf{n}), \qquad (7.16)$$

equals the total electric charge of the nucleus. In addition, we require the proton charge radius $\simeq 0.84$ fm [13, 14], which is added into r_{EM} in quadrature. Similarly, the quadrupole moment is given by

$$Q \equiv Q_{zz} = \sum_{\mathbf{n}} (3z^2 - \mathbf{r}^2) f_{pp}(\mathbf{n}), \qquad (7.17)$$

in terms of the z-component of the quadrupole tensor. As is apparent from Fig. 7.5, the MC calculation of $f_{pp}(\mathbf{n})$ involves a significant statistical uncertainty which needs to be propagated to the reported values of r_{EM} and Q. This can be done, for instance, using so-called jackknife resampling. We shall discuss this method further in the context of our treatment of α-α scattering. We point out again that with the pinhole algorithm discussed in Chap. 8, a much improved description of these observables is possible. In Table 7.1, we present results at LO for the rms charge radii and quadrupole moments of the even-parity states of ^{12}C and ^{16}O, with experimental values given where available. The rms radii computed at LO in NLEFT tend to be smaller than the empirical ones, in particular for ^{16}O the LO result for the ground state is $\simeq 20\%$ smaller than the empirical value, which is consistent with the overbinding at that order. At higher orders, the radii are expected to increase due to the repulsive NLO interactions. The quadrupole moment is known only for the 2_1^+ state of ^{12}C, which is well reproduced at LO.

7.1.4 Electromagnetic Transitions

The LO results for the electromagnetic transitions among the even-parity states of ^{12}C and ^{16}O are also shown in Table 7.1. The dominant decay modes that

Table 7.1 NLEFT results at LO, and experimental values for root-mean-square charge radii r_{EM}, quadrupole moments Q, and electromagnetic transition matrix elements among the low-lying even-parity states of ^{12}C and ^{16}O

Observable	Unit	^{12}C (LO)	^{12}C (Exp)	^{16}O (LO)	^{16}O (Exp)
$r_{EM}(0_1^+)$	[fm]	2.2(2)	2.47(2) [15]	2.3(1)	2.710(15) [16]
$r_{EM}(2_1^+)$	[fm]	2.2(2)	–	2.3(1)	–
$Q(2_1^+)$	[$e\,fm^2$]	6(2)	6(3) [17]	15(3)^a	–
$r_{EM}(0_2^+)$	[fm]	2.4(2)	–	2.3(1)	–
$r_{EM}(2_2^+)$	[fm]	2.4(2)	–	–	–
$Q(2_2^+)$	[$e\,fm^2$]	$-7(2)$	–	–	–
$B(E2, 2_1^+ \to 0_1^+)$	[$e^2\,fm^4$]	5(2)	7.6(4) [18]	6.2(1.6)^a	7.4(2) [19]
$B(E2, 2_1^+ \to 0_2^+)$	[$e^2\,fm^4$]	1.5(7)	2.6(4) [18]	46(8)^a	65(7) [20]
$B(E2, 2_2^+ \to 0_1^+)$	[$e^2\,fm^4$]	2(1)	–	–	–
$B(E2, 2_2^+ \to 0_2^+)$	[$e^2\,fm^4$]	6(2)	–	–	–
$m(E0, 0_2^+ \to 0_1^+)$	[$e\,fm^2$]	3(1)	5.5(1) [21]	3.0(1.4)^a	3.6(2) [22]

^{a} "Rescaled" LO results, which account approximately for the deviation from the empirical charge radius at LO. The uncertainties are 1σ estimates which include both statistical and extrapolation errors. The sign differences in Q for the $J = 2$ states reflect the oblate shape of the 2_1^+ state and the prolate shape of the 2_2^+ state

contribute to transitions between the 2^+ and 0^+ states are those driven by the $E2$ term in the multipole expansion (see e.g. Ref. [23] for a detailed treatment of the multipole expansion of transition amplitudes). Empirical results for $E2$ transitions are conventionally given in terms of the transition matrix element

$$B(E2, A \to B) \equiv \sum_{\mu m_B} \left| \langle J_B, m_B | \mathcal{M}(E2, \mu) | J_A, m_A \rangle \right|^2, \tag{7.18}$$

where

$$\mathcal{M}(E2, \mu) = \sum_{\mathbf{n}} \mathbf{r}^2 Y_{2,\mu}(\hat{n}) \, \tilde{f}_{pp}(\mathbf{n}), \tag{7.19}$$

in terms of the correlation function

$$\tilde{f}_{pp}(\mathbf{n}) \equiv \frac{L^3}{Z-1} \langle \Psi_B | : \rho_p^{a^\dagger, a}(\mathbf{n}) \rho_p^{a^\dagger, a}(0) : | \Psi_A \rangle, \tag{7.20}$$

which is computed by a coupled-channel MC simulation, where A and B denote the initial and final state wave functions, obtained by Euclidean time projection. In addition to the $E2$ transitions, decay modes proportional to the $E0$ amplitude also contribute measurably to the widths of the low-lying 0^+ resonances. It should be noted that direct, single-photon $E0$ transitions driven by the electric monopole operator are strictly forbidden by angular momentum conservation. However, when the photon energy exceeds twice the electron mass, the $E0$ transition can proceed through electron-positron pair creation. This transition, which is conventionally described by the matrix element

$$m(E0, A \rightarrow B) \equiv \sum_{\mathbf{n}} \mathbf{r}^2 \, \tilde{f}_{pp}(\mathbf{n}), \qquad (7.21)$$

is, in most cases, a highly suppressed process relative to the $E2$ transition.

Since the LO charge radius r_{EM} of the ^{16}O ground state is smaller than the empirical value r_{exp}, a systematic deviation appears, which arises from the overall size of the second moment of the charge distribution. To compensate for this overall scaling mismatch, one can also calculate "rescaled" quantities multiplied by powers of the ratio $r_{\text{EM}}/r_{\text{exp}}$, according to the length dimension of each observable. With the scaling factor included, we find that the NLEFT predictions for the $E2$ and $E0$ transitions are in good agreement with the experimental values. In particular, NLEFT is able to explain the empirical value of $B(E2, 2_1^+ \rightarrow 0_2^+)$ in ^{16}O, which is $\simeq 30$ times larger than the Weisskopf single-particle shell model estimate [23]. This supports the interpretation of the 2_1^+ state as a rotational excitation of the 0_2^+ state. As we shall see, such an enhancement (also referred to as "strongly collective behavior") of $B(E2, 2_1^+ \rightarrow 0_2^+)$ in ^{12}C is also critical for the generation of sufficient ^{12}C to account for the observed abundances of carbon and oxygen in the universe.

7.1.5   Alpha Clustering

So far, we have considered the spectra, charge density distributions, and transition amplitudes of the low-lying states in ^{12}C and ^{16}O, and concluded that these provide tentative evidence for the clustering of α particles in such nuclei. A direct calculation of the full three-dimensional matter and charge distribution inside the nucleus is computationally very costly. Therefore, in Chap. 8, we shall introduce the "pinhole algorithm", which provides a more effective way to probe the underlying structure of nuclei. Nevertheless, much information about the structure of nuclei can already be obtained by considering the degree of overlap the various states have with carefully chosen trial wave functions. As we will see, this gives a lot of insight into the mysterious phenomenon of α clustering, which is highly prevalent in light nuclei with $N = Z$ and A a multiple of four. The α particle enjoys a special role in nuclear clustering because it is spin- and isospin-saturated, is very deeply bound and hence very difficult to perturb, as its first excited state has an energy of $\simeq$20 MeV above the ground state. For a recent review on α clustering in light nuclei, we refer the reader to Ref. [24].

Let us first consider the case of ^{12}C, illustrated in Fig. 7.6. The initial states labeled A, B, C and D consist of plane-wave (delocalized) single-nucleon orbitals with a strong SU(4) symmetric interaction applied for a projection time t'. Hence, the nucleons are allowed to self-organize into a nucleus, and one thus does not build in any prior assumptions about α clustering. On the other hand, the trial states Δ

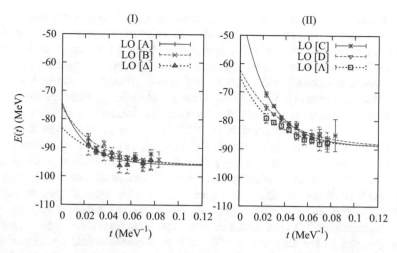

Fig. 7.6 Euclidean projection MC results for the ^{12}C spectrum. The left-hand panel shows the LO energy using initial states A, B and Δ, each of which approaches the ground state energy. The right-hand panel shows the LO energy using initial states C, D and Λ, which trace out an intermediate plateau $\simeq 7\,MeV$ above the ground state

and Λ consist of three α clusters each, arranged in such a way as to recover the same states found using the plane-wave trial states. For such α cluster states, the SU(4) symmetric interaction is taken to be weak, in order to allow these states to retain their basic structure during the Euclidean time evolution. For initial state Δ, the α clusters are formed by Gaussian wave packets centered on the vertices of a compact triangle. In order to construct eigenstates of total momentum and cubic lattice rotations, all possible translations and rotations (12 equivalent orientations) of the initial state are considered. Apparently, Δ is the only trial state composed of α clusters, which provides rapid convergence to the ^{12}C ground state. As shown in the left panel of Fig. 7.6, this coincides with the plane-wave trial states A and B, where all three α clusters are injected with zero total momentum. Hence, we may conclude that the 0_1^+ ground state of ^{12}C has the strongest overlap with trial state Δ, which consists of an isosceles right triangle of α clusters. In the absence of lattice artifacts, Δ would describe an equilateral triangular arrangement. It is equally important to see that very different initial states trace out the ground state with essentially the same energy. This can be used to estimate some (but not all) systematic uncertainties.

In the right-hand panel of Fig. 7.6, we consider plane-wave initial state C, where we take four nucleons with momenta $(0, 0, 0)$, four with $(2\pi/L, 2\pi/L, 2\pi/L)$, and four with $(-2\pi/L, -2\pi/L, -2\pi/L)$. For plane-wave initial state D, we instead take the second and third quadruplets of nucleons with momenta $(2\pi/L, 2\pi/L, 0)$ and $(-2\pi/L, -2\pi/L, 0)$, respectively. The trial state Λ is similar to Δ, except that the α clusters are centered on vertices in a "bent-arm", or obtuse, triangular configuration (with 24 equivalent orientations), reminiscent of an ozone molecule.

We note that in the limit of infinite Euclidean projection time, these states would eventually also converge to the ground state energy. However, as the practical extent of t is limited by sign oscillations and computational scaling, each of these trial states are found instead to approach an intermediate, metastable plateau at $-89(2)$ MeV, which is identified as the 0_2^+ Hoyle state. It should be noted that each of the trial states C, D, and Λ produce a state with an extended (or prolate) geometry, in contrast to the oblate triangular configuration of the ground state.

We shall also briefly comment on the NLEFT studies of ^{16}O, where the ground state was found to consist of α clusters arranged in a tetrahedral configuration. For the first excited 0^+ state (the analog of the Hoyle state in ^{16}O), a predominantly square configuration of α clusters was found. It should be noted that since the early work of Wheeler [25], many theoretical studies have suggested such α cluster structure [26–30], along with some experimental evidence for strong α-like correlations in ^{16}O [31]. While such models have been able to describe some of the puzzles in the structure of ^{16}O on a phenomenological (or geometrical) level, NLEFT is the first a priori framework which provides clear support for the α clustering hypothesis. We will return to such issues related to α clustering in Chap. 8.

7.2   Simulating Other Universes

The effects of variation of the fundamental parameters of nature is a question generating much interest way beyond the nuclear and particle physics communities (for reviews relates to these topics, see e.g. Refs. [32, 33]). In NLEFT, we can indeed calculate nuclear structure and reactions for values of the fundamental parameters that differ from the observed ones, which can loosely be called "simulating other universes". So far, we have alluded to the possibility that the ^{12}C spectrum may be fine-tuned with respect to the level ordering and the vicinity of the first excited 0^+ state (the so-called "Hoyle state") to the ^8Be-α threshold. In fact, the 3α process, whereby three α particles fuse to create ^{12}C, is among the most delicate reactions in nuclear astrophysics. To understand why this is so, we note that big bang nucleosynthesis (BBN) cannot proceed past $A = 7$ (^7Li) because of the lack of stable nuclei for $A = 5$ and $A = 8$. In order to circumvent this gap between BBN and stellar nucleosynthesis, two α particles must first form the unstable ^8Be resonance (with a lifetime of $\sim 10^{-16}$ s), followed by the reaction $^8Be(\alpha,\gamma)^{12}C$. The first reaction also exhibits a high level of fine-tuning, since the ^8Be nucleus is unbound by a mere 90 keV, and thus its lifetime is much longer than the expected $\sim 10^{-23}$ s for reactions driven by the strong interaction. Once carbon has been formed, the nucleosynthesis proceeds in massive stars by the burning of C, O, Ne, and Si up to Fe and Ni, followed by an explosive synthesis of heavier elements. The conditions at which the 3α process takes place (in particular the temperature) are found to play an important role in determining the mass and composition of the C/O core at the end of the He burning stage. This has a large impact on the subsequent evolution phases of massive stars, as it controls the mass of the Fe/Ni core which forms prior to supernova explosions. In turn, the temperature at which the 3α reaction proceeds is very sensitive to the position of the Hoyle state relative to the 3α threshold energy, see e.g. the textbooks [34, 35].

As noted above, the second step in the $^4\text{He}(\alpha\alpha,\gamma)^{12}\text{C}$ process is the capture of a third α particle into the 0^+ Hoyle state resonance, which is located 7.65 MeV above the ground state of ^{12}C. It is noteworthy that the existence of this resonance was first postulated by Hoyle in 1954 [36] in order to enhance the cross section during the He burning phase by orders of magnitude, i.e. to a level sufficient to account for the observed abundances of C and O in the universe. The Hoyle state then decays electromagnetically through $E2$ photon emission to the 2^+ state, located 4.44 MeV above the ground state of ^{12}C. The direct $E0$ transition to the ground state is suppressed relative to the $E2$ route through the 2^+ state. Provided that the ^4He and ^8Be nuclei are in thermal equilibrium, and that the sharp resonance approximation can be applied to the α capture on ^8Be, the reaction rate for $^4\text{He}(\alpha\alpha,\gamma)^{12}\text{C}$ can be expressed as [37]

$$r_{3\alpha} = 3^{\frac{3}{2}} N_\alpha^3 \left(\frac{2\pi\hbar^2}{|E_4|k_B T} \right)^3 \frac{\Gamma_\gamma}{\hbar} \exp\left(-\frac{\varepsilon}{k_B T} \right), \qquad (7.22)$$

where $|E_4| = M_\alpha$ is the mass and N_α is the number density of α particles in a stellar plasma at temperature T. Here, we have exhibited explicit factors of $\hbar$ and the Boltzmann constant k_B, which are usually set to one. The energy of the Hoyle state $\varepsilon = 379.47(18) \simeq 380\,\text{keV}$ relative to the 3α threshold is given by

$$\varepsilon \equiv \Delta E_b + \Delta E_h = E_{12}^\star - 3E_4, \qquad (7.23)$$

where $E_{12}^\star$ is the energy of the Hoyle state. Also, $\Delta E_b \equiv E_8 - 2E_4$ is the energy of the ^8Be ground state relative to the 2α threshold, and $\Delta E_h \equiv E_{12}^\star - E_8 - E_4$. All of the states involved in Eq. (7.22) can be computed in NLEFT using MC methods, along with their sensitivity to small changes in the fundamental parameters of nature, in particular the average light quark mass $m_q = (m_u + m_d)/2$ and the EM fine-structure constant α_{EM}. Clearly, the main effect of shifting ε is to change the temperature T at which He burning in massive stars takes place. The 3α reaction rate is much less sensitive to changes in the radiative width Γ_γ of the Hoyle state, when compared to the exponential dependence on ε. Nevertheless, the strongly enhanced $E2$ transition to the lowest 2^+ state is critical for allowing the Hoyle state to decay electromagnetically before disintegrating back into three α particles.

7.2.1 Sensitivity to the Fundamental Constants of Nature

During the He burning phase in massive stars, the only other significant reaction besides the 3α process is $^{12}\text{C}(\alpha,\gamma)^{16}\text{O}$. Hence, the competition of these two processes is what determines the C and O abundances at the end of the He burning stage. At present, the empirical understanding of $^{12}C(\alpha,\gamma)^{16}O$ relies on extrapolations down to astrophysical energies of $\sim$300 keV, which is the energy of the Gamow peak [34]. However, as this reaction is dominated by broad resonances, its sensitivity to small changes in the spectrum is much less than for the 3α process. From Eq. (7.22), one can draw the somewhat counter-intuitive conclusion that increasing ε will lead to the rapid processing of all ^{12}C to ^{16}O, as the star needs to increase its core temperature for the 3α process to be triggered. On the other hand, if ε is decreased, the He burning proceeds at a lower temperature, and hence very little ^{16}O is produced. While the range in ε compatible with life in the universe (as we know it) can be inferred from astrophysical models of stellar evolution up to the asymptotic giant branch stage, we should turn to NLEFT in order to relate such ad hoc changes to shifts in the fundamental parameters.

In order to gauge the sensitivity of the 3α reaction rate to shifts in the fundamental constants of nature [38, 39], we shall first consider the m_q-dependence or, equivalently, the M_π-dependence of the energies E_i, such as E_4, E_8, E_{12} and $E_{12}^\star$, which are involved in the 3α process. We remind the reader that in Chap. 2 we had discussed the Gell-Mann–Oakes–Renner relation, that in QCD is fulfilled to a few percent accuracy and allows one to translate the light quark mass dependence into the pion mass dependence, as done in the following. From the various energies, we can then obtain the sensitivities of ΔE_b and ΔE_h, and ultimately of ε. If we restrict ourselves to small shifts in M_π around the physical pion mass, denoted as M_π^{ph}, then it is sufficient to consider

$$\delta E_i \simeq \left.\frac{\partial E_i}{\partial M_\pi}\right|_{M_\pi^{\mathrm{ph}}} \delta M_\pi + \left.\frac{\partial E_i}{\partial \alpha_{\mathrm{EM}}}\right|_{\alpha_{\mathrm{EM}}^{\mathrm{ph}}} \delta \alpha_{\mathrm{EM}}, \tag{7.24}$$

with independent variation of M_π and α_{EM}. A useful way to express the sensitivity of an observable X to a parameter y is the dimensionless "K-factor"

$$K_X^i \equiv \left.\frac{y}{X}\frac{\partial X}{\partial y}\right|_{y^{\mathrm{ph}}}, \tag{7.25}$$

where we use the superscript $i = \{q, \pi, \alpha\}$ for $y = \{m_q, M_\pi, \alpha_{\mathrm{EM}}\}$. As an example, we can obtain K_X^q (i.e. the sensitivity of X to changes in m_q) in terms of K_X^π by means of the relation

$$K_X^q = K_X^\pi K_{M_\pi}^q, \tag{7.26}$$

where we take $K_{M_\pi}^q = 0.494^{+0.009}_{-0.013}$ [40], see also Ref. [41].

In addition to shifts in m_q, the effects of shifts in α_{EM} are also of interest. The sensitivity of the energies E_i to variations in α_{EM} can be obtained by computing the shifts $\Delta E_i(\alpha_{\mathrm{EM}})$ and $\Delta E_i(\tilde{C}_{pp})$. The former is due to the long-range Coulomb interaction, and the latter to the pp contact operator (see Chap. 4). We define

$$Q_{EM}(E_i) \equiv \Delta E_i(\tilde{C}_{pp}) x_{pp} + \Delta E_i(\alpha_{EM}), \tag{7.27}$$

where x_{pp} can be fixed by the accurately known EM contribution to the energy of ^{4}He. We take

$$K_X^\alpha = \frac{Q_{EM}(X)}{X} + \frac{\Delta M_\pi^{EM}}{X} \left. \frac{\partial X}{\partial M_\pi} \right|_{M_\pi^{ph}} + \dots, \tag{7.28}$$

where the second term denotes the EM shift of M_π. However, as such effects are expected to amount to a mere $\sim$4% correction [39], they are not considered further.

7.2.2 Quark Mass Dependence of the Nuclear Hamiltonian

We continue our analysis by identifying the different sources of the M_π-dependence in the NLEFT Hamiltonian, and for simplicity we shall restrict ourselves to the LO expressions. The explicit dependence of the energies E_i on M_π is due to the M_π-dependence of the nucleon mass m_N in the kinetic energy term of the NLEFT Hamiltonian, as well as to the obvious M_π-dependence of the OPE contribution. For the M_π-dependence of the nucleon mass, we define

$$x_1 \equiv \left. \frac{\partial m_N}{\partial M_\pi} \right|_{M_\pi^{ph}}, \tag{7.29}$$

which has been analyzed by combining ChPT with Lattice QCD, see e.g. the discussion in the review [42]. At $\mathcal{O}(p^2)$ in ChPT, $m_N(M_\pi)$ is given by

$$m_N = m_0 - 4c_1 M_\pi^2 + \mathcal{O}(p^3), \tag{7.30}$$

where m_0 denotes the value of m_N in the chiral limit. The value $x_1 \simeq 0.57$ was obtained in Ref. [40], and corresponds to the $\mathcal{O}(p^3)$ heavy-baryon (HB) ChPT result with $c_1 = -0.81\,\text{GeV}^{-1}$. Alternatively, one may determine x_1 from the pion-nucleon sigma term

$$\sigma_{\pi N} \equiv \left. M_\pi^2 \frac{\partial m_N}{\partial M_\pi^2} \right|_{M_\pi^{ph}}, \tag{7.31}$$

by means of the Feynman-Hellmann theorem [43–45]. Based on the theoretical error inherent in such analyses, we take $x_1 = 0.57\dots0.97$ as a conservative estimate. This value should be updated using more recent values of m_N and $\sigma_{\pi N}$, but as we will show later, its contribution is so small that the conservative range used here can be considered satisfactory. We leave it as an exercise to the reader to update these and the following values based on the most recent Lattice QCD results.

In addition to the explicit dependence on the pion mass in the OPE propagator, we also account for the M_π-dependence of the coupling $\tilde{g}_{\pi N} \equiv g_A/(2F_\pi)$, by defining

$$x_2 \equiv \left.\frac{\partial \tilde{g}_{\pi N}}{\partial M_\pi}\right|_{M_\pi^{ph}} = \left.\frac{1}{2F_\pi}\frac{\partial g_A}{\partial M_\pi}\right|_{M_\pi^{ph}} - \left.\frac{g_A}{2F_\pi^2}\frac{\partial F_\pi}{\partial M_\pi}\right|_{M_\pi^{ph}}, \tag{7.32}$$

where both contributions to x_2 have been studied by means of CHPT and Lattice QCD [40, 42, 46]. As for m_N in Eq. (7.30), we obtain

$$F_\pi = F\left(1 + \frac{M_\pi^2}{16\pi^2 F^2}\bar{l}_4 + \mathcal{O}(M_\pi^4)\right), \tag{7.33}$$

where $\bar{l}_4 \simeq 4.3$ [47], which agrees with modern lattice determinations [48]. Also, $F \simeq 86.2\,\mathrm{MeV}$ is the value of F_π in the chiral limit. As an alternative to the sub-leading order ChPT result, one may take $K_{F_\pi}^q = 0.048 \pm 0.012$ [40], which is based on a combination of ChPT and Lattice QCD. This gives

$$\left.\frac{\partial F_\pi}{\partial M_\pi}\right|_{M_\pi^{ph}} = \frac{F_\pi}{M_\pi}\frac{K_{F_\pi}^q}{K_{M_\pi}^q} \simeq 0.066, \tag{7.34}$$

using the central value $K_{M_\pi}^q \simeq 0.494$. While the chiral expansion of F_π shows good convergence, the equivalent expression for g_A is known to converge much slower. Constraints from $\pi N \to \pi\pi N$ yield a range of $x_2 = -0.056 \ldots 0.008$ [40], in units of the lattice spacing a, which is a relatively large uncertainty. As in the case of x_1, this large uncertainty does not matter, as the corresponding contribution will turn out to be small.

Next, we turn to the short-range part of the nuclear force that depends (implicitly) on the pion mass M_π. This dependence is more difficult to control within chiral EFT. Since the aim is a model-independent determination of the M_π-dependence of the nuclear energies E_i, we refrain from a chiral expansion of the short-range part of the nuclear force. Rather, we parameterize the M_π-dependence of the LO contact interactions in terms of two coefficients, C and C_{I^2}. As we shall see, this can be achieved in terms of the slopes of the inverse np S-wave scattering lengths

$$\bar{A}_s \equiv \left.\frac{\partial a_s^{-1}}{\partial M_\pi}\right|_{M_\pi^{ph}}, \quad \bar{A}_t \equiv \left.\frac{\partial a_t^{-1}}{\partial M_\pi}\right|_{M_\pi^{ph}}, \tag{7.35}$$

where we have used the subscripts s and t for the spin-singlet ($^1S_0$) and spin-triplet ($^3S_1$) np partial waves, respectively. For the purpose of our analysis, $\bar{A}_s$ and $\bar{A}_t$ are regarded as input parameters. Out final results shall be expressed in terms of these.

Given the sources of the pion mass dependence discussed so far, we may express the dependence of the energies E_i on M_π as

$$E_i = E_i(\tilde{M}_\pi, m_N(M_\pi), \tilde{g}_{\pi N}(M_\pi), C(M_\pi), C_{I^2}(M_\pi)), \tag{7.36}$$

where $\tilde{M}_\pi$ refers to the explicit M_π-dependence from the OPE propagator. It should be noted that we are neglecting the effects of the smearing of the LO contact interactions, as these are technically contributions of higher order. This is consistent, since we are treating the sensitivity to small changes in M_π as a first-order perturbation. The unperturbed LO energies E_i are computed in NLEFT with full account of the smearing effects. In order to assess the sensitivity of the 3α process (and of the various energy levels involved in that process) to shifts in M_π, we will compute quantities of the form $\partial E_i / \partial M_\pi$ at the physical point. Given Eq. (7.36), we find

$$\left. \frac{\partial E_i}{\partial M_\pi} \right|_{M_\pi^{\text{ph}}} = \left. \frac{\partial E_i}{\partial \tilde{M}_\pi} \right|_{M_\pi^{\text{ph}}} + x_1 \left. \frac{\partial E_i}{\partial m_N} \right|_{m_N^{\text{ph}}} + x_2 \left. \frac{\partial E_i}{\partial \tilde{g}_{\pi N}} \right|_{\tilde{g}_{\pi N}^{\text{ph}}}$$

$$+ x_3 \left. \frac{\partial E_i}{\partial C} \right|_{C^{\text{ph}}} + x_4 \left. \frac{\partial E_i}{\partial C_{I^2}} \right|_{C_{I^2}^{\text{ph}}}, \tag{7.37}$$

where we defined

$$x_3 \equiv \left. \frac{\partial C}{\partial M_\pi} \right|_{M_\pi^{\text{ph}}}, \qquad x_4 \equiv \left. \frac{\partial C_{I^2}}{\partial M_\pi} \right|_{M_\pi^{\text{ph}}}, \tag{7.38}$$

for the short-range components of the LO amplitude. Next, we show how the two-nucleon problem can be used to re-express the scheme-dependent parameters x_3 and x_4 in terms of available knowledge of the derivatives $\bar{A}_s$ and $\bar{A}_t$. Once the dependence of the S-wave np scattering lengths in the spin-singlet and triplet channels on M_π is known, x_3 and x_4 can be straightforwardly obtained. We begin by recalling from Chap. 5 the finite-volume formula due to Lüscher,

$$p \cot \delta = \frac{1}{\pi L} S(\eta) \approx -\frac{1}{a}, \qquad \eta \equiv m_N E \left(\frac{L}{2\pi} \right)^2, \tag{7.39}$$

which can be differentiated with respect to M_π. This yields

$$\frac{\partial a^{-1}}{\partial M_\pi} = -\frac{1}{\pi L} S'(\eta) \frac{\partial \eta}{\partial M_\pi}, \tag{7.40}$$

where the function $S(\eta)$ and its derivative may be computed numerically by means of a Taylor series expansion (see Chap. 5). The two-nucleon energy in the singlet channel is denoted by E_s, and that in the triplet channel by E_t. If we define

$$\zeta_{s,t} \equiv \frac{m_N L}{4\pi^3} S'(\eta_{s,t}), \qquad q_{s,t} \equiv \left.\frac{\partial E_{s,t}}{\partial C}\right|_{C^{\mathrm{ph}}}, \tag{7.41}$$

we find the relations

$$-\zeta_s^{-1} \bar{A}_s = \left.\frac{\partial E_s}{\partial \tilde{M}_\pi}\right|_{M_\pi^{\mathrm{ph}}} + x_1 \left(\frac{E_s}{m_N} + \left.\frac{\partial E_s}{\partial m_N}\right|_{m_N^{\mathrm{ph}}}\right) + x_2 \left.\frac{\partial E_s}{\partial \tilde{g}_{\pi N}}\right|_{\tilde{g}_{\pi N}^{\mathrm{ph}}}$$
$$+ (x_3 + x_4)\, q_s, \tag{7.42}$$

$$-\zeta_t^{-1} \bar{A}_t = \left.\frac{\partial E_t}{\partial \tilde{M}_\pi}\right|_{M_\pi^{\mathrm{ph}}} + x_1 \left(\frac{E_t}{m_N} + \left.\frac{\partial E_t}{\partial m_N}\right|_{m_N^{\mathrm{ph}}}\right) + x_2 \left.\frac{\partial E_t}{\partial \tilde{g}_{\pi N}}\right|_{\tilde{g}_{\pi N}^{\mathrm{ph}}}$$
$$+ (x_3 - 3x_4)\, q_t, \tag{7.43}$$

where the energies $E_{s,t}$ and their derivatives can be computed by exact numerical solution of the two-nucleon problem on a spatial lattice (see Chap. 5 for details). However, we may also eliminate x_3 and x_4 in favor of $\bar{A}_s$ and $\bar{A}_t$, to find

$$\left.\frac{\partial E_i}{\partial M_\pi}\right|_{M_\pi^{\mathrm{ph}}} \equiv -\frac{Q_s^{\mathrm{MC}}}{\zeta_s} \bar{A}_s - \frac{Q_t^{\mathrm{MC}}}{\zeta_t} \bar{A}_t - Q_s^{\mathrm{MC}} R_s(x_1, x_2)$$
$$- Q_t^{\mathrm{MC}} R_t(x_1, x_2) + R_{\mathrm{MC}}(x_1, x_2), \tag{7.44}$$

which is equivalent to Eq. (7.37). Here, the label "MC" indicates which quantities refer to the A-nucleon problem that is treated with Euclidean time projection and MC simulations. In Eq. (7.44), we have introduced

$$Q_s^{\mathrm{MC}} \equiv \frac{3}{4q_s} \left.\frac{\partial E_i}{\partial C}\right|_{C^{\mathrm{ph}}} + \frac{1}{4q_s} \left.\frac{\partial E_i}{\partial C_{I^2}}\right|_{C_{I^2}^{\mathrm{ph}}}, \tag{7.45}$$

$$Q_t^{\mathrm{MC}} \equiv \frac{1}{4q_t} \left.\frac{\partial E_i}{\partial C}\right|_{C^{\mathrm{ph}}} - \frac{1}{4q_t} \left.\frac{\partial E_i}{\partial C_{I^2}}\right|_{C_{I^2}^{\mathrm{ph}}}, \tag{7.46}$$

for which the error is given entirely by the statistical uncertainty of the MC calculation. We also have

$$R_{s,t}(x_1, x_2) \equiv \left.\frac{\partial E_{s,t}}{\partial \tilde{M}_\pi}\right|_{M_\pi^{\mathrm{ph}}} + x_1 \left(\frac{E_{s,t}}{m_N} + \left.\frac{\partial E_{s,t}}{\partial m_N}\right|_{m_N^{\mathrm{ph}}}\right) + x_2 \left.\frac{\partial E_{s,t}}{\partial \tilde{g}_{\pi N}}\right|_{\tilde{g}_{\pi N}^{\mathrm{ph}}}, \tag{7.47}$$

and

$$R_{\mathrm{MC}}(x_1, x_2) \equiv \left.\frac{\partial E_i}{\partial \tilde{M}_\pi}\right|_{M_\pi^{\mathrm{ph}}} + x_1 \left.\frac{\partial E_i}{\partial m_N}\right|_{m_N^{\mathrm{ph}}} + x_2 \left.\frac{\partial E_i}{\partial \tilde{g}_{\pi N}}\right|_{\tilde{g}_{\pi N}^{\mathrm{ph}}}, \tag{7.48}$$

where the dominant sources of uncertainty are due to x_1 and x_2.

While the most convenient way to proceed is by means of Eq. (7.44), it is also possible to solve Eqs. (7.42) and (7.43) directly for x_3 and x_4, and to express these as functions of $\bar{A}_{s,t}$ and $x_{1,2}$. This provides the relations

$$x_3 = 4.8470 \times 10^{-2} + 6.7127 \times 10^{-2} x_1$$
$$- 0.25101 \, x_2 - 0.37652 \, \bar{A}_s - 0.20467 \, \bar{A}_t \,, \tag{7.49}$$

$$x_4 = 4.9901 \times 10^{-3} - 1.8998 \times 10^{-3} x_1$$
$$- 1.2532 \times 10^{-2} x_2 - 0.12551 \, \bar{A}_s + 0.20467 \, \bar{A}_t \,, \tag{7.50}$$

for an $N = 32$ lattice ($L = 63.14\,\mathrm{fm}$). Here, the dimensionful quantities x_2, x_3 and x_4 are expressed in units of the inverse (spatial) lattice spacing. Results for $N = 24$ and $N = 32$ are found to be practically indistinguishable, which shows that finite volume effects in the Lüscher analysis are under control.

7.2.3   Sensitivity to the Light Quark Mass

We may now combine the MC results with the two-nucleon scattering analysis, in order to obtain predictions for the M_π-dependence of the various states featuring in the 3α process. We express the results as a function of $\bar{A}_s$ and $\bar{A}_t$. In this way, we obtain the following results for the M_π-dependence of the energy levels involved in the 3α process (here and in what follows, the strength of the EM interaction is not varied)

$$\left.\frac{\partial E_4}{\partial M_\pi}\right|_{M_\pi^{\mathrm{ph}}} = -0.339(5) \, \bar{A}_s - 0.697(4) \, \bar{A}_t + 0.0380(14)^{+0.008}_{-0.006}, \tag{7.51}$$

$$\left.\frac{\partial E_8}{\partial M_\pi}\right|_{M_\pi^{\mathrm{ph}}} = -0.794(32) \, \bar{A}_s - 1.584(23) \, \bar{A}_t + 0.089(9)^{+0.017}_{-0.011}, \tag{7.52}$$

$$\left.\frac{\partial E_{12}}{\partial M_\pi}\right|_{M_\pi^{\mathrm{ph}}} = -1.52(3) \, \bar{A}_s \quad 2.88(2) \, \bar{A}_t + 0.159(7)^{|\,0.023}_{-0.018}, \tag{7.53}$$

$$\left.\frac{\partial E_{12}^\star}{\partial M_\pi}\right|_{M_\pi^{\mathrm{ph}}} = -1.588(11) \, \bar{A}_s - 3.025(8) \, \bar{A}_t + 0.178(4)^{+0.026}_{-0.021}, \tag{7.54}$$

where the error receives contributions both from the statistical error of the MC calculation (given in parentheses) as well as from the uncertainties in x_1 and

x_2 (explicit positive and negative bounds given). It is noteworthy that x_1 and x_2 only affect the constant terms in the above results, and therefore the (sizable) uncertainties in these coefficients have a relatively minor impact. We can also assess the sensitivity to small shifts in M_π by computing the "K-factors" as defined in Eq. (7.25). For this purpose, we take $M_\pi = 138.0\,\text{MeV}$ as the isospin-averaged pion mass, and the empirical values $E_4^{\exp} = -28.30\,\text{MeV}$, $E_8^{\exp} = -56.50\,\text{MeV}$, $E_{12}^{\exp} = -92.16\,\text{MeV}$, and $E_{12}^{*\exp} = -84.51\,\text{MeV}$ for the E_i. This yields

$$K_{E_4}^\pi = 1.652(25)\,\bar{A}_s + 3.401(21)\,\bar{A}_t - 0.185(7)_{-0.039}^{+0.029}, \tag{7.55}$$

$$K_{E_8}^\pi = 1.94(8)\,\bar{A}_s + 3.87(6)\,\bar{A}_t - 0.217(21)_{-0.041}^{+0.027}, \tag{7.56}$$

$$K_{E_{12}}^\pi = 2.27(4)\,\bar{A}_s + 4.32(3)\,\bar{A}_t - 0.239(11)_{-0.034}^{+0.026}, \tag{7.57}$$

$$K_{E_{12}^*}^\pi = 2.593(19)\,\bar{A}_s + 4.940(13)\,\bar{A}_t - 0.291(7)_{-0.043}^{+0.034}, \tag{7.58}$$

where the same conventions for the errors have been applied. Having calculated the shifts of the individual energy levels involved in the 3α process, we may combine these and obtain similar predictions for the energy differences ΔE_b, ΔE_h and in particular for ε, which is the control parameter for the 3α reaction rate $r_{3\alpha}$. We find

$$\left.\frac{\partial \Delta E_b}{\partial M_\pi}\right|_{M_\pi^{\text{ph}}} = -0.117(34)\,\bar{A}_s - 0.189(24)\,\bar{A}_t + 0.013(9)_{-0.002}^{+0.003}, \tag{7.59}$$

$$\left.\frac{\partial \Delta E_h}{\partial M_\pi}\right|_{M_\pi^{\text{ph}}} = -0.455(35)\,\bar{A}_s - 0.744(24)\,\bar{A}_t + 0.051(10)_{-0.009}^{+0.008}, \tag{7.60}$$

$$\left.\frac{\partial \varepsilon}{\partial M_\pi}\right|_{M_\pi^{\text{ph}}} = -0.572(19)\,\bar{A}_s - 0.933(15)\,\bar{A}_t + 0.064(6)_{-0.009}^{+0.010}, \tag{7.61}$$

and

$$K_{\Delta E_b}^\pi = -175(51)\,\bar{A}_s - 284(36)\,\bar{A}_t + 19(13)_{-3.0}^{+4.5}, \tag{7.62}$$

$$K_{\Delta E_h}^\pi = -217(16)\,\bar{A}_s - 355(12)\,\bar{A}_t + 25(5)_{-4.5}^{+4.0}, \tag{7.63}$$

$$K_\varepsilon^\pi = -208(7)\,\bar{A}_s - 339(5)\,\bar{A}_t + 23(2)_{-3.4}^{+3.7}, \tag{7.64}$$

where we have used the empirical values $\Delta E_b^{\exp} = 92\,\text{keV}$, $\Delta E_h^{\exp} = 289\,\text{keV}$, and $\varepsilon = 380\,\text{keV}$.

7.2.4 Sensitivity to Electromagnetic Shifts

We now turn our attention to the MC results for the EM shifts, keeping the average light quark mass at its physical value. For that, we compute the energy shifts

$Q_{EM}(E_i)$ defined in Eq. (7.27), which involves the unknown parameter x_{pp}. This determines the relative strength of the proton-proton contact interaction that emerges from the lattice regularization of the long-range Coulomb force. We may fix x_{pp} by means of the known contribution of the Coulomb force to the binding energy of ^{4}He, which is (A. Nogga, private communication)

$$Q_{EM}(E_4) = 0.78(3)\,\text{MeV}, \tag{7.65}$$

where the quoted error reflects the model dependence. This is determined from the range of values corresponding to different phenomenological two- and three-nucleon potentials, as well as nuclear forces derived in chiral EFT.

We note that MC calculations for $\Delta E_i(\alpha_{EM})$ and $\Delta E_i(\tilde{C}_{pp})$ give

$$Q_{EM}(E_4) = 0.433(3)\,\text{MeV} \times x_{pp} + 0.613(2)\,\text{MeV}$$
$$\overset{!}{=} 0.78(3)\,\text{MeV} \;\to\; x_{pp} \simeq 0.39(5), \tag{7.66}$$

for ^{4}He, which enables us to predict

$$Q_{EM}(E_8) = 1.02(3)\,\text{MeV} \times x_{pp} + 2.35(2)\,\text{MeV}$$
$$= 2.75(8)\,\text{MeV}, \tag{7.67}$$

for ^{8}Be. Now, using $x_{pp} = 0.39(5)$ and the MC results of NLEFT, one predicts

$$Q_{EM}(E_{12}^\star) = 2.032(10)\,\text{MeV} \times x_{pp} + 5.54(2)\,\text{MeV}$$
$$= 6.33(6)\,\text{MeV}, \tag{7.68}$$

for the Hoyle state, and

$$Q_{EM}(E_{12}) = 1.95(2)\,\text{MeV} \times x_{pp} + 5.67(2)\,\text{MeV}$$
$$= 6.43(6)\,\text{MeV}, \tag{7.69}$$

for the ground state of ^{12}C. We are now in the position to predict the EM shifts of the energy differences relevant to the 3α process,

$$Q_{EM}(\Delta E_h) = 2.80(10)\,\text{MeV},$$
$$Q_{EM}(\Delta E_b) = 1.19(8)\,\text{MeV}, \tag{7.70}$$
$$Q_{EM}(\varepsilon) = 3.99(9)\,\text{MeV}.$$

for ΔE_b, ΔE_h and ε, respectively.

7.2.5   Correlations in Nuclear Binding Energies

Let us now return to the issue of the sensitivity of the excitation energy ΔE_c of the Hoyle state to small changes in M_π and $\alpha_{\rm EM}$. We compute ΔE_c from

$$\Delta E_c \equiv E_{12}^\star - E_{12}, \tag{7.71}$$

for which we find

$$\left. \frac{\partial \Delta E_c}{\partial M_\pi} \right|_{M_\pi^{\rm ph}} = -0.07(3)\, \bar{A}_s - 0.14(2)\, \bar{A}_t + 0.019(9)^{+0.004}_{-0.003}\,. \tag{7.72}$$

This yields

$$K_{\Delta E_c}^\pi = -1.3(5)\, \bar{A}_s - 2.6(4)\, \bar{A}_t + 0.34(15)^{+0.07}_{-0.05}\,, \tag{7.73}$$

for the sensitivity to small changes in M_π. The above result corresponds to the empirical value of $\Delta E_c^{\rm exp} = 7.65$ MeV. Finally, we have

$$Q_{\rm EM}(\Delta E_c) = 0.10(7)\ {\rm MeV}, \tag{7.74}$$

for the EM shift in the excitation energy of the Hoyle state.

We are also in a position to draw conclusions concerning the individual energies E_i and the associated energy differences. The first interesting observation is that the energy differences ΔE_h, ΔE_b and ε are, by themselves, extremely sensitive to changes in M_π as could be expected. Such a conclusion follows from the unnaturally large coefficients in Eqs. (7.62)–(7.64). Notice that the fact that ΔE_h, ΔE_b and ε are much smaller than the individual E_i does not, by itself, imply a strong fine-tuning. For example, the sensitivity of the Hoyle state excitation energy ΔE_c in Eq. (7.73) is of a natural size, in spite of $\Delta E_c^{\rm exp} \simeq 7.65$ MeV being almost an order of magnitude smaller than $|E_{12}^{\rm exp}| \simeq 92.16$ MeV. While ΔE_h, ΔE_b and ε are by themselves extremely sensitive to variations in M_π, the approximate relations

$$\frac{K_{\Delta E_h}^\pi}{K_{\Delta E_b}^\pi} \simeq 1.25, \qquad \frac{K_{\Delta E_h}^\pi}{K_\varepsilon^\pi} \simeq 1.05, \tag{7.75}$$

are satisfied for the central values of the individual terms in Eqs. (7.62)–(7.64) at the level of a few percent. This suggests that ΔE_h, ΔE_b and ε cannot be independently varied (or fine-tuned) by changing the singlet and triplet NN scattering lengths (or, equivalently, by changing the strength of the short-range NN force in the ${}^1S_0$ and ${}^3S_1$ channels). Moreover, it is apparent from Eqs. (7.55) to (7.58) that the energies E_i of the individual states are strongly correlated in a similar manner. In conclusion, the scenario of independent shifts of the energy levels pertinent to the 3α process under changes in the fundamental parameters appears strongly disfavored. In fact, it has been suggested earlier that the fine-tunings in the 3α process are correlated, which means that ε and ΔE_b cannot be adjusted independently of each other [49]. Only with the powerful machinery of NLEFT this can been shown explicitly, as demonstrated here.

7.2.6 Anthropic Considerations

The closeness of the ^{8}Be+^{4}He threshold to the Hoyle state has often been heralded as a prime example of the so-called anthropic principle (AP). Loosely defined, the AP states that the observed values of all physical and cosmological quantities are not equally probable. Rather, they take on values restricted by the requirement that there exist sites where carbon-oxygen based life can evolve, and by the requirement that the universe be old enough for it to have already done so [50]. However, see Ref. [51] for a critical assessment of the AP. The reader is also referred to a more detailed discussion of the AP and related issues in Ref. [52].

The correct physics question to be asked in this context is the following: Assuming that calculations of stellar evolution constrain the ad hoc variation in the reaction rate of the 3α process which is compatible with carbon-oxygen based life, how strongly does the reaction rate of the 3α process depend on shifts in the fundamental parameters, in particular m_q and α_{EM}? This, without any doubts, can be answered by nuclear lattice simulations, as the stellar modeling calculations of Refs. [37, 53] reveal how sensitively the production of carbon and oxygen in evolved stars depends on the proximity of the Hoyle state to the ^{8}Be+^{4}He threshold. These results indicate that sufficient abundances of both carbon and oxygen can be maintained within an envelope of $\pm 100\,$keV around the empirical value of $\varepsilon = 379.47(18)\,$keV. We note that very recent stellar simulations appear to soften this envelope [54]. For small variations $|\delta\alpha_{EM}/\alpha_{EM}| \ll 1$ and $|\delta m_q/m_q| \ll 1$, the resulting change in ε can be expressed as

$$\delta(\varepsilon) \approx \left.\frac{\partial\varepsilon}{\partial M_\pi}\right|_{M_\pi^{ph}} \delta M_\pi + \left.\frac{\partial\varepsilon}{\partial\alpha_{EM}}\right|_{\alpha_{EM}^{ph}} \delta\alpha_{EM} \qquad (7.76)$$

$$= \left.\frac{\partial\varepsilon}{\partial M_\pi}\right|_{M_\pi^{ph}} K_{M_\pi}^q M_\pi \left(\frac{\delta m_q}{m_q}\right) + Q_{EM}(\varepsilon)\left(\frac{\delta\alpha_{EM}}{\alpha_{EM}}\right),$$

where $K_{M_\pi}^q = 0.494^{+0.009}_{-0.013}$ [40]. Thus, the condition $|\delta(\varepsilon)| < 100\,$keV together with Eq. (7.70) leads to the predicted tolerance $|\delta\alpha_{EM}/\alpha_{EM}| \simeq 2.5\%$ of carbon-oxygen based life to shifts in α_{EM} when the quark masses are kept at their physical values.

The dependence on variations of the quark masses m_q for α_{EM} fixed at its physical value is more subtle. From Eq. (7.76) we find

$$\left|\left[0.572(19)\,\bar{A}_s + 0.933(15)\,\bar{A}_t - 0.064(6)\right]\left(\frac{\delta m_q}{m_q}\right)\right| < 0.15\%, \qquad (7.77)$$

using Eq. (7.61), where we have neglected the (relatively insignificant) errors due to x_1 and x_2. In the most generic scenario, assuming that both of the dimensionless quantities $\bar{A}_s$ and $\bar{A}_t$ are $\sim \mathcal{O}(1)$, and therefore $0.572(19)\,\bar{A}_s + 0.933(15)\,\bar{A}_t \sim \mathcal{O}(1)$, our results imply that a change in m_q of as little as $\simeq 0.15\%$ would suffice to render carbon-oxygen based life unlikely to exist. It should be noted that in such a scenario, one can approximate

$$\left.\frac{\partial \varepsilon}{\partial M_\pi}\right|_{M_\pi^{ph}} \approx 1.5 \left.\frac{\partial E_4}{\partial M_\pi}\right|_{M_\pi^{ph}}, \qquad (7.78)$$

which implies that the binding energy of ^{4}He should be fine-tuned under variation of m_q to its empirical value at the level of $\simeq 0.25\%$ in order to fulfill the condition $|\delta(\varepsilon)| < 100$ keV. Nevertheless, there clearly also exists a special value for the ratio of $\bar{A}_s$ to $\bar{A}_t$, given by

$$\bar{A}_s/\bar{A}_t \simeq -1.5, \qquad (7.79)$$

for which the dependence of ΔE_h, ΔE_b and ε on M_π becomes vanishingly small (compared to the statistical uncertainties of the MC calculation), such that the factor

$$0.572(19)\,\bar{A}_s + 0.933(15)\,\bar{A}_t - 0.064(6) \ll 1, \qquad (7.80)$$

in Eq. (7.77). In this case, one would conclude that the reaction rate of the 3α process were completely insensitive to shifts in m_q. As the realistic scenario is likely to be found somewhere in between these extreme cases, it becomes important to consider the available constraints on $\bar{A}_s$ and $\bar{A}_t$.

It should be noted that the quark mass dependence of the S-wave NN scattering lengths has been extensively analyzed within chiral EFT. As we have also found, the problem common to these calculations is the lack of knowledge about the m_q-dependence of the NN contact interactions. Estimates of the size of the corresponding LECs by means of dimensional analysis typically lead to large uncertainties for chiral extrapolations of the scattering lengths. For example, the NLO calculation of Refs. [9, 10] resulted in

$$K_{a_s}^q = 5 \pm 5, \qquad K_{a_t}^q = 1.1 \pm 0.9, \qquad (7.81)$$

which are nevertheless consistent with the NLO analysis of Ref. [11, 12], which yielded

$$K_{a_s}^q = 2.4 \pm 3.0, \qquad K_{a_t}^q = 3.0 \pm 3.5, \tag{7.82}$$

based on a perturbative treatment of the OPE. One may also attempt to combine chiral EFT with Lattice QCD. In particular, the NPLQCD collaboration has determined the regions for the S-wave scattering lengths consistent with their lattice results, $a_s = (0.63 \pm 0.50)$ fm and $a_t = (0.63 \pm 0.74)$ fm, obtained for $M_\pi = 353.7 \pm 2.1$ MeV [55]. It should be noted that these results remain under scrutiny [56].

Recently, the m_q-dependence of NN observables has been analyzed to NNLO in chiral EFT [40]. To overcome the difficulties due to the poorly known m_q-dependence of the short-range NN interactions, the fact that the LECs accompanying the NN contact interactions are saturated by exchanges of heavy mesons [57] was exploited. By means of a "unitarized" version of ChPT combined with Lattice QCD, the m_q-dependence of the meson resonances saturating these LECs can be described. It then becomes possible to analyze the m_q-dependence of the NN observables to NNLO, without relying on a chiral expansion of the NN contact interactions. The results of this procedure can be summarized by

$$K_{a_s}^q = 2.3_{-1.8}^{+1.9}, \qquad K_{a_t}^q = 0.32_{-0.18}^{+0.17}, \tag{7.83}$$

and

$$\bar{A}_s = -\frac{1}{a_s M_\pi} \frac{K_{a_s}^q}{K_{M_\pi}^q} \simeq 0.29_{-0.23}^{+0.25},$$

$$\bar{A}_t = -\frac{1}{a_t M_\pi} \frac{K_{a_t}^q}{K_{M_\pi}^q} \simeq -0.18_{-0.10}^{+0.10}, \tag{7.84}$$

which are roughly consistent with other parameter-free chiral EFT calculations [58]. Notably, direct application of the central values gives $\bar{A}_s/\bar{A}_t \simeq -1.6$, which leads to a strong cancellation of the dependence of ΔE_h, ΔE_b and ε on $\bar{A}_s$ and $\bar{A}_t$, and hence also to a mild dependence on M_π. As shown in Refs. [38, 39], taking into account the uncertainties in $\bar{A}_s$ and $\bar{A}_t$, one may conclude that changes of 2–3% in the quark masses are not detrimental for carbon-oxygen based life. However, in the future, more precise values of $\bar{A}_s$ and $\bar{A}_t$ will be obtained from lattice QCD, and this will help to lower this uncertainty. For the present state-of-the-art, see e.g. Ref. [56].

7.3   Alpha-Alpha Scattering

We shall conclude the chapter on the application of NLEFT to light and medium-mass nuclei by considering the problem of two α particles scattering on the lattice. While ab initio descriptions of nuclear scattering and reactions of light nuclei is a rapidly developing field [59–63], most numerical methods suffer from an adverse computational scaling with increased number of nucleons in the projectile (either exponential or factorial). Examples of processes that remain challenging are α-α scattering, α-^{12}C scattering and radiative capture, as well as the carbon and oxygen burning stages in the evolution of massive stars [64, 65]. Here, we describe how the adiabatic projection method, introduced in Chap. 6, can be combined with MC simulations to yield a computational scaling of $\sim(A_1 + A_2)^2$ for the two-nucleus scattering problem. Such scaling is sufficiently mild to allow for the ab initio calculation of α processes, and we shall illustrate this by considering α-α scattering to NNLO in NLEFT. In particular, we compute NLEFT predictions for the S-wave and D-wave α-α scattering phase shifts, for which experimental data exist [66–68].

7.3.1 Adiabatic Projection of Cluster States

We consider a set of initial two-α states $|\mathbf{R}\rangle$, labeled by the spatial separation vector $\mathbf{R}$. A suitable lattice size for the α-α calculations is $L \simeq 16$ fm, ignoring for a moment the Coulomb force between protons. The initial α wave functions are taken to be Gaussian wave packets which factorize into a product of two individual α clusters, according to

$$|\mathbf{R}\rangle = \sum_{\mathbf{r}} |\mathbf{r} + \mathbf{R}\rangle_1 \otimes |\mathbf{r}\rangle_2, \qquad (7.85)$$

for large $|\mathbf{R}|$. The summation over $\mathbf{r}$ is required to produce states with total momentum equal to zero. As we have done in Chap. 6 for NN scattering, we project onto the spherical harmonics using

$$|R\rangle^{l,m} = \sum_{\mathbf{R}'} Y_{l,m}(\hat{R}')\delta_{R,|\mathbf{R}'|}|\mathbf{R}'\rangle, \qquad (7.86)$$

with angular momentum quantum numbers (l, m), and we only consider $|\mathbf{R}| < L/2$. Here, the Kronecker delta $\delta_{R,|\mathbf{R}|}$ selects lattice points where $|\mathbf{R}| = R$. Hence, we are able to apply a partial-wave projection method similar to the one introduced for NN scattering in Chap. 5. This reduces the calculational complexity, and hence increases the expected accuracy of the α-α scattering observables.

In order to form "dressed" cluster states that approximately span the set of low-energy α-α scattering states in the periodic box, we compute the product

$$|R\rangle^{l,m}_t = \exp(-H_{\mathrm{LO}}t) \times \exp(-H_{\mathrm{SU(4)}}t')|R\rangle^{l,m}, \qquad (7.87)$$

such that effects of deformation and polarization of the α clusters are accounted for once the Euclidean projection time t is sufficiently large. Since we work at finite temporal lattice spacing δ, we are in practice computing

$$|R\rangle_t^{l,m} = M_{\text{LO}}^{N_t} \times M_{\text{SU}(4)}^{N_{t'}} |R\rangle^{l,m}, \tag{7.88}$$

in terms of the SU(4)-symmetric and full LO transfer matrices, which is applied for N_t time steps. Once the dressed cluster states have been obtained, we have the radial transfer matrix

$$\left[\mathcal{M}_t\right]_{R',R}^{l,m} = {}_t^{l,m}\langle R'|M|R\rangle_t^{l,m}, \tag{7.89}$$

for $N_t + 1$ Euclidean time steps, and

$$\left[\mathcal{N}_t\right]_{R',R}^{l,m} = {}_t^{l,m}\langle R'|R\rangle_t^{l,m}, \tag{7.90}$$

for the norm matrix. Then, we find

$$\left[\mathcal{M}_t^a\right]_{R',R}^{l,m} = \left[\mathcal{N}_t^{-1/2} \times \mathcal{M}_t \times \mathcal{N}_t^{-1/2}\right]_{R',R}^{l,m}, \tag{7.91}$$

for the radial adiabatic transfer matrix. As for the case of NN scattering, we impose a spherical hard wall boundary at a suitably chosen radius R_W. For large N_t, the standing waves of the radial adiabatic transfer matrix (7.91) are used to determine the elastic phase shifts for α-α scattering. We shall return momentarily to the issue of the long-range Coulomb interaction in the adiabatic formalism.

The auxiliary fields are updated using the HMC algorithm. Interspersed with the HMC updates, we perform Metropolis updates of the locations of the Gaussian wave packets $\mathbf{R}_1, \mathbf{R}_2$ in the initial state, and $\mathbf{R}_3, \mathbf{R}_4$ in the final state. Such Metropolis updates are accepted or rejected based on the absolute value of the LO amplitude

$$Z(\mathbf{R}_3, \mathbf{R}_4; \mathbf{R}_1, \mathbf{R}_2; 2t) \equiv {}_{t'}\langle \mathbf{R}_3, \mathbf{R}_4| \exp(-2H_{\text{LO}}t)|\mathbf{R}_1, \mathbf{R}_2\rangle_{t'}, \tag{7.92}$$

for the updated positions of the wave packets. In practical calculations with finite δ, the exponentials of the Hamiltonians are to be understood as transfer matrices. In Eq. (7.92), the trial states are

$$|\mathbf{R}_1, \mathbf{R}_2\rangle_{t'} \equiv \exp(-H_{\text{SU}(4)}t')|\mathbf{R}_1, \mathbf{R}_2\rangle, \tag{7.93}$$

where the two α clusters described by Gaussian wave packets, centered on lattice sites $\mathbf{R}_1$ and $\mathbf{R}_2$, respectively. We have formally given the projection time as $2t$ rather than t, as the total Euclidean projection time for the initial and final dressed cluster states taken together is $2t$.

As usual in NLEFT, the NLO and NNLO corrections, including Coulomb and isospin-breaking effects, are computed as first-order perturbations to (7.89), which gives a corresponding correction to the adiabatic transfer matrix (7.91). It should be noted that a non-perturbative treatment of the repulsive Coulomb contribution would cause increased sign oscillations, and decrease the quality of the MC data. Moreover, in the power counting of NLEFT the electric charge e is regarded as a small parameter. Nevertheless, an important effect of the Coulomb interaction is the change in the asymptotic behavior of the two-cluster wave function from spherical Bessel and Neumann functions to the equivalent Coulomb functions, as shown in Chap. 5.

7.3.2 Scattering Phase Shift

Within the adiabatic projection method, there are several ways in which scattering phase shifts can be obtained [69]. Firstly, the "Lüscher energy spectrum method" makes exclusive use of the energy of a scattering state. Secondly, in the "Lüscher wave function method", the computed wave function in the periodic box is fitted to its (known) asymptotic form. While the latter method works well for a one-dimensional system, the three-dimensional system was found to be much more problematic due to systematic errors from lattice artifacts, in particular for $l \geq 1$. The third method is the familiar spherical wall method, whereby scattering phase shifts are extracted by imposing a hard, spherical wall at a suitably chosen radius R_w. As shown in Chap. 5, a large range of scattering momenta can be accessed by variation of R_w, and by introduction of an adjustable "auxiliary potential" in the vicinity of the wall radius. In Ref. [69], the spherical wall method has been shown to be the method of choice for the type of reactions discussed here, because of its accuracy and applicability to arbitrary orbital angular momenta.

As we have found in Chap. 5, the Coulomb interaction modifies the asymptotic behavior of the scattering wave function. For simplicity, we shall here consider the case where the Coulomb potential is absent, noting that it can be straightforwardly accounted for by replacing the spherical Hankel functions with their Coulomb equivalents (for $\eta \neq 0$, where η is the Sommerfeld parameter). In the presence of a hard spherical wall, and at distances where the nuclear potential is negligible, the wave function of the two-cluster system is given by

$$\psi_{l,m}^{(p)} = A_l Y_{l,m}(\theta, \phi) \left[\cos \delta_l(p) j_l(pr) - \sin \delta_l(p) n_l(pr) \right], \tag{7.94}$$

in terms of the spherical Bessel and Neumann functions (or the Coulomb functions for $\eta = 0$), where p is the relative momentum of the α clusters, and A_l is a normalization coefficient. For $l = 0$, the use of Eq. (7.94) presents no problems, as the angular part is merely a constant. Each individual fit to a given wave function provides A_l, p and $\delta_l(p)$. As the MC data has some scatter due to the statistical error (see Fig. 7.7), the error can be propagated to the fitted parameters by resampling the input wave function data. For instance, a jackknife analysis can be performed, whereby individual radial shells are randomly omitted from the fit. When such a set of resampled fits is taken together, the mean and variance of the mean of each

Fig. 7.7 *S*-wave α-α radial scattering wave functions, for an adiabatic transfer matrix accurate to NNLO in NLEFT, and a hard spherical wall at a cluster separation radius of $r \simeq 36$ fm. The phase shift is extracted by fitting the MC data points using Coulomb functions (dashed lines)

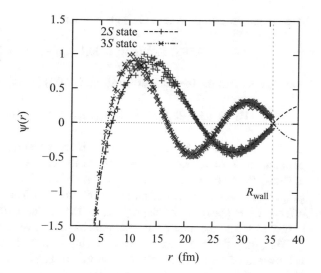

fitted parameter can be estimated. In order to account for *D*-wave α-α scattering, we note that the case of $l = 2$ is more difficult to analyze. In addition to the non-trivial angular dependence of the wave function, the angular momentum multiplet of $2l + 1$ states is broken up according to the irreducible representations of the cubic rotation group. For $l \neq 0$, instead of the wave function we consider the probability distribution

$$\sum_{m=-l}^{l} \left| \psi_{l,m}^{(p)} \right|^2 = \frac{2l + 1}{4\pi} |A_l|^2 \left[\cos \delta_l(p) j_l(pr) - \sin \delta_l(p) n_l(pr) \right]^2 , \qquad (7.95)$$

summed over m, which can again be fitted to MC data. Examples of radial wave functions for *S*-wave α-α scattering states are shown in Fig. 7.7. For a detailed study of fermion-dimer scattering within the adiabatic formalism, where the Lüscher energy spectrum and wave function methods are benchmarked against the spherical wall method, see Ref. [69].

We compute the radial adiabatic transfer matrix from Eq. (7.91) using a smaller lattice of $L \sim 10$–20 fm, after which the calculation is extended to much larger volumes of $L \sim 100$–200 fm. At large separations, the α clusters are effectively non-interacting, except for the Coulomb interaction. Matrix elements of the radial adiabatic transfer matrix are computed from MC simulations of a single α particle. We refer to this as the "trivial" two-cluster Hamiltonian [70]. Here, we shall briefly describe the construction of the trivial two-cluster transfer matrix (or Hamiltonian), into which the (infinite-range) Coulomb interaction is explicitly included. Again, we evolve an anti-symmetrized product of single-nucleon states comprising one α cluster located on lattice site $\mathbf{R}_1$ using the SU(4)-symmetric Hamiltonian

$$|\mathbf{R}_1\rangle_{t'} \equiv \exp(-H_{\text{SU(4)}} t')|\mathbf{R}_1\rangle, \qquad (7.96)$$

after which we compute the LO single-cluster amplitude

$$Z(\mathbf{R}_3; \mathbf{R}_1; 2t) \equiv {}_{t'}\langle \mathbf{R}_3| \exp(-2H_{\mathrm{LO}}t)|\mathbf{R}_1\rangle_{t'}, \tag{7.97}$$

and we note that the two-cluster amplitude (7.92) factorizes according to

$$Z(\mathbf{R}_3, \mathbf{R}_4; \mathbf{R}_1, \mathbf{R}_2; 2t) \rightarrow Z(\mathbf{R}_3; \mathbf{R}_1; 2t) \times Z(\mathbf{R}_4; \mathbf{R}_2; 2t), \tag{7.98}$$

in the limit where $\mathbf{R}_1$ and $\mathbf{R}_3$ are widely separated from $\mathbf{R}_2$ and $\mathbf{R}_4$. The generalization of the above formalism to the case of transfer matrices at finite δ is straightforward. It should be noted that the trivial Hamiltonian does not account properly for antisymmetrization between nucleons from different α clusters. Such effects are expected to be negligible, as the trivial Hamiltonian is only used for distances where the individual α clusters have little overlap.

The phase shift for S-wave α-α scattering, obtained from an adiabatic projection calculation of NLEFT to NNLO, is shown in Fig. 7.8 as a function of the energy E_{LAB} in the laboratory frame, together with experimental data [66–68], and a comparison with the halo EFT prediction of Ref. [71]. We note that

$$E_{\mathrm{LAB}} = \frac{p_{\mathrm{CM}}^2}{\mu} = 2E_{\mathrm{CM}}, \tag{7.99}$$

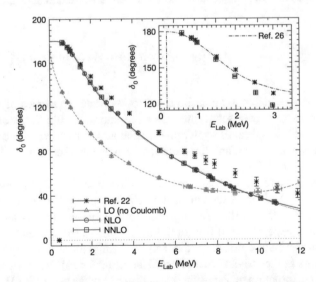

Fig. 7.8 S-wave phase shifts for α-α scattering at LO (green triangles), NLO (blue circles), and NNLO (red squares), compared with experimental data [66–68] (black asterisks). The error bars represent the uncertainty due to statistical MC errors, and errors from extrapolation in Euclidean projection time. The blue short-dashed (NLO) and red solid (NNLO) lines are determined from fits to MC data using a Coulomb-modified effective range expansion. The green dashed (LO) line corresponds to an effective range expansion without Coulomb. The black dot-dashed line in the inset shows the prediction of halo EFT with point-like α particles [71]

where

$$E_{CM} \equiv \frac{p_{CM}^2}{2\mu}, \quad \mu = \hat{M}_\alpha/2, \quad (7.100)$$

in terms of the momenta and energies defined in the CM frame, where $\hat{M}_\alpha$ is the "effective" mass of the α particle, as measured from the lattice dispersion relation. The phase shifts obtained from the MC calculation have been fitted with an effective range expansion (ERE), which has been modified to account for the long-range Coulomb interaction [70], see also Chap. 5. As Coulomb effects are absent at LO, the behavior of the phase shift in the vicinity of the α-α scattering threshold is qualitatively different at LO. The data in Fig. 7.8 correspond to MC simulations performed for a number of Euclidean time steps between $N_t = 4$ and $N_t = 10$, with the limit $N_t \rightarrow \infty$ taken by means of extrapolation. The residual dependence on Euclidean time was described by an exponential fit involving one excited state. In Ref. [70], resonance energies E_R and widths Γ_R were determined by analysis of the ERE phase shift in terms of a Breit-Wigner resonance, or more sophisticated methods [72]. The observed energy E_R of the S-wave resonance in the CM frame is 0.092 MeV above the α-α threshold. On the lattice, the ground state was found at $E_R = 0.79(9)$ MeV below threshold at LO, and at $E_R = 0.11(1)$ MeV below threshold at NLO and NNLO. Two remarks are in order: Firstly, the accuracy of $\simeq 200$ keV with respect to the location of the ^{8}Be pole, reflects the uncertainty at this order in EFT. Secondly, the very tiny NNLO correction appears to be an artifact of the relatively coarse lattice.

The NLEFT calculation of the D-wave α-α phase shift is shown in Fig. 7.9 (with the same conventions as for Fig. 7.8). In this case, the NLO and NNLO results are

Fig. 7.9 D-wave phase shifts for α-α scattering at LO (green triangles), NLO (blue circles), and NNLO (red squares), compared with experimental data [66–68] (black asterisks). Notation and conventions are as for Fig. 7.8

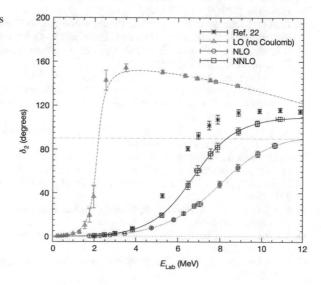

distinct, showing a clear improvement from NLO to NNLO. The dominant feature of the D-wave data is a wide resonance, located at $E_R = 2.92(18)$ MeV (in the CM frame), with a width of $\Gamma_R \simeq 1.34(50)$ MeV [68]. While the determination of parameters for such a wide resonance is expected to involve some model-dependence, the NLEFT calculations at NNLO give $E_R = 3.27(12)$ MeV and $\Gamma_R = 2.09(16)$ MeV, consistent with the empirical determinations. Clearly, there is room for improvement left, but we emphasize again that this is a truly ab initio calculation, without adjustable parameters.

Within the adiabatic projection framework, the ab initio calculation of the "holy grail of nuclear astrophysics" [73], i.e. the reaction $^{12}C(\alpha, \gamma)^{16}O$ at stellar energies, becomes feasible. Our present knowledge of this reaction involves significant uncertainties, as it is derived from extrapolations of experimental data down to stellar energies, where the cross section is too small to be measured directly. The computational time required is estimated to be roughly eight times that of α-α scattering, a factor of four since we have twice as many nucleons, and another factor of two since we need to consider two coupled channels (with and without the photon in the final state). This is clearly within reach, given present computing capabilities. Nevertheless, further refinement of the action is required, as discussed in Chap. 8, so as to achieve a desirable level of accuracy.

References

1. G. Hagen, M. Hjorth-Jensen, G.R. Jansen, R. Machleidt, T. Papenbrock, Evolution of shell structure in neutron-rich calcium isotopes. Phys. Rev. Lett. **109**, 032502 (2012)
2. E.D. Jurgenson, P. Maris, R.J. Furnstahl, P. Navrátil, W.E. Ormand, J.P. Vary, Structure of p-shell nuclei using three-nucleon interactions evolved with the similarity renormalization group. Phys. Rev. C **87**(5), 054312 (2013)
3. R. Roth, J. Langhammer, A. Calci, S. Binder, P. Navrátil, Similarity-transformed chiral NN+3N interactions for the ab initio description of 12-C and 16-O. Phys. Rev. Lett. **107**, 072501 (2011)
4. E. Epelbaum, H. Krebs, D. Lee, U.G. Meißner, Ab initio calculation of the Hoyle state. Phys. Rev. Lett. **106**, 192501 (2011)
5. T.A. Lähde, E. Epelbaum, H. Krebs, D. Lee, U.-G. Meißner, G. Rupak, Uncertainties of Euclidean time extrapolation in lattice effective field theory. J. Phys. G **42**, 034012 (2015)
6. P. Navrátil, J.P. Vary, B.R. Barrett, Properties of C-12 in the ab initio nuclear shell model. Phys. Rev. Lett. **84**, 5728 (2000)
7. P. Maris, J.P. Vary, A.M. Shirokov, Ab initio no-core full configuration calculations of light nuclei. Phys. Rev. C **79**, 014308 (2009)
8. S. Gandolfi, F. Pederiva, S. Fantoni, K.E. Schmidt, Auxiliary field diffusion Monte Carlo calculation of nuclei with A ≤ 40 with tensor interactions. Phys. Rev. Lett. **99**, 022507 (2007)
9. E. Epelbaum, U.-G. Meißner, W. Glöckle, Nuclear forces in the chiral limit. Nucl. Phys. A **714**, 535 (2003)
10. E. Epelbaum, U.-G. Meißner, W. Glöckle, Further comments on nuclear forces in the chiral limit (2002). arXiv: nucl-th/0208040
11. S.R. Beane, M.J. Savage, Variation of fundamental couplings and nuclear forces. Nucl. Phys. A **713**, 148 (2003)
12. S.R. Beane, M.J. Savage, The Quark mass dependence of two nucleon systems. Nucl. Phys. A **717**, 91 (2003)

13. R. Pohl et al., The size of the proton. Nature **466**, 213 (2010)
14. I.T. Lorenz, H.-W. Hammer, U.-G. Meißner, The size of the proton – closing in on the radius puzzle. Eur. Phys. J. A **48**, 151 (2012)
15. I.A. Schaller, L. Schellenberg, T.Q. Phan, G. Piller, A. Ruetschi, H. Schneuwly, Nuclear charge radii of the carbon isotopes C-12, C-13 and C-14. Nucl. Phys. A **379**, 523 (1982)
16. J.C. Kim, R.S. Hicks, R. Yen, I.P. Auer, H.S. Caplan, J.C. Bergstrom, Electron scattering from O-17. Nucl. Phys. A **297**, 301 (1978)
17. W.J. Vermeer, M.T. Esat, J.A. Kuehner, R.H. Spear, A.M. Baxter, S. Hinds, Electric quadrupole moment of the first excited state of 12 C. Phys. Lett. **122B**, 23 (1983)
18. F. Ajzenberg-Selove, Energy levels of light nuclei A = 11-12. Nucl. Phys. A **506**, 1 (1990)
19. R. Moreh, W.C. Sellyey, D. Sutton, R. Vodhanel, Widths of the 6.92 and 7.12 MeV levels in O-16 and the influence of the effective temperature. Phys. Rev. C **31**, 2314 (1985)
20. F. Ajzenberg-Selove, Energy levels of light nuclei A = 16-17. Nucl. Phys. A **460**, 1 (1986)
21. M. Chernykh, H. Feldmeier, T. Neff, P. von Neumann-Cosel, A. Richter, Pair decay width of the Hoyle state and carbon production in stars. Phys. Rev. Lett. **105**, 022501 (2010)
22. H. Miska et al., High resolution inelastic electron scattering and radiation widths of levels in 16 O. Phys. Lett. **58B**, 155 (1975)
23. A. Bohr, B.R. Mottelson, *Nuclear Structure*, vol. 1 (W. A. Benjamin, New York, 1969)
24. M. Freer, H. Horiuchi, Y. Kanada-En'yo, D. Lee, U.-G. Meißner, Microscopic clustering in light nuclei. Rev. Mod. Phys. **90**, 035004 (2018)
25. J.A. Wheeler, Molecular viewpoints in nuclear structure. Phys. Rev. **52**, 1083 (1937)
26. D.M. Dennison, Energy levels of the O-16 nucleus. Phys. Rev. **96**, 378 (1954)
27. D. Robson, Evidence for the tetrahedral nature of O-16. Phys. Rev. Lett. **42**, 876 (1979)
28. W. Bauhoff, H. Schultheis, R. Schultheis, Alpha cluster model and the spectrum of O-16. Phys. Rev. C **29**, 1046 (1984)
29. A. Tohsaki, H. Horiuchi, P. Schuck, G. Ropke, Alpha cluster condensation in C-12 and O-16. Phys. Rev. Lett. **87**, 192501 (2001)
30. R. Bijker, Algebraic cluster model with tetrahedral symmetry. AIP Conf. Proc. **1323**, 28 (2010)
31. M. Freer [CHARISSA Collaboration], Alpha-particle states in O-16 and Ne-20. J. Phys. G **31**, S1795 (2005)
32. C.J. Hogan, Why the universe is just so. Rev. Mod. Phys. **72**, 1149 (2000)
33. A.N. Schellekens, Life at the interface of particle physics and string theory. Rev. Mod. Phys. **85**, 1491 (2013)
34. C.E. Rolfs, W.S. Rodney, *Cauldrons in the Cosmos* (The University of Chicago Press, Chicago, 1988)
35. D. Arnett, *Supernovae and Nucleosynthesis* (Princeton University Press, Princeton, 1996)
36. F. Hoyle, On nuclear reactions occuring in very hot stars. 1. The synthesis of elements from carbon to nickel. Astrophys. J. Suppl. **1**, 121 (1954)
37. H. Oberhummer, A. Csoto, H. Schlattl, Stellar production rates of carbon and its abundance in the universe. Science **289**, 88 (2000)
38. E. Epelbaum, H. Krebs, T.A. Lähde, D. Lee, U.-G. Meißner, Viability of carbon-based life as a function of the light quark mass. Phys. Rev. Lett. **110**, 112502 (2013)
39. E. Epelbaum, H. Krebs, T.A. Lähde, D. Lee, U.-G. Meißner, Dependence of the triple-alpha process on the fundamental constants of nature. Eur. Phys. J. A **49**, 82 (2013)
40. J.C. Berengut, E. Epelbaum, V.V. Flambaum, C. Hanhart, U.-G. Meißner, J. Nebreda, J.R. Pelaez, Varying the light quark mass: impact on the nuclear force and Big Bang nucleosynthesis. Phys. Rev. D **87**, 085018 (2013)
41. P.F. Bedaque, T.I uu, L. Platter, Quark mass variation constraints from Big Bang nucleosynthesis. Phys. Rev. C **83**, 045803 (2011)
42. V. Bernard, Chiral perturbation theory and baryon properties. Prog. Part. Nucl. Phys. **60**, 82 (2008)
43. H. Hellmann, *Einführung in die Quantenchemie* (Deuticke, Leipzig, 1937)
44. R.P. Feynman, Forces in molecules. Phys. Rev. **56**, 340 (1939)

45. M. Frink, U.-G. Meißner, I. Scheller, Baryon masses, chiral extrapolations, and all that. Eur. Phys. J. A **24**, 395 (2005)
46. G. Colangelo et al., Review of lattice results concerning low energy particle physics. Eur. Phys. J. C **71**, 1695 (2011)
47. J. Gasser, H. Leutwyler, Chiral perturbation theory to one loop. Ann. Phys. **158**, 142 (1984)
48. J. Bijnens, G. Ecker, Mesonic low-energy constants. Ann. Rev. Nucl. Part. Sci. **64**, 149 (2014)
49. S. Weinberg, *Facing Up* (Harvard University Press, Cambridge, 2001)
50. B. Carter, in *Confrontation of Cosmological Theories with Observation*, ed. by M.S. Longair (Reidel, Dordrecht, 1974)
51. H. Kragh, An anthropic myth: Fred Hoyle's carbon-12 resonance level. Arch. Hist. Exact Sci. **64**, 721 (2010)
52. U.-G. Meißner, Anthropic considerations in nuclear physics. Sci. Bull. **60**, 43 (2015)
53. H. Oberhummer, A. Csoto, H. Schlattl, Fine tuning carbon based life in the universe by the triple alpha process in red giants, in *The Future of the Universe and the Future of Our Civilization* (World Scientific, Singapore, 2000), pp. 197–205
54. L. Huang, F.C. Adams, E. Grohs, Sensitivity of carbon and oxygen yields to the triple-alpha resonance in massive stars. Astropart. Phys. **105**, 103 (2019)
55. S.R. Beane, P.F. Bedaque, K. Orginos, M.J. Savage, Nucleon-nucleon scattering from fully-dynamical lattice QCD. Phys. Rev. Lett. **97**, 012001 (2006)
56. V. Baru, E. Epelbaum, A.A. Filin, J. Gegelia, Low-energy theorems for nucleon-nucleon scattering at unphysical pion masses. Phys. Rev. C **92**, 014001 (2015)
57. E. Epelbaum, U.-G. Meißner, W. Glöckle, C. Elster, Resonance saturation for four nucleon operators. Phys. Rev. C **65**, 044001 (2002)
58. E. Epelbaum, J. Gegelia, The two-nucleon problem in EFT reformulated: pion and nucleon masses as soft and hard scales. PoS CD **12**, 090 (2013)
59. K.M. Nollett, S.C. Pieper, R.B. Wiringa, J. Carlson, G.M. Hale, Quantum Monte Carlo calculations of neutron-alpha scattering. Phys. Rev. Lett. **99**, 022502 (2007)
60. S. Quaglioni, P. Navrátil, Ab initio many-body calculations of n-H-3, n-He-4, p-He-3,4,and and n-Be-10 scattering. Phys. Rev. Lett. **101**, 092501 (2008)
61. P. Navrátil, S. Quaglioni, Ab initio many-body calculations of the 3H(d,n)4He and 3He(d,p)4He fusion. Phys. Rev. Lett. **108**, 042503 (2012)
62. P. Navrátil, R. Roth, S. Quaglioni, Ab initio many-body calculation of the ^{7}Be$(p, \gamma)^8$B radiative capture. Phys. Lett. B **704**, 379 (2011)
63. G. Hupin, J. Langhammer, P. Navrátil, S. Quaglioni, A. Calci, R. Roth, Ab initio many-body calculations of nucleon-4He scattering with three-nucleon forces. Phys. Rev. C **88**, 054622 (2013)
64. M. Fink, F.K. Roepke, W. Hillebrandt, I.R. Seitenzahl, S.A. Sim, M. Kromer, Double-detonation sub-Chandrasekhar supernovae: can minimum helium shell masses detonate the core? Astron. Astrophys. **514**, A53 (2010)
65. K.J. Shen, L. Bildsten, The ignition of carbon detonations via converging shock waves in white dwarfs. Astrophys. J. **785**, 61 (2014)
66. N.P. Heydenburg, G.M. Temmer, Alpha-alpha scattering at low energies. Phys. Rev. **104**, 123 (1956)
67. R. Nilson, W.K. Jentschke, G.R. Briggs, R.O. Kerman, J.N. Snyder, Investigation of excited states in Be-8 by alpha-particle scattering from He. Phys. Rev. **109**, 850 (1958)
68. S.A. Afzal, A.A.Z. Ahmad, S. Ali, Systematic survey of the alpha-alpha interaction. Rev. Mod. Phys. **41**, 247 (1969)
69. A. Rokash, M. Pine, S. Elhatisari, D. Lee, E. Epelbaum, H. Krebs, Scattering cluster wave functions on the lattice using the adiabatic projection method. Phys. Rev. C **92**(5), 054612 (2015)
70. S. Elhatisari, D. Lee, G. Rupak, E. Epelbaum, H. Krebs, T.A. Lähde, T. Luu, U.-G. Meißner, Ab initio alpha-alpha scattering. Nature **528**, 111 (2015)
71. R. Higa, H.-W. Hammer, U. van Kolck, alpha alpha Scattering in halo effective field theory. Nucl. Phys. A **809**, 171 (2008)

72. G. Hupin, S. Quaglioni, P. Navrátil, Predictive theory for elastic scattering and recoil of protons from ^{4}He. Phys. Rev. C **90**(6), 061601 (2014)
73. W.A. Fowler, Experimental and theoretical nuclear astrophysics: the quest for the origin of the elements. Rev. Mod. Phys. **56**, 149 (1984)

Chapter 8
Further Developments

In this chapter, we provide an overview of NLEFT methods that have been recently developed. These methods allow for more detailed investigations than discussed so far, and therefore extend the toolbox of NLEFT to larger systems, as well as to more complex and accurate studies of nuclear structure and reactions. It should be noted that these methods are, as of the time of writing, still on-going developments. Hence, our emphasis here is to sketch the general directions of development, and to discuss the physical insights obtained so far.

We start in Sect. 8.1 by describing the symmetry-sign extrapolation (SSE) method. There, the approximate Wigner SU(4) symmetry of the nuclear forces is used to construct a family of Hamiltonians, for which the sign problem is mild. The full physical Hamiltonian may then be reconstructed by extrapolation in one parameter. Then, in Sect. 8.2, we consider non-local smearing, where the nucleon creation and annihilation operators are distributed over more than one lattice site. Such operators can be used to define non-local interactions. While such non-local smearing was originally thought to be merely a technical tool for the further suppression of sign oscillations, new insight into the formation of atomic nuclei could be gained by constructing a family of interactions with both local and non-local components, see Sect. 8.3. A new method to investigate and quantify clustering in nuclei is presented in Sect. 8.4, and in Sect. 8.5, the "pinhole algorithm" is introduced, which greatly facilitates the calculation of A-nucleon correlation functions in nuclei, and gives access to particle densities in lattice calculations. We comment briefly on detailed investigations of the lattice spacing dependence in few-nucleon systems in Sect. 8.6, and on NLEFT calculations beyond NNLO in Sect. 8.7. In Sect. 8.8, we discuss methods that seek to minimize the lattice artifacts due to the temporal lattice spacing δ. Finally, we give some concluding remarks in Sect. 8.9.

© Springer Nature Switzerland AG 2019
T. A. Lähde, U.-G. Meißner, *Nuclear Lattice Effective Field Theory*,
Lecture Notes in Physics 957, https://doi.org/10.1007/978-3-030-14189-9_8

8.1 The Symmetry-Sign Extrapolation Method

In our introduction of the symmetry-sign extrapolation (SSE) method [1], we consider the LO partition function of NLEFT along the lines of Ref. [2]. If we employ the notation and conventions of Chap. 4, we have

$$
\mathscr{Z}_{\mathrm{LO}} = \int \mathscr{D}\pi'_I \, \exp\left[-S_{\pi\pi}(\pi'_I) \right] \mathrm{Tr}\left\{ M_{\mathrm{LO}}^{(N_t-1)}(\pi'_I) \cdots M_{\mathrm{LO}}^{(0)}(\pi'_I) \right\}, \qquad (8.1)
$$

where we recall that π'_I is the (rescaled) pion field, and $S_{\pi\pi}$ the free pion lattice action. As in Chap. 4, we take the normal-ordered LO transfer matrix to be

$$
\begin{aligned}
M_{\mathrm{LO}}^{(t)}(\pi'_I) = \; : \exp\Bigg(&- \tilde{H}_{\mathrm{free}}\alpha_t \\
&- \frac{\alpha_t}{2N^3} \sum_{\mathbf{q}} D_s(\mathbf{q})\Big[\tilde{C}\, \rho^{a^\dagger,a}(\mathbf{q})\rho^{a^\dagger,a}(-\mathbf{q}) \\
&\qquad\qquad\qquad + \tilde{C}_{S^2}\sum_{S=1}^{3} \rho_S^{a^\dagger,a}(\mathbf{q})\rho_S^{a^\dagger,a}(-\mathbf{q}) \Big] \\
&- \frac{\alpha_t}{2N^3} \sum_{\mathbf{q}} D_s(\mathbf{q})\Big[\tilde{C}_{I^2}\sum_{I=1}^{3} \rho_I^{a^\dagger,a}(\mathbf{q})\rho_I^{a^\dagger,a}(-\mathbf{q}) \\
&\qquad\qquad\qquad + \tilde{C}_{S,I}\sum_{I,S=1}^{3} \rho_{S,I}^{a^\dagger,a}(\mathbf{q})\rho_{S,I}^{a^\dagger,a}(-\mathbf{q}) \Big] \\
&- \frac{g_A \alpha_t}{2F_\pi \sqrt{q_\pi}} \sum_{I,S=1}^{3} \sum_{\mathbf{n}} \Big[\nabla_{S,(\nu)}\pi'_I(\mathbf{n},t) \Big] \rho_{S,I}^{a^\dagger,a}(\mathbf{n}) \Bigg) :, \qquad (8.2)
\end{aligned}
$$

for Euclidean time steps $t = 0, \ldots, N_t - 1$. Again, the LO operator is improved in that the contact interactions depend on the momentum transfer $\mathbf{q}$ through a smooth smearing function $D_s(\mathbf{q})$. In Eq. (8.2), the various density operators for spin and isospin are defined in Chap. 4. We denote by $\alpha_t \equiv \delta/a$ the ratio of temporal and spatial lattice spacings, and by $\nabla_{S,(\nu)}$ the improved derivative coupling of the pions. From Chap. 4, we recall that $q_\pi \equiv \alpha_t(M_\pi^2 + 6)$ for $\nu = 2$. The coupling constants of Eq. (8.2) satisfy

$$
\tilde{C} = -3\tilde{C}_{S,I} = -\frac{3}{2}(\tilde{C}_{S^2} + \tilde{C}_{I^2}), \qquad (8.3)
$$

such that the smeared contact interactions only contribute to even-parity channels where we have antisymmetry in spin and symmetry in isospin, or vice versa. We recall that a tilde indicates that a dimensionful quantity, expressed in lattice units by multiplication with appropriate powers of a.

As we have seen in Chaps. 4 and 6, MC simulations are performed using a transfer matrix with auxiliary fields coupled to each of the densities in Eq. (8.2). From Chap. 4, we recall that the LO auxiliary-field transfer matrix at Euclidean time step t is

$$
\begin{aligned}
M_{\text{LO,aux}}^{(t)} & (\pi'_I, s, s_S, s_I, s_{S,I}) \\
=: \exp \Bigg(& - \tilde{H}_{\text{free}} \alpha_t + \sqrt{-\tilde{C} \alpha_t} \sum_{\mathbf{n}} s(\mathbf{n}, t) \, \rho^{a^\dagger, a}(\mathbf{n}) \\
& + i \sqrt{\tilde{C}_{S^2} \alpha_t} \sum_{S=1}^{3} \sum_{\mathbf{n}} s_S(\mathbf{n}, t) \rho_S^{a^\dagger, a}(\mathbf{n}) \\
& + i \sqrt{\tilde{C}_{I^2} \alpha_t} \sum_{I=1}^{3} \sum_{\mathbf{n}} s_I(\mathbf{n}, t) \rho_I^{a^\dagger, a}(\mathbf{n}) \\
& + i \sqrt{\tilde{C}_{S,I} \alpha_t} \sum_{I,S=1}^{3} \sum_{\mathbf{n}} s_{S,I}(\mathbf{n}, t) \rho_{S,I}^{a^\dagger, a}(\mathbf{n}) \\
& - \frac{g_A \alpha_t}{2 F_\pi \sqrt{q_\pi}} \sum_{I,S=1}^{3} \sum_{\mathbf{n}} \Big[\nabla_{S,(v)} \pi'_I(\mathbf{n}, t) \Big] \rho_{S,I}^{a^\dagger, a}(\mathbf{n}) \Bigg) :, \quad (8.4)
\end{aligned}
$$

where the physical values of the LO coupling constants are such that all factors inside the square roots in Eq. (8.4) are positive. Further, let us define an SU(4)-symmetric transfer matrix,

$$
M_4^{(t)} =: \exp \left(- \tilde{H}_{\text{free}} \alpha_t - \frac{\tilde{C}_4 \alpha_t}{2N^3} \sum_{\mathbf{q}} D_s(\mathbf{q}) \, \rho^{a^\dagger, a}(\mathbf{q}) \rho^{a^\dagger, a}(-\mathbf{q}) \right) :, \quad (8.5)
$$

where $\tilde{C}_4$ is the associated coupling constant, such that $\tilde{C}_4 < 0$. As before, the interaction can be rewritten using auxiliary fields, giving

$$
M_{4,\text{aux}}^{(t)}(s) =: \exp \left(- \tilde{H}_{\text{free}} \alpha_t + \sqrt{-\tilde{C}_4 \alpha_t} \sum_{\mathbf{n}} s(\mathbf{n}, t) \, \rho^{a^\dagger, a}(\mathbf{n}) \right) :, \quad (8.6)
$$

which is similar to the SU(4)-symmetric approximate transfer matrix, which we used as a low-energy filter in Chap. 6 to accelerate the convergence to the ground state as a function of Euclidean projection time. In the SSE method, we shall take this method one step further and consider a smooth transition between the full LO and SU(4)-symmetric interactions.

Let us consider the projection amplitude for an A-nucleon Slater determinant initial state, which is of the form

$$|\Psi\rangle \equiv |\psi_1\rangle \otimes |\psi_2\rangle \otimes \cdots \otimes |\psi_A\rangle, \tag{8.7}$$

taken together with an auxiliary-field transfer matrix $M_{\text{aux}}^{(t)}$. The Euclidean time projection amplitude is given by $\det(M)$, where $M_{i,j}$ is the $A \times A$ matrix obtained from the single-nucleon amplitudes

$$M_{i,j} \equiv \langle \psi_i | M_{\text{aux}}^{(t-1)} \cdots M_{\text{aux}}^{(0)} | \psi_j \rangle, \tag{8.8}$$

along the lines of Chap. 6. Let us define $\mathscr{U}[M]$ as the set of unitary matrices such that

$$U^\dagger M U = M^*, \tag{8.9}$$

for which it can be shown that $\det(M)$ is positive semi-definite, if there exists an antisymmetric matrix $U \in \mathscr{U}[M]$. The proof follows from the fact that the spectra of M and M^* must coincide, and the (real-valued) spectrum of M must be doubly degenerate, as a result of the antisymmetry of U. For details of this proof, see Refs. [3, 4]. From this, it follows straightforwardly that the projection amplitude $\det(M)$ is positive semi-definite, if there exists a unitary operator $U \in \mathscr{U}[M_{\text{aux}}^{(t)}]$ for all $t = 0, \cdots, N_t - 1$, and if the action of U on the single-particle states $|\psi_i\rangle$ can be represented as an antisymmetric $A \times A$ matrix.

We note that the set $\mathscr{U}[M_{4,\text{aux}}^{(t)}]$ contains the spin- and isospin matrices σ_2 and τ_2. Projection Monte Carlo (PMC) calculations with the SU(4)-symmetric theory are then free from sign oscillations, whenever the initial single-nucleon states are paired into spin-singlets or isospin-singlets, as shown in Refs. [4, 5]. To gain further gain insight into the violation of the SU(4) symmetry and the resulting sign oscillations, we note that the LO auxiliary-field transfer matrix $M_{\text{LO,aux}}^{(t)}$ contains pion interactions with the matrix structure $\sigma_S \tau_I$ acting on single-nucleon states, and smeared contact interactions proportional to $\mathbb{1}$, $i\sigma_S$, $i\tau_I$, and $i\sigma_S \tau_I$. Since

$$\sigma_2 \in \mathscr{U}[\sigma_S \tau_2], \mathscr{U}[i\sigma_S], \mathscr{U}[i\tau_2], \mathscr{U}[i\sigma_S \tau_1], \mathscr{U}[i\sigma_S \tau_3], \tag{8.10}$$

and

$$\sigma_2 \tau_3 \in \mathscr{U}[\sigma_S \tau_1], \mathscr{U}[i\sigma_S], \mathscr{U}[i\tau_1], \mathscr{U}[i\sigma_S \tau_2], \tag{8.11}$$

we note that for an initial state with an even number of neutrons paired into spin-singlets, and an even (but in general different) number of protons paired into spin-singlets, there will exist an antisymmetric representation for both σ_2 and $\sigma_2 \tau_3$ on the single nucleon states. The only structures in $M_{\text{LO,aux}}^{(t)}$ not included in both Eqs. (8.10) and (8.11) are $i\tau_3$ and $\sigma_S \tau_3$. Hence, the sign oscillations in $\det(M)$ are produced by contributions proportional to $i\tau_3$ and $\sigma_S\tau_3$, and the interference between the two sets of interactions in Eqs. (8.10) and (8.11). For initial states where the neutrons and protons cannot be paired into spin-singlets, there will be additional sign oscillations due to such unpaired nucleons. However, the number of such unpaired nucleons can often be kept to a minimum.

Armed with this insight, we can now develop the SSE method. At the heart of this method lies the definition of the "interpolating Hamiltonian" H_{int} as

$$H_{\text{int}} \equiv d_h H_{\text{LO}} + (1 - d_h)H_4, \quad 0 \le d_h \le 1, \tag{8.12}$$

where d_h is a real-valued parameter, and C_4 is the coupling constant of the (unphysical) SU(4)-symmetric theory. This can also be viewed as giving the couplings of the full LO Hamiltonian a linear dependence on d_h, according to

$$C(d_h) \equiv d_h C + (1 - d_h)C_4, \tag{8.13}$$

for the spin- and isospin-independent term, and

$$C_\alpha(d_h) \equiv d_h C_\alpha, \quad \alpha \in \{S^2; I^2; S, I\}, \tag{8.14}$$

for the spin- and isospin-dependent contributions. Similarly, the OPE contribution

$$g_A(d_h) \equiv d_h g_A, \tag{8.15}$$

vanishes for $d_h \to 0$. By taking $d_h < 1$, we can always decrease the sign problem to a tolerable level, while simultaneously tuning $\tilde{C}_4$ to a value favorable for an extrapolation $d_h \to 1$. Most significantly, one can use the constraint that the physical result at $d_h = 1$ should be independent of $\tilde{C}_4$. Furthermore, the dependence of calculated matrix elements on d_h is smooth in the vicinity of $d_h = 1$ (assuming that no level crossing occurs which has so far never been observed in the systems that have been investigated). Using the SSE, the sign problem could be arbitrarily ameliorated, at the price of introducing an extrapolation in a control parameter $d_h \to 1$. In practice, this means that the SSE method is only useful as long as the extrapolation errors can be kept under control. These observations are illustrated in Fig. 8.1.

The properties of H_{int} for various nuclei of physical interest were explored in Ref. [1]. There, it was investigated to what extent SSE can be used to suppress the sign problem, which for $d_h = 1$ becomes exponentially severe in the limit of large Euclidean time. Here, we only discuss briefly the most important results of this study. In Ref. [1], the ^{6}He and ^{12}C systems were extensively studied for a wide range of d_h and C_4. For ^{6}He, an extrapolation of all results to infinite Euclidean time was performed first, followed by an extrapolation $d_h \to 1$. For ^{12}C, the extrapolation $d_h \to 1$ was performed directly for data at finite Euclidean projection time. In both cases, similar behavior as a function of d_h was found, with excellent linearity in the vicinity of $d_h = 1$. Which ordering of the limits is preferable depends on how

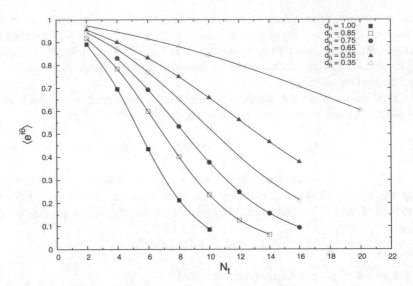

Fig. 8.1 The mean value of the exponential of the complex phase $\langle\exp(i\theta)\rangle$ of $\det(M)$ as a function of Euclidean time step N_t and SSE parameter d_h, for $C_4 = -4.8 \times 10^{-5}$ MeV^{-2} in the $A = 6$ system. Without the SSE method, practical MC simulations would be limited to $N_t \leq 10$

severe the sign problem is for a given system at $d_h = 1$. When the sign problem is severe, the method of first taking the limit $d_h \to 1$ followed by an extrapolation in Euclidean time, is likely to be more reliable. As a demonstration of the suppression of the sign problem using SSE, the dependence of the complex phase $\langle\exp(i\theta)\rangle$ on the number of Euclidean time steps N_t and the SSE parameter d_h, is shown in Fig. 8.1 for the $A = 6$ system. We find that for $d_h = 1$, $\langle\exp(i\theta)\rangle$ quickly approaches zero as N_t is increased, indicating that the sign problem becomes severe at rather modest Euclidean projection times. When d_h is decreased, the sign problem is successively diminished, allowing PMC to be extended to significantly larger N_t.

Finally, we note that a technique similar to SSE has been used in pioneering Shell Model MC calculations [6, 7]. There, the extrapolation in the average sign is performed by decomposing the Hamiltonian into "good sign" and "bad sign" parts, denoted H_G and H_B, respectively. The MC calculations are then performed by multiplying the coefficients of H_B by a parameter g, and by extrapolating from $g < 0$, where the simulations are free from sign oscillations, to the physical point $g = 1$. For SSE, the analysis in terms of "good" and "bad" signs is not the entire story. Most of the NLEFT interactions can be divided into two groups which are sign-free by themselves. As we have seen, a large portion of the sign oscillations is due to interference between the different underlying symmetries of the two groups of interactions. Since this effect is quadratic in the interfering interaction coefficients, the growth of the sign problem is more gradual. In all cases studied so far, one is therefore able to extrapolate from values not so far away from the physical point $d_h = 1$. Clearly, further studies are necessary to find out how useful the SSE method really is.

8.2 Nonlocal Smearing and Interactions

As we have found in Chap. 4, an important ingredient in NLEFT is smearing. This means that the smeared contact interactions are not strictly point-like, but distributed over a range of nearby lattice sites. Such methods are common in Lattice QCD, where they are used to enhance the strength of a given quark source, see e.g. Refs. [8–13]. In Chap. 4, we have already encountered one reason for operator smearing in NLEFT. Specifically, the four-nucleon interaction terms were smeared with a smooth function of Gaussian type, such that any additional parameters were fixed by the (averaged) nucleon-nucleon S-wave effective ranges. This type of (local) smearing is required to avoid excess binding due to configurations with four nucleons on one lattice site. Such smearing also provides the added value that the important effective range corrections are treated non-perturbatively, rather than perturbatively. Hence, some sizable higher-order corrections are also included in the smeared LO terms. We shall now consider a new type of non-local smearing, which was introduced in Ref. [14], with the original intention to further alleviate the sign problem. Such suppression can be understood from our earlier discussion of the residual sign oscillations. Consider a contribution, which has two nucleons on a lattice site, and involving some operator structure which breaks SU(4) symmetry. If we now distribute this two-nucleon state over the neighboring lattice sites, we are effectively increasing the interaction range. This has the effect of diminishing the sign oscillations as the "bad" configurations become smeared out, similarly to the method discussed in Chap. 4.

Let us now define non-local smearing. For this purpose, we introduce the notation $\langle \mathbf{n}', \mathbf{n} \rangle$ to denote the summation over the nearest-neighbor sites $\mathbf{n}'$ of site $\mathbf{n}$, as shown in Fig. 8.2. As our lattice geometry is a cubic lattice with N^3 sites and periodic boundaries, the summations over $\mathbf{n}'$ are defined in terms of periodic boundary conditions. For each lattice site $\mathbf{n}$, one can now define non-local annihilation and

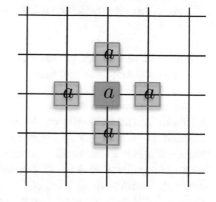

Fig. 8.2 Two-dimensional sketch of the non-local smearing, for an annihilation operator $a(\mathbf{n})$. The smeared operator is no longer point-like, but has support on the neighboring sites in every direction on the lattice

creation operators for each spin- and isospin component of the nucleon,

$$a_{\mathrm{NL}}(\mathbf{n}) \equiv a(\mathbf{n}) + s_{\mathrm{NL}} \sum_{\langle \mathbf{n}', \mathbf{n} \rangle} a(\mathbf{n}'), \tag{8.16}$$

$$a_{\mathrm{NL}}^{\dagger}(\mathbf{n}) \equiv a^{\dagger}(\mathbf{n}) + s_{\mathrm{NL}} \sum_{\langle \mathbf{n}', \mathbf{n} \rangle} a^{\dagger}(\mathbf{n}'), \tag{8.17}$$

where s_{NL} is a real-valued parameter which can be determined, for instance, by a fit to the lowest np partial waves. The introduction of such non-local smearing of operators makes it possible to construct wider classes of local, as well as and non-local interactions. As in Chap. 4, we define the smeared non-local densities,

$$\rho_{\mathrm{NL}}(\mathbf{n}) \equiv a_{\mathrm{NL}}^{\dagger}(\mathbf{n}) a_{\mathrm{NL}}(\mathbf{n}), \tag{8.18}$$

$$\rho_{S,\mathrm{NL}}(\mathbf{n}) \equiv a_{\mathrm{NL}}^{\dagger}(\mathbf{n}) [\sigma_S] \, a_{\mathrm{NL}}(\mathbf{n}), \tag{8.19}$$

$$\rho_{I,\mathrm{NL}}(\mathbf{n}) \equiv a_{\mathrm{NL}}^{\dagger}(\mathbf{n}) [\tau_I] \, a_{\mathrm{NL}}(\mathbf{n}), \tag{8.20}$$

$$\rho_{S,I,\mathrm{NL}}(\mathbf{n}) \equiv a_{\mathrm{NL}}^{\dagger}(\mathbf{n}) [\sigma_S \otimes \tau_I] \, a_{\mathrm{NL}}(\mathbf{n}), \tag{8.21}$$

and the smeared local densities

$$\rho_{\mathrm{L}}(\mathbf{n}) \equiv a^{\dagger}(\mathbf{n}) a(\mathbf{n}) + s_{\mathrm{L}} \sum_{\langle \mathbf{n}', \mathbf{n} \rangle} a^{\dagger}(\mathbf{n}') a(\mathbf{n}'), \tag{8.22}$$

$$\rho_{S,\mathrm{L}}(\mathbf{n}) \equiv a^{\dagger}(\mathbf{n}) [\sigma_S] a(\mathbf{n}) + s_{\mathrm{L}} \sum_{\langle \mathbf{n}', \mathbf{n} \rangle} a^{\dagger}(\mathbf{n}') [\sigma_S] a(\mathbf{n}'), \tag{8.23}$$

$$\rho_{I,\mathrm{L}}(\mathbf{n}) \equiv a^{\dagger}(\mathbf{n}) [\tau_I] a(\mathbf{n}) + s_{\mathrm{L}} \sum_{\langle \mathbf{n}', \mathbf{n} \rangle} a^{\dagger}(\mathbf{n}') [\tau_I] a(\mathbf{n}'), \tag{8.24}$$

$$\rho_{S,I,\mathrm{L}}(\mathbf{n}) \equiv a^{\dagger}(\mathbf{n}) [\sigma_S \otimes \tau_I] a(\mathbf{n}) + s_{\mathrm{L}} \sum_{\langle \mathbf{n}', \mathbf{n} \rangle} a^{\dagger}(\mathbf{n}') [\sigma_S \otimes \tau_I] a(\mathbf{n}'), \tag{8.25}$$

in terms of a real-valued parameter s_{L}. We note that as far the additional suppression of sign oscillations is concerned, either one of these smeared interactions will suffice.

We are now in a position to construct local and non-local LO short-range interactions. Let us briefly recall our definition of locality. Consider the nucleon-nucleon interaction $V(\mathbf{r}', \mathbf{r})$, where $\mathbf{r}$ is the spatial separation of the two incoming nucleons and $\mathbf{r}'$ is the spatial separation of the two outgoing nucleons. The short-range components in $V(\mathbf{r}', \mathbf{r})$ are called "non-local" if $\mathbf{r}' \neq \mathbf{r}$, in general. In contrast, the interaction is called "local" when $\mathbf{r}' = \mathbf{r}$ is imposed. We expect that the physics should not depend on the precise form of the short-distance interactions. However, as we are dealing with an EFT, the construction of non-local interactions effectively amounts to a resummation of some higher-order terms. Hence, at a given order in the chiral expansion, differences should be expected, as we shall discuss in more detail in Sect. 8.3. To be more precise, we write the non-local short-range in the standard which we encountered in Chap. 4, in terms of the non-local densities defined in Eqs. (8.18)–(8.21),

$$
V_{NL} \equiv \frac{C_{NL}}{2} \sum_{\mathbf{n}} : \rho_{NL}(\mathbf{n})\rho_{NL}(\mathbf{n}) : + \frac{C_{I^2,NL}}{2} \sum_{I=1}^{3}\sum_{\mathbf{n}} : \rho_{I,NL}(\mathbf{n})\rho_{I,NL}(\mathbf{n}) :,
$$

$$(8.26)$$

while the short-range interactions with local smearing are based on the densities defined in Eqs. (8.22)–(8.25),

$$
V_{L} \equiv \frac{C_{L}}{2} \sum_{\mathbf{n}} : \rho_{L}(\mathbf{n})\rho_{L}(\mathbf{n}) : + \frac{C_{S^2,L}}{2} \sum_{S=1}^{3}\sum_{\mathbf{n}} : \rho_{S,L}(\mathbf{n})\rho_{S,L}(\mathbf{n}) :
$$

$$
+ \frac{C_{I^2,L}}{2} \sum_{I=1}^{3}\sum_{\mathbf{n}} : \rho_{I,L}(\mathbf{n})\rho_{I,L}(\mathbf{n}) : + \frac{C_{S,I,L}}{2} \sum_{I,S=1}^{3}\sum_{\mathbf{n}} : \rho_{S,I,L}(\mathbf{n})\rho_{S,I,L}(\mathbf{n}) :,
$$

$$(8.27)$$

where, as before, we take special combinations of the LO coupling constants, such that the interaction in odd partial waves vanishes completely, cf. Eq. (8.3). To a first approximation, the strength of the local short-range interactions can be set equal in the S-wave singlet and triplet channels. Thus,

$$
C_{S^2,L} = C_{I^2,L} = C_{S,I,L} = -\frac{1}{3}C_{L},
$$

$$(8.28)$$

and only one independent coefficient remains. In future work, it will clearly be useful to consider relaxing this condition.

8.3 Insights into Nuclear Binding

Let us now consider what physical insights can be gained from the local and non-local smearing of the LO contact interactions. One of the central questions in nuclear physics is how protons and neutrons bind to form nuclei. Recently, some insight into this question was obtained using NLEFT in Ref. [14], where LO chiral EFT was considered with two variants of the short-range interactions. Admittedly, this is only a rough approximation to the full nuclear Hamiltonian, but it may nevertheless capture some features of nuclear binding on a qualitative level. We shall now provide a brief discussion of these results.

The starting point of Ref. [14] is two LO lattice interactions, referred to as "interaction A" and "interaction B". By design, these are similar to each other, consisting of smeared OPE and short-distance interactions. While in interaction A, only non-local terms, as in Eq. (8.26), are considered, interaction B is supplemented by local terms, as in Eq. (8.27). In both cases, the Coulomb potential between protons is included, although this is formally counted as an NLO effect. The parameters of interaction A are fitted to the low-energy np phase shifts and the deuteron binding energy, while the additional parameters of interaction B also allow for a description of S-wave α-α scattering at low energies, along the lines of Chap. 7. We refer the reader to Ref. [14] for further details.

Given our established Euclidean PMC methods, the ground-state energies of nuclei such as ^{3}H, ^{3}He, ^{4}He, ^{8}Be, ^{12}C, ^{16}O and ^{20}Ne, can be calculated. These results, shown in Table 8.1, reveal some striking phenomena. On the one hand, for interaction B the description of the nuclear binding energies is in general agreement with experiment. On the other hand, for interaction A, we note that ^{8}Be is significantly underbound, and this effect increases systematically up to ^{20}Ne, and presumably beyond. What is even more revealing is the calculated ratio of the LO energy for each of the α-like nuclei to that of the α particle (for the same interaction, without Coulomb). For interaction A, one finds

$$\frac{E(^8Be)}{E(^4He)} = 1.997(6), \quad \frac{E(^{12}C)}{E(^4He)} = 3.00(1), \quad \frac{E(^{16}O)}{E(^4He)} = 4.00(2), \quad \frac{E(^{20}Ne)}{E(^4He)} = 5.03(3),$$
(8.29)

which shows that for interaction A, the ground state in each case is simply a weakly-interacting Bose gas of α particles. The inability to generate more that non-interacting α particles indicates that some crucial physics is missing. Obviously, the missing physics is provided by interaction B, at least for the α-type nuclei. In fact, another problem becomes apparent when more neutrons (or protons) are added to a given nucleus. Then, interaction B provides too little binding, due to the repulsion in the $^1S_0$ partial wave. We shall return to this issue in Sect. 8.4.

Table 8.1 Ground state energies of nuclei from ^{3}H to ^{20}Ne for interactions A and B (as defined in the text)

Nucleus	A (LO)	B (LO)	A (LO + Coulomb)	B (LO + Coulomb)	Experiment
^{3}H	−7.82(5)	−7.78(12)	−7.82(5)	−7.78(12)	−8.482
^{3}He	−7.82(5)	−7.78(12)	−7.08(5)	−7.09(12)	−7.718
^{4}He	−29.36(4)	−29.19(6)	−28.62(4)	−28.45(6)	−28.296
^{8}Be	−58.61(14)	−59.73(6)	−56.51(14)	−57.29(7)	−56.591
^{12}C	−88.2(3)	−95.0(5)	−84.0(3)	−89.9(5)	−92.162
^{16}O	−117.5(6)	−135.4(7)	−110.5(6)	−126.0(7)	−127.619
^{20}Ne	−148(1)	−178(1)	−137(1)	−164(1)	−160.645

We show LO and LO + Coulomb results, along with the experimental values. All energies are given in units of MeV. The errors given in parentheses denote one-standard-deviation errors due to statistical (MC) and extrapolation uncertainties

The results shown in Table 8.1 can be used to draw some interesting conclusions on the many-body limit of nuclear physics (for a more general discussion, see Ref. [15]). For that purpose, we switch off the Coulomb interaction (as otherwise, nuclear matter would be unstable) and define a one-parameter family of interactions,

$$V_\lambda \equiv (1 - \lambda)V_A + \lambda V_B, \qquad 0 \le \lambda \le 1, \qquad (8.30)$$

where λ is a real-valued parameter. While the properties of systems of nucleons up to $A = 4$ vary only slightly with λ, the many-body ground state of V_λ undergoes a quantum phase transition from a Bose-condensed gas to a nuclear liquid. To understand this, consider the zero-temperature phase diagram. A phase transition occurs when the S-wave α-α scattering length $a_{\alpha\alpha}$ crosses zero, which happens for $\lambda \simeq 0$, and the Bose gas collapses due to the attractive interactions [16, 17]. At slightly larger values of λ, finite α-like nuclei also become bound, starting with the largest nuclei. The last α-like nucleus to be bound is ^{8}Be, which happens at the so-called unitarity limit where $|a_{\alpha\alpha}| \to \infty$. One finds that the quantum phase transition from the Bose-condensed gas to the nuclear liquid occurs at $\lambda_\infty = 0.0(1)$, where the uncertainty is due to the slow dependence of the energy levels on λ in the vicinity of $\lambda = 0$. In more detail, the critical point for the binding of ^{20}Ne occurs at $\lambda_{20} = 0.2(1)$. For the binding of the other α-type nuclei, we obtain $\lambda_{16} = 0.2(1)$ for ^{16}O, $\lambda_{12} = 0.3(1)$ for ^{12}C, and $\lambda_8 = 0.7(1)$ for ^{8}Be. A sudden change is found in the nucleon-nucleon density correlations at long distances when λ crosses the critical point, at the transition from a continuum state to a self-bound system. As λ increases further, the nucleus becomes more tightly bound, gradually losing its α-cluster substructure and becoming more like a nuclear liquid droplet. The quantum phase transition at $\lambda_\infty = 0.0(1)$ is the corresponding phenomenon in the many-body system, a first-order phase transition occurring for infinite matter. Clearly, these conclusions should be solidified by higher-order investigations. We note that similar results have been found based on density functional theory, see Refs. [18, 19].

We have found that the description of the nuclear ground-state energies are already rather good for the LO interaction B, supplemented by the Coulomb interaction. This lets us conclude that α-α scattering is a sensitive indicator, which is correlated with the binding energies of the medium-mass nuclei. Stated differently, by fitting the parameters of NLEFT to α-α scattering as well, the higher-order corrections due to the three- and four-body forces can be minimized. We note the general principle that precision studies of nuclear structure benefit whenever the multi-nucleon forces are rendered perturbative, see e.g. Ref. [20].

8.4 More on Clustering in Nuclei

The precise location of the so-called drip lines is one of the most interesting problems in nuclear physics. Loosely spoken, these are the boundaries in the nuclear chart, beyond which nuclei decay by the emission of a proton or neutron. Stated differently, how many neutrons or protons can be added to a given nucleus (such as ^{12}C), such that the system remains self-bound? As noted in Sect. 8.3, the LO action used to investigate the zero-temperature phase diagram of nuclear matter works well for α-type nuclei. However, if extra neutrons (or protons) are added, the precision of this interaction deteriorates quickly. Such a problem was also observed with the NNLO action, which was used in Chap. 7 to analyze the spectra of ^{12}C and ^{16}O, as well as α-α scattering. For example, using the action discussed in Sect. 8.3, the binding energies of ^{14}C and ^{12}C differ by a mere $\simeq 2.2$ MeV, much less than the experimental value $\simeq 13.1$ MeV. This effect can be largely traced back to the repulsion in the $^{1}S_0$ channel of np scattering. To overcome this, a radical step was taken in Ref. [21]. The LO contribution was reduced to the SU(4)-symmetric term of the smeared nonlocal action of Eqs. (8.18)–(8.21) supplemented by smeared OPE. Clearly, the description of the S-waves in np scattering suffer significantly from such a simplification, which must be corrected for by higher-order contributions. Nevertheless, some very interesting physics emerges from this simplified LO action. The three remaining parameters were fitted to the average np scattering lengths and effective ranges, as well as to the finite-volume energy of ^{8}Be, which is sensitive to the α-α scattering length. Then, the ground-state energies of the neutron-rich, even atomic number chains of He, Be, C and O were found to agree with experiment with a maximal deviation of $\simeq 0.7$ MeV per nucleon. While such a high level of agreement is encouraging, given the rather rough approximation to the nuclear Hamiltonian, these results must certainly be scrutinized by systematic higher-order calculations.

In Ref. [21], an important tool was developed for the quantitative measurement of cluster structures in nuclei. One of the most central nuclear clustering phenomena is the α-clustering observed in light nuclei. We have already encountered such clustering effects in our treatment of ^{12}C and ^{16}O in Chap. 7. The dominant role of α-clustering can be understood in terms of the large binding energy and energy gap of the ^{4}He nucleus. If we denote the ground state by 0^+_1 and the first excited state (also with zero total angular momentum and positive parity) by 0^+_2, we have $E(0^+_2) - E(0^+_1) \simeq 20.21$ MeV, and hence the basic structure of the α particle is rather difficult to perturb. For a recent review, see e.g. Ref. [22]. While α-clustering can be visualized in density plots obtained from many-body methods such as, Green's function MC, see e.g. Ref. [23], density functional theory, see e.g. Ref. [18], and relativistic mean field calculations, see e.g. Ref. [24], such results are not quantitative, in the sense that they do not provide a precise measure of clustering effects. In NLEFT, this problem has been addressed by consideration of the normal-ordered quantities

$$\rho_N \equiv \frac{1}{N!} \sum_{\mathbf{n}} : \rho^N(\mathbf{n}) :, \quad N = 3, 4 \tag{8.31}$$

where the summation extends over the entire lattice. We note that ρ_4 is clearly a probe of α-clustering. Furthermore, due to the energetics in even-even nuclei, there are no ^{3}H or ^{3}He clusters, such that ρ_3 can also be taken as a measure of

α-clustering. More precisely, while ρ_4 is sensitive to the center of the α-cluster (as it couples to four nucleons), ρ_3 receives contributions from a wider portion of the α-cluster wave function. However, defined in this way, ρ_N is scale- and scheme-dependent. Most obviously, ρ_N will depend on the lattice regulator (the lattice spacing a). Therefore, if $\rho_{3,\alpha}$ and $\rho_{4,\alpha}$ denote the corresponding values for the α particle, then the ratios

$$\rho_N^{\text{norm}} \equiv \frac{\rho_N}{\rho_{N,\alpha}}, \quad N = 3, 4 \tag{8.32}$$

are free of short-distance divergences, and are model-independent quantities (up to contributions from higher-dimensional operators in an operator product expansion). A detailed derivation of these statements can be found in Ref. [21].

The normalized ratios of Eq. (8.32) for a given isotope are found, to a good approximation, to be the same for $N = 3$ and $N = 4$. Also, ρ_N^{norm} is found to increase significantly with the mass of the nucleus. As ρ_3^{norm} is a measure of the effective number of α clusters N_α, it follows that any deviation from an integer-valued N_α indicates quantum entanglement. This can be understood easily. Simple counting gives $N_\alpha = 1, 2, 3, 4$ for (neutron-rich isotopes of) He, Be, C and O, respectively. However, the α clusters are not free, but immersed in a complex many-body system comprising the atomic nucleus, which leads to quantum entanglement. It is useful to quantify the entanglement of the nucleons comprising each α cluster with the outside medium. For that, we introduce the ρ_3-entanglement, defined as

$$\frac{\Delta_\alpha^{\rho_3}}{N_\alpha} \equiv \frac{\rho_3^{\text{norm}}}{N_\alpha} - 1, \tag{8.33}$$

which quantifies the deviation of the nuclear wave function from a pure product state of α-clusters, and thus provides a measure of the entanglement. The results for isotopic chains with an even number of neutrons are shown in Table 8.2.

We note that for Be isotopes, the deviation from $N_\alpha = 2$ is small. However, the ρ_3-entanglement grows steadily with A, and reaches a value of $\simeq 3/4$ at the end of the O chain. For such high values of the ρ_3-entanglement, the simple picture in terms of α-clusters and excess neutrons is no longer valid. With the ρ_3-entanglement, we have established a model-independent, quantitative measure of nuclear cluster formation in terms of entanglement of the wave function. This shows that the transition from cluster-like states in light systems, to nuclear liquid-like states in heavier systems, should not be viewed as a simple suppression of short-distance

Table 8.2 ρ_3-Entanglement for isotopic chains of α-cluster nuclei

Nucleus	$^{4,6,8}\text{He}$	$^{8,10,12,14}\text{Be}$	$^{12,14,16,18,20,22}\text{C}$	$^{16,18,20,22,24,26}\text{O}$
$\Delta_\alpha^{\rho_3}/N_\alpha$	0.00–0.03	0.20–0.35	0.25–0.50	0.50–0.75

Definitions are given in the main text

multi-nucleon correlations, but rather as an increasing entanglement of the nucleons involved in the multi-nucleon system. Similar conclusions can be drawn from the analysis of ρ_4-entanglement. Another way of looking at these results is in terms the "swelling" of α-clusters as the system becomes saturated with excess neutrons, an effect that has also been observed in ^{6}He and ^{8}He in the framework of Green's function MC calculations [23].

8.5   Pinhole Algorithm

In spite of the many strengths of the PMC methods presented so far, we note that the computation of density distributions remain difficult. This is because the MC simulations involve quantum states, which are superpositions of many different center-of-mass (CM) positions. This problem does not affect the calculation of moments of density distributions (such as the charge radius), which we have encountered in Chap. 7. However, in order to calculate the nucleon density distribution directly, a new method was developed in Ref. [21], referred to as the "pinhole algorithm".

To describe the pinhole algorithm, let us first consider the density operator for nucleons with spin i and isospin j at lattice site $\mathbf{n}$,

$$\rho_{i,j}(\mathbf{n}) \equiv a_{i,j}^\dagger(\mathbf{n})a_{i,j}(\mathbf{n}), \tag{8.34}$$

from which we construct the normal-ordered A-nucleon density operator

$$\rho_{i_1,j_1,\ldots,i_A,j_A}(\mathbf{n}_1,\ldots,\mathbf{n}_A) \equiv \, : \rho_{i_1,j_1}(\mathbf{n}_1)\cdots\rho_{i_A,j_A}(\mathbf{n}_A) :, \tag{8.35}$$

which satisfies the completeness (closure) relation

$$\sum_{i_1,j_1,\ldots,i_A,j_A} \sum_{\mathbf{n}_1,\ldots,\mathbf{n}_A} \rho_{i_1,j_1,\ldots,i_A,j_A}(\mathbf{n}_1,\ldots,\mathbf{n}_A) = A!, \tag{8.36}$$

in the A-nucleon subspace. We then consider the Euclidean time projection amplitudes

$$Z_{f,i} \equiv \langle \Psi_f | M_*^{N_{t'}} M^{N_t} M_*^{N_{t'}} | \Psi_i \rangle, \tag{8.37}$$

where the full NLEFT transfer matrix M is applied N_t times, and the (computationally inexpensive) SU(4) symmetric transfer matrix M_* is applied $N_{t'}$ times to the initial and final states.

We obtain the pinhole algorithm by including the A-nucleon density operator in the projection amplitudes, according to

$$Z_{f,i}^{\text{pinhole}} \equiv Z_{f,i}(i_1, j_1, \ldots, i_A, j_A; \mathbf{n}_1, \ldots, \mathbf{n}_A; N_t)$$

$$= \langle \Psi_f | M_*^{N_{t'}} M^{N_t/2} \rho_{i_1, j_1, \ldots, i_A, j_A}(\mathbf{n}_1, \ldots, \mathbf{n}_A) M^{N_t/2} M_*^{N_{t'}} | \Psi_i \rangle, \qquad (8.38)$$

which amounts to the insertion of a "screen", with pinholes located at the positions $\mathbf{n}_1, \ldots, \mathbf{n}_A$ and spin-isospin indices $i_1, j_1, \ldots, i_A, j_A$, at the midpoint of the Euclidean time evolution. This procedure is illustrated in Fig. 8.3. Due to the completeness relation of Eq. (8.36), the sum of this expectation value over the lattice points $\mathbf{n}_1, \ldots, \mathbf{n}_A$ and the spin-isospin labels $i_1, j_1, \ldots, i_A, j_A$ equals $A!$ times the amplitude of Eq. (8.37). As we have established many times, Eqs. (8.37) and (8.38) are calculated using auxiliary-field MC simulations. The path integral representation of Eq. (8.38) is given by

$$Z_{f,i}^{\text{pinhole}} = \int \mathscr{D}\sigma \, \langle \Phi_f[\sigma] | \rho_{i_1, j_1, \ldots, i_A, j_A}(\mathbf{n}_1, \ldots, \mathbf{n}_A) | \Phi_i[\sigma] \rangle, \qquad (8.39)$$

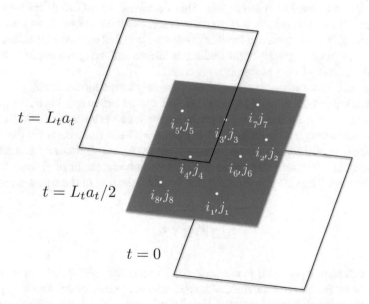

Fig. 8.3 Illustration of pinhole locations and spin-isospin indices, inserted at the mid-time point in the Euclidean time evolution. Note that the positions and indices of the pinholes are updated at regular intervals during a MC simulation

where $\mathscr{D}\sigma$ denotes the path integral measure of the auxiliary and the pion fields. In Eq. (8.39), the trial states which have been evolved in Euclidean time are

$$|\Phi_i[\sigma]\rangle \equiv M^{(N_{t'}+N_t/2-1)} \cdots M^{(N_{t'})} M_*^{(N_{t'}-1)} \cdots M_*^{(0)} |\Psi_i\rangle, \qquad (8.40)$$

$$\langle\Phi_f[\sigma]| \equiv \langle\Psi_f| M_*^{(2N_{t'}+N_t-1)} \cdots M_*^{(N_{t'}+N_t)} M^{(N_{t'}+N_t-1)} \cdots M^{(N_{t'}+N_t/2)}, \qquad (8.41)$$

and we recall that the transfer matrix at each Euclidean time slice depends on one or more components of the auxiliary and pion fields, collectively denoted by σ. The importance sampling of Eq. (8.39) proceeds according to the absolute value

$$A(\{\sigma\}; i_1, j_1, \ldots, i_A, j_A; \mathbf{n}_1, \ldots, \mathbf{n}_A; N_t)$$
$$\equiv \left| \langle\Phi_f[\sigma]| \rho_{i_1, j_1, \ldots, i_A, j_A} (\mathbf{n}_1, \ldots, \mathbf{n}_A) |\Phi_i[\sigma]\rangle \right|, \qquad (8.42)$$

as in the reweighting procedure discussed in Chap. 6. As for the phase of the pinhole amplitude, it can be positive as well as negative (if there are remaining sign oscillations). The configurations with positive and negative sign can be considered separately, and added after averaging. This increases the MC statistics (instead of just considering ensembles with a definite sign). For fixed spin-isospin indices and pinhole locations, the pion and auxiliary fields are updated using the HMC algorithm of Chap. 6, whereas updates of the pinhole locations and the spin-isospin indices are updated according to the Metropolis algorithm.

Let us conclude our discussion of the pinhole algorithm by considering the spatial resolution we can obtain with this method. For spatial lattice spacing a, the position $\mathbf{r}_i$ of each nucleon on the lattice is an integer-valued vector $\mathbf{n}_i$ times a. Since the CM is a mass-weighted average of A nucleons with the same mass (in the isospin limit), the CM position $\mathbf{r}_{\text{CM}}$ is an integer-valued vector $\mathbf{n}_{\text{CM}}$ times a/A. Therefore, the density distribution has a resolution scale which is a factor of A finer than the lattice spacing. The CM position is found by minimization of the squared radius

$$\sum_{i=1}^A |\mathbf{r}_{\text{CM}} - \mathbf{r}_i|^2, \qquad (8.43)$$

where each term $|\mathbf{r}_{\text{CM}} - \mathbf{r}_i|$ is minimized with respect to all periodic copies of the separation on the lattice. One the CM has been fixed, density distributions of protons and neutrons (or higher-order correlations) can be determined. We refer to Ref. [21] for results on the proton and neutron densities in the carbon isotopes ^{12}C, ^{14}C, and ^{16}C, along with visualizations of α-clusters in these isotopes, which can be made much more quantitative in comparison with earlier methods (compare Chap. 7 and Ref. [25]).

8.6 Lattice Spacing Dependence of Few-Nucleon Systems

In Chap. 4, we discussed the need for an "effective" four-nucleon contact term, in order to correctly reproduce the physical binding energy of ^{4}He with the standard, coarse-grained NNLO action. The original assumption was that the four-nucleon term absorbs artifacts due to the large lattice spacing $a \simeq 2$ fm. In order to clarify this issue, a detailed study was undertaken in Ref. [26]. There, NNLO calculations including all isospin breaking and electromagnetic effects, the complete TPE potential, and the three-nucleon forces were performed for lattice spacings $a = 1.97$ fm, $a = 1.64$ fm, and $a = 1.32$ fm. It was found that the so-called Tjon line (or more accurately, the Tjon band) was recovered as a was decreased. The Tjon band represents a correlation between the triton and the ^{4}He binding energies, which was first noted for local interactions [27]. This correlation was later confirmed using high-precision two- and three-nucleon potentials in Ref. [28], where it was also concluded that there is little room left for four-nucleon forces. Within the framework of pionless EFT, it could be shown that the same three-nucleon force required to renormalize the three-nucleon system [29] also renormalizes the four-nucleon system [30]. Based on these observations, the Tjon band arises naturally [31], see also Ref. [32]. Because of theoretical uncertainties, we refer here to the Tjon band instead of the Tjon line.

Let us return once more to the three- and four-nucleon system in NLEFT. Without the effective four-nucleon contact term of Chap. 4, the NLEFT result at NNLO for the coarse lattice spacing is outside the Tjon band. However, this is an artifact of the large value of the lattice spacing, as the results of Ref. [26] indicate,

$$a = 1.97 \text{ fm} \rightarrow E(^4\text{He}) = -34.55(18)(3) \text{ MeV}, \tag{8.44}$$

$$a = 1.64 \text{ fm} \rightarrow E(^4\text{He}) = -31.09(7)(6) \text{ MeV}, \tag{8.45}$$

$$a = 1.32 \text{ fm} \rightarrow E(^4\text{He}) = -28.37(28)(3) \text{ MeV}, \tag{8.46}$$

where the first parenthesis gives the error due to MC statistics, and the second parenthesis gives the error due to the uncertainties of the pertinent LECs up to NNLO. Note that for each value of a, the triton binding energy is fixed to its physical value by adjustment of the coupling constant E, which appears in the contact term of the NNLO three-nucleon force (see Chap. 4). Clearly, the Tjon band is recovered as a is decreased. The differences in the ^{4}He binding energies can largely be traced back to remaining discrepancies in the $^1S_0$ and $^3S_1$ partial waves at NNLO (see also Chap. 5). These results suggest that in an NNNLO (or N3LO) calculation for $A = 2$ and $A = 3$ the remaining lattice artifacts are removed, and the use of an effective four-nucleon contact term can be avoided.

8.7   First Results from NNNLO Calculations

The NLEFT formalism and results we have presented have, in most cases, involved a truncation of the EFT expansion at NNLO. A first step towards a systematic analysis of nuclei at NNNLO (or N3LO) has been taken in Ref. [33]. There, a new projection method was developed, which allows for an easier decomposition of the chiral NN force into partial waves. The end result is that lattice operators specific to each partial wave is obtained. This makes the fitting of the required LECs to scattering data much simpler. Previously, most of the low partial waves had to be fitted simultaneously, which (as we have found in Chap. 5) can be problematic due to the different accuracy of the PWA in each channel. With the new decomposition, each partial wave can be fitted independently, with the exception of partial waves coupled by the tensor interaction.

In Ref. [33], the phase shifts and mixing angles in the np system were computed at four different lattice spacings, namely $a = 1.97$ fm, $a = 1.64$ fm, $a = 1.32$ fm, and $a = 0.99$ fm. The results for the S-wave channels are shown in Fig. 8.4. At this high order in NLEFT, the phase shifts in the S-, P-, and D-waves, as well as the mixing angles ε_1 and ε_2, are indistinguishable from the Nijmegen PWA for nucleon CM momenta $\lesssim 200$ MeV. An error analysis based on the method described in Chap. 2 has also been performed. It remains to be seen how the description of nuclear structure and reactions will improve when based on these more precise and accurate nuclear forces.

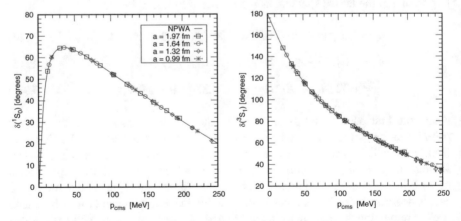

Fig. 8.4 S-wave phase shifts for np scattering at NNNLO, as a function of the CM momentum. Results are shown for (spatial) lattice spacings of $a = 0.99$ fm (boxes), $a = 1.23$ fm (circles), $a = 1.64$ fm (diamonds), and $a = 1.97$ fm (crosses). The solid lines give the Nijmegen PWA (NPWA) [34]. In the left/right panel, the $^1S_0/^3S_1$ phase shift is shown. Figure courtesy of Ning Li

8.8 Calculations in the Hamiltonian Limit

So far, most of our discussion concerning lattice artifacts has been focused on effects due to the (spatial) lattice spacing a. However, as pointed out already in Chap. 3, some dependence on the temporal lattice spacing δ is also expected to remain, which in general is dealt with by extrapolation to the Hamiltonian limit $\delta \to 0$. For an NLEFT study of such effects in the np system, see Ref. [35]. Here, we briefly outline the "Hamiltonian method", which was developed in S. Elhatisari et al. (private communication) and found to essentially remove lattice artifacts in δ. In the NLEFT calculations that we have described so far, the temporal lattice spacing used is $\delta = (150\,\mathrm{MeV})^{-1}$, and the typical number of Euclidean time slices is $N_t \simeq 15$.

Let us now discuss an alternative way to calculate ground-state energies. For simplicity, we consider the LO order Hamiltonian

$$H_{\mathrm{LO}} \equiv H_{\mathrm{free}} + H_{\mathrm{SU(4)}} + V_{\mathrm{OPE}} = H_4 + V_{\mathrm{OPE}}, \tag{8.47}$$

in the standard notation which we have established earlier. Let us now project the initial and final states in Euclidean time, using the auxiliary field transfer matrix, for a very fine temporal lattice spacing of $\delta = (1000\,\mathrm{MeV})^{-1}$ and a much large number of time slices $N_t \simeq 200$. These projected states are denoted

$$|\Psi_i(N_t/2)\rangle \equiv M_{\mathrm{LO}}^{N_t/2}|\Psi_i\rangle, \qquad \langle\Psi_f(N_t/2)| \equiv \langle\Psi_f|M_{\mathrm{LO}}^{N_t/2}, \tag{8.48}$$

in terms of which we can directly calculate energies, such as

$$E_{\mathrm{kin}}(N_t + 1) \equiv \frac{\langle\Psi_f(N_t/2)|H_{\mathrm{free}}|\Psi_i(N_t/2)\rangle}{\langle\Psi_f(N_t/2)|\Psi_i(N_t/2)\rangle}, \tag{8.49}$$

$$E_{\mathrm{SU(4)}}(N_t + 1) \equiv \frac{\langle\Psi_f(N_t/2)|H_{\mathrm{SU(4)}}|\Psi_i(N_t/2)\rangle}{\langle\Psi_f(N_t/2)|\Psi_i(N_t/2)\rangle}, \tag{8.50}$$

$$E_{\mathrm{OPE}}(N_t + 1) \equiv \frac{\langle\Psi_f(N_t/2)|V_{\mathrm{OPE}}|\Psi_i(N_t/2)\rangle}{\langle\Psi_f(N_t/2)|\Psi_i(N_t/2)\rangle}, \tag{8.51}$$

using the exact Hamiltonian formalism. Due to the fact that the inserted operators H_{free}, $H_{\mathrm{SU(4)}}$, and V_{OPE} in Eqs. (8.49)–(8.51) are exact and no longer defined in terms auxiliary fields (see Appendix F), their MC uncertainties are also greatly reduced. The key advantage of this exact calculation is that any lattice artifact due to the larger temporal lattice spacing δ becomes negligible.

8.9   Concluding Remarks

Let us conclude our presentation with some remarks on the present and future status of NLEFT. From Chaps. 3–7, we have shown that NLEFT has matured into a powerful and versatile tool of theoretical nuclear physics, which combines the well-established EFT methods for the nuclear forces with the advantageous computational scaling of Euclidean time projection and MC simulations for multi-nucleon systems. Furthermore, NLEFT enables the treatment of nuclear structure and reactions on the same footing, in terms of a unified calculational framework. In this last chapter, we have attempted to provide an overview of the many exciting, recent developments in NLEFT, and to give the reader a feeling for the future potential of the NLEFT framework. We look forward to many more insights into nuclear physics based on NLEFT in the years to come.

References

1. T.A. Lähde, T. Luu, D. Lee, U.-G. Meißner, E. Epelbaum, H. Krebs, G. Rupak, Nuclear lattice simulations using symmetry-sign extrapolation. Eur. Phys. J. A **51**, 92 (2015)
2. E. Epelbaum, H. Krebs, D. Lee, U.-G. Meißner, Lattice chiral effective field theory with three-body interactions at next-to-next-to-leading order. Eur. Phys. J. A **41**, 125 (2009)
3. D. Lee, Inequalities for low-energy symmetric nuclear matter. Phys. Rev. C **70**, 064002 (2004)
4. J.W. Chen, D. Lee, T. Schäfer, Inequalities for light nuclei in the Wigner symmetry limit. Phys. Rev. Lett. **93**, 242302 (2004)
5. D. Lee, Spectral convexity for attractive SU(2N) fermions. Phys. Rev. Lett. **98**, 182501 (2007)
6. Y. Alhassid, D.J. Dean, S.E. Koonin, G. Lang, W.E. Ormand, Practical solution to the Monte Carlo sign problem: realistic calculations of Fe-54. Phys. Rev. Lett. **72**, 613 (1994)
7. S.E. Koonin, D.J. Dean, K. Langanke, Shell model Monte Carlo methods. Phys. Rep. **278**, 1 (1997)
8. S. Gusken, A study of smearing techniques for hadron correlation functions. Nucl. Phys. Proc. Suppl. **17**, 361 (1990)
9. D. Daniel, R. Gupta, G.W. Kilcup, A. Patel, S.R. Sharpe, Phenomenology with Wilson fermions using smeared sources. Phys. Rev. D **46**, 3130 (1992)
10. C.R. Allton et al., UKQCD Collaboration, Gauge invariant smearing and matrix correlators using Wilson fermions at Beta = 6.2. Phys. Rev. D **47**, 5128 (1993)
11. C. Morningstar, M.J. Peardon, Analytic smearing of SU(3) link variables in lattice QCD. Phys. Rev. D **69**, 054501 (2004)
12. A. Hasenfratz, R. Hoffmann, S. Schaefer, Hypercubic smeared links for dynamical fermions. J. High Energy Phys. **0705**, 029 (2007)
13. R.G. Edwards, B. Joo, H.W. Lin, Tuning for three-flavors of anisotropic clover fermions with stout-link smearing. Phys. Rev. D **78**, 054501 (2008)
14. S. Elhatisari et al., Nuclear binding near a quantum phase transition. Phys. Rev. Lett. **117**, 132501 (2016)
15. D.J. Dean, Viewpoint: uncovering a quantum phase transition in nuclei. Physics **9**, 106 (2016)
16. H.T.C. Stoof, Atomic Bose gas with a negative scattering length. Phys. Rev. A **49**, 3824 (1994)
17. Y. Kagan, A.E. Muryshev, G.V. Shlyapnikov, Collapse and Bose-Einstein condensation in a trapped Bose gas with negative scattering length. Phys. Rev. Lett. **81**, 933 (1998)
18. J.-P. Ebran, E. Khan, T. Niksic, D. Vretenar, How atomic nuclei cluster. Nature **487**, 341 (2012)

19. J.P. Ebran, E. Khan, T. Niksic, D. Vretenar, Cluster-liquid transition in finite saturated fermionic systems. Phys. Rev. C **89**(3), 031303 (2014)
20. S.K. Bogner, A. Schwenk, R.J. Furnstahl, A. Nogga, Is nuclear matter perturbative with low-momentum interactions? Nucl. Phys. A **763**, 59 (2005)
21. S. Elhatisari et al., Ab initio calculations of the isotopic dependence of nuclear clustering. Phys. Rev. Lett. **119**, 222505 (2017)
22. M. Freer, H. Horiuchi, Y. Kanada-En'yo, D. Lee, U.-G. Meißner, Microscopic clustering in light nuclei. Rev. Mod. Phys. **90**, 035004 (2018)
23. R.B. Wiringa, S.C. Pieper, J. Carlson, V.R. Pandharipande, Quantum Monte Carlo calculations of A = 8 nuclei. Phys. Rev. C **62**, 014001 (2000)
24. P.W. Zhao, N. Itagaki, J. Meng, Rod-shaped nuclei at extreme spin and isospin. Phys. Rev. Lett. **115**, 022501 (2015)
25. E. Epelbaum, H. Krebs, T.A. Lähde, D. Lee, U.-G. Meißner, Structure and rotations of the Hoyle state. Phys. Rev. Lett. **109**, 252501 (2012)
26. N. Klein, S. Elhatisari, T.A. Lähde, D. Lee, U.-G. Meißner, The Tjon band in nuclear lattice effective field theory. Eur. Phys. J. A **54**, 121 (2018)
27. J.A. Tjon, Bound states of 4 He with local interactions. Phys. Lett. **56B**, 217 (1975)
28. A. Nogga, H. Kamada, W. Gloeckle, Modern nuclear force predictions for the alpha particle. Phys. Rev. Lett. **85**, 944 (2000)
29. P.F. Bedaque, H.W. Hammer, U. van Kolck, Renormalization of the three-body system with short range interactions. Phys. Rev. Lett. **82**, 463 (1999)
30. L. Platter, H.W. Hammer, U.-G. Meißner, The four boson system with short range interactions. Phys. Rev. A **70**, 052101 (2004)
31. L. Platter, H.-W. Hammer, U.-G. Meißner, On the correlation between the binding energies of the triton and the alpha-particle. Phys. Lett. B **607**, 254 (2005)
32. J. Kirscher, H.W. Griesshammer, D. Shukla, H.M. Hofmann, Universal correlations in pionless EFT with the resonating group model: three and four nucleons. Eur. Phys. J. A **44**, 239 (2010)
33. N. Li, S. Elhatisari, E. Epelbaum, D. Lee, B.N. Lu, U.-G. Meißner, Neutron-proton scattering with lattice chiral effective field theory at next-to-next-to-next-to-leading order. Phys. Rev. C **98**, 044002 (2018)
34. V.G.J. Stoks, R.A.M. Klomp, M.C.M. Rentmeester, J.J. de Swart, Partial wave analysis of all nucleon-nucleon scattering data below 350-MeV. Phys. Rev. C **48**, 792 (1993)
35. J.M. Alarcón, Recent developments in neutron-proton scattering with lattice effective field theory. PoS (CD15) 091 (2016)

Appendix A
Notations and Conventions

Throughout this book, we work in natural units, by setting

$$\hbar = c = k_B = 1, \tag{A.1}$$

where $\hbar \equiv h/2\pi$ is the reduced Planck constant, c is the speed of light in vacuum, and k_B is the Boltzmann constant. In a few occasions, these constants are explicitly displayed. Up-to-date values can be found at the Particle Data Group website, http://pdg.lbl.gov/. The metric tensor in Minkowski space is $g_{\mu\nu} \equiv \mathrm{diag}(1, -1, -1, -1)$, and in Euclidean space $g_{\mu\nu}^E \equiv \mathrm{diag}(-1, -1, -1, -1)$, with $\mu, \nu \in 0, 1, 2, 3$.

The natural length scale in nuclear physics is 1 Fermi = 1 femtometer = 1 fm = 10^{-15} m. The natural energy scale is Mega-electronvolts (MeV) or Giga-electronvolts (GeV), with $1\,\mathrm{GeV} = 1000\,\mathrm{MeV}$. The conversion factor from energy to length due to Eq. (A.1) is

$$\hbar c = 197.328\ \mathrm{MeV} \cdot \mathrm{fm}, \tag{A.2}$$

which we frequently use to express the lattice spacing a in units of $[\mathrm{MeV}^{-1}]$. Throughout, we work on a four-dimensional hypercubic lattice with periodic boundary conditions, of physical volume

$$V \equiv L \times L \times L \times L_t = L^3 \times L_t, \tag{A.3}$$

where L is the spatial and L_t the temporal extent of the lattice, in units of [fm]. In terms of the fundamental length scales on the lattice, we have

$$L \equiv Na, \quad L_t \equiv N_t \delta, \quad \alpha_t \equiv \delta/a, \tag{A.4}$$

where a is the (spatial) lattice spacing and δ the temporal lattice spacing, such that $N, N_t \in \mathbb{N}$. In most NLEFT calculations, we have $N = 6 \ldots 8$, with N_t of $\mathcal{O}(10)$.

© Springer Nature Switzerland AG 2019
T. A. Lähde, U.-G. Meißner, *Nuclear Lattice Effective Field Theory*,
Lecture Notes in Physics 957, https://doi.org/10.1007/978-3-030-14189-9

On the lattice, dimensionful quantities such as masses and coupling constants are expressed in dimensionless units by multiplication with appropriate powers of a.

A.1   Pauli Matrices

As the interactions between nucleons depend on spin and isospin, one encounters the Pauli matrices for spin and isospin. For spin degrees of freedom, these 2×2 matrices are defined by

$$[\sigma_i, \sigma_j] = 2i\varepsilon_{ijk}\sigma_k, \tag{A.5}$$

and

$$\sigma_i\sigma_j = \delta_{ij}\mathbb{1} + i\varepsilon_{ijk}\sigma_k, \tag{A.6}$$

for $i, j, k \in 1, 2, 3$. The equivalent matrices for isospin are denoted τ_i. The totally antisymmetric Levi-Civita tensor is given by $\varepsilon_{ijk} = 1$ for even permutations of (i, j, k) and $\varepsilon_{ijk} = -1$ for odd permutations. Also, the Kronecker delta $\delta_{ij} = 1$ for $i = j$, and zero otherwise. The explicit representation of the σ matrices is

$$\sigma_1 \equiv \begin{pmatrix} 0 & 1 \\ 1 & 0 \end{pmatrix}, \quad \sigma_2 \equiv \begin{pmatrix} 0 & -i \\ i & 0 \end{pmatrix}, \quad \sigma_3 \equiv \begin{pmatrix} 1 & 0 \\ 0 & -1 \end{pmatrix}, \tag{A.7}$$

which act in either spin- or isospin space. For notational convenience, we include the 2×2 unit matrix

$$\mathbb{1} \equiv \sigma_0 = \begin{pmatrix} 1 & 0 \\ 0 & 1 \end{pmatrix}, \tag{A.8}$$

which we occasionally denote as σ_0.

A.2   Spherical Harmonics

The spherical harmonics $Y_{l,m}(\theta, \phi)$ appear when the variables of the Laplace equation are separated in spherical coordinates, for instance in our partial-wave projection of the Schrödinger equation in Chap. 5. For orbital angular momentum l and magnetic quantum number m, we take the $Y_{l,m}(\theta, \phi)$ to be

$$Y_{l,m}(\theta, \phi) \equiv (-1)^m \sqrt{\frac{(2l+1)}{4\pi}\frac{(l-m)!}{(l+m)!}}\, P_{l,m}(\cos\theta)\exp(im\phi), \tag{A.9}$$

in terms of the associated Legendre polynomials

$$P_{l,m}(x) \equiv (1 - x^2)^{m/2} \frac{d^m}{dx^m} P_l(x), \tag{A.10}$$

where

$$P_l(x) \equiv \frac{1}{2^l l!} \frac{d^l}{dx^l} (x^2 - 1)^l, \tag{A.11}$$

are the Legendre polynomials. Note that we have defined the $P_{l,m}(x)$ without the Condon-Shortley phase $(-1)^m$, which we have instead included in Eq. (A.9). We note the orthonormality

$$\int d\Omega \, Y_{l,m}(\Omega) Y^*_{l,m'}(\Omega) = \delta_{l,l'} \delta_{m,m'}, \tag{A.12}$$

with $d\Omega \equiv \sin\theta d\theta d\phi$, and the property

$$Y^*_{l,m}(\theta, \phi) = (-1)^m \, Y_{l,-m}(\theta, \phi), \tag{A.13}$$

under complex conjugation. Special cases are

$$Y_{l,0}(\theta, \phi) = \sqrt{\frac{2l + 1}{4\pi}} P_l(\cos\theta), \tag{A.14}$$

for $m = 0$, and

$$Y_{l,\pm l}(\theta, \phi) = \frac{(\mp 1)^l}{2^l l!} \sqrt{\frac{(2l + 1)!}{4\pi}} \sin^l\theta \exp(\pm il\phi), \tag{A.15}$$

for $m = \pm l$.

A.3   Clebsch-Gordan Coefficients

The Clebsch-Gordan (CG) coefficients appear whenever spin and orbital angular momenta are coupled to total angular momentum J, such as in the spin-orbit or tensor components of the nuclear potential, or in the construction of shell-model trial states for Euclidean time projection. The CG coefficients appear in the resolution of identity

$$|J J_z\rangle = \sum_{m_1=-j_1}^{j_1} \sum_{m_2=-j_2}^{j_2} |j_1 m_1; j_2 m_2\rangle \langle j_1 m_1; j_2 m_2 | J J_z\rangle, \tag{A.16}$$

where angular momenta (j_1, m_1) and (j_2, m_2) are coupled to total angular momentum (J, J_z). These satisfy the orthogonality relations

$$\sum_{J=|j_1-j_2|}^{j_1+j_2} \sum_{J_z=-J}^{J} \langle j_1 m_1; j_2 m_2 | J J_z \rangle \langle J J_z | j_1 m_1'; j_2 m_2' \rangle$$

$$= \langle j_1 m_1; j_2 m_2 | j_1 m_1'; j_2 m_2' \rangle = \delta_{m_1, m_1'} \delta_{m_2, m_2'}, \tag{A.17}$$

and

$$\sum_{m_1, m_2} \langle J J_z | j_1 m_1; j_2 m_2 \rangle \langle j_1 m_1; j_2 m_2 | J' J_z' \rangle = \langle J J_z | J' J_z' \rangle = \delta_{J, J'} \delta_{J_z, J_z'}, \tag{A.18}$$

and we note that the CG coefficients vanish unless $J_z = m_1 + m_2$. The CG coefficients are typically tabulated for common values of j_1, j_2 and J, as the general expression is somewhat cumbersome. We note the special cases

$$\langle j_1 m_1; j_2 m_2 | 00 \rangle = \frac{(-1)^{j_1 - m_1}}{\sqrt{2j_1 - 1}} \delta_{j_1, j_2} \delta_{m_1, -m_2}, \tag{A.19}$$

for $J = 0$, and

$$\langle j_1 j_1; j_2 j_2 | (j_1 + j_2)(j_1 + j_2) \rangle = 1, \tag{A.20}$$

for $J = j_1 + j_2$ and $J = J_z$.

In the treatment of the Lüscher method in Chap. 5, some identities are expressed in terms of the Wigner 3-j symbols, which have more convenient symmetry relations than the equivalent CG coefficients. We note that the CG coefficients and the Wigner 3-j symbols are related by

$$\langle j_1 m_1; j_2 m_2 | J J_z \rangle = (-1)^{-j_1 + j_2 - J_z} \sqrt{2J + 1} \begin{pmatrix} j_1 & j_2 & J \\ m_1 & m_2 & -J_z \end{pmatrix} \tag{A.21}$$

and we also have the integral relation

$$\int d\Omega \, Y_{l_1, m_1}^*(\Omega) Y_{l_2, m_2}^*(\Omega) Y_{l, m}(\Omega)$$

$$= \sqrt{\frac{(2l_1 + 1)(2l_2 + 1)}{4\pi(2l + 1)}} \langle l_1 0; l_2 0 | l 0 \rangle \langle l_1 m_1; l_2 m_2 | l m \rangle, \tag{A.22}$$

which allows us to express the CG coefficients as integrals over the spherical harmonics. It should be noted that many more identities exist which involve the CG coefficients, apart from the ones given here.

A.4 Bessel Functions

In the treatment of the radial Schrödinger equation for the spherical wall method in Chap. 5 and the adiabatic projection method in Chap. 6, different types of Bessel functions appear. These are solutions to the differential equation

$$x^2\frac{d^2y}{dx^2} + x\frac{dy}{dx} + (x^2 - \alpha^2)y = 0, \tag{A.23}$$

where α is an arbitrary, complex-valued parameter. The most frequently encountered Bessel functions are those with integer-valued α, known as "cylinder functions" or ordinary Bessel functions, as they appear in the solution to the Laplace equation in cylindrical coordinates.

The Bessel functions with half-integer α, or spherical Bessel functions, are encountered when the Helmholtz equation is solved by separation of variables in spherical coordinates. Specifically, the radial equation is

$$x^2\frac{d^2y}{dx^2} + 2x\frac{dy}{dx} + (x^2 - n(n-1))y = 0, \tag{A.24}$$

which is solved by

$$j_n(x) \equiv \sqrt{\frac{\pi}{2x}}J_{n+1/2}(x), \tag{A.25}$$

$$y_n(x) \equiv \sqrt{\frac{\pi}{2x}}Y_{n+1/2}(x) = (-1)^{n+1}\sqrt{\frac{\pi}{2x}}J_{-n-1/2}(x), \tag{A.26}$$

where $J_n(x)$ and $Y_n(x)$ are the ordinary Bessel functions. Here, $j_n(x)$ is the regular spherical Bessel function, and $y_n(x)$ the irregular spherical Bessel function (or Neumann function). We also have the spherical Hankel functions

$$h_n^{(1)}(x) \equiv j_n(x) + iy_n(x), \tag{A.27}$$

$$h_n^{(2)}(x) \equiv j_n(x) - iy_n(x), \tag{A.28}$$

of the first and second kind. These describe the propagation of spherical waves, as in the problem of the scattering phase shift in Chap. 5. The spherical Bessel functions can be obtained from

$$j_n(x) = (-x)^n\left(\frac{1}{x}\frac{d}{dx}\right)^n\frac{\sin x}{x}, \tag{A.29}$$

and

$$y_n(x) = -(-x)^n \left(\frac{1}{x}\frac{d}{dx}\right)^n \frac{\cos x}{x}, \tag{A.30}$$

known as Rayleigh's formulas. Hence, the spherical Bessel functions have simple expressions in terms of polynomials and trigonometric functions.

Closely related are the Riccati-Bessel functions,

$$S_n(x) \equiv x j_n(x) = \sqrt{\frac{\pi x}{2}} J_{n+1/2}(x), \tag{A.31}$$

$$C_n(x) \equiv -x y_n(x) = -\sqrt{\frac{\pi x}{2}} Y_{n+1/2}(x), \tag{A.32}$$

$$\xi_n(x) \equiv x h_n^{(1)}(x) = \sqrt{\frac{\pi x}{2}} H_{n+1/2}^{(1)}(x) = S_n(x) - i C_n(x), \tag{A.33}$$

$$\zeta_n(x) \equiv x h_n^{(2)}(x) = \sqrt{\frac{\pi x}{2}} H_{n+1/2}^{(2)}(x) = S_n(x) + i C_n(x), \tag{A.34}$$

which satisfy the differential equation

$$x^2 \frac{d^2 y}{dx^2} + (x^2 - n(n-1))y = 0, \tag{A.35}$$

which is of the same form as the radial Schrödinger equation with a centrifugal barrier, and appears in the spherical wall problem of Chap. 5.

A.5   Confluent Hypergeometric Functions

When generalizing the nucleon-nucleon scattering problem to involve the long-range Coulomb interaction in Chap. 5, the spherical Bessel functions should be replaced by the equivalent Coulomb functions for Sommerfeld parameter η. We also found that the Coulomb functions could be expressed in terms of the Kummer confluent hypergeometric functions. These are solutions to the Kummer differential equation

$$z^2 \frac{d^2 w}{dz^2} + (b - z)\frac{dw}{dz} - aw = 0, \tag{A.36}$$

with a regular singularity at $z = 0$ and an irregular one at $z = \infty$. This equation has (in most cases) two linearly independent solutions.

The first solution to Eq. (A.36) is the Kummer function $M(a, b, z)$, or the confluent hypergeometric function of the first kind,

$$M(a, b, z) \equiv \sum_{n=0}^{\infty} \frac{a^{(n)} z^n}{b^{(n)} n!} = {}_1F_1(a; b; z), \tag{A.37}$$

where

$$a^{(n)} \equiv a(a + 1)(a + 2) \cdots (a + n - 1), \qquad a^{(0)} = 1, \tag{A.38}$$

is the ascending factorial. We note that $M(a, b, z)$ is analytic as a function of b, except for poles at non-positive integer values of b. Also,

$$_1F_1(a; c; z) = \lim_{b \to \infty} {}_2F_1(a; b; c; z/b), \tag{A.39}$$

such that the confluent hypergeometric function can be thought of as the limit $b \to \infty$ of the hypergeometric function.

A second linearly independent solution to Eq. (A.36) is the Tricomi function $U(a, b, z)$, or the confluent hypergeometric function of the second kind,

$$U(a, b, z) \equiv \frac{\Gamma(1 - b)}{\Gamma(a + 1 - b)} M(a, b, z) + \frac{\Gamma(b - 1)}{\Gamma(a)} z^{1-b} M(a + 1 - b, 2 - b, z), \tag{A.40}$$

where $\Gamma(z)$ is the Euler gamma function,

$$\Gamma(z) = \int_0^{\infty} t^{z-1} e^{-t} dt \,. \tag{A.41}$$

Note that $U(a, b, z)$ is undefined for integer-valued b, but can be extended to any integer b by continuity. Unlike $M(a, b, z)$ which is an entire function of z, $U(a, b, z)$ has (in most cases) a singularity at $z = 0$.

Several useful integral relations exist for the $M(a, b, z)$ and $U(a, b, z)$. For instance, we have

$$M(a, b, z) = \frac{\Gamma(b)}{\Gamma(a)\Gamma(b - a)} \int_0^1 du \, \exp(zu) u^{a-1} (1 - u)^{b-a-1}, \tag{A.42}$$

for $\text{Re } b > \text{Re } a > 0$, and

$$U(a, b, z) = \frac{1}{\Gamma(a)} \int_0^{\infty} dt \, \exp(-zt) t^{a-1} (1 + t)^{b-a-1}, \tag{A.43}$$

for Re $a > 0$. As can be expected from the relation of the $M(a, b, z)$ and $U(a, b, z)$ to the hypergeometric functions, a very large number of relations exist between Kummer and Tricomi functions for various arguments and their derivatives. For instance,

$$M(a, b, z) = \exp(z)M(b - a, b, -z), \qquad (A.44)$$

and

$$U(a, b, z) = z^{1-b}U(1 + a - b, 2 - b, z), \qquad (A.45)$$

which are known as the Kummer transformations.

Finally, we note that a large number of elementary functions can be obtained as special cases of the confluent hypergeometric functions, such as $M(0, b, z) = 1$, $U(0, b, z) = 1$, $M(b, b, z) = \exp(z)$, and the case of $b = 2a$ can be related to the Bessel functions.

Appendix B
Basics of the Nucleon-Nucleon Interaction

The most general structure of a non-relativistic two-nucleon potential can be expressed in terms a few operators only. This potential can be viewed as an operator acting in the position-, spin- and isospin spaces of the nucleons. This is due to the fact that the strong interactions are invariant under parity- and time reversal transformations, and (to a good approximation) also under isospin transformations. As the problem is non-relativistic, one further has invariance under translations, rotations and Galilean transformations. In fact, the approximate equality of proton-proton and neutron-proton forces (after Coulomb effects have been removed) led Heisenberg to introduce the concept of isospin [1]. As discussed earlier, isospin symmetry is broken in QCD through the different masses of the light quarks.

It is instructive to consider the isospin structure of the 2N interaction separately from the operators acting in the position-spin space. According to the classification of Ref. [2], the isospin structure of the two-nucleon force (2NF) falls into the four different classes

$$
\begin{aligned}
\text{Class I:} \quad & V_{\mathrm{I}} = \alpha_{\mathrm{I}} + \beta_{\mathrm{I}} \boldsymbol{\tau}_1 \cdot \boldsymbol{\tau}_2, \\
\text{Class II:} \quad & V_{\mathrm{II}} = \alpha_{\mathrm{II}} \tau_1^3 \tau_2^3, \\
\text{Class III:} \quad & V_{\mathrm{III}} = \alpha_{\mathrm{III}} (\tau_1^3 + \tau_2^3), \\
\text{Class IV:} \quad & V_{\mathrm{IV}} = \alpha_{\mathrm{IV}} (\tau_1^3 - \tau_2^3) + \beta_{\mathrm{IV}} [\boldsymbol{\tau}_1 \times \boldsymbol{\tau}_2]^3,
\end{aligned}
\tag{B.1}
$$

where the α_k and β_k are position-spin operators, and the τ_i are Pauli isospin matrices of nucleon i. While Class I forces are isospin-invariant, all the other classes are isospin-breaking. Class II forces maintain the so-called charge symmetry but break charge independence. They are usually referred to as charge independence breaking (CIB) forces. Here, charge symmetry represents the invariance under reflection about the $1 - 2$ plane in charge space. The charge symmetry operator P_{cs} transforms

© Springer Nature Switzerland AG 2019
T. A. Lähde, U.-G. Meißner, *Nuclear Lattice Effective Field Theory*,
Lecture Notes in Physics 957, https://doi.org/10.1007/978-3-030-14189-9

proton and neutron states into each other and is given by

$$P_{cs} \equiv \exp(i\pi T_2), \qquad \mathbf{T} \equiv \frac{\tau_1 + \tau_2}{2}, \tag{B.2}$$

where $\mathbf{T}$ denotes the total isospin operator. Class III forces break charge symmetry but do not generate isospin mixing in the NN system, i.e. they do not give rise to transitions between isospin-singlet and isospin-triplet NN states. Finally, Class IV forces break charge symmetry (CSB) and cause isospin mixing in the NN system. We note that the operator β_{IV} has to be odd under a time-reversal transformation. In what follows, we shall mostly deal with the isospin-conserving Class I forces.

Let us now consider the position-spin structure of the potential in the isospin-invariant case, along the lines of Ref. [3]. The available vectors are the position, momentum and spin operators for each nucleon, namely $\mathbf{r}_i$, $\mathbf{p}_i$, and $\boldsymbol{\sigma}_i$, with $i = 1, 2$. Due to translational and Galilean invariance, the potential should only depend on the relative distance,

$$\mathbf{r} \equiv \mathbf{r}_1 - \mathbf{r}_2, \tag{B.3}$$

and the relative momentum,

$$\mathbf{p} \equiv \frac{\mathbf{p}_1 - \mathbf{p}_2}{2}, \tag{B.4}$$

of the nucleons. Instead of separate proton and neutrons masses, we work with the nucleon mass

$$m_N \equiv \frac{m_p + m_n}{2}, \tag{B.5}$$

as we are considering the isospin-symmetric case. The most general operator basis, in which to construct an NN potential consistent with rotational symmetry, parity, time reversal, hermiticity, as well as invariance with respect to permutation $1 \leftrightarrow 2$ of the nucleon labels, is given by

$$\left\{ \mathbb{1}_{\text{spin}}, \boldsymbol{\sigma}_1 \cdot \boldsymbol{\sigma}_2, S_{12}(\mathbf{r}), S_{12}(\mathbf{p}), \mathbf{L} \cdot \mathbf{S}, (\mathbf{L} \cdot \mathbf{S})^2 \right\} \otimes \left\{ \mathbb{1}_{\text{isospin}}, \boldsymbol{\tau}_1 \cdot \boldsymbol{\tau}_2 \right\}, \tag{B.6}$$

where the first bracket refers to position and spin, and the second to isospin degrees of freedom. We define orbital angular momentum,

$$\mathbf{L} \equiv \mathbf{r} \times \mathbf{p}, \tag{B.7}$$

total spin,

$$\mathbf{S} \equiv \frac{\boldsymbol{\sigma}_1 + \boldsymbol{\sigma}_2}{2}, \tag{B.8}$$

and tensor operators

$$S_{12}(\mathbf{r}) = 3\boldsymbol{\sigma}_1 \cdot \hat{r}\,\boldsymbol{\sigma}_2 \cdot \hat{r} - \boldsymbol{\sigma}_1 \cdot \boldsymbol{\sigma}_2, \qquad \hat{r} \equiv \frac{\mathbf{r}}{|\mathbf{r}|}, \tag{B.9}$$

which form a complete basis of 12 operators in total. Each of these operators is multiplied by a scalar function, which depends on r^2, p^2, and L^2. These scalar functions are often referred to as the central, spin-spin, tensor, velocity-dependent tensor, spin-orbit and quadratic spin-orbit potentials, respectively.

We note that NN observables are often computed by solving the Lippmann-Schwinger (LS) equation in momentum space. The representation of the potential in momentum space is given by

$$V(\mathbf{p}', \mathbf{p}) \equiv \langle \mathbf{p}'|V|\mathbf{p}\rangle, \tag{B.10}$$

where $\mathbf{p}$ and $\mathbf{p}'$ denote the NN center-of-mass momenta in the initial and the final state, respectively. The most general form of the potential in momentum space then follows as

$$\left\{ \mathbb{1}_{\text{spin}}, \boldsymbol{\sigma}_1 \cdot \boldsymbol{\sigma}_2, S_{12}(\mathbf{q}), S_{12}(\mathbf{k}), i\mathbf{S} \cdot \mathbf{q} \times \mathbf{k}, \boldsymbol{\sigma}_1 \cdot \mathbf{q} \times \mathbf{k}\,\boldsymbol{\sigma}_2 \cdot \mathbf{q} \times \mathbf{k} \right\}$$

$$\otimes \left\{ \mathbb{1}_{\text{isospin}}, \boldsymbol{\tau}_1 \cdot \boldsymbol{\tau}_2 \right\}, \tag{B.11}$$

where

$$\mathbf{q} \equiv \mathbf{p}' - \mathbf{p}, \qquad \mathbf{k} \equiv \frac{\mathbf{p}' + \mathbf{p}}{2}, \tag{B.12}$$

such that each operator is again multiplied with a scalar function, which depends on p^2, p'^2 and $\mathbf{p} \cdot \mathbf{p}'$.

For low-energy processes (such as those relevant for NLEFT), it is convenient to switch to the partial wave basis $|\mathbf{p}\rangle \to |plm_l\rangle$. A two-nucleon state $|p(ls)jm_j\rangle$ in the partial-wave basis is characterized by the orbital angular momentum l, the spin s, and the total angular momentum quantum number j, as well as the magnetic quantum number m_j. The partial-wave decomposition of the potential in Eq. (B.11) is then

$$\langle p'(l's')j'm'_j|V|p(ls)jm_j\rangle \equiv \delta_{j'j}\delta_{m'_jm_j}\delta_{s's}\,V^{sj}_{l'l}(p', p), \tag{B.13}$$

where

$$V_{l'l}^{sj}(p', p) = \sum_{m'_l, m_l} \int d\hat{p}' d\hat{p} \, \langle l', m'_l; s, m_j - m'_l | j, m_j \rangle \langle l, m_l; s, m_j - m_l | j, m_j \rangle$$

$$\times Y_{l', m'_l}^*(\hat{p}') Y_{l, m_l}(\hat{p}) \, \langle s(m_j - m'_l) | V(\mathbf{p}', \mathbf{p}) | s(m_j - m_l) \rangle, \qquad (B.14)$$

in terms of the Clebsch-Gordan (CG) coefficients and the spherical harmonics $Y_{l, m_l}(\hat{p})$. The first two Kronecker δ's in Eq. (B.13) reflect conservation of the total angular momentum. The rotational invariance of the potential makes the matrix elements independent of m_j. In the case of conserved isospin, the conservation of the total spin of the NN system can be deduced. However, in the more general case of broken isospin symmetry, transitions between the spin-singlet and spin-triplet channels are possible. For each individual operator in Eq. (B.11), the expression (B.14) can be simplified. Effectively, one is left with an integral over $\hat{p} \cdot \hat{p}'$, and an integrand expressed in terms of the corresponding scalar potential function and Legendre polynomials. For explicit formulae, see e.g. Refs. [4, 5].

The LS equation is commonly used to calculate scattering states and bound states. In the partial-wave basis, we have

$$T_{l'l}^{sj}(p', p) = V_{l'l}^{sj}(p', p) + \sum_{l''} \int_0^\infty \frac{dp'' \, p''^2}{(2\pi)^3} V_{l'l''}^{sj}(p', p'') \frac{m_N}{p^2 - p''^2 + i\varepsilon} T_{l''l}^{sj}(p'', p),$$

$$(B.15)$$

for the half-off-shell T-matrix, with $\varepsilon \to 0^+$. In the case of uncoupled channels, orbital angular momentum l is conserved. The relation between the on-shell S-matrix and T-matrix is given by

$$S_{l'l}^{sj}(p) = \delta_{l'l} - \frac{i}{8\pi^2} p m_N T_{l'l}^{sj}(p), \qquad (B.16)$$

with standard normalization conventions. Let us also briefly consider the deuteron, which is the only NN bound state. The deuteron wave function and binding energy E_d are obtained from the homogeneous part of Eq. (B.15),

$$\phi_l(p) \equiv \frac{1}{E_d - p^2/m_N} \sum_{l'} \int_0^\infty \frac{dp' \, p'^2}{(2\pi)^3} V_{ll'}^{sj}(p, p') \phi_{l'}(p'), \qquad (B.17)$$

for $s = j = 1$ and $l = l' = 0, 2$. Once the wave function is known, any deuteron observable can be calculated, see e.g. Ref. [5].

As in our treatment of scattering in Chap. 5, the phase shifts for uncoupled channels are found from

$$S_{jj}^{0j}(p) = \exp(2i\delta_j^{0j}(p)), \quad S_{jj}^{1j}(p) = \exp(2i\delta_j^{1j}(p)), \qquad (B.18)$$

where we have used the notation δ_l^{sj} for the phase shift in each partial-wave channel. For scattering channels coupled by the tensor interaction, we recall the Stapp parameterization [6] of Chap. 5. Hence, the S-matrix for coupled channels ($j > 0$) is given by

$$S = \begin{pmatrix} S_{j-1\,j-1}^{1j} & S_{j-1\,j+1}^{1j} \\ S_{j+1\,j-1}^{1j} & S_{j+1\,j+1}^{1j} \end{pmatrix} \tag{B.19}$$

$$= \begin{pmatrix} \cos(2\epsilon)\exp(2i\delta_{j-1}^{1j}) & i\sin(2\epsilon)\exp(i\delta_{j-1}^{1j} + i\delta_{j+1}^{1j}) \\ i\sin(2\epsilon)\exp(i\delta_{j-1}^{1j} + i\delta_{j+1}^{1j}) & \cos(2\epsilon)\exp(2i\delta_{j+1}^{1j}) \end{pmatrix}, \tag{B.20}$$

where the momentum argument p has been suppressed for brevity. It should be noted that another frequently used parameterization is that of Blatt and Biedenharn [7]. If we denote the Blatt-Biedenharn phase shifts and mixing angles by $\tilde{\delta}$ and $\tilde{\epsilon}$ respectively, one finds the relations

$$\delta_{j-1} + \delta_{j+1} = \tilde{\delta}_{j-1} + \tilde{\delta}_{j+1}, \tag{B.21}$$

and

$$\frac{\tan(2\epsilon)}{\tan(2\tilde{\epsilon})} = \sin(\delta_{j-1} - \delta_{j+1}), \tag{B.22}$$

$$\frac{\sin(2\epsilon)}{\sin(2\tilde{\epsilon})} = \sin(\tilde{\delta}_{j-1} - \tilde{\delta}_{j+1}), \tag{B.23}$$

to those of the Stapp parameterization. Once the phase shifts and mixing angles are known, NN scattering observables can be computed in a standard way, see e.g. Refs. [8, 9].

Finally, a few remarks on the electromagnetic interaction between two nucleons are in order. The inclusion of the Coulomb interaction requires special care when scattering observables are calculated, due to its infinite (unscreened) range. In particular, the S-matrix should be formulated in terms of asymptotic Coulomb states, as detailed in Chap. 5. The electromagnetic interaction between the nucleons is dominated by the Coulomb force, with small corrections from magnetic moment interactions and vacuum polarization. Note that the expansion of the scattering amplitude in partial waves converges very slowly in the presence of magnetic moment interactions. For explicit expressions and a detailed discussion of their implementation when calculating NN observables, we refer to Ref. [10].

Appendix C
Study of Rotational Symmetry Breaking Effects in an α Cluster Model

In Chap. 3, we have described how the full rotational symmetry group SO(3) is reduced on the lattice to the finite group of cubic rotations SO(3,Z). Hence, the eigenfunctions of the lattice Hamiltonian do not unambiguously belong to a particular (continuum) quantum number [11–13]. In the continuum limit and the limit of infinite lattice volume, quantum bound states with total angular momentum J form a degenerate multiplet of $2J + 1$ members. At finite lattice spacing, these energy levels are split into sub-groups according to the different irreducible representations or "irreps" of SO(3,Z). This is an unphysical effect, which makes it more difficult to obtain definite NLEFT predictions at finite lattice spacing. The sizes of such energy splittings depend on the lattice spacing, as well as the lattice volume and the choice of boundary conditions. Here, we will illustrate the effects of rotational symmetry breaking in terms of a simplified α cluster model. We shall focus on the ^{8}Be and ^{12}C nuclei, as these are composed of two and three α particles (or ^{4}He nuclei), respectively. This type of calculation is computationally much less demanding than the corresponding full NLEFT calculations, and moreover do not require Monte Carlo methods. Yet, we shall find that important insights can be gained from such a "toy model" calculation, including some useful improvements with which the effects of rotational symmetry breaking on observables (such as transition matrix elements) can be minimized.

C.1   Alpha Cluster Hamiltonian

Our starting point is the continuum Hamiltonian

$$H \equiv -\frac{\nabla^2}{2\mu} + V(r),$$

(C.1)

© Springer Nature Switzerland AG 2019
T. A. Lähde, U.-G. Meißner, *Nuclear Lattice Effective Field Theory*,
Lecture Notes in Physics 957, https://doi.org/10.1007/978-3-030-14189-9

where $M_\alpha = 3727.0$ MeV is the mass of the α particle, and $\mu \equiv M_\alpha/2$ is the reduced mass of the α-α system. In Eq. (C.1), $V(r) \equiv V_N(r) + V_C(r)$ is the α-α potential, as a function of the distance $r = \sqrt{x^2 + y^2 + z^2}$ between the two α particles [14]. $V_N(r)$ denotes the nuclear component, and $V_C(r)$ the Coulomb component of the α-α potential. For $V_N(r)$, we use the ansatz

$$V_N(r) \equiv V_0 \exp\left(-\eta_0^2 r^2\right) + V_1 \exp\left(-\eta_1^2 r^2\right), \tag{C.2}$$

of the Ali-Bodmer type [15], with parameters $V_0 = -216.0$ MeV, $V_1 = 354.0$ MeV, $\eta_0 = 0.436$ fm^{-1}, and $\eta_1 = 0.529$ fm^{-1}. These have been determined by fixing the S-wave and D-wave α-α scattering lengths to their experimental values. The Coulomb potential between two α particles is given by

$$V_C(r) \equiv \frac{4e^2}{r} \mathrm{erf}\left(\frac{\sqrt{3}r}{2R_\alpha}\right), \tag{C.3}$$

where the error function is

$$\mathrm{erf}(x) \equiv \frac{2}{\sqrt{\pi}} \int_0^x dy \, \exp(-y^2), \tag{C.4}$$

$R_\alpha \simeq 1.44$ fm is the (rms) radius of the α particle, and e is the fundamental unit of charge. Since we shall consider systems with more than two α particles, we should also include an explicit three-body interaction. Hence, we take [16]

$$V(\mathbf{r}_1, \mathbf{r}_2, \mathbf{r}_3) \equiv V_0 \exp\left[-\lambda(r_{12}^2 + r_{13}^2 + r_{23}^2)\right], \tag{C.5}$$

as the three-α potential. Here, the vectors $\mathbf{r}_i$ ($i = 1, 2, 3$) denote the coordinates of the α particles, and the r_{ij} denote the distances between the pairs (i, j). In Ref. [16], the width parameter λ was taken to be $\lambda = 0.00506$ fm^{-2}, and we set $V_0 = -4.41$ MeV in order to reproduce the empirical binding energy of the ^{12}C ground state.

We solve the Schrödinger equation

$$H\psi_i(\mathbf{r}_1, \mathbf{r}_2, \ldots, \mathbf{r}_{N-1}) = E_i\psi_i(\mathbf{r}_1, \mathbf{r}_2, \ldots, \mathbf{r}_{N-1}), \tag{C.6}$$

for the $3(N-1)$ relative coordinates of a system of N α particles. As the spatial vectors $\mathbf{r}$ assume discrete values on the lattice, the Hamiltonian H can be expressed in terms of a matrix. In a cubic box of dimension L, we impose the periodic boundary conditions

$$\psi(\mathbf{r} + \hat{n}_i L) = \psi(\mathbf{r}), \tag{C.7}$$

on the wave functions, where the $\hat{n}_i$ are unit vectors along the three spatial coordinate axes $i = x, y, z$. As we have discussed in Chap. 5, Monte Carlo methods are usually not required for few-body systems up to $N = 3$. Instead, the energy eigenvalues and wave functions of ^{8}Be ($N = 2$) and ^{12}C ($N = 3$) can be found by diagonalization of a Hamiltonian matrix of dimension $(L/a)^3 \times (L/a)^3$. As in Chap. 4, we replace the kinetic energy term of the Hamiltonian (C.1) by a finite-difference approximation. In one dimension, the second derivative can be expressed as

$$f''(x) \approx c_0^{(q)} f(x) + \sum_{k=1}^{q} c_k^{(q)} [f(x+ka) + f(x-ka)], \qquad (\text{C.8})$$

where a is the (spatial) lattice spacing and the $c_k^{(q)}$ denote a fixed set of coefficients. The formula of $\mathcal{O}(q)$ involves $2q+1$ lattice points, with a truncation error of $\mathcal{O}(a^{2q})$. If we denote

$$f_{na}(x) \equiv f(x - na) + f(x + na), \qquad (\text{C.9})$$

we may construct increasingly accurate approximations for $f''(x)$ according to

$$\frac{1}{a^2} f_a(x) - \frac{2}{a^2} f(x), \qquad (\text{C.10})$$

for $q = 1$,

$$-\frac{1}{12a^2} f_{2a}(x) + \frac{4}{3a^2} f_a(x) - \frac{5}{2a^2} f(x), \qquad (\text{C.11})$$

for $q = 2$,

$$\frac{1}{90a^2} f_{3a}(x) - \frac{3}{20a^2} f_{2a}(x) + \frac{3}{2a^2} f_a(x) - \frac{49}{18a^2} f(x), \qquad (\text{C.12})$$

for $q = 3$, and

$$-\frac{1}{560a^2} f_{4a}(x) + \frac{8}{315a^2} f_{3a}(x) - \frac{1}{5a^2} f_{2a}(x) + \frac{8}{5a^2} f_a(x) - \frac{205}{72a^2} f(x), \qquad (\text{C.13})$$

for $q = 4$. For a detailed derivation of such formulas for arbitrary q, see Ref. [14]. In our α cluster calculations for ^{8}Be and ^{12}C, we express the lattice Laplacian using Eq. (C.13). This choice is sufficiently accurate to remove most rotational symmetry breaking effects due to the kinetic energy operator.

C.2   Effects of Finite Volume and Lattice Spacing

Empirically, the simplest α-cluster nucleus (^{8}Be) is found $\simeq 90$ keV above the α-α threshold. It should be noted that (as in nature) the ^{8}Be nucleus is unbound with the parameters of Eq. (C.2). Since our objective is to study bound-state properties, we (arbitrarily) increase the magnitude of V_0 by 30%. With this adjustment, we find the ^{8}Be ground state at $E(0^+) = -10.8$ MeV, and one excited state at $E(2^+) = -3.3$ MeV, again relative to the α-α threshold. On the lattice, the 2^+ state splits into two multiplets corresponding to the E and T_2 representations of SO(3,Z), see Chap. 3 for notation and further details. These splittings are due to finite-volume as well as discretization effects.

We shall now make some phenomenological observations based on the model calculations of the 2α and 3α systems. While we focus our discussion on the ^{8}Be nucleus, we note that qualitatively similar results are obtained for ^{12}C nucleus, once the three-body interaction of Eq. (C.5) is included to obtain the physical binding energy of the ^{12}C ground state. In order to study effects due to the finite lattice volume L, we use a small lattice spacing of $a = 1.0$ fm. For small L, the 2^+_T state is pushed upward, while the 2^+_E state is pushed downward. As L is increased, these states are found to merge for $L \geq 15$ fm, and converge to the infinite-volume (continuum) result. This confirms the expectation that discretization errors are negligible at $a = 1.0$ fm, such that the observed energy splittings are entirely due to finite-volume effects. Moreover, the results shown in Ref. [14] were found to be consistent with the finite-volume energy shift formulas derived in Refs. [17, 18]. These observations suggest that the "multiplet-averaged" energy

$$E(2^+_A) \equiv \frac{1}{5} \left(2E(2^+_E) + 3E(2^+_T) \right), \tag{C.14}$$

would be an advantageous quantity to compute. Here, the weight factors 2 and 3 denote the number of members in each representation of SO(3,Z). The definition of Eq. (C.14) is further suggested by the fact that averaging over all elements of the spin multiplets simplifies the L-dependence of the finite volume corrections for two-body bound states [17, 18]. Also, the multiplet-averaged energy was found to be much closer, at any given L, to the infinite-volume continuum result than either of the E or T_2 branches. Another useful observable is the multiplet-averaged excitation gap,

$$E_{ex}(2^+) \equiv E(2^+_A) - E(0^+), \tag{C.15}$$

which is found to converge faster than the energy splitting,

$$\Delta E(2^+) = |E(2^+_E) - E(2^+_T)|, \tag{C.16}$$

to its value in the $L \to \infty$ limit. Note that the use of the magnitude of the energy splitting is merely a matter of convenience. For instance, at $L = 10$ fm, the excitation energy $E_{ex}(2^+)$ was found to be accurate to $\simeq 100$ keV, while the $\Delta E(2^+)$ energy splitting still exceeds 1 MeV.

Next, we shall consider systematic errors due to the finite lattice spacing a. In Fig. C.1, we show the 0^+ and 2^+ levels, $E(2_A^+)$, $E_{ex}(2^+)$, and $\Delta E(2^+)$ as a function of the lattice spacing a. In all cases, we take $L \geq 15$ fm and $q = 4$. While the energy splittings are negligible for $a \leq 1.0$ fm, an oscillatory behavior is found at larger a, such that the splittings become as large as 10 MeV for $a \geq 2.5$ fm. In Ref. [14], the cause of the oscillatory behavior was found to be related to the different spatial structure of the probability density distributions $|\psi(\mathbf{r})|^2$ for states in the E and T_2 representations, which in turn minimizes the energies of the respective states at different values of a. In spite of such large effects, it should be noted that the error of the multiplet-averaged excitation gap $E_{ex}(2^+)$, shown in the lower panel of Fig. C.1, does not exceed $\simeq 0.5$ MeV even for $a \leq 2.5$ fm. Hence, we conclude that the calculation of multiplet-averaged excitation gaps is a useful tool for reducing systematic errors in any lattice calculation of bound-state energies.

So far, we have only considered improvements to the lattice dispersion relation, and kept the same functional form for the interaction, which was inherited from the (smooth) continuum potential. When short-ranged interactions are involved (such as cutoff-dependent contact interactions of NLEFT), one should also consider improvements to the short-range interaction operators themselves. There are, however, practical reasons for restricting our attention to improvements of the lattice

Fig. C.1 Upper panel: The lowest 0^+ and 2^+ levels of the ^{8}Be nucleus as a function of (spatial) lattice spacing a, for $L \geq 15$ fm and $q = 4$ ("$N = 4$ formula"). The dashed line gives the weighted average $E(2_A^+)$. Lower panel: Energy splitting $\Delta E(2^+)$ (open circles) and average excitation gap $E_{ex}(2^+)$ (full circles) as a function of a. The dotted line shows the result for $E_{ex}(2^+)$ in the limits $L \to \infty$ and $a \to 0$

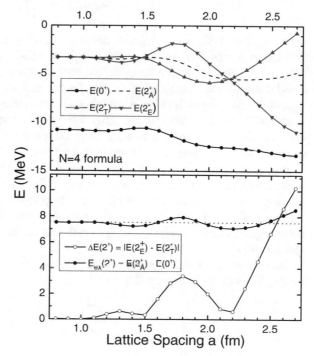

dispersion relation. As α clusters are built from protons and neutrons in ab initio nuclear lattice simulations, the details of the interactions between the emergent α clusters are difficult to compute and control in terms of the underlying nucleon lattice action. It is much easier to compute and modify the α dispersion relation via the underlying nucleon lattice dispersion relation. This is why we have focused our attention on improvements to the lattice dispersion relation.

C.3   Lattice Wave Functions

The continuum Hamiltonian H of Eq. (C.1) is invariant under spatial rotations. As a result, the bound states of H form degenerate angular momentum multiplets. We denote the bound-state wave functions by ϕ_{lm}, with angular momentum l and $-l \leq m \leq l$, such that the angular dependence of these wave functions is given by the spherical harmonics $Y_{l,m}$. Note that for systems with more than one bound state with the same value of l, additional radial quantum numbers would be required.

On the lattice, the multiplets of angular momentum l are split into *irreps* of the SO(3,Z) group, and we recall that the splitting patterns of the multiplets for $l \leq 8$ were given in Chap. 5. In order to specify which wave functions belong to the same *irrep*, we define a quantum number k valid on the lattice through the relation

$$R_z \left(\frac{\pi}{2} \right) \equiv \exp \left(-i \frac{\pi}{2} k \right), \tag{C.17}$$

following Ref. [19]. Here, $R_z(\pi/2)$ denotes a rotation around the z-axis by $\pi/2$. The integers k (which equal m modulo 4), are non-degenerate for each *irrep* of SO(3,Z). We use the notation $\psi_{l\tau k}$ for the wave functions, classified according to l, k and the *irrep* τ it belongs to. If the angular momentum l contains more than one "branch" belonging to the same *irrep*, we distinguish them by adding primes to the names of the *irreps*. For instance, $\psi_{6T_2'1}$ refers to a wave function with $l = 6$ and $k = 1$, which belongs to the second T_2 *irrep*. Note that the quantum number l is only approximate on the lattice, in the sense that a wave function labeled by l can have overlap with an infinite number of *irreps* of the rotational group carrying angular momenta other than l. However, such mixing effects are suppressed by powers of the lattice spacing. In the continuum limit, the $\psi_{l\tau k}$ form a complete basis for the subspace of bound states, and the corresponding energies are degenerate for the same angular momentum l. In contrast, on the lattice the energies depend on both l and the *irrep* τ.

In the continuum limit $a \to 0$, we can write down unitary transformations from the ϕ_{lm} basis to the $\psi_{l\tau k}$ basis [20],

$$\phi_{lm} \equiv \sum_{\tau k} U_{lm\tau k} \psi_{l\tau k}, \tag{C.18}$$

and vice versa,

$$\psi_{l\tau k} \equiv \sum_m U^{-1}_{l\tau km} \phi_{lm}, \tag{C.19}$$

which we illustrate below for the case of $l = 2$. The wave functions

$$\psi_{2E0} = \phi_{20}, \qquad \psi_{2E2} = \sqrt{\frac{1}{2}}(\phi_{22} + \phi_{2\bar{2}}), \tag{C.20}$$

belong to *irrep E*, and

$$\psi_{2T_21} = \phi_{21}, \qquad \psi_{2T_22} = -i\sqrt{\frac{1}{2}}(\phi_{22} - \phi_{2\bar{2}}), \qquad \psi_{2T_23} = \phi_{2\bar{1}}, \tag{C.21}$$

to *irrep T_2* [20]. In order to avoid confusing notation, we have used $\bar{m}$ to denote $-m$. On the lattice, the bound state wave functions $\psi_{l\tau k}$ are obtained by simultaneous diagonalization of the lattice Hamiltonian (or transfer matrix) and $R_z(\pi/2)$. Since the full rotational symmetry is broken, l should be viewed as a label for the angular momentum multiplet, which is obtained by continuously taking the limit $a \to 0$. Still, the unitary transformation of Eq. (C.18) can be used to define the wave functions ϕ_{lm} for $a \neq 0$. We do this even though the wave functions ϕ_{lm} are in general no longer exact eigenstates of H when $a \neq 0$.

We use φ_{lm} to denote the continuum limit ($a \to 0$) of the lattice wave functions ϕ_{lm}. For observables, we use $(f|O|i)$ with (round) parentheses to denote matrix elements computed by summation over lattice sites. Also, we use the notation $\langle f|O|i \rangle$ with brackets for matrix elements computed by integration over continuous space. Let us now consider the bound-state wave function ϕ_{00} for a zero angular momentum state, and ϕ_{lm} for a general angular momentum state. Then, the matrix element

$$C_m \equiv (lm|r^l Y_{l,m}|00) = \sum_{\mathbf{n}} \phi^*_{lm}(\mathbf{n}a)|\mathbf{n}a|^l Y_{l,m}(\hat{n})\phi_{00}(\mathbf{n}a), \tag{C.22}$$

of the multipole operator $r^l Y_{l,m}$ should become independent of m in the limit $a \to 0$. When using Eq. (C.18), the requirement that C_m satisfy this constraint provides a convenient check that the phases of ϕ_{lm} and $\psi_{l\tau k}$ agree with standard conventions [20].

C.4   Factorization of Matrix Elements

Let us now consider a pair of wave functions $\phi_{l_1 m_1}(\mathbf{r})$ and $\phi_{l_2 m_2}(\mathbf{r})$. On the one hand, the lattice matrix element of the multipole moment operator $r^{l'} Y_{l,m}$ is

$$(l_1 m_1 | r^{l'} Y_{l,m} | l_2 m_2) = \sum_{\mathbf{n}} \phi^*_{l_1 m_1}(\mathbf{n}a) | \mathbf{n}a |^{l'} Y_{l,m}(\hat{n}) \phi_{l_2 m_2}(\mathbf{n}a), \qquad (C.23)$$

where the summation runs over all lattice sites. Here, we have considered independent integers l and l' in the multipole moment operator, in order to keep the radial and angular degrees of freedom independent. This makes our conclusions sufficiently general and applicable to all irreducible tensor operators. On the other hand, the continuum version of Eq. (C.23) is

$$\langle l_1 m_1 | r^{l'} Y_{l,m} | l_2 m_2 \rangle = \int d^3 r \, \varphi^*_{l_1 m_1}(\mathbf{r}) r^{l'} Y_{l,m}(\Omega) \varphi_{l_2 m_2}(\mathbf{r}), \qquad (C.24)$$

where $\varphi_{l_1 m_1}(\mathbf{r})$ and $\varphi_{l_2 m_2}(\mathbf{r})$ denote the wave functions in the continuum limit. As we have found in Chap. 7, matrix elements of the form (C.24) occur frequently in the calculation of nuclear observables, such as rms radii, quadrupole moments, and transition probabilities. Here, we focus on lattice artifacts that produce differences between the numerical values of Eqs. (C.23) and (C.24) at a given lattice spacing a, and methods for removing them.

According to the Wigner-Eckart theorem, the matrix element (C.24) can be expressed as the product

$$\langle l_1 m_1 | r^{l'} Y_{l,m} | l_2 m_2 \rangle = \langle l_1 | r^{l'} | l_2 \rangle \, Q^{l_1 m_1}_{l_2 m_2, lm}, \qquad (C.25)$$

where the "reduced" matrix element

$$\langle l_1 | r^{l'} | l_2 \rangle \equiv \int dr \, r^{l'+2} R^*_1(r) R_2(r), \qquad (C.26)$$

encodes the dynamics of the problem, and

$$Q^{l_1 m_1}_{l_2 m_2, lm} \equiv \int d\Omega \, Y^*_{l_1, m_1}(\Omega) Y_{l,m}(\Omega) Y_{l_2, m_2}(\Omega), \qquad (C.27)$$

which can be expressed in terms of Clebsch-Gordan (CG) coefficients. In Eq. (C.26), $R_1(r)$ and $R_2(r)$ denote the radial parts of the wave functions $\varphi_{l_1 m_1}(\mathbf{r})$ and $\varphi_{l_2 m_2}(\mathbf{r})$, respectively. The radial integral in Eq. (C.26) represents the matrix element of the moment operator of order l', and is independent of m, m_1 and m_2. We can express

$$Q^{l_1 m_1}_{l_2 m_2, lm} = \sqrt{\frac{(2l+1)(2l_2+1)}{4\pi(2l_1+1)}} \, C^{l_1 0}_{l_2 0, l0} \, C^{l_1 m_1}_{l_2 m_2, lm}, \qquad (C.28)$$

Table C.1 The factors $Q^{l_1 m_1}_{l_2 m_2, lm}$ defined in Eq. (C.28) for $l_1 = l_2 = 2$ and $0 \le l \le 4$

l_1	l	l_2	m_1	m	m_2	$Q^{l_1 m_1}_{l_2 m_2, lm}$	m_1	m	m_2	$Q^{l_1 m_1}_{l_2 m_2, lm}$
2	0	2	0	0	0	$\frac{1}{2\sqrt{\pi}}$	1	0	1	$\frac{1}{2\sqrt{\pi}}$
			2	0	2	$\frac{1}{2\sqrt{\pi}}$				
2	2	2	0	0	0	$\frac{1}{7}\sqrt{\frac{5}{\pi}}$	2	1	1	$\sqrt{\frac{15}{14\pi}}$
			2	0	2	$-\frac{1}{7}\sqrt{\frac{5}{\pi}}$	1	1	0	$\frac{1}{14}\sqrt{\frac{5}{\pi}}$
			2	2	0	$-\frac{1}{7}\sqrt{\frac{5}{\pi}}$	1	0	1	$\frac{1}{14}\sqrt{\frac{5}{\pi}}$
			1	2	$\bar{1}$	$-\sqrt{\frac{15}{14\pi}}$				
2	4	2	0	0	0	$\frac{3}{7\sqrt{\pi}}$	1	3	$\bar{2}$	$\frac{1}{2}\sqrt{\frac{5}{7\pi}}$
			2	0	2	$\frac{1}{14\sqrt{\pi}}$	1	2	$\bar{1}$	$-\frac{1}{7}\sqrt{\frac{10}{\pi}}$
			2	1	1	$-\frac{1}{14}\sqrt{\frac{5}{\pi}}$	1	0	1	$-\frac{2}{7\sqrt{\pi}}$
			1	1	0	$\frac{1}{7}\sqrt{\frac{15}{2\pi}}$	2	2	0	$\frac{1}{14}\sqrt{\frac{15}{\pi}}$
			2	4	$\bar{2}$	$\sqrt{\frac{5}{14\pi}}$				

Those not given here can be obtained by means of standard tables of CG coefficients

in terms of the CG coefficients

$$C^{l_1 m_1}_{l_2 m_2, lm} \equiv \langle l_2, m_2; l, m | l_1, m_1 \rangle, \tag{C.29}$$

such that the dependence on m, m_1, m_2 and l is entirely absorbed into Eq. (C.28). Examples of the $Q^{l_1 m_1}_{l_2 m_2, lm}$ for $l_1 = l_2 = 2$ and $0 \le l \le 4$ are given in Table C.1, and we note that others can be computed using standard tables of CG coefficients.

For $a \ne 0$, a decomposition of the wave functions into products of radial and angular components is precluded due to rotational symmetry breaking. However, the Wigner-Eckart theorem is still applicable to each *irrep* of SO(3,Z). Thus, the lattice matrix elements that belong to the same *irrep* are related by the CG coefficients of the cubic group, which can be straightforwardly computed using decompositions into spherical harmonics [20, 21]. For instance,

$$(2T_2 1 | r^2 Y_{2E0} | 2T_2 1) = -\sqrt{\frac{1}{3}} (2T_2 1 | r^2 Y_{2E2} | 2T_2 \bar{1}), \tag{C.30}$$

where

$$Y_{2E0} \equiv Y_{20}, \qquad Y_{2E2} \equiv \sqrt{\frac{1}{2}} (Y_{22} + Y_{2\bar{2}}), \tag{C.31}$$

analogously to the relations in Eqs. (C.20) and (C.21). The factor $-\sqrt{1/3}$ is independent of the lattice spacing, box size, and details of the potential. Even though the factorization (C.25) is not exact on the lattice, we shall nevertheless divide the lattice matrix elements (C.23) by the $Q^{l_1 m_1}_{l_2 m_2, lm}$ of Eq. (C.28), giving

$$(l_1 m_1 \| r^{l'} Y_{l,m} \| l_2 m_2) \equiv \frac{(l_1 m_1 | r^{l'} Y_{l,m} | l_2 m_2)}{Q^{l_1 m_1}_{l_2 m_2, lm}}, \qquad Q^{l_1 m_1}_{l_2 m_2, lm} \neq 0, \tag{C.32}$$

and we note that such "reduced lattice matrix elements" (denoted by double vertical lines) all converge to $\langle l_1 \| r^{l'} \| l_2 \rangle$ for $a \to 0$. For $a \neq 0$, some dependence on m, m_1, and m_2 will remain. Hence, the splittings between the components of Eq. (C.32) can be used to quantify the degree to which rotational symmetry is broken.

C.5   Isotropic Average

We shall now develop some strategies for the removal of spatial anisotropies associated with the orientation of the lattice wave functions relative to the lattice axes. We start with the continuum wave functions $\varphi_{l_1 m_1}$ and $\varphi_{l_2 m_2}$ and define the "skewed" matrix element

$$\langle l_1 m_1 | r^{l'} Y_{l,m} | l_2 m_2 \rangle_\Lambda \equiv \int d^3 r \, \varphi^*_{l_1 m_1} (R(\Lambda)\mathbf{r}) \, r^{l'} Y_{l,m} (R(\Lambda)\Omega) \, \varphi_{l_2 m_2} (R(\Lambda)\mathbf{r}), \tag{C.33}$$

where $\Lambda \equiv (\alpha, \beta, \gamma)$ is a set of Euler angles and $R(\Lambda)$ is an element of the SO(3) rotation group. In the continuum limit, rotational invariance of the integral measure guarantees the equality of $\langle l_1 m_1 | r^{l'} Y_{l,m} | l_2 m_2 \rangle$ and $\langle l_1 m_1 | r^{l'} Y_{l,m} | l_2 m_2 \rangle_\Lambda$. For $a \neq 0$ and $l \neq 0$, the skewed matrix element will depend on Λ. This unphysical orientation dependence should be eliminated as far as possible. We define the "isotropically averaged" matrix element

$$\langle l_1 m_1 | r^{l'} Y_{l,m} | l_2 m_2 \rangle_\circ \equiv \int d^3 \Lambda \, \langle l_1 m_1 | r^{l'} Y_{l,m} | l_2 m_2 \rangle_\Lambda$$

$$= C^{l_1 m_1}_{l_2 m_2, lm} \left[\frac{1}{2 l_1 + 1} \sum_{m', m'_1, m'_2} C^{l_1 m'_1}_{l_2 m'_2, lm'} \langle l_1 m'_1 | r^{l'} Y_{l,m'} | l_2 m'_2 \rangle \right], \tag{C.34}$$

where the circle "∘" denotes averaging over all possible orientations, with $d^3 \Lambda$ the normalized invariant measure on the SO(3) group space. Given Eq. (C.34), we take

$$(l_1 m_1 |r^{l'} Y_{l,m}| l_2 m_2)_\circ \equiv C^{l_1 m_1}_{l_2 m_2, lm} \left[\frac{1}{2l_1 + 1} \sum_{m', m'_1, m'_2} C^{l_1 m'_1}_{l_2 m'_2, lm'} (l_1 m'_1 |r^{l'} Y_{l,m'}| l_2 m'_2) \right],$$

$$(C.35)$$

as the "isotropically averaged lattice matrix element". As in Eq. (C.32), we obtain

$$(l_1 \| r^{l'} Y_l \| l_2)_\circ \equiv \frac{(l_1 m_1 |r^{l'} Y_{l,m}| l_2 m_2)_\circ}{Q^{l_1 m_1}_{l_2 m_2, lm}}, \qquad Q^{l_1 m_1}_{l_2 m_2, lm} \neq 0, \qquad (C.36)$$

as the "isotropically averaged reduced lattice matrix element". Again, $(l_1 \| r^{l'} Y_l \| l_2)_\circ$ reduces to the radial matrix element $\langle l_1 | r^{l'} | l_2 \rangle$ when $a \to 0$. It should be noted that $\langle l_1 | r^{l'} | l_2 \rangle$ is not only independent of m, m_1 and m_2, but also independent of l. Then, a non-trivial test of rotational symmetry restoration is to check that $(l_1 \| r^{l'} Y_l \| l_2)_\circ$, as defined in Eq. (C.36), is independent of l. If $(l_1 \| r^{l'} Y_l \| l_2)_\circ$ is independent of l to a good approximation, then we have succeeded in (approximately) factorizing the radial and angular parts of the lattice wave function by means of isotropic averaging. In what follows, we will test this numerically for the α-cluster model.

C.6 Lattice Spacing Dependence of Transition Operators

Let us consider the rms radius operator for the ^{8}Be nucleus, which is obtained by setting $l = 0$ and $l' = 2$ in Eqs. (C.23) and (C.24). In the upper panel of Fig. C.2, we show $\langle r^2 \rangle$ for the lowest 2^+ states of ^{8}Be, as a function of the lattice spacing a. The wave functions $\psi_{2\tau k}$ are obtained by simultaneous diagonalization of the lattice Hamiltonian H and $R_z(\pi/2)$. Finite volume effects are suppressed by taking the box size $L \geq 16$ fm. The linear combinations ϕ_{2m} are then constructed according to Eqs. (C.20) and (C.21). Only three cases with $m \geq 0$ are shown, since time reversal symmetry ensures equal results for m and $-m$. The three branches shown in the upper panel of Fig. C.2 are not linearly independent, because of the cubic symmetries on the lattice. According to Eqs. (C.20) and (C.21), the wave functions ϕ_{21} and ϕ_{20} belong to *irreps* E and T_2, respectively. Similarly, ϕ_{22} is a mixture of the *irreps* E and T_2, with equal weights. We use the simplified notation $(m \| U \| m)$ as an abbreviation for $(2m \| r^2 Y_{00} \| 2m)$. Thus, $(2 \| 0 \| 2)$ equals the arithmetic average of $(0 \| 0 \| 0)$ from *irrep* E and $(1 \| 0 \| 1)$ from *irrep* T_2. Hence, the isotropically averaged reduced lattice matrix element $(2 \| r^2 Y_0 \| 2)_\circ$ is given by

$$(2 \| r^2 Y_0 \| 2)_\circ = \frac{3}{5}(1 \| 0 \| 1) + \frac{2}{5}(0 \| 0 \| 0), \qquad (C.37)$$

where the factors 3 and 2 in the numerators are given by the multiplicities of the cubic representations. It is easy to verify that the weighted average formula is applicable for any angular momentum, provided that the factors 3 and 2 in Eq. (C.37) are replaced by the multiplicities of the appropriate *irreps*. We now turn to the case of $l = 2$ and $l' = 2$, likewise for ^{8}Be. Again, we use $(\alpha\|\beta\|\gamma)$ as an abbreviation for $(2\alpha\|r^2 Y_{2\beta}\|2\gamma)$. Here, the subscripts α, β and γ assume values from -2 to 2 (we only show the components with $\alpha = \beta + \gamma$). As the $2l + 1$ wave function components in an angular momentum multiplet mix on the lattice, some components with $\alpha \neq \beta + \gamma$ survive for large a. However, as the corresponding CG coefficients vanish, such components do not contribute to the isotropic average.

Given the results shown in Fig. C.2, we can compare the components $(m\|0\|m)$ to the continuum limit and determine which ones exhibit the least dependence on a. Clearly, for the scalar operator r^2, the arithmetic average $(2\|0\|2)$ and the multiplet-weighted average both remain accurate over a wide range of lattice spacings. The latter is theoretically preferable, because of its clear physical interpretation as

Fig. C.2 Expectation values of $\langle r^2 \rangle$ for the lowest 2^+ states of the ^{8}Be nucleus as a function of a. Upper panel: Scalar operator r^2. Lower panel: Insertion of the spherical harmonic $Y_{2\beta}$. The solid lines show the isotropic averages according to Eq. (C.36). We maintain $L \geq 16$ fm for all values of a, in order to suppress finite-volume effects. Note the very large and unphysical spread of the individual branches for $a \simeq 2$ fm

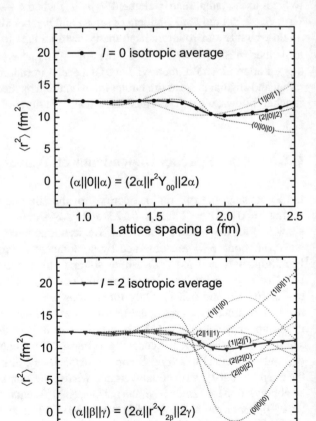

isotropic averaging. The insertion of the spherical harmonic $Y_{2\beta}$ makes the situation much more complicated, as evidenced in the lower panel of Fig. C.2. Firstly, as for the scalar operators, one finds that $(2\|0\|2)$ equals the arithmetic average of $(1\|0\|1)$ and $(0\|0\|0)$, while $(2\|2\|0)$ equals that of $(1\|1\|0)$ and $(0\|0\|0)$. Secondly, applying the Wigner-Eckart theorem for the cubic group, we obtain multiple linear identities between the lattice matrix elements. These involve not only the components shown in Fig. C.2, but also those that vanish when $a \to 0$.

Without isotropic averaging, the operator expectation value of any given branch $(\alpha\|\beta\|\gamma)$ at $a \simeq 2$ fm could hardly be considered a reliable estimator of the limit $a \to 0$, because of the large oscillations shown in Fig. C.2. As noted already for the binding energies, such oscillations are associated with the commensurability of the lattice with the size and shape of the lattice wave functions. The lattice wave functions receive large contributions from lattice vectors which form the corresponding representation of the cubic rotational group, and are closest in length to the average (continuum) separation R between constituent particles. Roughly speaking, if the lattice vectors closest in length to R are shorter than R, then $\langle r^2 \rangle$ falls below its continuum value. Conversely, if the lattice separation vectors closest in length to R are longer than R, then $\langle r^2 \rangle$ exceeds its continuum value. We find that isotropic averaging leads to an almost complete elimination of such oscillations.

C.7 Summary

Here, we have studied the breaking of rotational symmetry due to lattice artifacts, with emphasis on how the degeneracy of multiplets of bound states with the same angular momentum is affected. As bound-state wave functions on the lattice are classified according to the irreducible representations of the cubic group instead of the full SO(3) rotational group, the treatment of observables represented by irreducible tensor operators becomes more complicated. We have used a simplified α-cluster model to study the lattice matrix elements of such operators, and found that their qualitative behavior as a function of the lattice spacing a is mainly determined by the angular momentum quantum numbers of the states and operators.

In order to minimize the effects of rotational symmetry breaking, we have introduced an "isotropic average", which consists of a linear combination of the components of the matrix element. The weight of a given component is given by the GC coefficient with the associated quantum numbers. We have shown that this method is equivalent to averaging over all possible lattice orientations. We have also found that isotropic averaging eliminates, to a good approximation, the anisotropy caused by lattice artifacts. This was illustrated by numerical calculations for the ^{8}Be nucleus within the α-cluster model. For instance, the isotropically averaged $\langle r^2 \rangle$ and $\langle r^4 \rangle$ were found to only slightly underestimate their continuum values in the region 1.7 fm $\leq a \leq 2.0$ fm, while the individual components could be wildly inaccurate. For $a < 1.7$ fm, the deviation from the continuum was found to be negligible. The overall agreement with the continuum results was excellent.

We note that the present conclusions have been obtained within a highly simplified α-cluster model, which is amenable to calculations without Monte Carlo methods. Nevertheless, the method of isotropic averaging is immediately applicable to ab initio lattice Monte Carlo results. For instance, in the lattice calculation of transition amplitudes between low-energy excited states and the ground state of a nucleus (as described for ^{12}C and ^{16}O in Chap. 7), this method is expected to provide an immediate improvement by removing ambiguities due to unphysical level splittings, and by minimizing the dependence on the lattice spacing.

Appendix D
Monte Carlo Sampling

Here, as well as in Appendices E and F, we provide technical details on various aspects of the NLEFT calculations, in the hope that they will be found helpful for the development of practical Monte Carlo (MC) codes. We use a "pseudocode" notation inspired by Fortran 90 conventions, where functions and subroutines are indicated with typewriter font. However, as our presentation does not strongly depend on the choice of programming language, we expect the ideas to be easily applicable to the reader's framework of choice. We are here assuming that NLEFT has been formulated along the lines of the LO_3 action as described in Chap. 4, to which the reader is referred to for proofs and derivations. Much of what we present here is concerned with the evolution of trial wave functions for N_f fermions in Euclidean time, by repeated application of the transfer matrix operator $\hat{M}$, which depends on a set of auxiliary fields $\{\phi\}$ defined on discrete time slices $t = 0 \ldots, N_t - 1$. Each spatial lattice dimension has sites $0, \ldots, N - 1$ with periodic boundaries. As before, we define the ratio of temporal to spatial lattice spacings as $\alpha_t \equiv \delta/a$. The trial states are defined in terms of single-particle wave functions for nucleons $p = 0, \ldots, N_f - 1$. In general, we also evolve more than one wave function, in which case we speak of "coupled-channel" simulations with $c = 0, \ldots, N_{ch} - 1$.

We shall now consider the MC evaluation of the NLEFT projection amplitude, as defined in Chap. 6. We recall that this can be expressed in terms of a path integral over auxiliary fields, with additional summations over channels and spatial locations of the nucleons in the trial state. Quantities that involve fermions (nucleons) are expressed in terms of

- "vectors" $Z(\mathbf{n}, t, s, i, p)$ which describe the Euclidean time evolution of nucleon p with spin s and isospin i in the initial state,
- "dual vectors" $Z^d(\mathbf{n}, t, s, i, p)$ which describe the Euclidean time evolution of nucleon p with spin s and isospin i in the final state,

© Springer Nature Switzerland AG 2019

T. A. Lähde, U.-G. Meißner, *Nuclear Lattice Effective Field Theory*,
Lecture Notes in Physics 957, https://doi.org/10.1007/978-3-030-14189-9

which we shall treat in detail in Appendix E. We also consider updates of the positions of the nucleons in the initial and final states, which we implement using the notation

- $x_1(p)$, $y_1(p)$, and $z_1(p)$ for translations of initial-state nucleon p in each lattice dimension,
- $x_2(p)$, $y_2(p)$, and $z_2(p)$ for translations of final-state nucleon p in each lattice dimension,
- x_{CM}, y_{CM}, and z_{CM} for translations in each lattice dimension of the center-of-mass of the nucleons in the final state,

all of which may (optionally) be fixed or updated during the MC calculation. Which of these positional updates are enabled during a given calculation depends of the type of trial state, and the specific aim of the calculation. Below, we shall illustrate this by means of specific examples. The auxiliary fields are always updated with the Hybrid Monte Carlo (HMC) algorithm, which we have encountered in Chap. 6. We describe the implementation of different trial states in Sect. D.1 and the different MC updates in Sect. D.2.

D.1   Examples of Trial States

We begin by specifying the spin-isospin part of the wave function for each nucleon, which can be either protons or neutrons with spin-up or spin-down. We take

$$s = 0, \quad i = 0, \tag{D.1}$$

for $type(p) = p+$,

$$s = 1, \quad i = 0, \tag{D.2}$$

for $type(p) = p-$,

$$s = 0, \quad i = 1, \tag{D.3}$$

for $type(p) = n+$, and

$$s = 1, \quad i = 1, \tag{D.4}$$

for $type(p) = n-$. We also need to specify the spatial part of the wave function for each nucleon. For simplicity, we make use of plane-wave components in a cubic box, and take

$$Z_{init}(\mathbf{n}, s, i, p) = 1, \tag{D.5}$$

for $\texttt{trig}(p) = \texttt{one}$,

$$Z_{\text{init}}(\mathbf{n}, s, i, p) = \sqrt{2}\cos\left(\frac{2\pi k_x(p)}{N}x + \frac{2\pi k_y(p)}{N}y + \frac{2\pi k_z(p)}{N}z\right), \tag{D.6}$$

for $\texttt{trig}(p) = \texttt{cos}$, and

$$Z_{\text{init}}(\mathbf{n}, s, i, p) = \sqrt{2}\sin\left(\frac{2\pi k_x(p)}{N}x + \frac{2\pi k_y(p)}{N}y + \frac{2\pi k_z(p)}{N}z\right). \tag{D.7}$$

and for $\texttt{trig}(p) = \texttt{sin}$. Thus, we are in a position to specify the spin, isospin, and the integers k_x, k_y and k_z for each nucleon in the trial state. Finally, we note that

$$Z_{\text{init}}^d(\mathbf{n}, s, i, p) = Z_{\text{init}}^*(\mathbf{n}, s, i, p), \tag{D.8}$$

for the dual Euclidean time projection vectors.

D.1.1 Helium-4 Using Zero-Momentum States

For this example, we define $\texttt{multichannel} = 0$, and $\texttt{nspatial} = 0$, as we are not performing a multichannel calculation or updating nucleon positions (such updates are only needed for nucleons defined as localized wave packets). For further details, see Sect. D.2. We set the nucleon translations (which are not updated) to $x_1(p) = y_1(p) = z_1(p) = 0$ and $x_2(p) = y_2(p) = z_2(p) = 0$. We also define the quantities

$$N_{\text{fill}}(p, c) = p, \tag{D.9}$$

where $p = 0, \ldots, N_f - 1$, and $c = 0, \ldots, N_{\text{ch}} - 1$, and

$$N_{\text{trim}}(p) = 0, \quad N_{\text{trim}}^d(p) = N_{\text{trim}}(p), \tag{D.10}$$

for $p = 0, \ldots, N_{\text{all}} - 1$, which will be needed in Appendix E in order to properly account for the Pauli principle in the calculation of Z and Z^d. Note that we allow $N_{\text{all}} \neq N_f$ for sufficient generality. For an Euclidean time projection calculation of the ground state of ^{4}He, we set $N_f = 4$, $N_{\text{all}} = 4$, $N_{\text{ch}} = 1$, and define

$$
\begin{aligned}
&\texttt{type}(0) = \texttt{p+}, \quad \texttt{trig}(0) = \texttt{one}, \quad k_x(0) = k_y(0) = k_z(0) = 0, \\
&\texttt{type}(1) = \texttt{n|}, \quad \texttt{trig}(1) = \texttt{one}, \quad k_x(1) = k_y(1) = k_z(1) = 0, \\
&\texttt{type}(2) = \texttt{p-}, \quad \texttt{trig}(2) = \texttt{one}, \quad k_x(2) = k_y(2) = k_z(2) = 0, \\
&\texttt{type}(3) = \texttt{n-}, \quad \texttt{trig}(3) = \texttt{one}, \quad k_x(3) = k_y(3) = k_z(3) = 0,
\end{aligned} \tag{D.11}
$$

which gives a ^{4}He trial state with zero total momentum.

D.1.2   Carbon-12 Using Zero-Momentum States

For ^{12}C, we take $N_f = 12$, $N_{\text{all}} = 12$, and $N_{\text{ch}} = 1$, with multichannel = 0, and nspatial = 0, and we proceed otherwise according to the example for ^{4}He. However, we now set

$$N_{\text{trim}}(0:3) = 0, \quad N_{\text{trim}}(4:7) = 1, \quad N_{\text{trim}}(8:11) = 2, \tag{D.12}$$

and

$$N_{\text{trim}}^d(p) = N_{\text{trim}}(p), \tag{D.13}$$

such that $N_{\text{trim}}(p)$ equals the slice t of Euclidean projection time on which nucleon p is "injected". We are therefore able to simply replicate the definitions (D.11) for nucleons 4–7 and 8–11, as the Pauli principle is circumvented. The extension of this approach to more than 12 nucleons in a state of zero total momentum is straightforward.

D.1.3   Carbon-12 Using States with Finite Momentum

Let us consider a different approach to the case of ^{12}C, whereby instead of injecting four nucleons at a time on successive time slices, we set $N_{\text{trim}}(p) = 0$ throughout. We are therefore unable to insert more than four nucleons into zero-momentum states. However, we may still resort to successively filling the lowest available modes in the cubic box, which gives

$$
\begin{aligned}
&\text{type}(0) = \text{p+}, &&\text{trig}(0) = \text{one}, &&k_x(0) = k_y(0) = k_z(0) = 0, \\
&\text{type}(1) = \text{n+}, &&\text{trig}(1) = \text{one}, &&k_x(1) = k_y(1) = k_z(1) = 0, \\
&\text{type}(2) = \text{p-}, &&\text{trig}(2) = \text{one}, &&k_x(2) = k_y(2) = k_z(2) = 0, \\
&\text{type}(3) = \text{n-}, &&\text{trig}(3) = \text{one}, &&k_x(3) = k_y(3) = k_z(3) = 0, \\
&\text{type}(4) = \text{p+}, &&\text{trig}(4) = \text{cos}, &&k_x(4) = k_y(4) = k_z(4) = 1, \\
&\text{type}(5) = \text{n+}, &&\text{trig}(5) = \text{cos}, &&k_x(5) = k_y(5) = k_z(5) = 1, \\
&\text{type}(6) = \text{p-}, &&\text{trig}(6) = \text{cos}, &&k_x(6) = k_y(6) = k_z(6) = 1, \\
&\text{type}(7) = \text{n-}, &&\text{trig}(7) = \text{cos}, &&k_x(7) = k_y(7) = k_z(7) = 1, \\
&\text{type}(8) = \text{p+}, &&\text{trig}(8) = \text{sin}, &&k_x(8) = k_y(8) = k_z(8) = 1, \\
&\text{type}(9) = \text{n+}, &&\text{trig}(9) = \text{sin}, &&k_x(9) = k_y(9) = k_z(9) = 1, \\
&\text{type}(10) = \text{p-}, &&\text{trig}(10) = \text{sin}, &&k_x(10) = k_y(10) = k_z(10) = 1, \\
&\text{type}(11) = \text{n-}, &&\text{trig}(11) = \text{sin}, &&k_x(11) = k_y(11) = k_z(11) = 1, \tag{D.14}
\end{aligned}
$$

and we recall from Chap. 7 that such a trial state has been extensively used in realistic studies of ^{12}C. However, it is found that such a finite-momentum trial state has a large overlap with the 0_2^+ "Hoyle state" of ^{12}C, and would require a very long Euclidean projection time in order to converge to the 0_1^+ ground state. Such long projection times are not feasible for the LO$_3$ action due to sign oscillations (note, however, the recent developments with non-locally smeared LO actions in Chap. 8). In order to find rapid convergence the ground state of ^{12}C, the zero-momentum trial state with $N_{\text{trim}}(p) \neq 0$ is more useful. This situation is similar for other nuclei with $A > 4$.

D.2 Types of Monte Carlo Updates

In each MC simulation, the auxiliary and pion fields $\{\phi\}$ are initialized using a "hot start", where each field is populated from a random Gaussian distribution, which is then allowed to "thermalize" as the simulation progresses (in principle, a "cold start" could also be used, where all auxiliary fields are initialized to zero). Random numbers can be obtained using any given high-quality generator of (pseudo-) random numbers. Each iteration of the MC simulation consists of three types of MC updates, specifically global updates of $\{\phi\}$, updates of the center-of-mass (CM) position in the final nucleon configuration, and updates of the relative separations between nucleons in both the initial and final states, in the following sequence:

1. Compute the boson (auxiliary field and pion) action and the fermion action.
2. For ndoCM $= 1$, update the CM position of the final state and the fermion action.
3. For multichannel $= 1$ and nspatial $= 0$, update the channels c_1 and c_2 and the fermion action.
4. For multichannel $= 1$ and nspatial $= 1$, perform a translational update of the initial state. the initial channel c_1, and the fermion action. The translational update is proposed by the routine spatial.
5. For multichannel $= 1$ and nspatial $= 1$, perform a translational update of the final state. the final channel c_2, and the fermion action. The translational update is proposed by the routine spatial.
6. Hybrid Monte Carlo (HMC) update of the auxiliary and pion fields.

We shall now briefly describe the implementation of each of these updates. In general, these require the numerical evaluation of the boson action, the fermion action, and various functional derivatives thereof. Details on their implementation are given in Appendix E.

D.2.1   Hybrid Monte Carlo Updates

As we have argued in Chap. 6, for theories with fermions, global Hybrid Monte Carlo (HMC) updates are often computationally superior to random local updates, especially when the number of spatial dimensions $d > 2$. For this purpose, we introduce a set of conjugate momentum fields $\{\pi\}$ corresponding to the auxiliary- and pion field components $\{\phi\}$. In Chap. 4, we introduced the method of accelerating the convergence of Euclidean time projection by means of a simplified, Wigner SU(4)-invariant LO Hamiltonian. In the "outer", Wigner SU(4)-invariant region, the HMC action is

$$S_{\text{HMC}} \equiv \frac{1}{2} \sum_{\mathbf{n},t} \pi^2(\mathbf{n}, t) + S_\phi + S_f, \tag{D.15}$$

for $t < N_t^{\text{outer}}$ and $t \geq N_t^{\text{outer}} + N_t^{\text{inner}}$, and

$$S_{\text{HMC}} \equiv \frac{1}{2} \sum_{\mathbf{n},t} \pi^2(\mathbf{n}, t) + \frac{1}{2} \sum_{k=1}^{3} \sum_{\mathbf{n},t} \pi_k^{S\,2}(\mathbf{n}, t) + \frac{1}{2} \sum_{l=1}^{3} \sum_{\mathbf{n},t} \pi_l^{I\,2}(\mathbf{n}, t) \tag{D.16}$$

$$+ \frac{1}{2} \sum_{k=1}^{3} \sum_{l=1}^{3} \sum_{\mathbf{n},t} \pi_{k,l}^{SI\,2}(\mathbf{n}, t) + \frac{1}{2} \sum_{l=1}^{3} \sum_{\mathbf{n},t} \pi_l^{\pi\,2}(\mathbf{n}, t) + S_\phi + S_\pi + S_f,$$

in the "inner" region where the full LO action is used, for $t \geq N_t^{\text{outer}}$ and $t < N_t^{\text{outer}} + N_t^{\text{inner}}$. The $\{\pi\}$ fields do not couple to the $\{\phi\}$ fields, and hence leave the dynamics unchanged. The summations over k and l account for spin- and isospin dependence of the auxiliary and pion fields (and their respective canonically conjugate momentum fields). The boson action is given by $S_\phi + S_\pi$, and the fermion action by S_f.

We are now in a position now evolve the $\{\phi\}$ fields in a fictitious "MC time" τ using the Hamiltonian equations of motion, along the lines of Chap. 6. During this global HMC (or "Molecular Dynamics") update of $\{\phi\}$ and $\{\pi\}$, S_{HMC} should remain unchanged. In practice, the Hamiltonian equations of motion are integrated using a finite time step ϵ. The total HMC evolution time is

$$\tau \equiv \epsilon N_{\text{HMC}}, \tag{D.17}$$

and we recall that S_{HMC} is exactly conserved only in the limit $\epsilon \to 0$. In Chap. 6, we established that an acceptable (for the purpose of maintaining detailed balance) integration method should be reversible and symplectic. Here, we shall implement the "leap-frog" integrator from Chap. 6. A given configuration $\{\pi\}$ and $\{\phi\}$ is then updated according to

$$\pi(\mathbf{n}, t) \rightarrow \pi(\mathbf{n}, t) - \epsilon \frac{\delta S_{\mathrm{HMC}}}{\delta \phi(\mathbf{n}, t)},$$

$$\pi_k^S(\mathbf{n}, t) \rightarrow \pi_k^S(\mathbf{n}, t) - \epsilon \frac{\delta S_{\mathrm{HMC}}}{\delta \phi_k^S(\mathbf{n}, t)},$$

$$\pi_l^I(\mathbf{n}, t) \rightarrow \pi_l^I(\mathbf{n}, t) - \epsilon \frac{\delta S_{\mathrm{HMC}}}{\delta \phi_l^I(\mathbf{n}, t)},$$

$$\pi_{k,l}^{SI}(\mathbf{n}, t) \rightarrow \pi_{k,l}^{SI}(\mathbf{n}, t) - \epsilon \frac{\delta S_{\mathrm{HMC}}}{\delta \phi_{k,l}^{SI}(\mathbf{n}, t)},$$

$$\pi_l^\pi(\mathbf{n}, t) \rightarrow \pi_l^\pi(\mathbf{n}, t) - \epsilon \frac{\delta S_{\mathrm{HMC}}}{\delta \phi_l^\pi(\mathbf{n}, t)}, \tag{D.18}$$

for the conjugate momentum fields $\{\pi\}$, and

$$\phi(\mathbf{n}, t) \rightarrow \phi(\mathbf{n}, t) + \epsilon \pi(\mathbf{n}, t),$$

$$\phi_k^S(\mathbf{n}, t) \rightarrow \phi_k^S(\mathbf{n}, t) + \epsilon \pi_k^S(\mathbf{n}, t),$$

$$\phi_l^I(\mathbf{n}, t) \rightarrow \phi_l^I(\mathbf{n}, t) + \epsilon \pi_l^I(\mathbf{n}, t),$$

$$\phi_{k,l}^{SI}(\mathbf{n}, t) \rightarrow \phi_{k,l}^{SI}(\mathbf{n}, t) + \epsilon \pi_{k,l}^{SI}(\mathbf{n}, t),$$

$$\phi_l^\pi(\mathbf{n}, t) \rightarrow \phi_l^\pi(\mathbf{n}, t) + \epsilon \pi_l^\pi(\mathbf{n}, t), \tag{D.19}$$

for the auxiliary and pion fields $\{\phi\}$. The update $\{\phi\} \rightarrow \{\phi'\}$ and $\{\pi\} \rightarrow \{\pi'\}$ or "trajectory", then proceeds according to the algorithm

1. Select $\{\pi\}$ from a random Gaussian distribution,
2. Compute $S_{\mathrm{HMC}}(\mathrm{old})$ using $\{\phi\}$ and $\{\pi\}$ (see Appendix E).
3. Compute the functional derivatives ("force terms") using $\{\phi\}$ (see Appendix E).
4. Perform an initial "half-step" $\epsilon \rightarrow \epsilon/2$ in $\{\pi\}$, according to Eq. (D.18).
5. Perform a "full step" ϵ in $\{\phi\}$, according to Eq. (D.19).
6. Recompute the "force terms" in Eq. (D.18) using the updated $\{\phi\}$.
7. Perform a "full step" ϵ in $\{\pi\}$, according to Eq. (D.18).
8. Return to step 3 and repeat for N_{HMC} updates of $\{\phi\}$.
9. Perform a final "half-step" $\epsilon \rightarrow \epsilon/2$ in $\{\pi\}$, according to Eq. (D.18).
10. Compute $S_{\mathrm{HMC}}(\mathrm{new})$ using $\{\phi'\}$ and $\{\pi'\}$.
11. Perform a Metropolis accept/reject test. Select a random number r in the range $[0:1]$, if

$$r < \min\left[1, \exp(-\Delta S)\right], \quad \Delta S \equiv S_{\mathrm{HMC}}(\mathrm{new}) - S_{\mathrm{HMC}}(\mathrm{old}), \tag{D.20}$$

then update $\{\phi\} \rightarrow \{\phi'\}$, else reject the configuration $\{\phi'\}$.
12. Return to step 1.

We emphasize that for the updated configuration $\{\phi'\}$ to be significantly different (decorrelated) from the initial configuration $\{\phi\}$, the HMC trajectories need to be sufficiently long. For a hybrid algorithm such as HMC, optimal performance and decorrelation is found when $\epsilon N_{HMC} \sim 1$ (we recall that the entirely stochastic Langevin method is recovered for $N_{HMC} = 1$). Since the computational cost per update is proportional to N_{HMC}, it is clearly advantageous to maximize ϵ. As we have argued in Chap. 6, for the leap-frog integrator ϵ should be adjusted such that a Metropolis acceptance rate of $\simeq 60\%$ is maintained.

D.2.2   Center-of-Mass Position Updates

In MC simulations where the trial states are localized (such as α-cluster trial states), we need to sum over all the (relative) locations of the center-of-mass (CM) in the initial and final states of the Euclidean projection amplitude. In practice, it suffices to update the position of the CM in the final state. This feature is enabled by setting ndoCM $= 1$. We illustrate the procedure by considering an update of the CM position of an α cluster in the final state. Let us assume that (x_{CM}, y_{CM}, z_{CM}) denotes the position of the CM. If we index the spatial lattice sites according to

$$n_r \equiv x + Ny + N^2 z, \tag{D.21}$$

then

$$N_{CM} \equiv x_{CM} + Ny_{CM} + N^2 z_{CM}, \tag{D.22}$$

gives the indexed position of the CM. We define the heat-bath probability measure

$$\Pi_{CM}(n_r) \equiv \Pi_0 \exp\left(-\frac{d^2(x, y, z)}{2\Gamma_{CM}^2}\right), \tag{D.23}$$

where the normalization is

$$\Pi_0 \equiv \sum_{n_r=0}^{N^3-1} \exp\left(-\frac{d^2(x, y, z)}{2\Gamma_{CM}^2}\right), \tag{D.24}$$

$\Gamma_{CM} = 1$ is a Gaussian width parameter (which can be adjusted to optimize the acceptance rate of the CM updates), and

$$d^2(x, y, z) \equiv \min\left[x^2, (N-x)^2\right] + \min\left[y^2, (N-y)^2\right] + \min\left[z^2, (N-z)^2\right], \tag{D.25}$$

is the squared distance on the lattice. We expect the Euclidean projection amplitude to be dominated by contributions where the CM positions of the initial and final states are close to each other. We also define the "probability sum"

$$\Pi_{\text{CM}}^{\text{sum}}(n_r) \equiv \Pi_{\text{CM}}^{\text{sum}}(n_r - 1) + \Pi_{\text{CM}}(n_r), \tag{D.26}$$

with the condition $\Pi_{\text{CM}}^{\text{sum}}(0) = \Pi_{\text{CM}}(0)$.

The update of the CM position of the final state proceeds according to the following algorithm:

1. Select a random number r in the range $[0:1]$, and loop over all lattice sites n_r. Once $\Pi_{\text{CM}}^{\text{sum}}(n_r) \geq r$, take $N_{\text{CM}}(\text{new}) = n_r$.
2. Compute (for details, see Appendix E) the ratio

$$\frac{\det(Z_{\text{end}})}{|\det(Z_{\text{corr}})|}, \tag{D.27}$$

with the numerator given by the `getendoverlap` routine using $N_{\text{CM}}(\text{new})$, according to Eq. (E.20). The denominator $|\det(Z_{\text{corr}})|$ should be computed using $N_{\text{CM}}(\text{old})$.
3. Select a random number r in the range $[0:1]$, if

$$r < \left\{ \frac{|\det(Z_{\text{end}})|}{|\det(Z_{\text{corr}})|} \times \frac{\Pi_{\text{CM}}(N_{\text{CM}}(\text{old}))}{\Pi_{\text{CM}}(N_{\text{CM}}(\text{new}))} \right\}, \tag{D.28}$$

then update the CM position $N_{\text{CM}}(\text{old}) = N_{\text{CM}}(\text{new})$, otherwise retain the old CM position.
4. If the CM position was updated, recompute the fermion action S_f and update the HMC action S_{HMC} accordingly.

D.2.3   Translational Updates

Let us now outline how the relative positions of the nucleons in the initial and final states can be updated. Such translational updates can concern either individual nucleons, or groups of nucleons (such as α-clusters). Examples of calculations where translational updates occur are simulations of the spatial structure of nuclear states (such as the Hoyle state of ^{12}C), simulations of adiabatic Hamiltonians (where different separations between α clusters need to be probed), and the "pinhole algorithm" described in Chap. 8. Clearly, the translational updates are highly problem-dependent, and depend (for instance) on whether we are considering a problem with two α-clusters (such as α-α scattering) or three α-clusters (such as ^{12}C structure calculations). In the former case, we are updating the position of one α-cluster, in the latter the positions of two clusters.

Here, we shall only give a rough outline of how algorithms for such translational updates can be constructed:

1. Consider a subroutine `spatial`, which takes as input the translations

$$(x_1(p), y_1(p), z_1(p)), \tag{D.29}$$

of each nucleon p, and the corresponding translations

$$(x_2(p), y_2(p), z_2(p)), \tag{D.30}$$

at the "opposite" boundary of the Euclidean time evolution. As output, updated translations

$$(\tilde{x}_1(p), \tilde{y}_1(p), \tilde{z}_1(p)), \tag{D.31}$$

and an updated channel index $\tilde{c}_1$ is generated, along with the amplitude

$$\Pi(x_{\alpha_1}, y_{\alpha_1}, z_{\alpha_1}, x_{\alpha_2}, y_{\alpha_2}, z_{\alpha_2}), \tag{D.32}$$

for the initial positions of the α-clusters, and

$$\Pi(\tilde{x}_{\alpha_1}, \tilde{y}_{\alpha_1}, \tilde{z}_{\alpha_1}, \tilde{x}_{\alpha_2}, \tilde{y}_{\alpha_2}, \tilde{z}_{\alpha_2}), \tag{D.33}$$

for their updated positions.
2. For an update of the final state configuration, we simply replace $1 \leftrightarrow 2$ in the subroutine `spatial`.
3. As the probability amplitude Π is highly problem-specific, we shall leave it unspecified. Assuming that Π and the new α-cluster positions have been generated, we take for a system of 12 nucleons

$$\tilde{x}_1(0:3) = \tilde{y}_1(0:3) = \tilde{z}_1(0:3) = 0, \tag{D.34}$$

and

$$\tilde{x}_1(4:7) = \tilde{x}_{\alpha_1}, \quad \tilde{y}_1(4:7) = \tilde{y}_{\alpha_1}, \quad \tilde{z}_1(4:7) = \tilde{z}_{\alpha_1}, \tag{D.35}$$

and

$$\tilde{x}_1(8:11) = \tilde{x}_{\alpha_2}, \quad \tilde{y}_1(8:11) = \tilde{y}_{\alpha_2}, \quad \tilde{z}_1(8:11) = \tilde{z}_{\alpha_2}, \tag{D.36}$$

for the updated translations of the nucleons, since we are moving α-clusters, not individual nucleons. For the old configuration of α-clusters, we take

$$x_{\alpha_1} = x_1(4), \quad y_{\alpha_1} = y_1(4), \quad z_{\alpha_1} = z_1(4), \tag{D.37}$$

and

$$x_{\alpha_2} = x_1(8), \quad y_{\alpha_2} = y_1(8), \quad z_{\alpha_2} = z_1(8), \tag{D.38}$$

such that we are effectively keeping one α-cluster fixed and proposing moves for the second and third cluster relative to the first.

4. As for the CM update, we use the `getendoverlap` routine to compute

$$\frac{\det(Z_{end})}{|\det(Z_{corr})|}, \tag{D.39}$$

according to Eq. (E.20) using the updated translations $(\tilde{x}_1(p), \tilde{y}_1(p), \tilde{z}_1(p))$ and channel index $\tilde{c}_1$. Again, the denominator $|\det(Z_{corr})|$ should be computed using the old nucleon translations and channel index.

5. Select a random number r in the range $[0:1]$, if

$$r < \left\{ \frac{|\det(Z_{end})|}{|\det(Z_{corr})|} \times \frac{\Pi(x_{\alpha_1}, y_{\alpha_1}, z_{\alpha_1}, x_{\alpha_2}, y_{\alpha_2}, z_{\alpha_2})}{\Pi(\tilde{x}_{\alpha_1}, \tilde{y}_{\alpha_1}, \tilde{z}_{\alpha_1}, \tilde{x}_{\alpha_2}, \tilde{y}_{\alpha_2}, \tilde{z}_{\alpha_2})} \right\}, \tag{D.40}$$

then the nucleon translations and channel index are updated, otherwise the old ones are retained.

6. If the nucleon translations were updated, recompute the fermion action S_f and update the HMC action S_{HMC} accordingly.

Appendix E
Hybrid Monte Carlo Action and Force

In Appendix D, we have given some practical examples of how the Monte Carlo (MC) sampling in the NLEFT calculations can be implemented. These MC methods require the repeated computation of the action and its functional derivatives with respect to the auxiliary fields. We begin with a discussion of the kinetic terms of the auxiliary and pion fields (the bosonic part of the action). The implementation of the fermionic part of the action is of particular interest, due to the many complicating factors such as spin, isospin, and smearing of the leading-order (LO) contact terms. Here, we shall mostly concern ourselves with the LO contributions, specifically along the lines of the LO$_3$ action, which has been introduced in Chap. 4. The ideas and techniques presented here should also be applicable to the later developments of the LO action as detailed in Chap. 8, where non-locally smeared contact terms are considered. As the higher-order terms in NLEFT are treated perturbatively in the MC simulations, we defer their treatment to Appendix F, where methods for the calculation of observables are presented.

E.1 Auxiliary Field Action

We describe here how the bosonic action of the auxiliary field components is computed. The auxiliary field action S_ϕ is given by

$$S_\phi \equiv \frac{1}{2} \sum_{\mathbf{n},t} \phi(\mathbf{n}, t) G_{\mathrm{SU}(4)}^{-1}(\mathbf{n}, t), \qquad (E.1)$$

© Springer Nature Switzerland AG 2019
T. A. Lähde, U.-G. Meißner, *Nuclear Lattice Effective Field Theory*,
Lecture Notes in Physics 957, https://doi.org/10.1007/978-3-030-14189-9

for $t < N_t^{\text{outer}}$ and $t \geq N_t^{\text{outer}} + N_t^{\text{inner}}$, and

$$
S_\phi \equiv \frac{1}{2} \sum_{\mathbf{n},t} \phi(\mathbf{n}, t) G^{-1}(\mathbf{n}, t) + \frac{1}{2} \sum_{k=1}^{3} \sum_{\mathbf{n},t} \phi_k^S(\mathbf{n}, t) G_{S,k}^{-1}(\mathbf{n}, t)
$$

$$
+ \frac{1}{2} \sum_{l=1}^{3} \sum_{\mathbf{n},t} \phi_l^I(\mathbf{n}, t) G_{I,l}^{-1}(\mathbf{n}, t) + \frac{1}{2} \sum_{k=1}^{3} \sum_{l=1}^{3} \sum_{\mathbf{n},t} \phi_{k,l}^{SI}(\mathbf{n}, t) G_{SI,k,l}^{-1}(\mathbf{n}, t),
$$
(E.2)

for $t \geq N_t^{\text{outer}}$ and $t < N_t^{\text{outer}} + N_t^{\text{inner}}$. For each auxiliary field component

$$
G_{SI,k,l}^{-1}(\mathbf{n}, t) \equiv \sum_{\mathbf{n}'} D_\phi^{-1}(\mathbf{n} - \mathbf{n}') \phi_{k,l}^{SI}(\mathbf{n}', t),
$$
(E.3)

accounts for the smearing of the two-nucleon contact terms. This smearing is identical for all auxiliary field components.

Expressions similar to Eq. (E.3) are advantageously computed with the FFTW implementation of the Fast Fourier Transform algorithm. For instance, for contributions due to $\phi_{k,l}^{SI}$, we have

$$
Z_{\text{mom}}(\mathbf{k}) \equiv \mathcal{F} \phi_{k,l}^{SI}(\mathbf{n}, t), \qquad G_{SI,k,l}^{-1}(\mathbf{n}, t) \equiv \mathcal{F}^{-1} Z_{\text{mom}}^P(\mathbf{k})
$$
(E.4)

where $\mathcal{F}$ and $\mathcal{F}^{-1}$ denote execution of the forward and backward FFTW transforms, respectively. We have

$$
Z_{\text{mom}}^P(k_1, k_2, k_3) = N^{-3} \frac{N_\phi}{\tilde{G}_\phi(k_1, k_2, k_3)} Z_{\text{mom}}(k_1, k_2, k_3),
$$
(E.5)

where we use the notation $f(\mathbf{k})$ and $f(k_1, k_2, k_3)$ interchangeably. The "smearing factor" is given by

$$
\tilde{G}_\phi(\mathbf{k}) \equiv \exp\left(-\frac{b_2}{2} k_\phi^2 - \frac{b_4}{4} k_\phi^4 - \frac{b_6}{8} k_\phi^6 - \frac{b_8}{16} k_\phi^8 - \ldots \right),
$$
(E.6)

for the full action, and by

$$
\tilde{G}_\phi(\mathbf{k}) \equiv \exp\left(-\frac{\tilde{b}_2}{2} k_\phi^2 \right),
$$
(E.7)

for the simplified SU(4)-invariant action, for which $\tilde{b}_2 \neq b_2$. Moreover, only $\tilde{b}_2$ and b_4 are non-zero in actual calculations. We take

$$
\frac{k_\phi^2}{2} \equiv 3\omega_{2,0}^\phi - \omega_{2,1}^\phi \sum_{i=1}^3 \cos\left(\frac{2\pi}{N}\hat{k}_i\right)
$$

$$
+ \omega_{2,2}^\phi \sum_{i=1}^3 \cos\left(\frac{4\pi}{N}\hat{k}_i\right) - \omega_{2,3}^\phi \sum_{i=1}^3 \cos\left(\frac{6\pi}{N}\hat{k}_i\right), \tag{E.8}
$$

in terms of the integer-valued lattice momentum components $\hat{k}_i$, and the normalization factor is given by

$$
N_\phi \equiv N^{-3} \sum_k \tilde{G}_\phi(\mathbf{k}), \tag{E.9}
$$

which needs to be computed only once, as it does not depend on any of the auxiliary or pion fields.

E.2   Pion Action

The action of the (rescaled) pion fields S_π is only computed for $t \geq N_t^{\text{outer}}$ and $t < N_t^{\text{outer}} + N_t^{\text{inner}}$, and is given by

$$
S_\pi \equiv \frac{1}{2} \sum_{l=1}^3 \sum_{\mathbf{n},t} \phi_l^{\pi\,2}(\mathbf{n}, t) + \frac{\alpha_t}{q_\pi} \sum_{l=1}^3 \sum_{\mathbf{n},t} \phi_l^\pi(\mathbf{n}, t)
$$

$$
\times \left[-\omega_{2,1}^\pi \sum_{k=1}^3 \phi_l^\pi(\mathrm{mod}(\mathbf{n} + \hat{e}_k, L), t) \right.
$$

$$
\left. + \omega_{2,2}^\pi \sum_{k=1}^3 \phi_l^\pi(\mathrm{mod}(\mathbf{n} + 2\hat{e}_k, L), t) - \omega_{2,3}^\pi \sum_{k=1}^3 \phi_l^\pi(\mathrm{mod}(\mathbf{n} + 3\hat{e}_k, L), t) \right],
$$

$$
\tag{E.10}
$$

where the "rescaling factor" q_π is given by

$$
q_\pi \equiv \alpha_t(M_\pi^2 + 6\omega_{2,0}^\pi), \tag{E.11}
$$

where Eq. (E.10) includes third-nearest-neighbor terms, as required for an $\mathcal{O}(a^4)$ improved lattice action. Periodic lattice boundary conditions have been applied in

Eq. (E.10). We also define

$$\tilde{G}_\pi^{-1}(\mathbf{k}) \equiv 1 + \frac{2\alpha_t}{q_\pi}$$

$$\times \left[-\omega_{2,1}^\pi \sum_{i=1}^{3} \cos\left(\frac{2\pi}{N}\hat{k}_i\right) \right.$$

$$\left. +\omega_{2,2}^\pi \sum_{i=1}^{3} \cos\left(\frac{4\pi}{N}\hat{k}_i\right) - \omega_{2,3}^\pi \sum_{i=1}^{3} \cos\left(\frac{6\pi}{N}\hat{k}_i\right) \right], \quad \text{(E.12)}$$

analogously to Eq. (E.6).

E.3   Fermion Correlators

The single-particle wave functions are defined on time slices from $t = 0$ to $t = N_t$, and the evolution in Euclidean time is effected by the operations

$$|\psi(t+1)\rangle_{p_1} = \hat{M}(t)|\psi(t)\rangle_{p_1} =: \exp(-\alpha_t \hat{H}(t)) : |\psi(t)\rangle_{p_1}, \quad \text{(E.13)}$$

and

$$_{p_2}\langle\psi(t-1)| = {}_{p_2}\langle\psi(t)|\hat{M}(t-1) = {}_{p_2}\langle\psi(t)| : \exp(-\alpha_t \hat{H}(t-1)) :, \quad \text{(E.14)}$$

where the time indices of $\hat{H}$ and $\hat{M}$ refer to the auxiliary and pion fields $\{\phi\}$, which are defined for $t = 0$ to $t = N_t - 1$. The boundary conditions are

$$Z(\mathbf{n}, t = 0, s, i, p_1) \equiv |\psi(0)\rangle_{p_1} = |\psi_{\text{init}}\rangle_{p_1}, \quad \text{(E.15)}$$

for particle p_1 in the initial nucleon configuration, and

$$_{p_2}\langle\psi_{\text{fin}}| = {}_{p_2}\langle\psi(N_t)| \equiv Z^d(\mathbf{n}, t = N_t, s, i, p_2), \quad \text{(E.16)}$$

for particle p_2 in the final nucleon configuration. Z and Z^d are obtained for all time slices using the getzvecs and getzdualvecs routines, respectively. From these, we form the "correlator matrix"

$$Z_{\text{corr}}(p_2, p_1) \equiv {}_{p_2}\langle\psi(t^*)|\psi(t^*)\rangle_{p_1}$$

$$= N^{-3} \sum_{s,i} \sum_{\mathbf{n}} Z^d(\mathbf{n}, t^*, s, i, p_2) Z(\mathbf{n}, t^*, s, i, p_1), \quad \text{(E.17)}$$

where $t^* \equiv N_t^{\text{outer}} + N_t^{\text{inner}}$ equals the maximal Euclidean evolution time. Note that p_1 has been evolved for $N_t^{\text{outer}} + N_t^{\text{inner}}$ steps, and p_2 for $N_t - t^* = N_t^{\text{outer}}$ steps. We compute the inverse and log-determinant of Eq. (E.17) with the `getinvcorr` routine. We also form the "mid-overlap matrix"

$$
\begin{aligned}
Z_{t_2,t_1}(p_2, p_1) &\equiv {}_{p_2}\langle \psi(t_2)|\psi(t_1)\rangle_{p_1} \\
&= N^{-3} \sum_{s,i} \sum_{\mathbf{n}} Z^d(\mathbf{n}, t_2, s, i, p_2) Z(\mathbf{n}, t_1, s, i, p_1),
\end{aligned}
\tag{E.18}
$$

and compute the ratio

$$
\frac{\det(Z_{t_2,t_1})}{|\det(Z_{\text{corr}})|},
\tag{E.19}
$$

using Eq. (E.18) with the `getmidoverlap` routine. Finally, we form the "end-overlap matrix"

$$
\begin{aligned}
Z_{\text{end}}(p_2, p_1) \\
\equiv {}_{p_2}\langle \psi(N_t)|\psi(N_t)\rangle_{p_1} \\
= N^{-3} \sum_{s,i} \sum_{\mathbf{n}} Z^d(\mathbf{n}, N_t, s, i, N_{\text{fill}}(p_2, c_2)) Z(\mathbf{n}, N_t - N_{\text{trim}}^d(p_2), s, i, p_1),
\end{aligned}
$$

$$
\tag{E.20}
$$

and compute the ratio

$$
\frac{\det(Z_{\text{end}})}{|\det(Z_{\text{corr}})|},
\tag{E.21}
$$

using Eq. (E.20) with the `getendoverlap` routine. The denominators of Eqs. (E.19) and (E.21) can be provided as an input argument to these routines. The injection of nucleons in the final state has been explicitly accounted for in Eq. (E.20).

We compute the inverses and determinants of these correlation matrices using LAPACK routines for LU decomposition, and it should be noted that these operations do not consume an appreciable amount of CPU time since the problem is of size $N_f \times N_f$ only. It is advantageous to perform all operations in terms of log-determinants, as this avoids problems with numerical under or overflow.

E.3.1   Fermion Action

The fermion action S_f is evaluated by computing the determinant of Eq. (E.17). Depending on details such as the choice of Hubbard-Stratonovich transformation, $\det(Z_{\text{corr}})$ is complex-valued (or at least of indefinite sign) as NLEFT, in general, has a non-vanishing sign problem. In the polar representation,

$$\det(Z_{\text{corr}}) \equiv |\det(Z_{\text{corr}})| \exp(i\varphi), \tag{E.22}$$

or

$$\log \det(Z_{\text{corr}}) = \log |\det(Z_{\text{corr}})| + i\varphi, \tag{E.23}$$

which gives

$$S_f \equiv -\log |\det(Z_{\text{corr}})| = -\text{Re} \log \det(Z_{\text{corr}}), \tag{E.24}$$

which we equate with the fermion action S_f. Hence, the MC probability weight is given by the absolute value of the determinant, and the complex phase

$$\exp(i\varphi) = \exp(i \, \text{Im} \log \det(Z_{\text{corr}})), \tag{E.25}$$

is treated as a part of the observable being computed. This equals the standard "reweighting" procedure for MC calculations with a sign problem, where importance sampling is performed according to the absolute value of the fermion determinant. This approach is feasible as long as $\exp(i\varphi)$ is not too strongly fluctuating. In practice, we compute

$$\exp(i\varphi) = \frac{\det(Z_{\text{corr}})}{|\det(Z_{\text{corr}})|}, \tag{E.26}$$

and form the (real-valued) ensemble averages. As described in Chap. 6 and Appendix F, we divide each amplitude by the ensemble average of Eq. (E.26), when computing matrix elements of operators.

E.3.2   Evolution in Euclidean Time

The routine `getzvecs` evolves all nucleons in the state $|\psi(t_1)\rangle$ by recursive application of the transfer matrix operator

$$\hat{M}(t) = \, : \exp(-\alpha_t \hat{H}(t)) : \, \approx 1 - \alpha_t \hat{H}(t), \tag{E.27}$$

to a time t_2, such that $t_2 > t_1$. Similarly, getzdualvecs evolves all nucleons in the "dual" state $\langle \psi(t_2)|$ to a time t_1, again such that $t_2 > t_1$. Here, we always set $t_1 = 0$ and $t_2 = N_t$. Note that we describe these routines as they are actually implemented, and hence the notation is not the most transparent. However, the method as described here is then also sufficiently general to deal with all different types of calculations we encounter.

The Euclidean time evolution is initialized by taking

$$
\begin{aligned}
Z(\mathbf{n}, t_1, s, i, p) \equiv Z_{\text{init}}(&\text{mod}(x - x_1(p) + N, N), \\
&\text{mod}(y - y_1(p) + N, N), \\
&\text{mod}(z - z_1(p) + N, N), \\
&s, i, N_{\text{fill}}(p, c_1)),
\end{aligned}
\tag{E.28}
$$

for the vectors Z, and

$$
\begin{aligned}
Z^d(\mathbf{n}, t_2, s, i, p) \equiv Z_{\text{init}}^d(&\text{mod}(x - x_{\text{CM}} - x_2(p) + 2N, N), \\
&\text{mod}(y - y_{\text{CM}} - y_2(p) + 2N, N), \\
&\text{mod}(z - z_{\text{CM}} - z_2(p) + 2N, N), \\
&s, i, N_{\text{fill}}(p, c_2)),
\end{aligned}
\tag{E.29}
$$

for the dual vectors Z^d. Note that the arguments $N_{\text{fill}}(p, c) \neq p$ for multi-channel calculations only. Here, we have explicitly accounted for translations of each individual nucleon in the initial and final states, as well as translations of the CM in the final state, as described in Appendix D.

We note that in getzvecs, for $t \leq N_{\text{trim}}(N_{\text{fill}}(p, c_1))$,

$$
Z(\mathbf{n}, t, s, i, p) = Z(\mathbf{n}, t - 1, s, i, p),
\tag{E.30}
$$

and in getzdualvecs, for $t - 1 \geq N_t - N_{\text{trim}}^d(N_{\text{fill}}(p, c_2))$,

$$
Z^d(\mathbf{n}, t - 1, s, i, p) = Z^d(\mathbf{n}, t, s, i, p),
\tag{E.31}
$$

gives the equivalent condition. This construction accounts for the injection of nucleons onto the lattice without violation of the constraints due to the Pauli principle.

E.3.3   Spin- and Isospin-Independent Contributions

The auxiliary fields are defined on temporal lattice slices $t = 0, \ldots, N_t - 1$, where the first and last N_t^{outer} slices are described by an SU(4) symmetric interaction, with N_t^{inner} slices of the full LO interaction. Hence, $N_t = 2N_t^{\text{outer}} + N_t^{\text{inner}}$.

For the getzvecs routine, the spin- and isospin-independent contributions are given by

$$Z(\mathbf{n}, t, s, i, p) = \left[1 + \sqrt{-\tilde{C}\alpha_t}\, \phi(\mathbf{n}, t - 1) - \alpha_t V_{\text{wall}}(\mathbf{n})\right] Z(\mathbf{n}, t - 1, s, i, p),$$
(E.32)

where for $t \leq N_t^{\text{outer}}$ and $t > N_t^{\text{outer}} + N_t^{\text{inner}}$, we set $\tilde{C} \to \tilde{C}_{\text{SU(4)}}$, where in general $\tilde{C} \neq \tilde{C}_{\text{SU(4)}}$. For the getzdualvecs routine, the equivalent expression is

$$Z^d(\mathbf{n}, t - 1, s, i, p) = \left[1 + \sqrt{-\tilde{C}\alpha_t}\, \phi(\mathbf{n}, t - 1) - \alpha_t V_{\text{wall}}(\mathbf{n})\right] Z^d(\mathbf{n}, t, s, i, p),$$
(E.33)

which illustrates the qualitative differences. The contributions from the nucleon kinetic energy term with $\mathcal{O}(a^4)$ improvement, are

$$Z(\mathbf{n}, t, s, i, p) = Z(\mathbf{n}, t, s, i, p) - 6h\omega_{2,0}^N Z(\mathbf{n}, t - 1, s, i, p)$$

$$+ h\omega_{2,1}^N \sum_{k=1}^{3} Z(\text{mod}(\mathbf{n} + \hat{e}_k, N), t - 1, s, i, p)$$

$$- h\omega_{2,2}^N \sum_{k=1}^{3} Z(\text{mod}(\mathbf{n} + 2\hat{e}_k, N), t - 1, s, i, p)$$

$$+ h\omega_{2,3}^N \sum_{k=1}^{3} Z(\text{mod}(\mathbf{n} + 3\hat{e}_k, N), t - 1, s, i, p), \quad (E.34)$$

and

$$Z^d(\mathbf{n}, t - 1, s, i, p) = Z^d(\mathbf{n}, t - 1, s, i, p) - 6h\omega_{2,0}^N Z^d(\mathbf{n}, t, s, i, p)$$

$$+ h\omega_{2,1}^N \sum_{k=1}^{3} Z^d(\text{mod}(\mathbf{n} + \hat{e}_k, N), t, s, i, p)$$

$$- h\omega_{2,2}^N \sum_{k=1}^{3} Z^d(\text{mod}(\mathbf{n} + 2\hat{e}_k, N), t, s, i, p)$$

$$+ h\omega_{2,3}^N \sum_{k=1}^{3} Z^d(\text{mod}(\mathbf{n} + 3\hat{e}_k, N), t, s, i, p), \quad (E.35)$$

where $h \equiv \alpha_t/(2m_N)$. As the "spherical wall" potential $V_{\text{wall}}(\mathbf{n})$ induces severe sign oscillations, it is typically not used in MC calculations. Strictly speaking, the spherical wall is only used in order to determine scattering phase shifts for a two-body system (such as the np or α-α scattering problems). See also Chap. 5 for further details on the spherical wall method.

In what follows, we shall only give explicit expressions for the `getzvecs` routine, whenever the appropriate modifications for the `getzdualvecs` routine can be straightforwardly inferred.

E.3.4 Spin- and Isospin-Dependent Contributions

These operations are only performed if $t \geq N_t^{\text{outer}} + 1$ and $t \leq N_t^{\text{outer}} + N_t^{\text{inner}}$. For the Pauli matrices, we introduce the notation

$$Z_{\sigma,k}^{2\times2}(i, j) \equiv [\sigma_k]_{i,j}, \qquad Z_{\tau,l}^{2\times2}(i, j) \equiv [\tau_l]_{i,j}, \tag{E.36}$$

where $k = 1, 2, 3$ and $l = 1, 2, 3$. We include the unit matrices in spin- and isospin space as the cases of $k = 0$ and $l = 0$. For each time slice t and each spatial lattice site $\mathbf{n}$, we begin by defining

$$\tilde{Z}_S^{2\times2}(s', s'') \equiv \sum_{k=1}^{3} \phi_k^S(\mathbf{n}, t - 1) Z_{\sigma,k}^{2\times2}(s', s''), \tag{E.37}$$

where we use the "tilde" notation to indicate that the degrees of freedom $(\mathbf{n}, t)$ have been suppressed. Explicitly, we find

$$\tilde{Z}_S^{2\times2}(0, 0) = \phi_3^S(\mathbf{n}, t - 1), \tag{E.38}$$

$$\tilde{Z}_S^{2\times2}(1, 1) = -\phi_3^S(\mathbf{n}, t - 1), \tag{E.39}$$

$$\tilde{Z}_S^{2\times2}(1, 0) = \phi_1^S(\mathbf{n}, t - 1) + i\phi_2^S(\mathbf{n}, t - 1), \tag{E.40}$$

$$\tilde{Z}_S^{2\times2}(0, 1) = \phi_1^S(\mathbf{n}, t - 1) - i\phi_2^S(\mathbf{n}, t - 1), \tag{E.41}$$

which gives the contribution

$$\tilde{Z}^{2\times2}(s', s'', i', i') = \tilde{Z}^{2\times2}(s', s'', i', i') + i\sqrt{\tilde{C}_{S^2}\alpha_t}\, \tilde{Z}_S^{2\times2}(s', s''), \tag{E.42}$$

from the auxiliary-field components ϕ_k^S with $k = 1, 2, 3$.

We then have, analogously,

$$\tilde{Z}_I^{2\times2}(i', i'') \equiv \sum_{l=1}^{3} \phi_l^I(\mathbf{n}, t - 1) Z_{\tau,l}^{2\times2}(i', i''), \tag{E.43}$$

explicitly

$$\tilde{Z}_I^{2\times2}(0,0) = \phi_3^I(\mathbf{n}, t-1), \tag{E.44}$$

$$\tilde{Z}_I^{2\times2}(1,1) = -\phi_3^I(\mathbf{n}, t-1), \tag{E.45}$$

$$\tilde{Z}_I^{2\times2}(1,0) = \phi_1^I(\mathbf{n}, t-1) + i\phi_2^I(\mathbf{n}, t-1), \tag{E.46}$$

$$\tilde{Z}_I^{2\times2}(0,1) = \phi_1^I(\mathbf{n}, t-1) - i\phi_2^I(\mathbf{n}, t-1), \tag{E.47}$$

which gives the contribution

$$\bar{Z}^{2\times2}(s', s', i', i'') = \tilde{Z}^{2\times2}(s', s', i', i'') + i\sqrt{\tilde{C}_{I^2}}\alpha_t\,\tilde{Z}_I^{2\times2}(i', i''), \tag{E.48}$$

from the auxiliary-field components ϕ_l^I with $l = 1, 2, 3$.

We also have the components $\phi_{k,l}^{SI}$ with $k = 1, 2, 3$ and $l = 1, 2, 3$, for which we define

$$\tilde{Z}_{SI}^{2\times2}(s', s'', i', i'') \equiv \sum_{k=1}^{3}\sum_{l=1}^{3} \phi_{k,l}^{SI}(\mathbf{n}, t-1) Z_{\sigma,k}^{2\times2}(s', s'') Z_{\tau,l}^{2\times2}(i', i''), \tag{E.49}$$

giving

$$\bar{Z}^{2\times2}(s', s'', i', i'') = \tilde{Z}^{2\times2}(s', s'', i', i'') + i\sqrt{\tilde{C}_{S,I}}\alpha_t\,\tilde{Z}_{SI}^{2\times2}(s', s'', i', i''). \tag{E.50}$$

For the (rescaled) OPE fields ϕ_l^π with $l = 1, 2, 3$, we find

$$\tilde{Z}_{SI,\pi}^{2\times2}(s', s'', i', i'') \equiv \sum_{k=1}^{3}\sum_{l=1}^{3} \Delta_k \phi_l^\pi(\mathbf{n}, t-1) Z_{\sigma,k}^{2\times2}(s', s'') Z_{\tau,l}^{2\times2}(i', i'')$$

$$= \sum_{l=1}^{3} \tilde{Z}_{\pi,l}^{2\times2}(s', s'') Z_{\tau,l}^{2\times2}(i', i''), \tag{E.51}$$

explicitly

$$\tilde{Z}_{\pi,l}^{2\times2}(0,0) = \Delta_3 \phi_l^\pi(\mathbf{n}, t-1), \tag{E.52}$$

$$\tilde{Z}_{\pi,l}^{2\times2}(1,1) = -\Delta_3 \phi_l^\pi(\mathbf{n}, t-1), \tag{E.53}$$

$$\tilde{Z}_{\pi,l}^{2\times2}(1,0) = \Delta_1 \phi_l^\pi(\mathbf{n}, t-1) + i\Delta_2 \phi_l^\pi(\mathbf{n}, t-1), \tag{E.54}$$

$$\tilde{Z}_{\pi,l}^{2\times2}(0,1) = \Delta_1 \phi_l^\pi(\mathbf{n}, t-1) - i\Delta_2 \phi_l^\pi(\mathbf{n}, t-1), \tag{E.55}$$

where

$$\Delta_k \phi_l^\pi (\mathbf{n}, t - 1) = \theta_{2,1} \, \phi_l^\pi (\text{mod}(\mathbf{n} + \hat{e}_k, N), t - 1)$$
$$- \theta_{2,1} \, \phi_l^\pi (\text{mod}(\mathbf{n} - \hat{e}_k + N, N), t - 1)$$
$$- \theta_{2,2} \, \phi_l^\pi (\text{mod}(\mathbf{n} + 2\hat{e}_k, N), t - 1)$$
$$+ \theta_{2,2} \, \phi_l^\pi (\text{mod}(\mathbf{n} - 2\hat{e}_k + N, N), t - 1)$$
$$+ \theta_{2,3} \, \phi_l^\pi (\text{mod}(\mathbf{n} + 3\hat{e}_k, N), t - 1)$$
$$- \theta_{2,3} \, \phi_l^\pi (\text{mod}(\mathbf{n} - 3\hat{e}_k + N, N), t - 1), \qquad (E.56)$$

such that

$$\tilde{Z}^{2\times2}(s', s'', i', i'') = \tilde{Z}^{2\times2}(s', s'', i', i'') - \frac{g_A \alpha_t}{2 F_\pi \sqrt{q_\pi}} \, \tilde{Z}^{2\times2}_{SI,\pi} (s', s'', i', i''),$$

$$(E.57)$$

where the rescaling factor q_π is defined in Eq. (E.11). Finally, the spin- and isospin-dependent contributions are all accounted for by taking

$$Z(\mathbf{n}, t, s, i, p) = Z(\mathbf{n}, t, s, i, p) + \sum_{s',i'} \tilde{Z}^{2\times2}(s, s', i, i') Z(\mathbf{n}, t - 1, s', i', p).$$

$$(E.58)$$

E.4 HMC Force Terms

Functional derivatives of the Hybrid Monte Carlo (HMC) action S_{HMC}, with respect to the auxiliary field components $\{\phi\}$, are needed for the global Molecular Dynamics (MD) updates. As we have seen in Chap. 6, such expressions occur when the MD update is performed by numerical integration of the Hamiltonian equations of motion, for instance with the "leap-frog" integrator. We illustrate the principle using the case of $\phi(\mathbf{n}, t)$, for which the functional derivative is

$$\frac{\delta S_{\text{HMC}}}{\delta \phi(\mathbf{n}, t)} \equiv -F_\phi(\mathbf{n}, t) = \frac{\delta S_\phi}{\delta \phi(\mathbf{n}, t)} + \frac{\delta S_f}{\delta \phi(\mathbf{n}, t)}$$

$$= G^{-1}(\mathbf{n}, t) - \text{Re} \sum_{p_1, p_2} \frac{\delta Z_{\text{corr}}(p_2, p_1)}{\delta \phi(\mathbf{n}, t)} Z_{\text{corr}}^{-1}(p_1, p_2),$$

$$(E.59)$$

where we note the sign convention of the force being the negative of the functional derivative. Using

$$\frac{\partial \det(M)}{\partial \lambda} = \det(M) \, \text{Tr}\left(\frac{\partial M}{\partial \lambda} M^{-1}\right),$$
(E.60)

known as Jacobi's formula, we can evaluate the functional derivative of the fermion action S_f. The "derivative correlator" is

$$\frac{\delta Z_{\text{corr}}(p_2, p_1)}{\delta \phi(\mathbf{n}, t)} = N^{-3}\sqrt{-\tilde{C}\alpha_t} \sum_{s,i} Z^d(\mathbf{n}, t+1, s, i, p_2) Z(\mathbf{n}, t, s, i, p_1),$$
(E.61)

where for $t < N_t^{\text{outer}}$ and $t \geq N_t^{\text{outer}} + N_t^{\text{inner}}$, we substitute $\tilde{C} \rightarrow \tilde{C}_{\text{SU(4)}}$, with $\tilde{C} \neq \tilde{C}_{\text{SU(4)}}$, for consistency with the treatment of S_f. We also note that Eq. (E.61) vanishes in the "outer" region of the Euclidean time evolution, unless

$$t \geq N_{\text{trim}}(N_{\text{fill}}(p_1, c_1)),$$
(E.62)

and

$$t \leq N_t - N_{\text{trim}}^d(N_{\text{fill}}(p_2, c_2)) - 1,$$
(E.63)

are simultaneously satisfied.

The spin- and isospin-dependent force terms are only included for $t \geq N_t^{\text{outer}}$ and $t < N_t^{\text{outer}} + N_t^{\text{inner}}$. We define

$$Z_{dV}^{\text{all}}(\mathbf{n}, t, k, l) \equiv N^{-3} \sum_{s,s',i,i'} Z_{\sigma,k}^{2\times 2}(s, s') Z_{\tau,l}^{2\times 2}(i, i')$$

$$\times \sum_{p_1, p_2} Z^d(\mathbf{n}, t+1, s, i, p_2) Z(\mathbf{n}, t, s', i', p_1) Z_{\text{corr}}^{-1}(p_1, p_2),$$
(E.64)

in terms of which we can conveniently express the remaining force terms. For the $\phi_k^S(\mathbf{n}, t)$ with $k = 1, 2, 3$ we find

$$\frac{\delta S_{\text{HMC}}}{\delta \phi_k^S(\mathbf{n}, t)} = \frac{\delta S_\phi}{\delta \phi_k^S(\mathbf{n}, t)} + \frac{\delta S_f}{\delta \phi_k^S(\mathbf{n}, t)}$$

$$= G_{S,k}^{-1}(\mathbf{n}, t) - \text{Re}\, i \sqrt{\tilde{C}_{S^2}\alpha_t} \, Z_{dV}^{\text{all}}(\mathbf{n}, t, k, 0),$$
(E.65)

for the $\phi_l^I(\mathbf{n}, t)$ with $l = 1, 2, 3$ we have

$$
\frac{\delta S_{\mathrm{HMC}}}{\delta \phi_l^I(\mathbf{n}, t)} = \frac{\delta S_\phi}{\delta \phi_l^I(\mathbf{n}, t)} + \frac{\delta S_f}{\delta \phi_l^I(\mathbf{n}, t)}
$$

$$
= G_{I,l}^{-1}(\mathbf{n}, t) - \operatorname{Re} i \sqrt{\tilde{C}_{I^2} \alpha_t} \, Z_{dV}^{\mathrm{all}}(\mathbf{n}, t, 0, l), \qquad (\text{E.66})
$$

for the $\phi_{k,l}^{SI}(\mathbf{n}, t)$ with $k = 1, 2, 3$ and $l = 1, 2, 3$, we have

$$
\frac{\delta S_{\mathrm{HMC}}}{\delta \phi_{k,l}^{SI}(\mathbf{n}, t)} = \frac{\delta S_\phi}{\delta \phi_{k,l}^{SI}(\mathbf{n}, t)} + \frac{\delta S_f}{\delta \phi_{k,l}^{SI}(\mathbf{n}, t)}
$$

$$
= G_{SI,k,l}^{-1}(\mathbf{n}, t) - \operatorname{Re} i \sqrt{\tilde{C}_{S,I} \alpha_t} \, Z_{dV}^{\mathrm{all}}(\mathbf{n}, t, k, l), \qquad (\text{E.67})
$$

and for the pion fields $\phi_l^\pi(\mathbf{n}, t)$ with $l = 1, 2, 3$, we have

$$
\frac{\delta S_{\mathrm{HMC}}}{\delta \phi_l^\pi(\mathbf{n}, t)} = \frac{\delta S_\pi}{\delta \phi_l^\pi(\mathbf{n}, t)} + \frac{\delta S_f}{\delta \phi_l^\pi(\mathbf{n}, t)}
$$

$$
= \phi_l^\pi(\mathbf{n}, t) + \frac{\alpha_t}{q_\pi} \Bigg\{ -\omega_{2,1}^\pi \sum_{k=1}^{3} \Big[\phi_l^\pi(\mathrm{mod}(\mathbf{n} + \hat{e}_k, N), t)
$$

$$
+ \phi_l^\pi(\mathrm{mod}(\mathbf{n} - \hat{e}_k + N, N), t) \Big]
$$

$$
+ \omega_{2,2}^\pi \sum_{k=1}^{3} \Big[\phi_l^\pi(\mathrm{mod}(\mathbf{n} + 2\hat{e}_k, N), t)
$$

$$
+ \phi_l^\pi(\mathrm{mod}(\mathbf{n} - 2\hat{e}_k + N, N), t) \Big]
$$

$$
- \omega_{2,3}^\pi \sum_{k=1}^{3} \Big[\phi_l^\pi(\mathrm{mod}(\mathbf{n} + 3\hat{e}_k, N), t)
$$

$$
+ \phi_l^\pi(\mathrm{mod}(\mathbf{n} - 3\hat{e}_k + N, N), t) \Big] \Bigg\}
$$

$$
- \frac{g_A \alpha_t}{2 F_\pi \sqrt{q_\pi}} \Bigg\{ -\frac{\theta_{2,1}}{2} \sum_{k=1}^{3} \Big[Z_{dV}^{\mathrm{all}}(\mathrm{mod}(\mathbf{n} - \ddot{e}_k + N, N), t, k, l)
$$

$$
- Z_{dV}^{\mathrm{all}}(\mathrm{mod}(\mathbf{n} + \hat{e}_k, N), t, k, l) \Big]
$$

$$+ \frac{\theta_{2,2}}{2} \sum_{k=1}^{3} \left[Z_{dV}^{\text{all}}(\text{mod}(\mathbf{n} - 2\hat{e}_k + N, N), t, k, l) \right.$$

$$\left. - Z_{dV}^{\text{all}}(\text{mod}(\mathbf{n} + 2\hat{e}_k, N), t, k, l) \right]$$

$$- \frac{\theta_{2,3}}{2} \sum_{k=1}^{3} \left[Z_{dV}^{\text{all}}(\text{mod}(\mathbf{n} - 3\hat{e}_k + N, N), t, k, l) \right.$$

$$\left. \left. - Z_{dV}^{\text{all}}(\text{mod}(\mathbf{n} + 3\hat{e}_k, N), t, k, l) \right] \right\},$$

$$(E.68)$$

where the rescaling factor q_π is given by Eq. (E.11). We have explicitly included periodic boundary conditions and improvement coefficients for both the pion dispersion relation and derivative coupling, sufficient for an $\mathcal{O}(a^4)$ treatment of both. See also Chap. 4 for a more detailed description.

Appendix F
Monte Carlo Calculation of Observables

We shall now consider the calculation of operator expectation values in the Monte Carlo (MC) simulations. We have already encountered such operators in the perturbative treatment of NLEFT beyond LO in Chaps. 4 and 5, in the development of the projection Monte Carlo (PMC) method in Chap. 6, as well as in the calculation of the nuclear charge distribution and transition matrix elements in Chap. 7. At regular intervals (typically after several consecutive MC updates), matrix elements of various operators are computed, or "measured" in the jargon of the MC community. As we have described in Chap. 6, these "snapshots" are then binned together over the course of several measurements, and the variance of the mean decreases as the square root of the number of snapshots, provided that successive measurements are properly decorrelated.

Here, we provide some further details on the MC calculation of operator matrix elements of the form

$$\langle \mathcal{O} \rangle \equiv \frac{\langle \psi(t_m) | \mathcal{O} | \psi(t_m) \rangle}{\langle \psi(t^*) | \psi(t^*) \rangle}, \tag{F.1}$$

where $t^* \equiv N_t^{\text{outer}} + N_t^{\text{inner}}$ is the maximal Euclidean projection time, and $t_m \equiv N_t^{\text{outer}} + N_t^{\text{inner}}/2$. We demonstrate the principle by means of the occupation number operators, and the "transient energy" of the Euclidean projection amplitude, from which the ground-state energy at LO can be extracted. We also show how the energy shifts due to the NLO and NNLO operators (including electromagnetic and isospin-breaking operators) can be computed, when considered as first-order perturbations.

© Springer Nature Switzerland AG 2019
T. A. Lähde, U.-G. Meißner, *Nuclear Lattice Effective Field Theory*,
Lecture Notes in Physics 957, https://doi.org/10.1007/978-3-030-14189-9

F.1   Occupation Number Operators

As a first example, we consider the average number of singly, doubly, triply and quadruply occupied lattice sites. The Pauli principle prohibits more than four nucleons from simultaneously occupying a given lattice site. We ignore complications due to multiple channels in the MC simulation, and take $t_m = \text{int}(N_t/2)$, where $N_t \equiv 2N_t^{\text{outer}} + N_t^{\text{inner}}$. We consider the matrix elements of the operators

$$\mathcal{O}_1(\mathbf{n}) \equiv \rho^{a^\dagger, a}(\mathbf{n}), \tag{F.2}$$

$$\mathcal{O}_2(\mathbf{n}) \equiv \frac{1}{2} : \rho^{a^\dagger, a}(\mathbf{n})\rho^{a^\dagger, a}(\mathbf{n}) :, \tag{F.3}$$

$$\mathcal{O}_3(\mathbf{n}) \equiv \frac{1}{6} : \rho^{a^\dagger, a}(\mathbf{n})\rho^{a^\dagger, a}(\mathbf{n})\rho^{a^\dagger, a}(\mathbf{n}) :, \tag{F.4}$$

$$\mathcal{O}_4(\mathbf{n}) \equiv \frac{1}{24} : \rho^{a^\dagger, a}(\mathbf{n})\rho^{a^\dagger, a}(\mathbf{n})\rho^{a^\dagger, a}(\mathbf{n})\rho^{a^\dagger, a}(\mathbf{n}) :, \tag{F.5}$$

according to Eq. (F.1). As outlined in Chap. 4, we represent each density operator using a functional derivative with respect to a "source field" $\varepsilon(\mathbf{n})$,

$$\rho^{a^\dagger, a}(\mathbf{n}) = \frac{\delta}{\delta\varepsilon(\mathbf{n})} \exp\left(\sum_{\mathbf{n}'} \varepsilon(\mathbf{n}')\rho^{a^\dagger, a}(\mathbf{n}')\right)\Bigg|_{\varepsilon \to 0}, \tag{F.6}$$

such that multiple density operators can be inserted (for instance) at time slice t_m, by successive functional derivatives with respect to $\varepsilon(\mathbf{n})$. These can be evaluated using a numerical finite difference formula, given a small (but finite) constant $\varepsilon(\mathbf{n}) = \varepsilon$.

Given a suitable finite difference formula, we need to evaluate the matrix element of the exponential in Eq. (F.6) for positive and negative multiples of ε. Let us first establish a suitable notation for the task. We define

$$Z_{\text{corr}}^{q+}(p_2, p_1; \mathbf{n}) \equiv N^{-3} \sum_{s,i} \sum_{\mathbf{n}'} Z^d(\mathbf{n}', t_m, s, i, p_2) Z(\mathbf{n}', t_m, s, i, p_1)$$

$$+ q\varepsilon N^{-3} \sum_{s,i} Z^d(\mathbf{n}, t_m, s, i, p_2) Z(\mathbf{n}, t_m, s, i, p_1), \quad \text{(F.7)}$$

for positive multiples of ε, and

$$Z_{\text{corr}}^{q-}(p_2, p_1; \mathbf{n}) \equiv N^{-3} \sum_{s,i} \sum_{\mathbf{n}'} Z^d(\mathbf{n}', t_m, s, i, p_2) Z(\mathbf{n}', t_m, s, i, p_1)$$

$$- q\varepsilon N^{-3} \sum_{s,i} Z^d(\mathbf{n}, t_m, s, i, p_2) Z(\mathbf{n}, t_m, s, i, p_1), \quad \text{(F.8)}$$

for negative multiples. These satisfy

$$Z_{corr}^{0+}(p_2, p_1; \mathbf{n}) = Z_{corr}^{0-}(p_2, p_1; \mathbf{n}) \equiv Z_{corr}^{0}(p_2, p_1), \tag{F.9}$$

and we compute

$$Z_{rdet}^{q+}(\mathbf{n}) \equiv \frac{\det(Z_{corr}^{q+}; \mathbf{n})}{|\det(Z_{corr})|}, \quad Z_{rdet}^{q-}(\mathbf{n}) \equiv \frac{\det(Z_{corr}^{q-}; \mathbf{n})}{|\det(Z_{corr})|}, \tag{F.10}$$

and

$$Z_{rdet}^{0} \equiv \frac{\det(Z_{corr}^{0})}{|\det(Z_{corr})|}, \tag{F.11}$$

and evaluate the finite differences

$$\langle \mathcal{O}_1(\mathbf{n}) \rangle = \frac{Z_{rdet}^{1+}(\mathbf{n}) - Z_{rdet}^{1-}(\mathbf{n})}{2\varepsilon}, \tag{F.12}$$

$$\langle \mathcal{O}_2(\mathbf{n}) \rangle = \frac{1}{2} \frac{Z_{rdet}^{1+}(\mathbf{n}) + Z_{rdet}^{1-}(\mathbf{n}) - 2Z_{rdet}^{0}}{\varepsilon^2}, \tag{F.13}$$

$$\langle \mathcal{O}_3(\mathbf{n}) \rangle = \frac{1}{6} \frac{Z_{rdet}^{3+}(\mathbf{n}) - 3Z_{rdet}^{1+}(\mathbf{n}) + 3Z_{rdet}^{1-}(\mathbf{n}) - Z_{rdet}^{3-}(\mathbf{n})}{8\varepsilon^3}, \tag{F.14}$$

$$\langle \mathcal{O}_4(\mathbf{n}) \rangle = \frac{1}{24} \frac{Z_{rdet}^{2+}(\mathbf{n}) - 4Z_{rdet}^{1+}(\mathbf{n}) + 6Z_{rdet}^{0} - 4Z_{rdet}^{1-}(\mathbf{n}) + Z_{rdet}^{2-}(\mathbf{n})}{\varepsilon^4}, \tag{F.15}$$

which, for sufficiently small ϵ, gives us the required matrix elements. Note that ϵ should nevertheless not be taken too small, or excessive numerical round-off error may result.

Let us consider one possible way to evaluate such matrix elements with parallel computing. Consider a system running a MC simulation with N_p MPI processes. For each configuration in the ensemble, a lattice site $\mathbf{n}$ is randomly chosen by each of the processes. Each of the MPI processes computes its own ensemble averages. At regular intervals, the results from each process may be merged and used to update cumulative averages. For the occupation numbers, this equals

$$\langle\langle N_i \rangle\rangle \equiv \frac{N^3}{\langle\langle \exp(i\varphi) \rangle\rangle} \times \frac{1}{N_p} \sum_{p=0}^{N_p-1} \langle\langle \mathrm{Re}\, \mathcal{O}_i(\mathbf{n}) \rangle\rangle_p, \tag{F.16}$$

for a calculation performed with $N_t \equiv 2N_t^{outer} + N_t^{inner}$ time slices. As in our discussion of the reweighting procedure for the fermion action in Chap. 6, we divide by the average phase factor due to sign oscillations. We note that MC errors can be added in quadrature, although more sophisticated methods such as Jackknife resampling give a more accurate estimate.

F.2   Transient Energy

As a second example, let us consider the calculation of the LO "transient energy" E_{LO}^* of Chap. 6, which converges to the ground state energy in the limit of large Euclidean projection time (provided that the trial state has non-zero overlap with the true ground state to begin with).

We consider the ratio of amplitudes

$$\frac{\langle \psi(t^* + \Delta t) | \psi(t^*) \rangle}{\langle \psi(t^*) | \psi(t^*) \rangle} = \exp(\alpha_t \Delta t \, E_{LO}^*(t^* - \Delta t/2)), \tag{F.17}$$

where we denote by Δt the number of removed time slices. We define

$$Z_{rdet}^{\Delta t}(t) \equiv \frac{\det(Z_{t+\Delta t,t})}{|\det(Z_{corr})|}, \tag{F.18}$$

and compute

$$Z_{over}(t^*, \Delta t) \equiv \frac{1}{N_t^{inner} - \Delta t + 1} \sum_{t=N_t^{outer}}^{N_t^{outer} + N_t^{inner} - \Delta t} Z_{rdet}^{\Delta t}(t), \tag{F.19}$$

where we have averaged over all possible ways of removing Δt time steps from the Euclidean time evolution. The transient energy is then found from

$$E_{LO}^*(t^* - \Delta t/2)) = \frac{1}{\alpha_t \Delta t} \times \log \frac{\langle\langle Z_{over}(t^*, \Delta t) \rangle\rangle}{\langle\langle \exp(i\varphi) \rangle\rangle}, \tag{F.20}$$

and the ground state energy is given by

$$E_{LO} \equiv \lim_{t^* \to \infty} E_{LO}^*(t^* - \Delta t/2)), \tag{F.21}$$

where Δt can in principle be freely chosen. This naturally raises the question of whether MC simulations only need to be performed for the maximal Euclidean projection time t^* allowed by the sign problem or other computational constraints, or whether more information can be obtained by additional simulations at smaller times. In practice, we find that MC data of high quality can only be obtained for $\Delta t \leq 2$. If further time slices are removed, the statistical error and correlation have been found to increase dramatically. Instead of removing time slices, we may also add "new" time slices, although the question then arises how the auxiliary fields on the new slices should be determined. For more details, see Sect. F.3.

F.3   Perturbative Contributions

We are now in a position to consider the energy shifts due to NLO and higher-order operators. We shall treat these as first-order perturbations, along the lines of Chap. 6. For instance, at NLO we have the ratio of amplitudes

$$\frac{\langle \psi(t_m) | \hat{M}_{\mathrm{NLO}} | \psi(t_m) \rangle}{\langle \psi(t_m) | \hat{M}_{\mathrm{LO}} | \psi(t_m) \rangle} = 1 - \alpha_t \Delta E_{\mathrm{NLO}}(t^*) + \ldots, \tag{F.22}$$

where we have ignored contributions which are of higher order in the NLO coupling constants. Similarly, we have

$$\frac{\langle \psi(t_m) | \hat{M}_{\mathrm{NNLO}} | \psi(t_m) \rangle}{\langle \psi(t_m) | \hat{M}_{\mathrm{LO}} | \psi(t_m) \rangle} = 1 - \alpha_t \Delta E_{\mathrm{NNLO}}(t^*) + \ldots, \tag{F.23}$$

for the NNLO contributions. The NLO transfer matrix operator is

$$\hat{M}_{\mathrm{NLO}} \equiv \hat{M}_{\mathrm{LO}} + \sum_i \Delta \hat{M}_{\mathrm{NLO},i}, \tag{F.24}$$

with a similar expression for $\hat{M}_{\mathrm{NNLO}}$. These receive contributions from higher-order contact terms as well as from terms that involve the pion fields, such as two-pion exchange (TPE) and contributions from the three-nucleon force. For instance, the NLO perturbations due to contact terms are of the form

$$\Delta \hat{M}_{\mathrm{NLO},i}(t) \propto -\alpha_t \sum_{\mathbf{n}} \frac{\delta}{\delta \varepsilon_j(\mathbf{n}, t)} \frac{\delta}{\delta \varepsilon_k(\mathbf{n}, t)} \hat{M}(\{\varepsilon\}, t) \Big|_{\{\varepsilon\} \to 0}, \tag{F.25}$$

where the functional derivatives and constants of proportionality are chosen appropriately in each case. The transfer matrix with source terms is

$$\hat{M}(\{\varepsilon\}, t) \equiv\, : \exp(-\alpha_t \hat{H}(t)) \exp(U(\{\varepsilon\}, t) + U_{I^2}(\{\varepsilon\}, t)) :$$
$$\approx 1 - \alpha_t \hat{H}(t) + U(\{\varepsilon\}, t) + U_{I^2}(\{\varepsilon\}, t), \tag{F.26}$$

such that

$$\lim_{\{\varepsilon\} \to 0} \hat{M}(\{\varepsilon\}, t) = \hat{M}_{\mathrm{LO}}(t), \tag{F.27}$$

and we recall that the source terms are given explicitly in Eqs. (4.120) and (4.121) of Chap. 4.

F.3.1   Conventions for Matrix Elements

When expressions of the form (F.25) are inserted at t_m, we need to account for the fact that these contain a transfer matrix operator, and hence require assigning them auxiliary and pion field configurations $\{\phi\}$ at t_m. We can either "eat" an extant time slice ($\Delta t = 1$) or insert a "new" time slice ($\Delta t = 0$) for $\hat{M}(\{\varepsilon\}, t)$. In the latter case, we need to specify how the corresponding auxiliary field is to be generated.

In practice, for N_t^{inner} even,

- $\Delta t = 0, t_m = N_t^{\mathrm{outer}} + N_t^{\mathrm{inner}}/2$, we insert a new time slice at t_m,
- $\Delta t = 1$, we randomly take $t_m = N_t^{\mathrm{outer}} + (N_t^{\mathrm{inner}} - 2)/2$ or $t_m = N_t^{\mathrm{outer}} + (N_t^{\mathrm{inner}} - 2)/2 + 1$, and use the existing time slice at t_m. For instance, for $N_t^{\mathrm{outer}} = 0$ and $N_t^{\mathrm{inner}} = 8$, we have $t_m = 3$ or $t_m = 4$, and use the operators M that effect the time evolutions $3 \leftrightarrow 4$ or $4 \leftrightarrow 5$,

and for N_t^{inner} odd,

- $\Delta t = 0, t_m = N_t^{\mathrm{outer}} + (N_t^{\mathrm{inner}} - 1)/2$, we insert a new time slice at t_m,
- $\Delta t = 1$, we take $t_m = N_t^{\mathrm{outer}} + (N_t^{\mathrm{inner}} - 1)/2$, and use the existing time slice at t_m. For instance, for $N_t^{\mathrm{outer}} = 0$ and $N_t^{\mathrm{inner}} = 7$, we have $t_m = 3$, and use the operator M that effects the time evolution $3 \leftrightarrow 4$,

such that for $\Delta t = 0$, we obtain

$$\frac{\langle \psi(t_m) | \Delta \hat{M}_{\mathrm{NLO}}(t_{\mathrm{step}}) | \psi(t_m) \rangle}{\langle \psi(t^*) | \psi(t^*) \rangle} = \langle \Delta E_{\mathrm{NLO}}^*(t^* + 1/2) \rangle, \tag{F.28}$$

and for $\Delta t = 1$, we find

$$\frac{\langle \psi(t_m + 1) | \Delta \hat{M}_{\mathrm{NLO}}(t_m) | \psi(t_m) \rangle}{\langle \psi(t^*) | \psi(t^*) \rangle} = \langle \Delta E_{\mathrm{NLO}}^*(t^* - 1/2) \rangle, \tag{F.29}$$

and as for the transient energy, we obtain ΔE_{NLO} in the limit $t^* \to \infty$, either by direct MC calculation, or (in most practical cases, due to the sign problem) by extrapolation. Examples of such extrapolations in Euclidean time are given in Chap. 6.

F.3.2   Auxiliary and Pion Field Steps

The perturbative treatment is very similar to the calculation of operator expectation values, except that we need to consider the contribution of the evolution operator $1 - \alpha_t \hat{H}(t)$ in Eq. (F.26), which we assign to a routine called dostep. This performs the operation

$$Z_{\text{step}}(\mathbf{n}, s, i, p) = \left[1 + \sqrt{-\tilde{C}\alpha_t}\, \phi_{\text{step}}(\mathbf{n}) - \alpha_t V_{\text{wall}}(\mathbf{n})\right] Z(\mathbf{n}, t_m, s, i, p), \qquad \text{(F.30)}$$

and we note that the inclusion of the nucleon kinetic energy terms, spin-isospin dependent terms, and pion-nucleon coupling terms is straightforward, and proceeds according to the getzvecs routine of Appendix E.

For $\Delta t = 1$, we take

$$\phi_{\text{step}}(\mathbf{n}) = \phi(\mathbf{n}, t), \qquad \phi^{\pi}_{l,\text{step}}(\mathbf{n}) = \phi^{\pi}_l(\mathbf{n}, t), \qquad \text{(F.31)}$$

i.e. we use an "extant" time slice. For the case of $\Delta t = 0$, no such time slice is available, so we generate a "new" time slice from Gaussian noise, similar to the "pseudofermion" method of Chap. 6, for the stochastic computation of fermion determinants. This equals

$$\phi_{\text{step}}(\mathbf{n}) = \sum_{\mathbf{n}'} D_{\phi}^{1/2}(\mathbf{n} - \mathbf{n}')\, G_{\text{step}}^{-1}(\mathbf{n}'), \qquad \text{(F.32)}$$

and

$$\phi^{\pi}_{l,\text{step}}(\mathbf{n}) = \sum_{\mathbf{n}'} D_{\pi}^{1/2}(\mathbf{n} - \mathbf{n}')\, G_{\pi,\text{step}}^{-1}(\mathbf{n}'), \qquad \text{(F.33)}$$

where G_{step}^{-1} and $G_{\pi,\text{step}}^{-1}$ are chosen from a random Gaussian distribution. As in Appendix E, we use FFTW for the efficient evaluation of the above expressions. For the auxiliary fields, this gives

$$\phi_{\text{step}}(\mathbf{n}) = \mathscr{F}^{-1} Z_{\text{mom}}^P(\mathbf{k}), \qquad Z_{\text{mom}}(\mathbf{k}) \equiv \mathscr{F} G_{\text{step}}^{-1}(\mathbf{n}), \qquad \text{(F.34)}$$

with

$$Z_{\text{mom}}^P(k_1, k_2, k_3) = N^{-3} \frac{\tilde{G}_{\phi}^{1/2}(k_1, k_2, k_3)}{N_{\phi}^{1/2}} Z_{\text{mom}}(k_1, k_2, k_3), \qquad \text{(F.35)}$$

according to Eq. (E.6), and for the pion fields, we have

$$\phi^{\pi}_{l,\text{step}}(\mathbf{n}) = \mathscr{F}^{-1} Z_{\text{mom}}^P(\mathbf{k}), \qquad Z_{\text{mom}}(\mathbf{k}) \equiv \mathscr{F} G_{\pi,\text{step}}^{-1}(\mathbf{n}), \qquad \text{(F.36)}$$

with

$$Z^p_{\text{mom}}(k_1, k_2, k_3) = N^{-3} \tilde{G}^{1/2}_\pi(k_1, k_2, k_3) Z_{\text{mom}}(k_1, k_2, k_3), \qquad (\text{F.37})$$

according to Eq. (E.12).

F.3.3   Energy Shifts from Finite Differences

In order to illustrate the perturbative calculation of higher-order contributions in NLEFT, we first consider the simplest operators at hand, namely ΔV and ΔV_{I^2}. These are formally similar to the LO operators without smearing, and are counted as NLO corrections. The operator ΔV is

$$\Delta V \equiv \frac{1}{2} \Delta \tilde{C} : \sum_{\mathbf{n}} \rho^{a^\dagger, a}(\mathbf{n}) \rho^{a^\dagger, a}(\mathbf{n}) :, \qquad (\text{F.38})$$

and contributes

$$\Delta \hat{M}(t) = -\frac{1}{2} \Delta \tilde{C} \alpha_t \sum_{\mathbf{n}} \frac{\delta}{\delta \varepsilon_1(\mathbf{n}, t)} \frac{\delta}{\delta \varepsilon_1(\mathbf{n}, t)} \hat{M}(\{\varepsilon\}, t) \bigg|_{\{\varepsilon\} \to 0}, \qquad (\text{F.39})$$

to Eq. (F.25), with

$$U(\{\varepsilon\}, t) \equiv \sum_{\mathbf{n}} \varepsilon_1(\mathbf{n}, t) \rho^{a^\dagger, a}(\mathbf{n}), \qquad (\text{F.40})$$

from Eq. (4.120). We are now in a position to evaluate the energy shift for $\Delta t = 0$ or $\Delta t = 1$. Given Z_{step}, we define

$$Z^0_{\text{corr}}(p_2, p_1) \equiv N^{-3} \sum_{s,i} \sum_{\mathbf{n}'} Z^d(\mathbf{n}', t_m + \Delta t, s, i, p_2) Z_{\text{step}}(\mathbf{n}', s, i, p_1), \qquad (\text{F.41})$$

and

$$Z^{q+}_{\text{corr}}(p_2, p_1; \mathbf{n}) \equiv Z^0_{\text{corr}}(p_2, p_1) + q\varepsilon N^{-3} \sum_{s,i} Z^d(\mathbf{n}, t_m + \Delta t, s, i, p_2)$$

$$\times Z(\mathbf{n}, t_m, s, i, p_1), \qquad (\text{F.42})$$

and

$$Z_{\text{corr}}^{q-}(p_2, p_1; \mathbf{n}) \equiv Z_{\text{corr}}^0(p_2, p_1) - q\varepsilon N^{-3} \sum_{s,i} Z^d(\mathbf{n}, t_m + \Delta t, s, i, p_2)$$

$$\times Z(\mathbf{n}, t_m, s, i, p_1), \tag{F.43}$$

where ε is an appropriately chosen small parameter. Note that for $\Delta t = 1$, one time slice has been "eaten" by the evolution operator in Eq. (F.25). We again compute the ratios

$$Z_{\text{rdet}}^{q+}(\mathbf{n}) \equiv \frac{\det(Z_{\text{corr}}^{q+}; \mathbf{n})}{|\det(Z_{\text{corr}})|}, \quad Z_{\text{rdet}}^{q-}(\mathbf{n}) \equiv \frac{\det(Z_{\text{corr}}^{q-}; \mathbf{n})}{|\det(Z_{\text{corr}})|}, \tag{F.44}$$

and

$$Z_{\text{rdet}}^0 \equiv \frac{\det(Z_{\text{corr}}^0)}{|\det(Z_{\text{corr}})|}, \tag{F.45}$$

which using the finite difference formula

$$\left.\frac{\partial^2 f(x)}{\partial x^2}\right|_{x_0=0} \approx \frac{f(\varepsilon) + f(-\varepsilon) - 2f(0)}{\varepsilon^2}, \tag{F.46}$$

gives

$$\langle \Delta V \rangle = \frac{1}{2}\Delta C \sum_{\mathbf{n}} \frac{Z_{\text{rdet}}^{1+}(\mathbf{n}) + Z_{\text{rdet}}^{1-}(\mathbf{n}) - 2Z_{\text{rdet}}^0}{\varepsilon^2}, \tag{F.47}$$

for which the binning and ensemble averaging proceeds similarly to the occupation number operators.

The case of ΔV_{I2} introduces a minor complication due to isospin, but is otherwise treated identically. We now have

$$\Delta V_{I2} \equiv \frac{1}{2}\Delta\tilde{C}_{I2} : \sum_{\mathbf{n},l} \rho_I^{a^\dagger,a}(\mathbf{n}, l)\rho_I^{a^\dagger,a}(\mathbf{n}, l) :, \tag{F.48}$$

with

$$\Delta\hat{M}(t) = -\frac{1}{2}\Delta\tilde{C}_{I2}\alpha_t \sum_{\mathbf{n},l} \frac{\delta}{\delta\varepsilon_9(\mathbf{n}, t, l)} \frac{\delta}{\delta\varepsilon_9(\mathbf{n}, t, l)} \hat{M}(\{\varepsilon\}, t)\bigg|_{\{\varepsilon\}\to 0}, \tag{F.49}$$

where

$$U_{I^2}(\{\varepsilon\}, t) \equiv \sum_{\mathbf{n},l} \varepsilon_9(\mathbf{n}, t, l) \rho_I^{a^\dagger, a}(\mathbf{n}, l), \tag{F.50}$$

from Eq. (4.121). The isospin degrees of freedom are then accounted for by

$$Z_{\text{corr}}^{q+}(p_2, p_1; \mathbf{n}, l) \equiv Z_{\text{corr}}^0(p_2, p_1) + q\varepsilon N^{-3} \sum_{s,i,i'} Z^d(\mathbf{n}, t_m + \Delta t, s, i, p_2)$$

$$\times Z_{\tau,l}^{2\times2}(i, i') Z(\mathbf{n}, t_m, s, i', p_1), \tag{F.51}$$

and

$$Z_{\text{corr}}^{q-}(p_2, p_1; \mathbf{n}, l) \equiv Z_{\text{corr}}^0(p_2, p_1) - q\varepsilon N^{-3} \sum_{s,i,i'} Z^d(\mathbf{n}, t_m + \Delta t, s, i, p_2)$$

$$\times Z_{\tau,l}^{2\times2}(i, i') Z(\mathbf{n}, t_m, s, i', p_1), \tag{F.52}$$

which gives

$$\langle \Delta V_{I^2} \rangle = \frac{1}{2} \Delta \tilde{C}_{I^2} \sum_{\mathbf{n},l} \frac{Z_{\text{rdet}}^{1+}(\mathbf{n}, l) + Z_{\text{rdet}}^{1-}(\mathbf{n}, l) - 2Z_{\text{rdet}}^0}{\varepsilon^2}, \tag{F.53}$$

as the desired result.

F.4   NLO Operators

From a computational point of view, the NLO operators can be classified into quadratic operators which contain factors of q^2 with various combinations of spin and isospin operators, tensor operators with various instances of operators of the form $(\boldsymbol{\sigma}_1 \cdot \mathbf{q})(\boldsymbol{\sigma}_2 \cdot \mathbf{q})$, and spin-orbit operators of the form $\mathbf{q} \times \mathbf{k}$. We hope that the material presented here shall be sufficiently general to allow the reader to construct the equivalent expressions for more complicated operators, such as those needed for the three-nucleon force at NNLO. It should also be noted that the finite-difference formulas used here are by no means unique, and should be viewed as a compromise between accuracy and convenience.

F.4.1   Quadratic Operators

We are now in a position to apply the methods developed for the operators ΔV and ΔV_{12}. We first consider the quadratic operator

$$V_{q^2} \equiv -\frac{1}{2}\tilde{C}_{q^2} : \sum_{\mathbf{n}}\sum_{j=1}^{3} \rho^{a^\dagger,a}(\mathbf{n})\nabla_j^2 \rho^{a^\dagger,a}(\mathbf{n}) :, \tag{F.54}$$

which we evaluate using the established method of numerical finite differences with respect to external source fields. We have

$$\Delta\hat{M}(t) = \frac{1}{2}\tilde{C}_{q^2}\alpha_t \sum_{\mathbf{n}}\sum_{j=1}^{3} \frac{\delta}{\delta\varepsilon_1(\mathbf{n},t)}\frac{\delta}{\delta\varepsilon_5(\mathbf{n},t,j)}\hat{M}(\{\varepsilon\},t)\Bigg|_{\{\varepsilon\}\to 0}, \tag{F.55}$$

for which the finite difference

$$\frac{\partial^2 f(x,y)}{\partial x\partial y}\Bigg|_{(x_0,y_0)=(0,0)} \approx \frac{f(\varepsilon,\varepsilon)-f(0,\varepsilon)-f(\varepsilon,0)+f(0,0)}{\varepsilon^2}, \tag{F.56}$$

gives

$$\langle V_{q^2}\rangle = -\frac{1}{2}\tilde{C}_{q^2}\sum_{\mathbf{n}} \frac{Z_{\text{rdet}}^{1,1}(\mathbf{n})-Z_{\text{rdet}}^{1,0}(\mathbf{n})-Z_{\text{rdet}}^{0,1}(\mathbf{n})+Z_{\text{rdet}}^0}{\varepsilon^2}, \tag{F.57}$$

and we again use the definition

$$Z_{\text{rdet}}^{p,q}(\mathbf{n}) \equiv \frac{\det(Z_{\text{corr}}^{p,q};\mathbf{n})}{|\det(Z_{\text{corr}})|}, \tag{F.58}$$

which we shall not hereafter repeat for similar quantities. We define

$$Z_{\text{corr}}^{p,q}(p_2,p_1;\mathbf{n}) \equiv Z_{\text{corr}}^0(p_2,p_1)$$

$$+ p\varepsilon N^{-3}\sum_{s,i} Z_1'(p_2,p_1;\mathbf{n},s,s,i,i)$$

$$+ q\varepsilon N^{-3}\sum_{s,i} Z_{qq}'(p_2,p_1;\mathbf{n},s,s,i,i), \tag{F.59}$$

with

$$Z_1'(p_2,p_1;\mathbf{n},s,s',i,i') \equiv Z^d(\mathbf{n},t_m+\Delta t,s,i,p_2)Z(\mathbf{n},t_m,s',i',p_1), \tag{F.60}$$

and

$$
\begin{aligned}
Z'_{\text{qq}}&(p_2, p_1; \mathbf{n}, s, s', i, i') \\
&\equiv \frac{1}{4}\Bigg\{ -6Z^d(\mathbf{n}, t_m + \Delta t, s, i, p_2)Z(\mathbf{n}, t_m, s', i', p_1) \\
&\quad + \sum_{j=1}^{3}\Bigg[Z^d(\text{mod}(\mathbf{n} + 2\hat{e}_j, N), t_m + \Delta t, s, i, p_2) \\
&\qquad\qquad \times Z(\text{mod}(\mathbf{n} + 2\hat{e}_j, N), t_m, s', i', p_1)\Bigg] \\
&\quad + \sum_{j=1}^{3}\Bigg[Z^d(\text{mod}(\mathbf{n} - 2\hat{e}_j + N, N), t_m + \Delta t, s, i, p_2) \\
&\qquad\qquad \times Z(\text{mod}(\mathbf{n} - 2\hat{e}_j + N, N), t_m, s', i', p_1)\Bigg]\Bigg\},
\end{aligned}
\tag{F.61}
$$

where the finite difference formula

$$
\left.\frac{\partial^2 f(x)}{\partial x^2}\right|_{x_0=0} \approx \frac{f(2h) - 2f(0) + f(-2h)}{4h^2},
\tag{F.62}
$$

has been used for ∇_j^2. It should be noted that the form usually given in the literature is obtained by setting $h' = 2h$.

The isospin-dependent quadratic term is

$$
V_{I^2, q^2} \equiv -\frac{1}{2}\tilde{C}_{I^2, q^2} : \sum_{\mathbf{n}, l}\sum_{j=1}^{3} \rho_I^{a^\dagger, a}(\mathbf{n}, l)\nabla_j^2\, \rho_I^{a^\dagger, a}(\mathbf{n}, l) :,
\tag{F.63}
$$

with

$$
\begin{aligned}
Z_{\text{corr}}^{p,q}(p_2, p_1; \mathbf{n}, l) &\equiv Z_{\text{corr}}^0(p_2, p_1) \\
&\quad + p\varepsilon N^{-3} \sum_{s,i,i'} Z_{\tau,l}^{2\times2}(i, i')Z_1'(p_2, p_1; \mathbf{n}, s, s, i, i') \\
&\quad + q\varepsilon N^{-3} \sum_{s,i,i'} Z_{\tau,l}^{2\times2}(i, i')Z_{\text{qq}}'(p_2, p_1; \mathbf{n}, s, s, i, i'),
\end{aligned}
\tag{F.64}
$$

and

$$
\langle V_{I^2, q^2}\rangle = -\frac{1}{2}\tilde{C}_{I^2, q^2} \sum_{\mathbf{n}, l} \frac{Z_{\text{rdet}}^{1,1}(\mathbf{n}, l) - Z_{\text{rdet}}^{1,0}(\mathbf{n}, l) - Z_{\text{rdet}}^{0,1}(\mathbf{n}, l) + Z_{\text{rdet}}^0}{\varepsilon^2}.
\tag{F.65}
$$

The spin-dependent quadratic term is

$$V_{S^2,q^2} \equiv -\frac{1}{2}\tilde{C}_{S^2,q^2} : \sum_{\mathbf{n},k}\sum_{j=1}^{3} \rho_S^{a^\dagger,a}(\mathbf{n},k)\nabla_j^2\,\rho_S^{a^\dagger,a}(\mathbf{n},k) :, \tag{F.66}$$

with

$$
\begin{aligned}
Z_{\text{corr}}^{p,q}(p_2,p_1;\mathbf{n},k) &\equiv Z_{\text{corr}}^0(p_2,p_1) \\
&\quad + p\varepsilon N^{-3}\sum_{s,s',i} Z_{\sigma,k}^{2\times2}(s,s')Z_1'(p_2,p_1;\mathbf{n},s,s',i,i) \\
&\quad + q\varepsilon N^{-3}\sum_{s,s',i} Z_{\sigma,k}^{2\times2}(s,s')Z_{qq}'(p_2,p_1;\mathbf{n},s,s',i,i),
\end{aligned}
\tag{F.67}
$$

and

$$\langle V_{S^2,q^2}\rangle = -\frac{1}{2}\tilde{C}_{S^2,q^2}\sum_{\mathbf{n},k}\frac{Z_{\text{rdet}}^{1,1}(\mathbf{n},k) - Z_{\text{rdet}}^{1,0}(\mathbf{n},k) - Z_{\text{rdet}}^{0,1}(\mathbf{n},k) + Z_{\text{rdet}}^0}{\varepsilon^2}, \tag{F.68}$$

The spin-isospin dependent quadratic term is

$$V_{S^2,I^2,q^2} \equiv -\frac{1}{2}\tilde{C}_{S^2,I^2,q^2} : \sum_{\mathbf{n},k,l}\sum_{j=1}^{3} \rho_{SI}^{a^\dagger,a}(\mathbf{n},k,l)\nabla_j^2\,\rho_{SI}^{a^\dagger,a}(\mathbf{n},k,l) :, \tag{F.69}$$

with

$$
\begin{aligned}
Z_{\text{corr}}^{p,q}(p_2,p_1;\mathbf{n},k,l) &\equiv Z_{\text{corr}}^0(p_2,p_1) \\
&\quad + p\varepsilon N^{-3}\sum_{s,s',i,i'} Z_{\sigma,k}^{2\times2}(s,s')Z_{\tau,l}^{2\times2}(i,i')Z_1'(p_2,p_1;\mathbf{n},s,s',i,i') \\
&\quad + q\varepsilon N^{-3}\sum_{s,s',i,i'} Z_{\sigma,k}^{2\times2}(s,s')Z_{\tau,l}^{2\times2}(i,i') \\
&\qquad \times Z_{qq}'(p_2,p_1;\mathbf{n},s,s',i,i'),
\end{aligned}
\tag{F.70}
$$

and

$$\langle V_{S^2,I^2,q^2}\rangle = -\frac{1}{2}C_{S^2,I^2,q^2}\sum_{\mathbf{n},k,l}\frac{Z_{\text{rdet}}^{1,1}(\mathbf{n},k,l) - Z_{\text{rdet}}^{1,0}(\mathbf{n},k,l) - Z_{\text{rdet}}^{0,1}(\mathbf{n},k,l) + Z_{\text{rdet}}^0}{\varepsilon^2}. \tag{F.71}$$

F.4.2   Tensor Operators

Next, we turn to the "tensor" operators which have a different structure from the quadratic operators considered above. The isospin-independent operator is

$$
V_{(q \cdot S)^2} \equiv \frac{1}{2} \tilde{C}_{(q \cdot S)^2} : \sum_{\mathbf{n}} \sum_{j=1}^{3} \Delta_j \rho_S^{a^{\dagger},a}(\mathbf{n}, j) \sum_{j'=1}^{3} \Delta_{j'} \rho_S^{a^{\dagger},a}(\mathbf{n}, j') :,
\qquad (\text{F.72})
$$

for which we use

$$
Z_{\text{corr}}^{q+}(p_2, p_1; \mathbf{n}) \equiv Z_{\text{corr}}^{0}(p_2, p_1) + q \varepsilon N^{-3} \sum_{s,s',i} Z_{\text{qs}}'(p_2, p_1; \mathbf{n}, s, s', i, i),
\qquad (\text{F.73})
$$

and

$$
Z_{\text{corr}}^{q-}(p_2, p_1; \mathbf{n}) \equiv Z_{\text{corr}}^{0}(p_2, p_1) - q \varepsilon N^{-3} \sum_{s,s',i} Z_{\text{qs}}'(p_2, p_1; \mathbf{n}, s, s', i, i),
\qquad (\text{F.74})
$$

with

$$
\begin{aligned}
Z_{\text{qs}}'&(p_2, p_1; \mathbf{n}, s, s', i, i') \\
&\equiv \frac{1}{2} \Bigg\{ \sum_{j=1}^{3} \Bigg[Z^d(\text{mod}(\mathbf{n} + \hat{e}_j, N), t_m + \Delta t, s, i, p_2) \\
&\qquad\qquad\qquad \times Z_{\sigma,j}^{2 \times 2}(s, s') Z(\text{mod}(\mathbf{n} + \hat{e}_j, N), t_m, s', i', p_1) \Bigg] \\
&\qquad - \sum_{j=1}^{3} \Bigg[Z^d(\text{mod}(\mathbf{n} - \hat{e}_j + N, N), t_m + \Delta t, s, i, p_2) \\
&\qquad\qquad\qquad \times Z_{\sigma,j}^{2 \times 2}(s, s') Z(\text{mod}(\mathbf{n} - \hat{e}_j + N, N), t_m, s', i', p_1) \Bigg] \Bigg\},
\end{aligned}
\qquad (\text{F.75})
$$

which gives

$$
\langle V_{(q \cdot S)^2} \rangle = \frac{1}{2} \tilde{C}_{(q \cdot S)^2} \sum_{\mathbf{n}} \frac{Z_{\text{rdet}}^{1+}(\mathbf{n}) + Z_{\text{rdet}}^{1-}(\mathbf{n}) - 2 Z_{\text{rdet}}^{0}}{\varepsilon^2}.
\qquad (\text{F.76})
$$

The equivalent isospin-dependent term is given by

$$
V_{I^2,(q \cdot S)^2} \equiv \frac{1}{2} \tilde{C}_{I^2,(q \cdot S)^2} : \sum_{\mathbf{n},l} \sum_{j=1}^{3} \Delta_j \rho_{SI}^{a^\dagger,a}(\mathbf{n},j,l) \sum_{j'=1}^{3} \Delta_{j'} \rho_{SI}^{a^\dagger,a}(\mathbf{n},j',l) :,
$$

(F.77)

for which

$$
Z_{\text{corr}}^{q+}(p_2, p_1; \mathbf{n})
$$
$$
\equiv Z_{\text{corr}}^{0}(p_2, p_1) + q\varepsilon N^{-3} \sum_{s,s',i,i'} Z_{\tau,l}^{2\times2}(i,i') Z_{\text{qs}}'(p_2, p_1; \mathbf{n}, s, s', i, i'),
$$

(F.78)

and

$$
Z_{\text{corr}}^{q-}(p_2, p_1; \mathbf{n})
$$
$$
\equiv Z_{\text{corr}}^{0}(p_2, p_1) - q\varepsilon N^{-3} \sum_{s,s',i,i'} Z_{\tau,l}^{2\times2}(i,i') Z_{\text{qs}}'(p_2, p_1; \mathbf{n}, s, s', i, i'),
$$

(F.79)

giving

$$
\langle V_{I^2,(q \cdot S)^2} \rangle = \frac{1}{2} \tilde{C}_{I^2,(q \cdot S)^2} \sum_{\mathbf{n},l} \frac{Z_{\text{rdet}}^{1+}(\mathbf{n},l) + Z_{\text{rdet}}^{1-}(\mathbf{n},l) - 2Z_{\text{rdet}}^{0}}{\varepsilon^2}.
$$

(F.80)

F.4.3 Rotational Symmetry Breaking Operators

On the lattice, we introduce two operators which are invariant under the cubic rotational group $SO(3,Z)$ but not under the full $SO(3)$ symmetry. The structure of these operators is similar to the tensor operators. The first one is

$$
V_{SSqq} \equiv -\frac{1}{2} \tilde{C}_{SSqq} : \sum_{\mathbf{n}} \sum_{j=1}^{3} \rho_S^{a^\dagger,a}(\mathbf{n},j) \nabla_j^2 \rho_S^{a^\dagger,a}(\mathbf{n},j) :,
$$

(F.81)

with

$$
\begin{aligned}
Z_{\text{corr}}^{p,q}(p_2, p_1; \mathbf{n}, j) &\equiv Z_{\text{corr}}^0(p_2, p_1) \\
&\quad + p\varepsilon N^{-3} \sum_{s,s',i} Z_{\sigma,j}^{2\times2}(s, s') Z_1'(p_2, p_1; \mathbf{n}, s, s', i, i) \\
&\quad + q\varepsilon N^{-3} \sum_{s,s',i} Z_{\sigma,j}^{2\times2}(s, s') Z_{\text{qq},j}'(p_2, p_1; \mathbf{n}, s, s', i, i),
\end{aligned}
$$

$$(\text{F.82})$$

where

$$
\begin{aligned}
Z_{\text{qq},j}'(p_2, p_1; \mathbf{n}, s, s', i, i') &\equiv \frac{1}{4}\Big[-2Z^d(\mathbf{n}, t_m + \Delta t, s, i, p_2) Z(\mathbf{n}, t_m, s', i', p_1) \\
&\quad + Z^d(\text{mod}(\mathbf{n} + 2\hat{e}_j, N), t_m + \Delta t, s, i, p_2) \\
&\quad \times Z(\text{mod}(\mathbf{n} + 2\hat{e}_j, N), t_m, s', i', p_1) \\
&\quad + Z^d(\text{mod}(\mathbf{n} - 2\hat{e}_j + N, N), t_m + \Delta t, s, i, p_2) \\
&\quad \times Z(\text{mod}(\mathbf{n} - 2\hat{e}_j + N, N), t_m, s', i', p_1) \Big],
\end{aligned}
$$

$$(\text{F.83})$$

which gives

$$
\langle V_{SSqq} \rangle = -\frac{1}{2}\tilde{C}_{SSqq} \sum_{\mathbf{n},j} \frac{Z_{\text{rdet}}^{1,1}(\mathbf{n}, j) - Z_{\text{rdet}}^{1,0}(\mathbf{n}, j) - Z_{\text{rdet}}^{0,1}(\mathbf{n}, j) + Z_{\text{rdet}}^0}{\varepsilon^2}.
$$

$$(\text{F.84})$$

The second operator is given by

$$
V_{I^2, SSqq} \equiv -\frac{1}{2}\tilde{C}_{I^2, SSqq} : \sum_{\mathbf{n},l} \sum_{j=1}^3 \rho_{SI}^{a^\dagger, a}(\mathbf{n}, j, l) \nabla_j^2 \rho_{SI}^{a^\dagger, a}(\mathbf{n}, j, l) :,
$$

$$(\text{F.85})$$

with

$$
\begin{aligned}
Z_{\text{corr}}^{p,q}(p_2, p_1; \mathbf{n}, j, l) &\\
&\equiv Z_{\text{corr}}^0(p_2, p_1) \\
&\quad + p\varepsilon N^{-3} \sum_{s,s',i,i'} Z_{\sigma,j}^{2\times2}(s, s') Z_{\tau,l}^{2\times2}(i, i') Z_1'(p_2, p_1; \mathbf{n}, s, s', i, i') \\
&\quad + q\varepsilon N^{-3} \sum_{s,s',i,i'} Z_{\sigma,j}^{2\times2}(s, s') Z_{\tau,l}^{2\times2}(i, i') Z_{\text{qq},j}'(p_2, p_1; \mathbf{n}, s, s', i, i'),
\end{aligned}
$$

$$(\text{F.86})$$

such that

$$\langle V_{I^2, SSqq} \rangle = -\frac{1}{2} \tilde{C}_{I^2, SSqq} \sum_{\mathbf{n}, j, l} \frac{Z_{\text{rdet}}^{1,1}(\mathbf{n}, j, l) - Z_{\text{rdet}}^{1,0}(\mathbf{n}, j, l) - Z_{\text{rdet}}^{0,1}(\mathbf{n}, j, l) + Z_{\text{rdet}}^{0}}{\varepsilon^2}.$$
$$\tag{F.87}$$

We adjust the $I = 0$ combination of these operators to eliminate the unphysical mixing of the $^3D_3$ partial wave in the $^3S(D)_1$ channel. For this purpose, we introduce

$$\tilde{C}_{SSqq} \equiv \frac{1}{4} \left(3\tilde{C}_{SSqq}^{I=1} + \tilde{C}_{SSqq}^{I=0} \right), \qquad \tilde{C}_{I^2, SSqq} \equiv \frac{1}{4} \left(\tilde{C}_{SSqq}^{I=1} - \tilde{C}_{SSqq}^{I=0} \right), \tag{F.88}$$

as the coupling constants for the $I = 0$ and $I = 1$ channels. If we define

$$V_{SSqq}^{I=0} \equiv V_{SSqq} + V_{I^2, SSqq}, \tag{F.89}$$

and set

$$\tilde{C}_{SSqq} \equiv \frac{1}{4} \tilde{C}_{SSqq}^{I=0}, \qquad \tilde{C}_{I^2, SSqq} \equiv -\frac{1}{4} \tilde{C}_{SSqq}^{I=0}, \tag{F.90}$$

we may tune $\tilde{C}_{SSqq}^{I=0}$ to eliminate the unphysical mixing of the $^3D_3$ partial wave. If necessary, we may also include the $I = 1$ combination in order to eliminate the unphysical mixing of the $^3F_4$ partial wave in the $^3P(F)_2$ channel.

F.4.4   Spin-Orbit Operators

The "spin-orbit" operators are the most complicated structures encountered at NLO. The isospin-independent spin-orbit operator is given by

$$V_{(iq \times S) \cdot k} \equiv -\frac{i}{2} \tilde{C}_{(iq \times S) \cdot k} : \sum_{\mathbf{n}, k'} \sum_{k=1}^{3} \sum_{j=1}^{3} \varepsilon_{k, k', j}$$

$$\times \left\{ \Pi^{a^{\dagger}, a}(\mathbf{n}, k) \Delta_j \rho_S^{a^{\dagger}, a}(\mathbf{n}, k') \right.$$

$$\left. + \Pi_S^{a^{\dagger}, a}(\mathbf{n}, k, k') \Delta_j \rho^{u^{\dagger}, a}(\mathbf{n}) \right\} :, \tag{F.91}$$

with two distinct contributions. If we define

$$Z'_{\text{p},j}(p_2, p_1; \mathbf{n}, s, s', i, i')$$

$$\equiv \frac{1}{2}\Big[-Z^d(\mathbf{n}, t_m + \Delta t, s, i, p_2)$$

$$\times Z(\text{mod}(\mathbf{n} - \hat{e}_j + N, N), t_m, s', i', p_1)$$

$$- Z^d(\text{mod}(\mathbf{n} + \hat{e}_j, N), t_m + \Delta t, s, i, p_2)$$

$$\times Z(\mathbf{n}, t_m, s', i', p_1)$$

$$+ Z^d(\mathbf{n}, t_m + \Delta t, s, i, p_2)$$

$$\times Z(\text{mod}(\mathbf{n} + \hat{e}_j, N), t_m, s', i', p_1)$$

$$+ Z^d(\text{mod}(\mathbf{n} - \hat{e}_j + N, N), t_m + \Delta t, s, i, p_2)$$

$$\times Z(\mathbf{n}, t_m, s', i', p_1)\Big], \tag{F.92}$$

and

$$Z'_{\text{q},j}(p_2, p_1; \mathbf{n}, s, s', i, i') \equiv \frac{1}{2}\Big[Z^d(\text{mod}(\mathbf{n} + \hat{e}_j, N), t_m + \Delta t, s, i, p_2)$$

$$\times Z(\text{mod}(\mathbf{n} + \hat{e}_j, N), t_m, s', i', p_1)$$

$$- Z^d(\text{mod}(\mathbf{n} - \hat{e}_j + N, N), t_m + \Delta t, s, i, p_2)$$

$$\times Z(\text{mod}(\mathbf{n} - \hat{e}_j + N, N), t_m, s', i', p_1)\Big], \tag{F.93}$$

then we can construct

$$Z_{\text{corr}}^{p,q}(p_2, p_1; \mathbf{n}, k) \equiv Z_{\text{corr}}^0(p_2, p_1)$$

$$+ p\varepsilon N^{-3} \sum_{s,s',i} Z_{\sigma,0}^{2\times2}(s, s') Z'_{\text{p},k}(p_2, p_1; \mathbf{n}, s, s', i, i)$$

$$+ q\varepsilon N^{-3} \sum_{s,s',i} Z_{\sigma,j'}^{2\times2}(s, s') Z'_{\text{q},j''}(p_2, p_1; \mathbf{n}, s, s', i, i)$$

$$- q\varepsilon N^{-3} \sum_{s,s',i} Z_{\sigma,j''}^{2\times2}(s, s') Z'_{\text{q},j'}(p_2, p_1; \mathbf{n}, s, s', i, i), \tag{F.94}$$

where

$$j' = \mathrm{mod}(k, 3) + 1, \qquad j'' = \mathrm{mod}(k + 1, 3) + 1, \tag{F.95}$$

for the first term in Eq. (F.91), and

$$
\begin{aligned}
Z_{\mathrm{corr}}^{p,q}(p_2, p_1; \mathbf{n}, j) &\equiv Z_{\mathrm{corr}}^{0}(p_2, p_1) \\
&\quad + p\varepsilon N^{-3} \sum_{s,s',i} Z_{\sigma,0}^{2\times 2}(s, s') Z_{\mathrm{q},j}'(p_2, p_1; \mathbf{n}, s, s', i, i) \\
&\quad - q\varepsilon N^{-3} \sum_{s,s',i} Z_{\sigma,j'}^{2\times 2}(s, s') Z_{\mathrm{p},j''}'(p_2, p_1; \mathbf{n}, s, s', i, i) \\
&\quad + q\varepsilon N^{-3} \sum_{s,s',i} Z_{\sigma,j''}^{2\times 2}(s, s') Z_{\mathrm{p},j'}'(p_2, p_1; \mathbf{n}, s, s', i, i),
\end{aligned}
\tag{F.96}
$$

where

$$j' = \mathrm{mod}(j, 3) + 1, \qquad j'' = \mathrm{mod}(j + 1, 3) + 1, \tag{F.97}$$

for the second term in Eq. (F.91). The contribution of the first term is given by

$$
\langle V_{(iq\times S)\cdot k} \rangle = -\frac{i}{2} \tilde{C}_{(iq\times S)\cdot k} \sum_{\mathbf{n},k} \frac{Z_{\mathrm{rdet}}^{1,1}(\mathbf{n}, k) - Z_{\mathrm{rdet}}^{1,0}(\mathbf{n}, k) - Z_{\mathrm{rdet}}^{0,1}(\mathbf{n}, k) + Z_{\mathrm{rdet}}^{0}}{\varepsilon^2},
\tag{F.98}
$$

and that of the second term in Eq. (F.91) is obtained by taking $k \to j$ in the ratios of determinants.

We note that in the continuum limit, the spin-orbit interaction vanishes unless $S = 1$, and that it is antisymmetric under the exchange of $\mathbf{q}$ and $\mathbf{k}$, thus Eq. (F.91) should contribute only in odd-parity channels with $S = 1$. However, at non-zero lattice spacing, the exact t-u channel antisymmetry of Eq. (F.91) is violated. Thus, we expect lattice artifacts to appear in even-parity channels with $S = 1$. We note that such artifacts may be eliminated by projecting onto $I = 1$, and for this purpose we include the isospin-dependent spin-orbit operator

$$
\begin{aligned}
V_{I^2,(iq\times S)\cdot k} &\equiv -\frac{i}{2} \tilde{C}_{I^2,(iq\times S)\cdot k} : \sum_{\mathbf{n},l,k'} \sum_{k=1}^{3} \sum_{j=1}^{3} \varepsilon_{k,k',j} \\
&\quad \times \Big\{ \Pi_I^{a^\dagger,a}(\mathbf{n}, k, l) \Delta_j \rho_{SI}^{a^\dagger,a}(\mathbf{n}, k', l) \\
&\quad + \Pi_{SI}^{a^\dagger,a}(\mathbf{n}, k, k', l) \Delta_j \rho_I^{a^\dagger,a}(\mathbf{n}, l) \Big\} :,
\end{aligned}
\tag{F.99}
$$

for which we construct

$$Z_{\text{corr}}^{p,q}(p_2, p_1; \mathbf{n}, k, l)$$

$$\equiv Z_{\text{corr}}^0(p_2, p_1)$$

$$+ p\varepsilon N^{-3} \sum_{s,s',i,i'} Z_{\sigma,0}^{2\times2}(s, s') Z_{\tau,l}^{2\times2}(i, i') Z_{\text{p},k}'(p_2, p_1; \mathbf{n}, s, s', i, i')$$

$$+ q\varepsilon N^{-3} \sum_{s,s',i,i'} Z_{\sigma,j'}^{2\times2}(s, s') Z_{\tau,l}^{2\times2}(i, i') Z_{\text{q},j''}'(p_2, p_1; \mathbf{n}, s, s', i, i')$$

$$- q\varepsilon N^{-3} \sum_{s,s',i,i'} Z_{\sigma,j''}^{2\times2}(s, s') Z_{\tau,l}^{2\times2}(i, i') Z_{\text{q},j'}'(p_2, p_1; \mathbf{n}, s, s', i, i'), \qquad \text{(F.100)}$$

where

$$j' = \text{mod}(k, 3) + 1, \qquad j'' = \text{mod}(k + 1, 3) + 1, \qquad \text{(F.101)}$$

for the first term in Eq. (F.99), and

$$Z_{\text{corr}}^{p,q}(p_2, p_1; \mathbf{n}, j, l)$$

$$\equiv Z_{\text{corr}}^0(p_2, p_1)$$

$$+ p\varepsilon N^{-3} \sum_{s,s',i,i'} Z_{\sigma,0}^{2\times2}(s, s') Z_{\tau,l}^{2\times2}(i, i') Z_{\text{q},j}'(p_2, p_1; \mathbf{n}, s, s', i, i')$$

$$- q\varepsilon N^{-3} \sum_{s,s',i,i'} Z_{\sigma,j'}^{2\times2}(s, s') Z_{\tau,l}^{2\times2}(i, i') Z_{\text{p},j''}'(p_2, p_1; \mathbf{n}, s, s', i, i')$$

$$+ q\varepsilon N^{-3} \sum_{s,s',i,i'} Z_{\sigma,j''}^{2\times2}(s, s') Z_{\tau,l}^{2\times2}(i, i') Z_{\text{p},j'}'(p_2, p_1; \mathbf{n}, s, s', i, i'), \qquad \text{(F.102)}$$

where

$$j' = \text{mod}(j, 3) + 1, \qquad j'' = \text{mod}(j + 1, 3) + 1, \qquad \text{(F.103)}$$

for the second term in Eq. (F.99). Again, we find

$$\langle V_{I^2,(iq\times S)\cdot k}\rangle = -\frac{i}{2}\tilde{C}_{I^2,(iq\times S)\cdot k}$$

$$\times \sum_{\mathbf{n},k,l} \frac{Z_{\text{rdet}}^{1,1}(\mathbf{n}, k, l) - Z_{\text{rdet}}^{1,0}(\mathbf{n}, k, l) - Z_{\text{rdet}}^{0,1}(\mathbf{n}, k, l) + Z_{\text{rdet}}^0}{\varepsilon^2},$$

$$\text{(F.104)}$$

with $k \to j$ for the second term in Eq. (F.99).

We introduce separate coupling constants for the $I = 0$ and $I = 1$ channels according to

$$\tilde{C}_{(iq \times S) \cdot k} \equiv \frac{1}{4}(3\tilde{C}^{I=1}_{(iq \times S) \cdot k} + \tilde{C}^{I=0}_{(iq \times S) \cdot k}), \qquad (F.105)$$

and

$$\tilde{C}_{I^2, (iq \times S) \cdot k} \equiv \frac{1}{4}(\tilde{C}^{I=1}_{(iq \times S) \cdot k} - \tilde{C}^{I=0}_{(iq \times S) \cdot k}), \qquad (F.106)$$

with

$$V^{I=1}_{(iq \times S) \cdot k} \equiv V_{(iq \times S) \cdot k} + V_{I^2, (iq \times S) \cdot k}, \qquad (F.107)$$

and if we set

$$\tilde{C}_{(iq \times S) \cdot k} \equiv \frac{3}{4}\tilde{C}^{I=1}_{(iq \times S) \cdot k}, \quad \tilde{C}_{I^2, (iq \times S) \cdot k} \equiv \frac{1}{4}\tilde{C}^{I=1}_{(iq \times S) \cdot k}, \qquad (F.108)$$

then the above-mentioned lattice artifacts due to the spin-orbit interaction for odd-parity channels with $S = 1$ are eliminated.

References

1. W. Heisenberg, On the structure of atomic nuclei. Z. Phys. **77**, 1 (1932)
2. E.M. Henley, G.A. Miller, Meson theory of charge dependent nuclear forces, in *Mesons in Nuclei*, ed. by M. Rho, D. Wilkinson, vol. I (North Holland, Amsterdam, 1979), pp. 405–434
3. S. Okubo, R.E. Marshak, Velocity dependence of the two-nucleon interaction. Ann. Phys. **4**, 166 (1958)
4. K. Erkelenz, Current status of the relativistic two nucleon one boson exchange potential. Phys. Rep. **13**, 191 (1974)
5. E. Epelbaum, W. Glöckle, U.-G. Meißner, The Two-nucleon system at next-to-next-to-next-to-leading order. Nucl. Phys. A **747**, 362 (2005)
6. H.P. Stapp, T.J. Ypsilantis, N. Metropolis, Phase shift analysis of 310-MeV proton–proton scattering experiments. Phys. Rev. **105**, 302 (1957)
7. J.M. Blatt, L.C. Biedenharn, Neutron-proton scattering with spin-orbit coupling. 1. General expressions. Phys. Rev. **86**, 399 (1952)
8. J. Bystricky, F. Lehar, P. Winternitz, Formalism of nucleon-nucleon elastic scattering experiments. J. Phys. (France) **39**, 1 (1978)
9. W. Glöckle, *The Quantum Mechanical Few-Body Problem* (Springer, Heidelberg, 1983)
10. V.G.J. Stoks, R.A.M. Klomp, M.C.M. Rentmeester, J.J. de Swart, Partial wave analysis of all nucleon-nucleon scattering data below 350-MeV. Phys. Rev. C **48**, 792 (1993)
11. R.C. Johnson, Angular momentum on a lattice. Phys. Lett. **114B**, 147 (1982)
12. B. Berg, A. Billoire, Glueball spectroscopy in four-dimensional SU(3) lattice gauge theory. 1. Nucl. Phys. B **221**, 109 (1983)

13. J.E. Mandula, G. Zweig, J. Govaerts, Representations of the rotation reflection symmetry group of the four-dimensional cubic lattice. Nucl. Phys. B **228**, 91 (1983)
14. B.N. Lu, T.A. Lähde, D. Lee, U.-G. Meißner, Breaking and restoration of rotational symmetry on the lattice for bound state multiplets. Phys. Rev. D **90**, 034507 (2014)
15. S. Ali, A.R. Bodmer, Phenomenological α-α potentials. Nucl. Phys. **80**, 99 (1966)
16. O. Portilho, D.A. Agrello, S.A. Coon, Three-body potential among alpha particles. Phys. Rev. C **27**, 2923 (1983)
17. S. Koenig, D. Lee, H.-W. Hammer, Volume dependence of bound states with angular momentum. Phys. Rev. Lett. **107**, 112001 (2011)
18. S. Koenig, D. Lee, H.-W. Hammer, Non-relativistic bound states in a finite volume. Ann. Phys. **327**, 1450 (2012)
19. B.N. Lu, T.A. Lähde, D. Lee, U.-G. Meißner, Breaking and restoration of rotational symmetry for irreducible tensor operators on the lattice. Phys. Rev. D **92**, 014506 (2015)
20. S.L. Altmann, A.P. Cracknell, Lattice harmonics 1. Cubic groups. Rev. Mod. Phys. **37**, 19 (1965)
21. K. Rykhlinskaya, S. Fritzsche, Generation of Clebsch-Gordan coefficients for the point and double groups. Comput. Phys. Commun. **174**, 903 (2006)

Index

© Springer Nature Switzerland AG 2019
T. A. Lähde, U.-G. Meißner, *Nuclear Lattice Effective Field Theory*,
Lecture Notes in Physics 957, https://doi.org/10.1007/978-3-030-14189-9

Printed in the United States
By Bookmasters